D0535837

THE REDISCOVERY OF THE EARTH

Stephen E. DWORNIK
Alastair G.W. CAMERON
Keith E. BULLEN
Stephen MOORBATH
Cyril PONNAMPERUMA
Peter M. MOLTON
Bartholomew NAGY
Martin F. GLAESSNER
Gordon A. GROSS
John Tuzo WILSON

THE REDISCOVERY OF THE EARTH

Takesi NAGATA
John F. DEWEY
Xavier LE PICHON
William R. DICKINSON
Stanley K. RUNCORN
Alan G. SMITH
Albert E.J. ENGEL
Rhodes W. FAIRBRIDGE
Edwin H. COLBERT
Vladimir V. BELOUSOV

(La Riscoperta della Terra)

Edited by
Lloyd Motz, Ph.D.
Professor of Astronomy
Columbia University

VNR **VAN NOSTRAND REINHOLD COMPANY**
NEW YORK CINCINNATI ATLANTA DALLAS SAN FRANCISCO
LONDON TORONTO MELBOURNE

Van Nostrand Reinhold Company Regional Offices:
New York Cincinnati Atlanta Dallas San Francisco

Van Nostrand Reinhold Company International Offices:
London Toronto Melbourne

Edizioni Scientifiche e Tecniche Mondadori
Editorial Director: Edgardo Macorini

Printed and bound in Italy by Arnoldo Mondadori Editore, Verona.

Library of Congress Catalog Card Number: 79-24650
ISBN 0-442-26779-7

Cover photo by NASA

Published by Van Nostrand Reinhold Company
A Division of Litton Educational Publishing, Inc.
135 West 50th Street, New York N.Y. 10020

16 15 14 13 12 11 10 9 8 7 6 5 4 3 2 1

Library of Congress Cataloging in Publication Data

Main entry under title:

The rediscovery of the earth.

Translation of La Riscoperta della terra.
Includes bibliographies and index.
1. Earth sciences. I. Motz, Lloyd, 1910–
QE26.2.R5713 550 79-24650
ISBN 0-442-26779-7

CONTENTS

1978 and the U.S. Pioneer Venus Orbiter and multiprobe missions which were launched on May 20, 1978 as a single package and which accumulated data over a 30-day period from December 4, 1978 into January 1979. These data dealt with the following features of Venus: its magnetic field, its ionosphere, its interaction with the solar wind, the thermal profile of its atmosphere, the composition and density of its upper and lower atmospheres, the nature of its cloud cover, its infrared emission, wind velocities, and surface conditions.

The remarkable amount and kind of information about Mars obtained from the U.S. Mariner 9 and 10 flights were greatly enhanced by the data collected by the U.S. Viking 1 and 2 orbiters and landers which were launched in August and September 1975 and went into synchronous orbits around Mars in June and August 1976. The landers from these two orbiters were deposited on the surface of Mars shortly after each orbiter had gone into orbit. In addition to transmitting direct photographs of the surface of Mars, showing the Martian landscape as it would look to a person standing on Mars, the landers carried out a series of chemical experiments to determine whether the Martian soil contained living organisms and whether the organic chemistry of the soil indicated the presence of past or present life. Even though these experiments showed an unusual amount of chemical activity in the Martian soil, they were inconclusive in so far as the presence of life is concerned.

Exciting and revealing as the photographs of Jupiter taken by Pioneer 10 and 11 are, they are orders of magnitude below those obtained by Voyager 1, which came to within 110,000 kms of Jupiter's cloud surface during February and March 1979 and transmitted to the Earth 15,000 photographs of Jupiter's rapidly moving and wildly turbulent bands of colored clouds, of its Great Red Spot, of its inner satellites, and of its previously unknown faint rings.

The article that follows by Stephen E. Dwornik, head of NASA's Planetology Program presents a survey of the planets as we now see them.

STEPHEN E. DWORNIK

Director of the Planetary Geology Program at NASA. He was born in Buffalo, New York in 1926. He joined NASA in 1965 after long experience in the field of mining at the U.S. Army Corps of Engineers' Research and Development Laboratory, which he joined just after receiving his degree in geology in 1951. At NASA he has supervised the development of the unmanned probes, beginning with the guidance of the Surveyor Program that, for the first time, carried five unmanned probes to a landing on the Moon, bringing back the first direct analyses of its surface.

THOUGH it may appear to be a paradox, the rediscovery of the planet Earth in the mid-seventies—in other words the formulation of a new overall description of the characteristics and life of our planet—must necessarily begin in space. This new starting point consists of the view obtained through space exploration, of Earth as a planet belonging to the solar system, and as a member of a family of heavenly bodies to which it is closely linked both by reason of its position in relation to them and its origin. We can now, for the first time in the history of the human race, compare our planet with its companions in the solar system, and at long last come to fairly reliable conclusions as to whether the Earth is unique in the solar system and, if this is so, for what reasons.

The history of unmanned space probes begins with the U.S. Rangers and the USSR's Zondi (1964–65) and with the pioneering TIROS modules (1960), which were little more than rudimentary cameras, sent respectively around the Moon to determine its morphology, and around the Earth to examine cloud coverage on the side facing the Sun. From then on, with an eye to manned spacecraft landing on the Moon, more and more sophisticated computation and remote control systems were developed, culminating in the successful soft landings of Russian and American automatic lunar probes (Luna 9 and Surveyor 1, 1966).

Meteorological satellites—the early Nimbus series (1964) and the later and more sophisticated ITOS (1970)—made it possible to keep an eye on cloud coverage even in night-side regions while remote analysis techniques reached a peak of perfection with ERTS-1 (1972) and ERTS-2 (1975) (or Landsat 1 and Landsat 2 as NASA later renamed them). These were able to monitor meteorological, ecological and oceanographic characteristics and to watch over vegetation on a planetary scale.Following the important planetary and lunar discoveries stemming from the U.S. and Soviet space probes from Sputnik to the Mariner 10 mission, a new series of unmanned flights, which revealed some new and startling planetary facts were initiated and successfully carried out by these two countries from 1974 to 1979. The USSR continued its Venera series of Venus probes with Venera 6 and 7 and then with the Venera 9, 10 orbiters and landers which achieved soft landings on Venus and, for the first time, transmitted to the Earth direct photographs of the Venusian surface, showing its rocks and soil. These were followed by the more recent Venera soft landing missions in

selves would also constitute both science and the history of science at the same time.

For the earth sciences the interval during which the yearbook project assumed concrete form was a classical age of discovery and driving enthusiasm that followed the realization that every individual research result was falling easily into place within a coherent and consistent new theory of planetary activity—the plate tectonics model. And every year the researchers found themselves confronting a planet that was totally unknown, unbelievably inconstant and variable. Phenomena that were apparently unrelated turned out to be different aspects of only one universal process of transformation. The past, then, would have to be completely reinterpreted. A clue would have to be followed through the labyrinth of Earth's history to an interpretation of its present form, and an evaluation of the consequences of this new and different reading of its past in terms of our knowledge of the evolution of life.

Finally, the great effort of research that had resulted in the construction of this new, coherent model of the planet revealed itself in its actual dimensions to be a total renovation that had transformed the appearance of the Earth before the very eyes of its human inhabitants. It was a cultural revolution with a scope that seemed no less wide than that initiated by Copernicus and the explorers who had redefined the geographical features of the planet. What made the revolution even broader was that the same years witnessed the development of the capacity to send men into space and to probe other planets—to scan the geography of Earth from a completely new point of view and compare it with those of the Moon, Mars, Mercury, and other members of the solar system. Mankind has essentially experienced and is still experiencing to a great extent both the excitement of discovery and the adventure of rediscovering the planet.

*This book is a collection of articles already published in the yearbooks (*Scienza e Tecnica*), articles selected from those regarded as most significant in their documentation of the new view of the planet as it has emerged in the last few years. The sequence of articles is intended to suggest the complex meaning of the new aspect assumed by the planet. First, we see how Earth finds its place in the solar system and how it compares with other heavenly bodies as photographed by space probes from various points. Second, we spend some time on what is known of the origin and primordial history of the planet and go into some detail on significant events like the appearance of the first crust and first forms of vegetable and animal life. Third, we examine the physical mechanism responsible for the changes the planet is now undergoing, from continental drift to the expansion of the ocean bottoms to plate tectonics, and, by projecting ourselves backwards in time trace the sutures that join past and present in the fields of geography, the environment, and ultimately the evolution of life. We also thought it would be appropriate to include the voice of a dissident, V. V. Belousov, who offers an alternate version of the phenomena in certain cases where we might otherwise consider matters to have been definitely resolved. The task of science is after all to dispute the opposition and not to ignore it.*

Each article is preceded by a brief introduction intended to fulfill several functions. The first is the obvious one of stressing the significance of the individual author's contribution and providing bridges between the individual contributions.

The second and perhaps more important function is to make up a primary level of approach to the book, which from this point of view is a completely new experiment in scientific communication. We have attempted in short to have the introductions maintain a level of communication sufficient to maximize our readership while at the same time in their totality summarizing the fundamental concepts of the individual articles. So a reader approaching geological discussion for the first time can, by simply reading all the introductions in order, gain a fairly complete idea of the themes discussed and of their cultural importance. At a second reading he can go more deeply into the subjects that interest him the most without necessarily following the order in which they are printed. Thus, the book offers two possible levels of reading, that of a first approach and that of profound and direct study—achieved by reading what has been written by the scientists themselves responsible for the increase in this field of knowledge.

EUGENIO DE ROSA

Preface

Human beings discovered the planet Earth for the first time when, following Copernicus (1543), we were able to assign it its true position within the solar system and precisely define its geographical features. The second stage in the "discovery" of the planet extends over a long period. If we understand "discovery" to include both the acquisition of new knowledge and its dissemination, we can probably locate the process between two limits in time—the 6th century B.C., when the Milesian, Anaximander, first translated the knowledge available to him of the then known world into symbolic form (constructing the first "map"), and December 16, 1911, when Roald Amundsen finally set foot for the first time on the most inaccessible spot on Earth, and the most remote from human habitation—the South Pole. The flag he raised at 0° south latitude was to become a symbol, the culmination of an endeavor that had continued almost as long as human history itself, with the main stages in Western culture coinciding with the appearance of successive fundamental modifications in the structure of the map of the Earth. For example, Anaximander's view was replaced by Ptolemy's (A.D. 150 the "Earth" for Europeans being practically confined to the limited number of known lands bordering the Mediterranean); the Portuguese navigators rounded the southernmost extremity of Africa (1497), and the Indian Ocean (which for Ptolemy had been an enclosed sea) was joined to the Atlantic; Columbus and Magellan added America and the Pacific to the map (1492 and 1520); James Cook added Australia (1770) and the Russian, von Bellingshausen, the Antarctic (1819–21). The Northwest Passage (McClure, 1851 and Amundsen, 1903–1906), the Northeast Passage (Nordenskjold, 1879), the conquest of the North Pole (Peary, 1909) and that of the South Pole (1911) all helped to fill in the details, and finally Earth began to assume the definite form we know today.

In the sixties and seventies of this century we have "rediscovered" our planet—we have looked back on it from outer space and have been able to compare it with other heavenly bodies. But above all, we have come to understand that Earth is subject to a continuous process of change, of which volcanism, seismicity, and mountain-building are only the most evident manifestations. We have come to understand that these phenomena are not separate and independent but actually different aspects of the same process—the dynamics of the terrestrial crust, which translates into the perennial wanderings of the continents, the creation and destruction of new crust, and the continuing modification of those geographic features it has worked so hard to form. As a result, we have had to reexamine these characteristics, to see them in a new light as ephemeral evidence, fleeting frames from a motion picture with a more complex plot than we had previously assumed. In short, our rediscovery has been of a new Earth, different from the one we had been familiar with.

This book is evidence of that cultural revolution as seen through the eyes of its protagonists, the scientists and researchers who were able to bring it about step by step. In 1967, the same Mondadori team that, under the leadership of Edgardo Macorini, had just finished the first edition of the largest Italian encyclopedia, the Enciclopedia della Scienza e della Tecnica, *was given the task of following yearly progress in the different sectors concerned with science and technology with the object of bringing out a series of yearbooks to keep the initial reference work up to date. These books are not, like the encyclopedia itself, organized alphabetically, but in a monograph format. This arrangement, which was the source of much perplexity at first, turned out over a number of years to be a determinant in the gathering and explication of the essential facts of a complex but basically ordered process like scientific research. What the team had to do, then, was to survey a vast panorama composed of the various fields of knowledge involved, and work from a number of options with respect to the emerging subject matter and protagonists, in whatever part of the world they might be.*

Once the main subjects were outlined, the team began to contact the scientists, researchers, and engineers who had made the most significant contributions to the fields, in order to provide first-hand accounts that would maintain the high level of information in the encyclopedia. In short, great care was taken to ensure that each article would represent a document with a value beyond that of merely updating the parent work. Each would factually and individually spotlight a particular moment in the very rapid development of science, and the actual sequence of reports from the eyewitnesses them-

THE REDISCOVERY OF THE EARTH

STEPHEN E. DWORNIK

The Solar System View in 1975

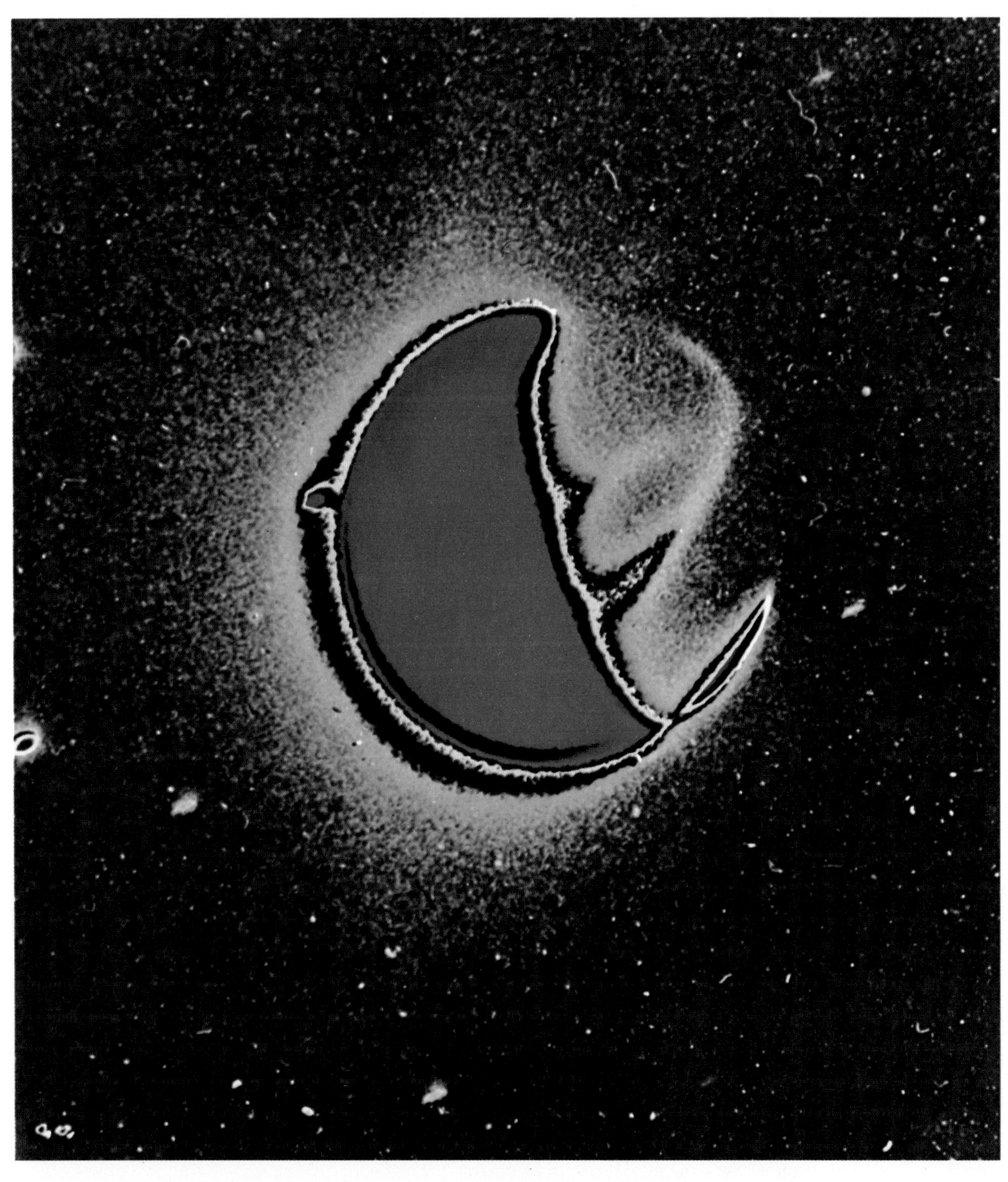

Overleaf: A most unusual view of the Earth and its atmosphere taken by the astronauts of the Apollo 16 mission. The photograph, taken by ultraviolet light, shows the oxygen in the Earth's atmosphere ionized by the ultraviolet radiation of the Sun (NASA)

Our solar system consists of planets, moons, comets, asteroids, meteorites, dust, all revolving around a star—the Sun. These objects range in size from a fraction of a millimeter, to the Sun itself which has a diameter of approximately 860,000 miles. Relative to our Earth, the Sun has an equatorial diameter about 110 times larger: Jupiter, the largest of the planets has a diameter about 11 times larger, and Mercury, the smallest, is less than 0.4 times as large.

While distances between bodies in our solar system happen to be great, it requires an imaginative mind to comprehend distances between our system and the nearest possible solar system, alpha centauri. Scientists measure astronomical distances in light-years. A light-year is the distance light travels in one year at the rate of 300,000 kilometers per second. In these terms, North America is about 3/100 of a light-second from central Europe: Earth is about 1 light-second from our moon and nearly 8 light-seconds from the sun. The overall diameter of our solar system is about 10 light-hours, the galaxy in which we are located is about 90,000 light-years in diameter. If you imagined our solar system to be but 1 meter wide, located in Rome, alpha centauri would be located about 300 km away near Florence.

The diagram shows the relative distances of the planets from the Sun as compared to a 100-meter long soccer field. On this scale, our own moon would be represented by a pinpoint on the circumference of the dot that represents the Earth. The exploration of the solar system up to 1979 has indeed been "close to home."

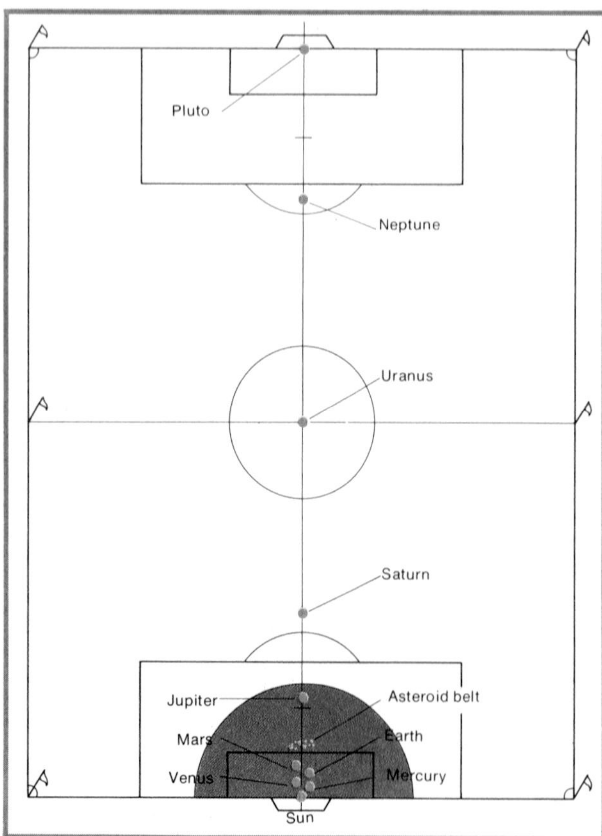

Relative distances of the planets from the Sun, as seen on the scale of a 100-meter long soccer field.

In the late seventies, we can view the solar system with instruments and by using techniques that did not exist before 1965. Space exploration has progressed from its initial stage of spacecraft weighing a few kilograms placed in orbit around the Earth, to the most complex, sophisticated technological achievement man has ever viewed—the landing of a man on the Moon and the return to Earth of a lunar sample. More recently in 1979 we have the amazing pictures of Jupiter and its satellites sent to the Earth by Voyager 1.

Space exploration can be divided into three general approaches:

1. Man and instrument on Earth
2. Man on Earth and instrument in space, landed on a solar system body
3. Man and instrument in space on a solar system body.

Each approach has its own unique advantages and disadvantages and each is necessary to carry out a successful space exploration program. Telescopic observations, however limited by the Earth's own atmosphere, provide valuable data. Development of Earth-based radar imaging techniques will be able to probe the depths of Venus and give us information on the surface of that mysterious planet that is shrouded from view by its dense clouds. Unmanned spacecraft will continue to carry man's instruments to the far reaches of our solar system, and even to the uncharted regions of the universe. Because distances such as these must be measured in terms of time, the journey of man out of our solar system perhaps may never be achieved.

The Results of the Viking Mission to Mars

Two Viking orbiters and landers reached Mars in the summer of 1976 and performed a series of very important experiments and collected significant observational data. The Viking 1 orbiter was placed in an elliptical synchronous orbit around Mars on June 20, 1976, with a period of 24.6 hours, and Viking 2 went into orbit on August 7 with a similar period. The two landers from these orbiters landed safely on the Martian surface in September and began to operate almost immediately. Operating with incredible success for several months, these two Viking missions carried out 13 scientific investigations transmitting back to the Earth important information about the Martian atmosphere and surface. Photographs of the surface from a height of 1500 km (from the orbiters) and directly from the ground (from the landers) reveal an incredible, wind-swept landscape of rocks and stones, partially covered with sand drifts. The atmospheric composition, the chemical composition of the soil and rocks, and the meteorological conditions were extensively measured and analyzed. The Viking missions continued their Martian investigations for about a year.

From the photographs taken by lander 2, starting September 3, one sees that the Martian surface is a boulder-strewn, reddish desert, threaded by troughs or fractures that form a complex polygonal network. The photographs from both landers show a pinkish colored sky, whose color is probably due to a fine suspension of the sandy surface material in the upper Martian atmosphere. The general appearance and structure of the surface rocks strongly indicate a volcanic origin. The inorganic analysis of the rocks and the soil shows very high abundances of silicon and iron, with significant concentrations of magnesium, aluminum, calcium and titanium. There is a much higher concentration of sulfur in the Martian soil than in the Earth's soil but a much smaller concentration of potassium.

The analysis of the Martian atmosphere shows the presence of krypton and xenon as well as argon and neon. In the height range from 120 to 200 km of the atmosphere carbon dioxide is the major constituent, but detectable concentrations of nitrogen, argon, carbon monoxide, molecular oxygen, atomic oxygen, and nitric oxide are also present. The temperature of the upper atmosphere is about 200°K.

The surface soil is relatively dense and rocky in some spots but loose and granular in others. From the depths to which the foot pads of the landers sank into the surface it is clear that the soil consists of weakly cohesive, fine-grained material which can flow around heavy objects deposited on it. This indicates that air (the atmospheric gas) is locked in the pores of the soil.

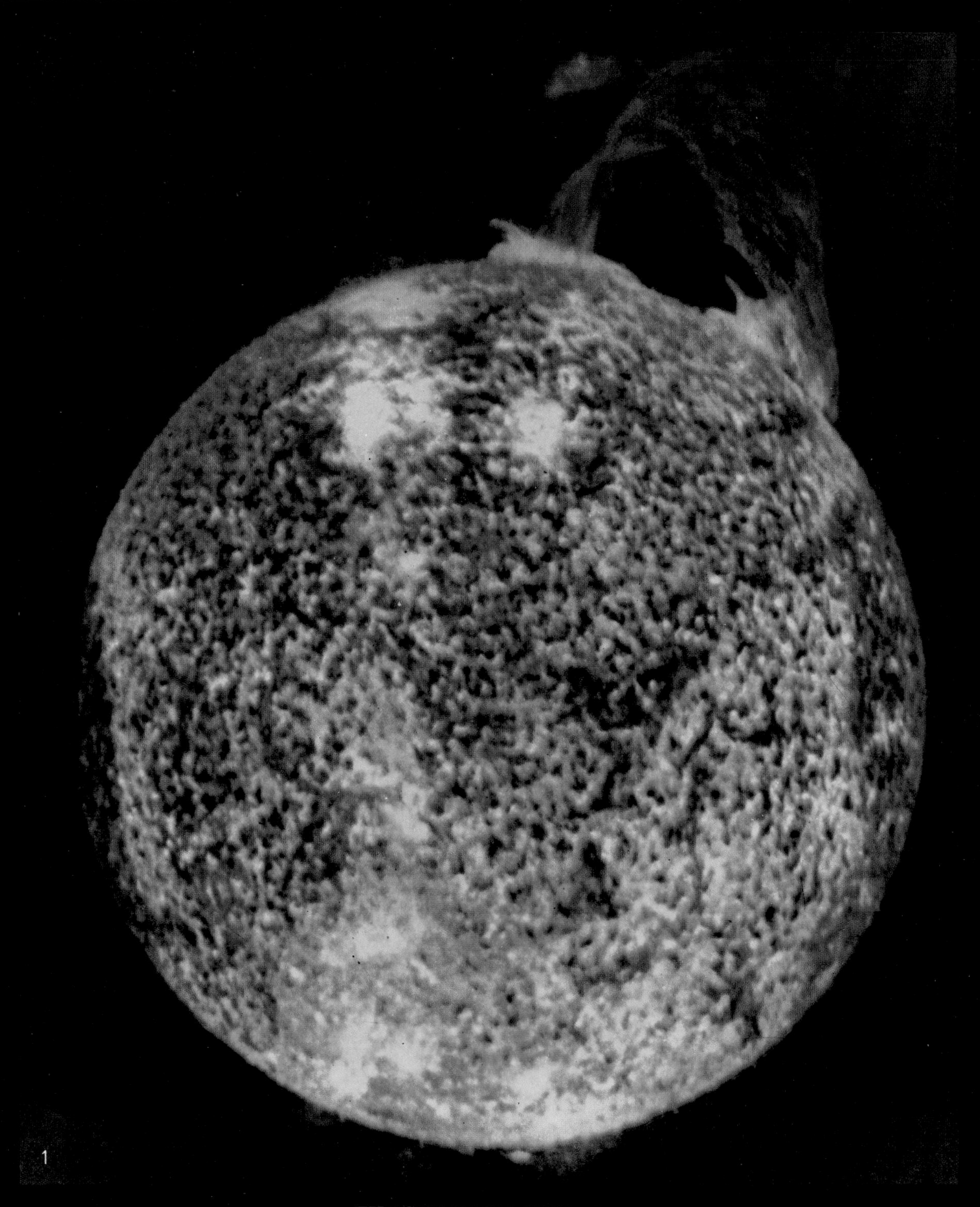

THE SUN FROM SKYLAB

Fig. 1-1. This photograph of the Sun, taken in 1973, by NASA's Skylab 4, shows one of the most spectacular solar flares ever recorded, covering more than 588,000 km of the solar surface. The previous pictures, taken some 17 hours earlier, showed this feature as a large quiescent prominence on the eastern side of the Sun. The flare looks like a twisted sheet of gas in the process of unwinding itself. Skylab photographs such as these may provide clues to the mechanism by which such quiescent features erupt from the Sun.

In this photograph, the solar poles are distinguished by the relative lack of a supergranulated network, and they have a much darker tone than the central portions of the disk. Several active regions are seen on the eastern side of the disk.

The photograph was taken using ionized helium light, by the extreme ultraviolet spectroheliographic instrument that registered the light from the ionosphere.

2

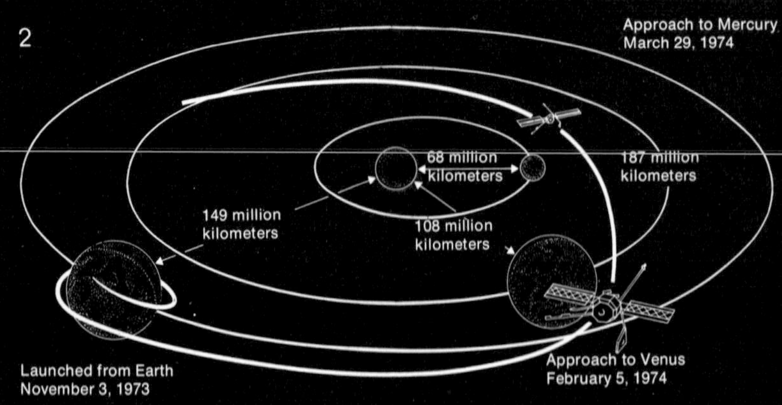

MARINER 10 MISSION TO VENUS AND MERCURY

Fig. 1-2. The first close-up pictures of Venus and Mercury were obtained by the NASA Mariner spacecraft, launched November 3, 1973 from Cape Kennedy. Its flight path took it to Venus, where the planetary gravity field altered its course toward Mercury. The spacecraft first encountered Mercury on March 29, 1974 (Figs. 1-3 to 1-8) and went into orbit around the Sun. On September 21, 1974, the spacecraft encountered Mercury for the second time (Figs. 1-9 to 1-12). It will remain in Sun-Mercury orbit until it is drawn into the Sun by the Sun's gravity field.

Fig. 1-3. Eighteen individual pictures from this photomosaic of Mercury. The pictures were taken from an altitude of about 200,000 km by the spacecraft as it approached the planet. About two-thirds of the visible surface is in the southern hemisphere. The smallest feature that can be seen is about 1 km in size; the largest crater is about 200 km in diameter. The bright crater near the equator is tentatively named "Kuiper" after scientist Gerard Kuiper. The bright streaks, or rays, emerging outward from this crater were caused by debris material that was ejected by the body that formed the crater. Craters are the predominant surface feature. They range in size from basins to shallow, barely discernible depressions. Some craters have lost their ejecta deposits and secondary crater fields. Some younger craters exhibit more recent features such as terraces on the side walls, central peaks and ring complexes. This view of Mercury does not show any of the widespread flood plain debris which is a characteristic of Earth's moon.

Mercury, the innermost planet of our solar system is small (less than half the diameter of Earth) hot, because of its proximity to the Sun, and has an unusual orbit around the Sun. For every two rotations around the Sun, it rotates on its own axis three times. This characteristic was first discovered in the late 1960s using ground-based radar antennas.

Prior to 1974, no information was available concerning surface characteristics. Astronomers were only able to plot faint planetary wide markings. Mariner 10 photographs showed that Mercury had basically three different types of terrain features: cratered, plains and scarps.

Cratered terrain. Craters are the predominant landform, ranging in size from basins as large as 1300 km across, down to pits that are barely detectable on the highest resolution photographs. The craters and basins are morphologically similar to lunar craters of the same size and exhibit the same stages of degradation as their lunar counterparts. The radial distance of continuous ejecta deposits, and areas of secondary craters surrounding the larger craters, do not extend as far around as similar size lunar craters; and the craters are shallower than their lunar counterparts. All three differences are consistent with Mercury's greater (2x) gravitational acceleration which would reduce the ballistic range of ejecta and cause a greater degree of postcratering collapse of the crater rim.

The most prominent feature on Mercury is the Caloris Basin and it is similar in appearance and size to the lunar Imbrium basin. The basin is bounded by a ring of mountains which form a scarp averaging 2 km in height. Extending outward from the main scarp for at least 1 km, the basin diameter is a radial system of linear hills which is best developed northeast of the basin. The visible portion of the basin is surrounded by the radial system of hills and beyond the hills by an area of level plains.

Plains. Nearly level surfaces, wh are relatively free of craters lar than 10 km, occur over the floo many large craters and basins, as w as around several large basins. rims of the enclosing craters as wel the surrounding heavily cratered rain, contain abundant craters present in the younger, smoot floors.

Plains material fills the Caloris Ba to within 2 km of the highest pe

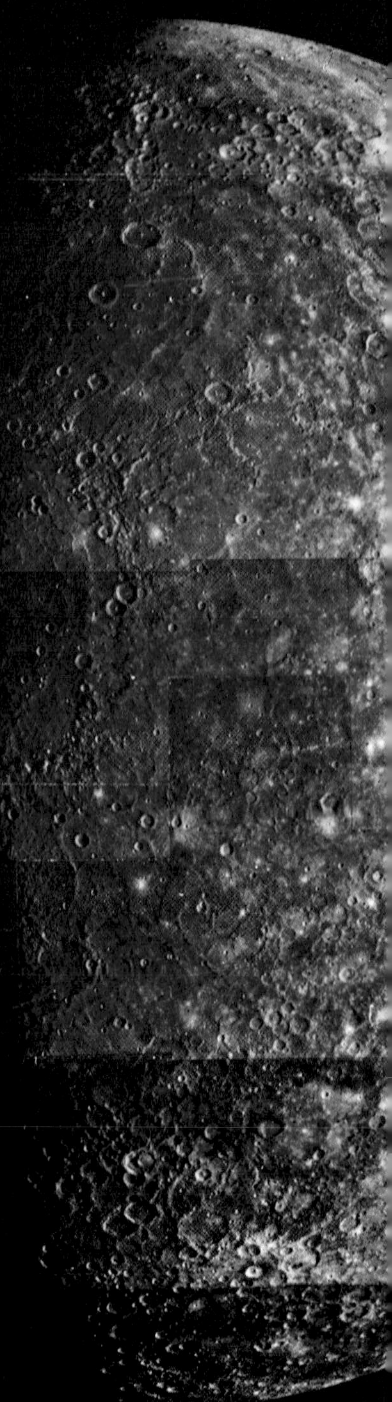

3

nd surrounds most of the basin in an rcuate band from 1000 to 1500 km ride. The plains within the basin con-ain numerous ridges and fractures, nd their extent and complexity is reater than on lunar Maria. Some nd to follow the direction of the frac-ures, suggesting that the structures re related. The Caloris Basin frac-ure pattern probably developed dur-ng subsidence of the central part of he basin floor after the plains mate-ials had been laid down.

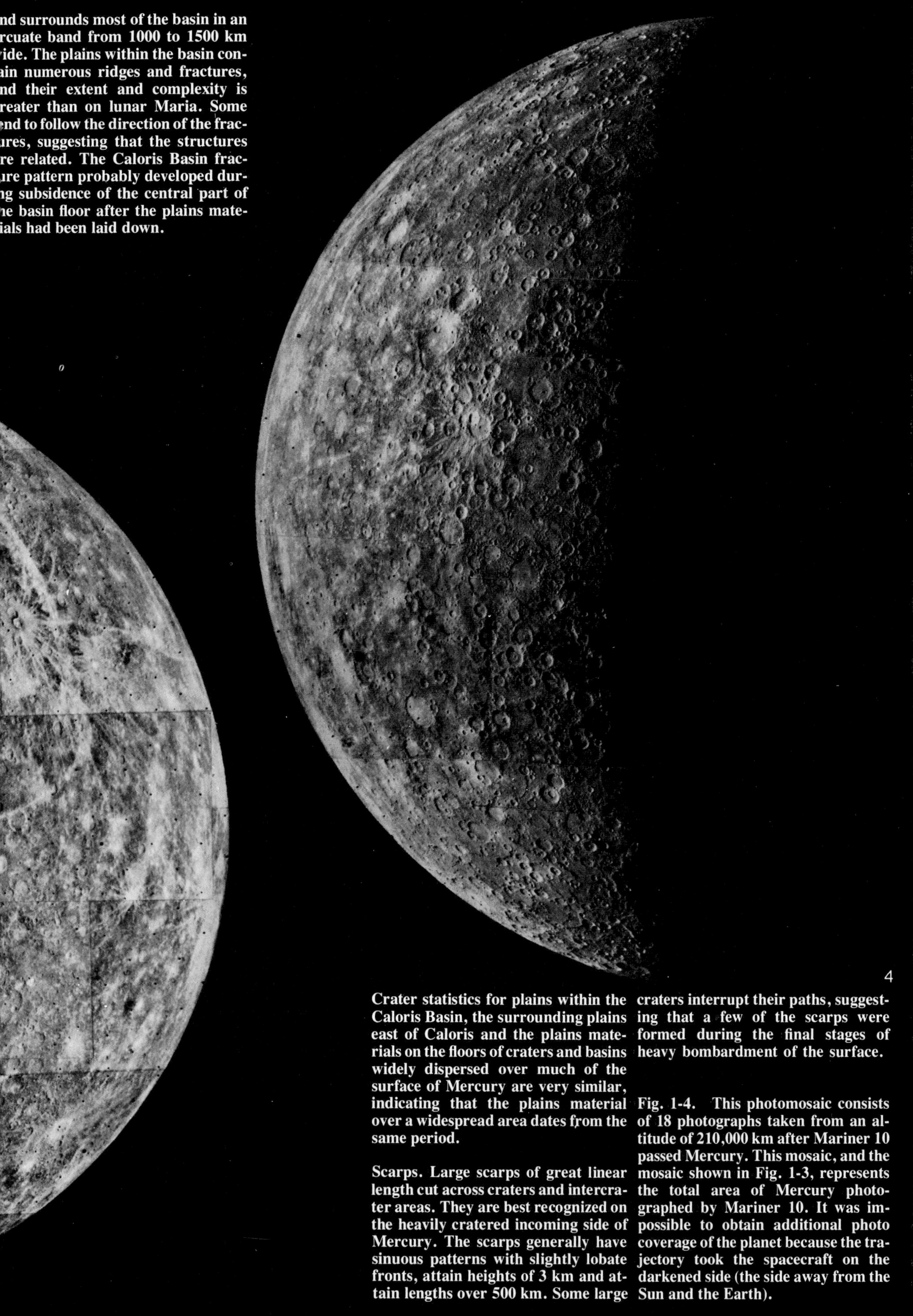

Crater statistics for plains within the Caloris Basin, the surrounding plains east of Caloris and the plains mate-rials on the floors of craters and basins widely dispersed over much of the surface of Mercury are very similar, indicating that the plains material over a widespread area dates from the same period.

Scarps. Large scarps of great linear length cut across craters and intercra-ter areas. They are best recognized on the heavily cratered incoming side of Mercury. The scarps generally have sinuous patterns with slightly lobate fronts, attain heights of 3 km and at-tain lengths over 500 km. Some large craters interrupt their paths, suggest-ing that a few of the scarps were formed during the final stages of heavy bombardment of the surface.

Fig. 1-4. This photomosaic consists of 18 photographs taken from an al-titude of 210,000 km after Mariner 10 passed Mercury. This mosaic, and the mosaic shown in Fig. 1-3, represents the total area of Mercury photo-graphed by Mariner 10. It was im-possible to obtain additional photo coverage of the planet because the tra-jectory took the spacecraft on the darkened side (the side away from the Sun and the Earth).

MERCURY'S LARGEST BASIN

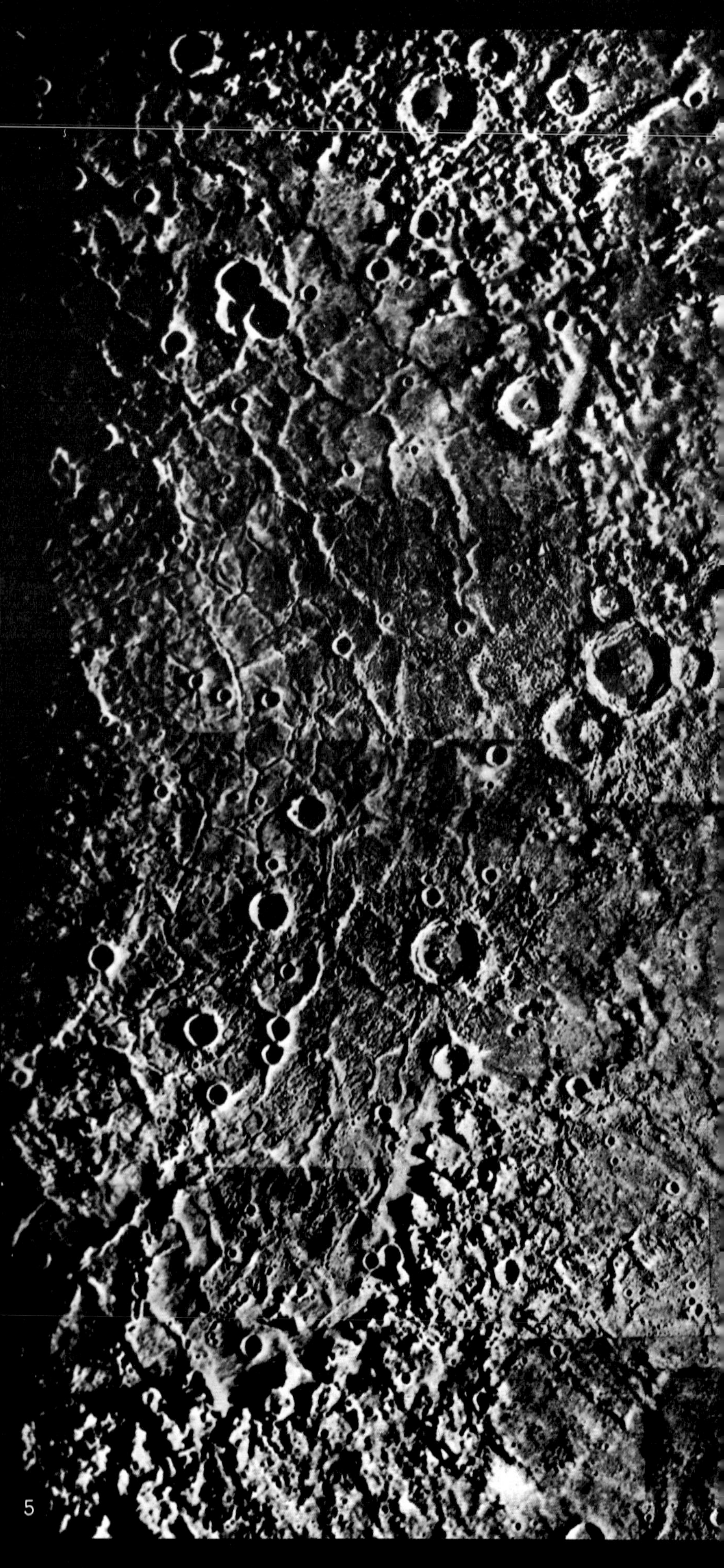

Fig. 1-5. Caloris Basin, the largest structural feature apparent in the Mariner 10 pictures, is similar in appearance and size to the lunar Imbrium Basin and undoubtedly was caused by the impact of a body tens of kilometers in diameter. The basin, 1300 km in diameter, is bounded by a ring of mountains which forms an irregular scarp rising to around a height of about 2 km above the basin floor. Between about 23° and 30°N, the scarp is much lower and appears to be mantled by plains material. In the northeastern part of the basin, a weak outer scarp occurs at a distance of about 150 km beyond the main scarp. Between these two scarps is a terrain characterized by relatively smooth hills or domes similar in appearance to the terrain adjacent to the Rook Mountains in the lunar Orientale Basin. Surrounding the main scarp and extending outward for a distance of 1300 km is a radial system of linear hills which is most developed to the northeast of the basin. The radial system is only weakly developed in the terrain between the two scarps; its main development begins beyond the outer scarp in this area. This radial system of hills is surrounded by smooth plains material at least over the visible eastern portion of the basin.

The plain inside the Caloris Basin contains numerous ridges and is intensely fractured. Ridges range from 1.5 to 13 km in width, have heights of about 300 m and lengths in excess of 300 km, and are superficially similar to the lunar mare ridges. The extent and complexity of the ridges and associated fracturing inside Caloris Basin are greater than they are in the lunar Maria Basin. Fractures are closely spaced, with some forming polygonal patterns; others are almost sinuous, although unlike lunar sinuous rills in having a detailed planimetric outline. They range in width from 6 km down to the best resolution of the photography of the basin floor (700 m). The widest fractures are flatfloored and grabenlike. Fractures transect, are parallel to, and even occur along the tops of ridges. The directions of fractures tend to follow the direction of the ridges, suggesting that the structures are related. The Caloris Basin fracture pattern seems consistent with the slight subsidence of the central part of the basin floor which apparently followed the creation of the plains. Subsidence has also affected lunar mare basins, but not to the same extent or in exactly the same pattern.

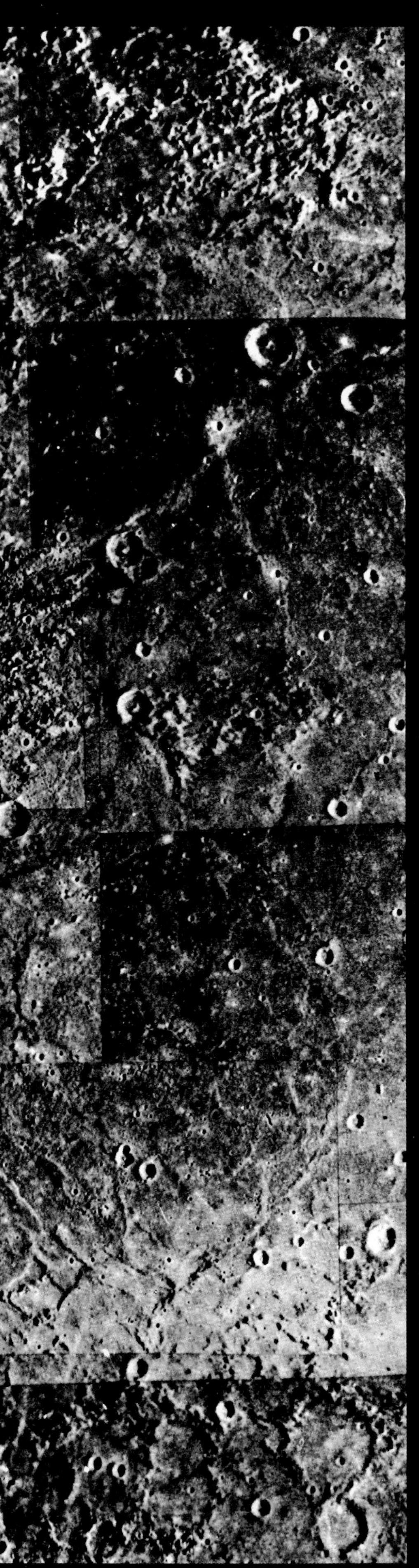

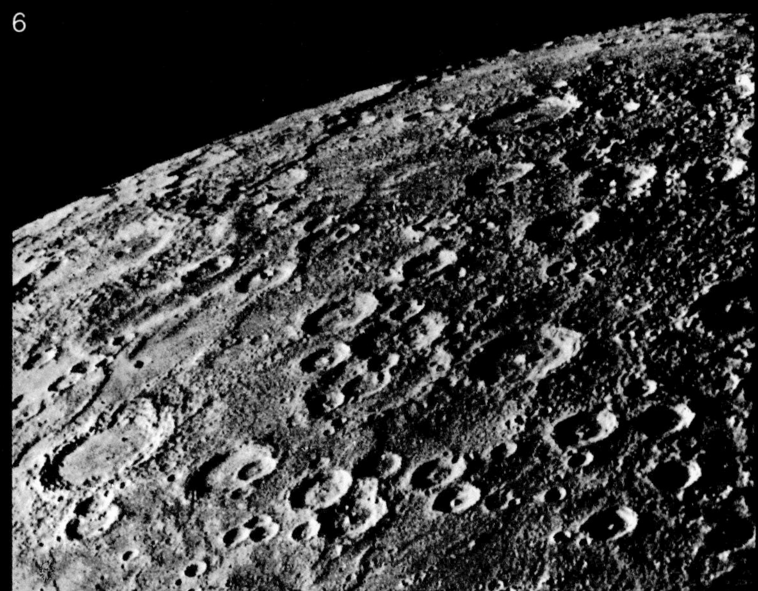

Fig. 1-6. This view of Mercury's northern section shows a prominent east-facing scarp extending from the section near the middle of the photo, southward for hundreds of kilometers. This is the first time that a scarp of this size has been detected on any planet. It is believed that it was formed as a result of crustal folding and fracturing. If this is true, Mercury underwent a period when tectonic activity was the predominant geological phenomena that sculptured its surface, and probably occurred at the same time that it was going through an overall global cooling-down period. The plains material evident on page 20 marked C and F, is thought to have been formed at the same time that scarp features were formed. Low sun angle illumination of the surface reveals craters of a few kilometers in size. The largest crater located near the limb on the left side of the photograph is approximately 300 km wide. The rims of the larger craters rise 1–2 km above the surrounding terrain.

Many craters have central peaks and rims. The majority of the craters were formed by the impact of ejecta thrown across the Martian landscape by the impact of a meteorite.

Fig. 1-7. This photograph, which covers an area 290 km by 270 km, shows a heavily cratered surface, low hills, and basins filled with plains material. The large valley that runs from the most prominent basin (80 km in diameter) near the lower center of the photograph is 7 km wide and more than 100 km long. Both the crater and valley are filled with volcanic material similar to the fill material in the Caloris Basin (marked D page 20). The surface of this plains material contains fewer craters and is therefore younger in age than the surrounding heavily cratered terrain. The plains materials are thought to have been formed during periods of local volcanic activity that followed formation of most of the basins, rather than from molten rocks resulting from a major impact cratering event. Patches of plains materials on the floors of craters and basins over the rest of Mercury are indistinguishable in age or morphology from the plains concentric to and inside Caloris Basin. Some of these smaller tracts of plains materials could perhaps be impact melt from nearby craters or basins, but for many there is no well-defined source crater (Fig.1-11). We have observed no direct evidence of volcanism such as cones, domes, or flow fronts.

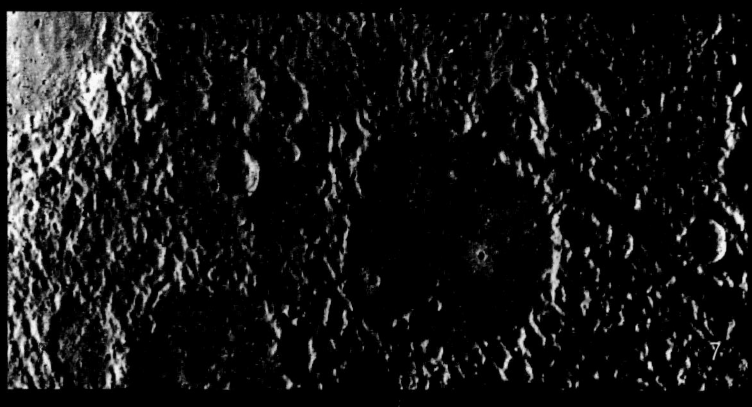

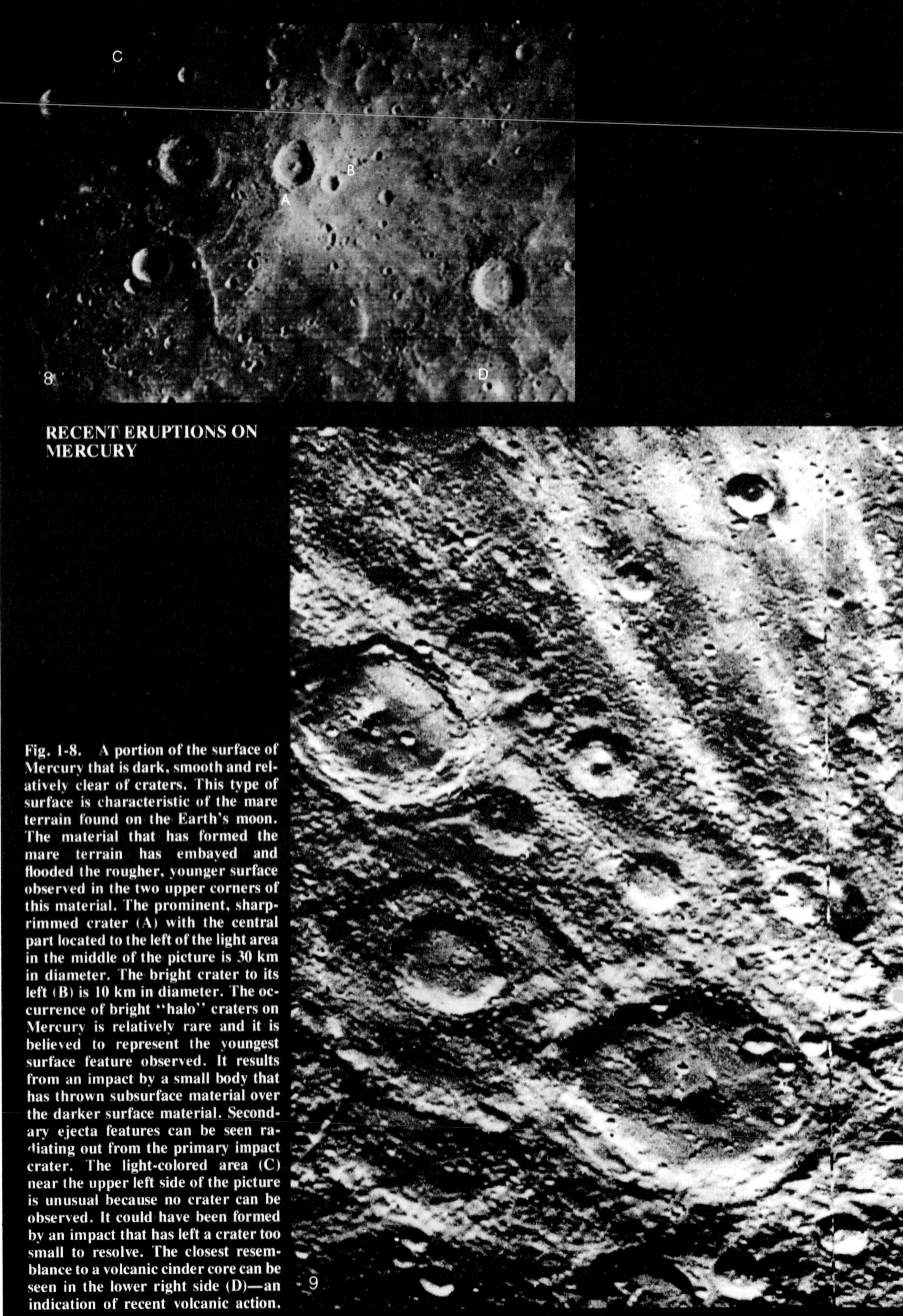

RECENT ERUPTIONS ON MERCURY

Fig. 1-8. A portion of the surface of Mercury that is dark, smooth and relatively clear of craters. This type of surface is characteristic of the mare terrain found on the Earth's moon. The material that has formed the mare terrain has embayed and flooded the rougher, younger surface observed in the two upper corners of this material. The prominent, sharp-rimmed crater (A) with the central part located to the left of the light area in the middle of the picture is 30 km in diameter. The bright crater to its left (B) is 10 km in diameter. The occurrence of bright "halo" craters on Mercury is relatively rare and it is believed to represent the youngest surface feature observed. It results from an impact by a small body that has thrown subsurface material over the darker surface material. Secondary ejecta features can be seen radiating out from the primary impact crater. The light-colored area (C) near the upper left side of the picture is unusual because no crater can be observed. It could have been formed by an impact that has left a crater too small to resolve. The closest resemblance to a volcanic cinder core can be seen in the lower right side (D)—an indication of recent volcanic action.

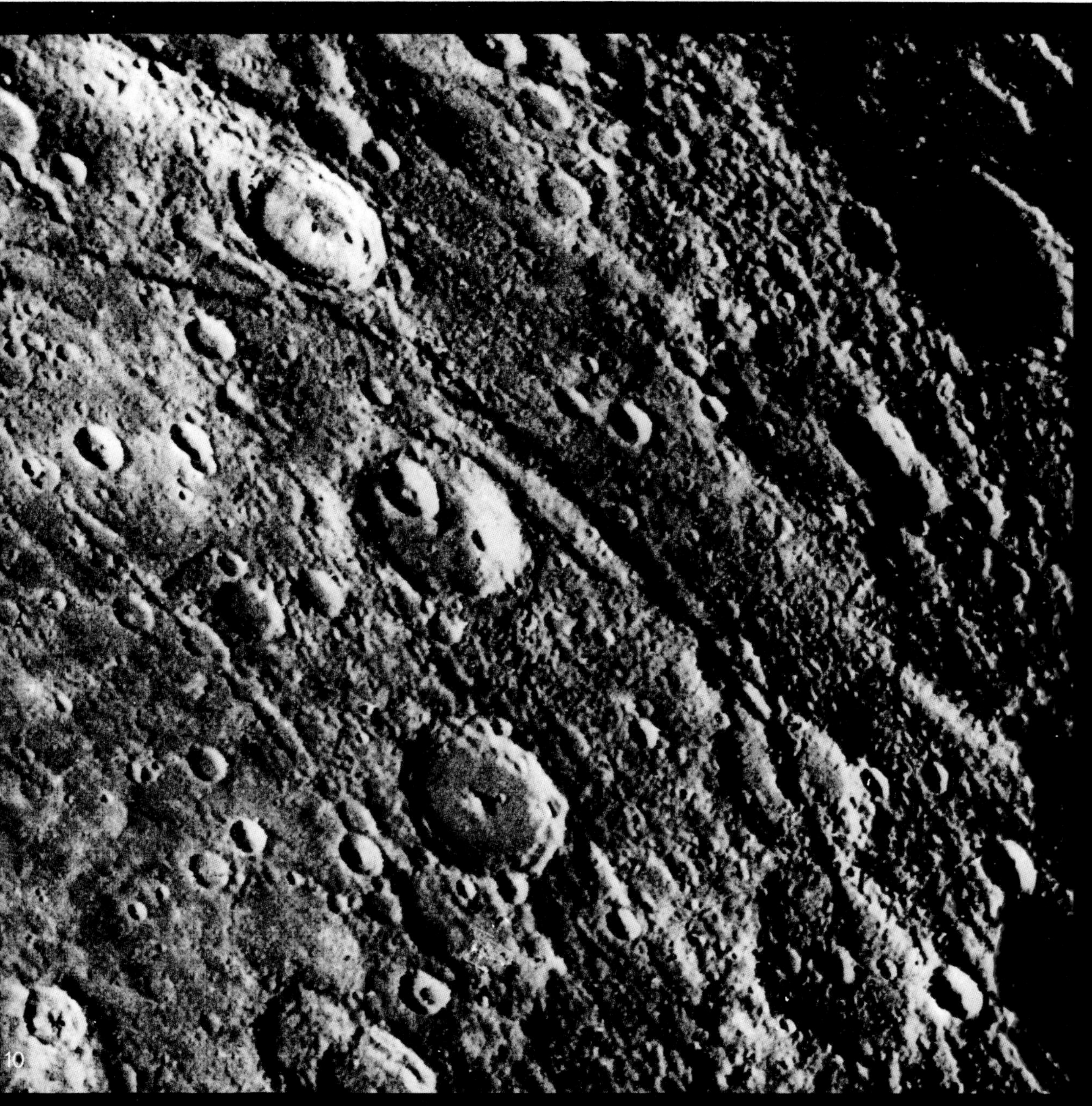

Fig. 1-9. A field of bright rays, created by ejecta from a crater, radiating to the north (top) from off camera (lower right) is seen in this view of Mercury taken in 1974 by Mariner 10. The source of the rays is a large new crater to the south, near Mercury's south pole. Mariner 10 was about 48,000 km from Mercury when the picture was taken at 2:01 p.m. PDT just 3 minutes after the spacecraft was closest to the planet. The largest crater shown is 100 km in diameter.

Fig. 1-10. A scarp or cliff, more than 460 kilometers long extends diagonally from upper left to lower right in this Mariner 10 picture. Numerous similar structures have been discovered by Mariner 10 during the television sequences on the spacecraft's second flyby of the planet. These structures are believed to be formed by compression caused by crustal shortening. The picture was taken from 64,500 km.

11

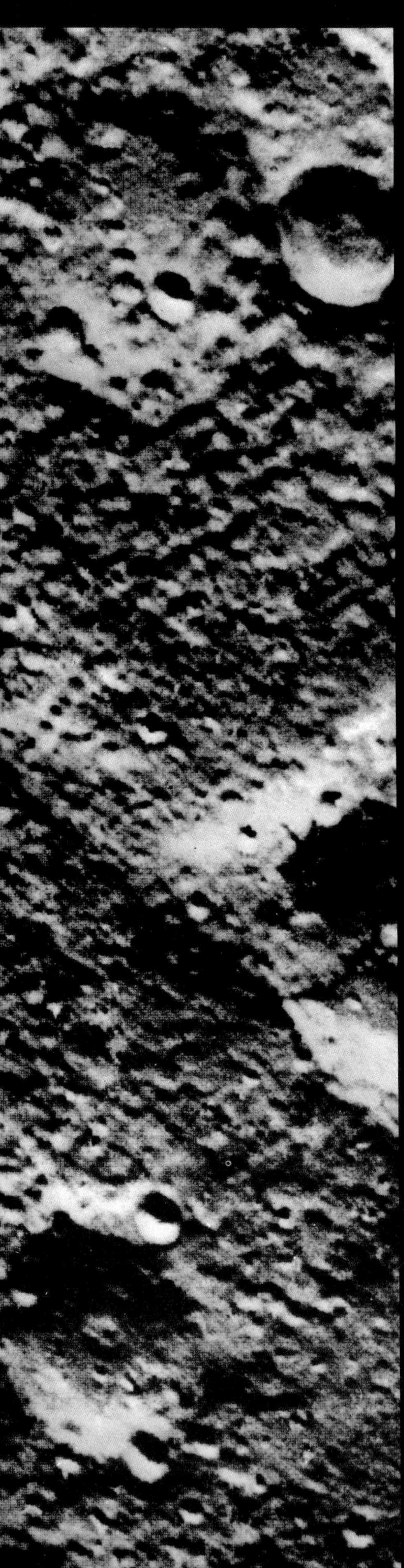

MERCURY'S LARGE CRATERS

Fig. 1-11. Taken about 40 minutes before Mariner 10 made its close approach to Mercury on September 21, 1974, this picture shows a large double-ringed basin in the center of the picture, 230km in diameter, located in the planet's south polar region—75° S. Lat. 120° W. Long. The basin was seen from a different angle on Mariner 10's first sweep past Mercury in March of the same year. This picture was taken from about 55,000 km. North is at upper left.

Fig. 1-12. Mercury's south pole was photographed by one of Mariner 10's TV cameras as the spacecraft made its second close-up flyby of the planet on September 21, 1974. The pole is located inside the large crater, 180 km in diameter on Mercury's limb (lower center). The crater floor is shadowed, and its far rim, illuminated by the Sun, appears to be disconnected from the edge of the planet. Just above and to the right of the south pole is a double ringed basin about 200 km in diameter. A bright ray system, splashed out of a 50 km crater is seen at the upper right. The stripe across the top occurred during computer processing. The picture was taken from a distance of 85,800 km less than 2 hours after Mariner 10 reached its closest point to the planet.

12

VENUS SEEN FROM MARINER 10 (see following p. 24)

Fig. 1-13. On February 5, 1974, Mariner 10, carrying two television cameras, crossed the Venus terminator from the dark side, swinging around the planet on a hyperbolic trajectory on its way to Mercury. The cameras were designed mainly to observe the surface details of Mercury, however, the optical design incorporated special filters, coating, and transmitting glass in order to image Venus in the ultraviolet.

Faint UV markings were discovered on Venus in 1926. Decades of subsequent UV observations from the Earth suggest a retrograde equatorial motion of 4-day duration. At least one feature (which takes the form of a dark horizontal Y) appears to be quasi-permanent.

The Mariner 10 pictures contain a surprising amount of information about the general circulation of this part of the atmosphere, which will enhance the value of ground-based observations as well as establishing a specific scientific framework for future entry probes and orbiters.

VENUS AS REVEALED BY THE USSR VENERA MISSIONS AND THE U.S. PIONEER ORBITER AND MULTIPROBE MISSIONS SINCE 1974

Fig. 1-14. The U.S. Mariner Venus flyby missions and the early USSR Venera missions, with their soft landings on the Venusian surface, revealed more about the climate and surface conditions on Venus than had ever been discovered before. In 1975 the Venera 9 and 10 flights dropped two landers on Venus which survived for approximately one hour on the planet's surface. During their slow descent through the relatively cool clouds and their very rapid descent through the hot atmosphere, the landers made important observations of the lower atmosphere and the surface. During their hour's life on the surface the landers transmitted panoramic photographs of the landing areas which reveal a rocky desert covered with sharp-edged stones of various sizes in the centimeter range. Their sharp edges indicate that they originated recently. The soil and rocks possess the natural radioactivity of uranium, thorium, and potassium.

The density of the rocks, as measured down to a depth of a few meters, is about 2.7 gms cm^{-3}, which is typical of crystalline basaltic rocks. The surface conditions on Venus are strong evidence that the same geochemical processes occurred (and are still occurring) on Venus as on the Moon, the Earth, and Mars, and that these processes subdivided these four bodies into their present shell structure, with their basalt crusts. The relative abundance of CO_2, as measured directly, exceeds 97% and the relative abundance of H_2O is 5×10^{-4}. The chemical analysis of the clouds carried out by Venera 9 and 10 confirms the previous evidence that the clouds consist of sulfuric acid droplets.

The U.S. Pioneer Venus Orbiter, with its multiprobe instruments, was launched in May 1978 and reached Venus in December 1978, going into a highly eccentric, near polar orbit on December 4, 1978, and the second Pioneer Venus Multiprobe was launched on August 8, 1978 and reached Venus on December 9, 1978. During this same period the USSR launched additional Venera vehicles which reached Venus shortly after the Pioneer missions did. The Pioneer Venus magnetometers revealed a very active Venusian ionosphere that is very responsive to the solar wind. The magnetic fields surrounding Venus are, on the whole, quite weak, amounting at most to a few gammas, but there are very rapid field fluctuations which may result in a 10-fold increase in field intensities at isolated regions.The electric field detectors on Pioneer Venus Orbiter showed that there is a strong, but variable interaction between the solar wind and the ionosphere which generates shock waves and electron plasma oscillations that emit strong electron beams. The ionosphere on the day side of Venus is rich in the ions of atomic oxygen (O^+), with C^+, N^+, H^+, and He^+ as secondary components. In the lower ionosphere the molecular ions O_2^+ dominate, with NO^+, CO^+, and CO_2^+ constituting less than 10% of the total abundance.One of the most surprising discoveries stemming from the measurements of the composition of the Venusian atmosphere is the relatively high concentrations of the argon-36 and argon-40 isotopes and the helium-4 isotope. Thus, at an altitude of 150 km there are 5×10^6 helium -4 atoms per cc compared to 6×10^9 CO_2 molecules in the Venusian atmosphere. The abundances of argon-36 and argon-40 are approximately 80 parts per million at an altitude of 135 km. This is about 300 times as great as in the Earth's atmosphere.

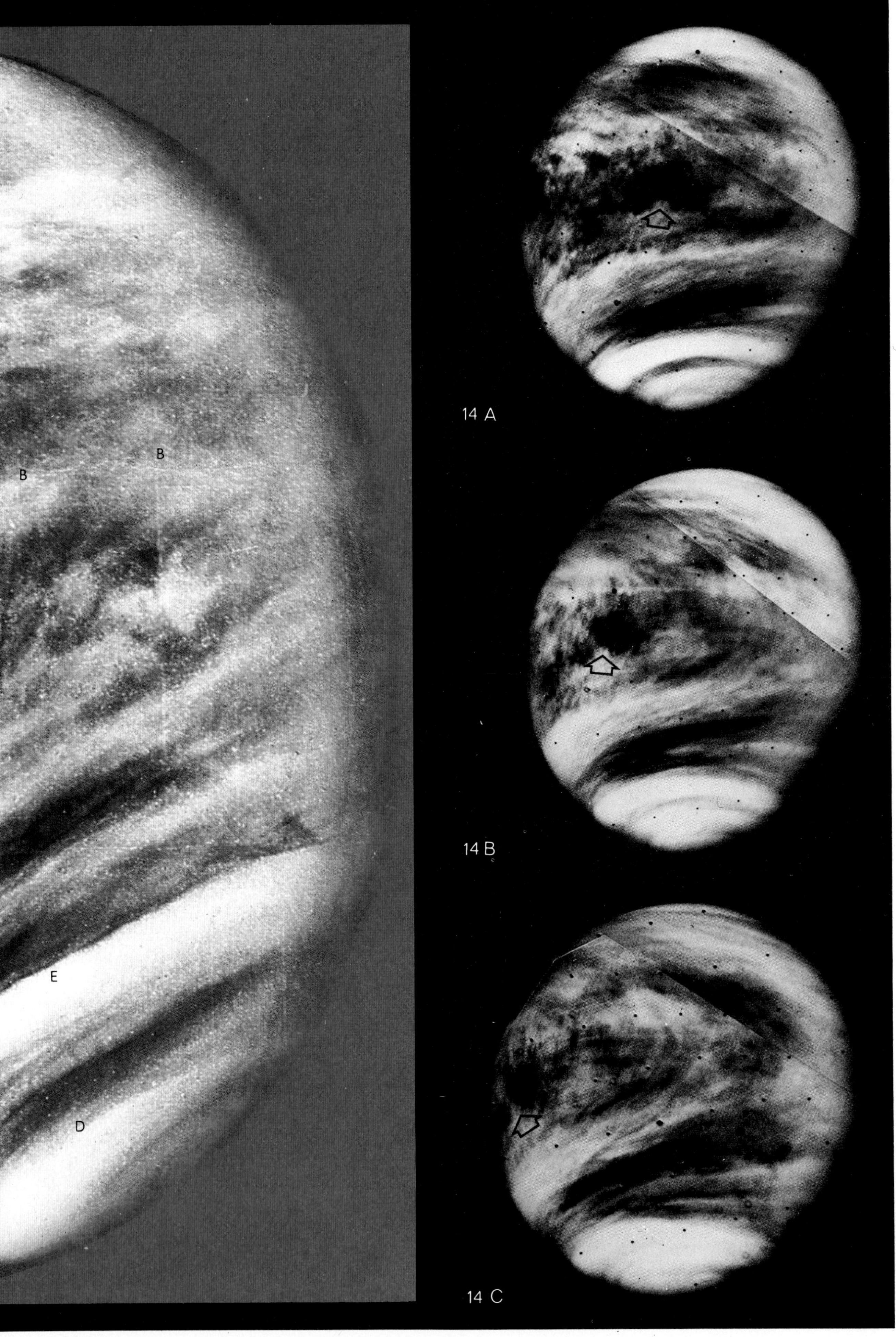
B
B
E
D
14 A
14 B
14 C

APOLLO MISSIONS ON THE MOON

Fig. 1-15. The Moon, Earth's closest neighbor, has long fascinated both scientists and laymen alike. Due to its close proximity, it appears as the brightest, largest body in the sky and, therefore, receives much attention. Long before the advent of space exploration, scientists were able to photograph its surface through ground-based telescopes. Geologists were able to interpret light colored areas as highlands and dark areas as lava-filled basins. Craters were interpreted to be of both volcanic and impact origin. But up until the mid-1960s, our understanding of the properties and characteristics of the Moon depended entirely upon ground-based observations. In the mid-sixties NASA conducted two programs to the Moon. The first was a lunar orbiter, which was successfully launched and put into lunar orbit, to photograph the surface. The second mission, Surveyor, successfully landed five spacecraft on the lunar surface to collect data regarding the chemical and physical properties of the surface material.

Extremely valuable information was derived from these missions, however, it remained to the Apollo program to send men to the Moon, to explore it, and return with samples, before we could understand enough to draw conclusions regarding its origin and evolution. To that end, the Apollo program met its goals. The splashdown of the Apollo 17 mission marked the end of the Apollo program.

Figs. 1-1 to 1-6 present just a few photographs of this mission which was launched on December 7, 1972.

The Taurus-Littrow landing site of Apollo 17 was picked because of the possibility of obtaining both older and younger rocks there than had previously been possible.

It was hoped that the discovery of younger basaltic rocks, differing in crystallization age from the 3.2 to 3.7 billion years of previous samples of mare basalts, would lead to an improved understanding both of volcanism and of the thermal history of the Moon. It was also hoped that the discovery of rocks formed earlier than 3.7 to 4.0 billion years ago would lead to further understanding both of the early lunar crust and of material present at the time of the formation of the Moon.

A close-up examination of lunar bedrock and subsurface rock structure was necessary to draw conclusions regarding its volcanic and thermal history. The Apollo program is now recorded as one of man's greatest technological achievements. However, in time, new missions to the Moon will provide answers to the questions that were raised by the Apollo exploration.

Apollo 17 scientist-astronaut Harrison S. Schmitt is photographed standing next to a huge, split lunar boulder during the third Apollo 17 extravehicular activity at the Taurus-Littrow landing site. Scoop marks in the soil at the base of the boulder and on the side of the boulder itself, mark the location of samples collected by Schmitt. The boulder is a breccia, a rock composed of fragments of other rocks and cemented together in a fine-grained matrix.

Fig. 1-16. The valley of Taurus-Littrow as seen from the lunar module (opposite page) Challenger while in orbit before the powered descent. The Command and Service Module (CSM) America can be seen crossing the base of South Massif. The valley is about 7 km at its narrowest point between the South and North massifs. The South Massif is the large mountain just beyond the CSM. The light colored material that extends north onto the valley floor from South Massif is the Rock Slide, and opposite the Rock Slide, the mountain on the north side of the landing site is North Massif. The crests of South Massif and North Massif are 2500 and 2100 m respectively, above the landing site.

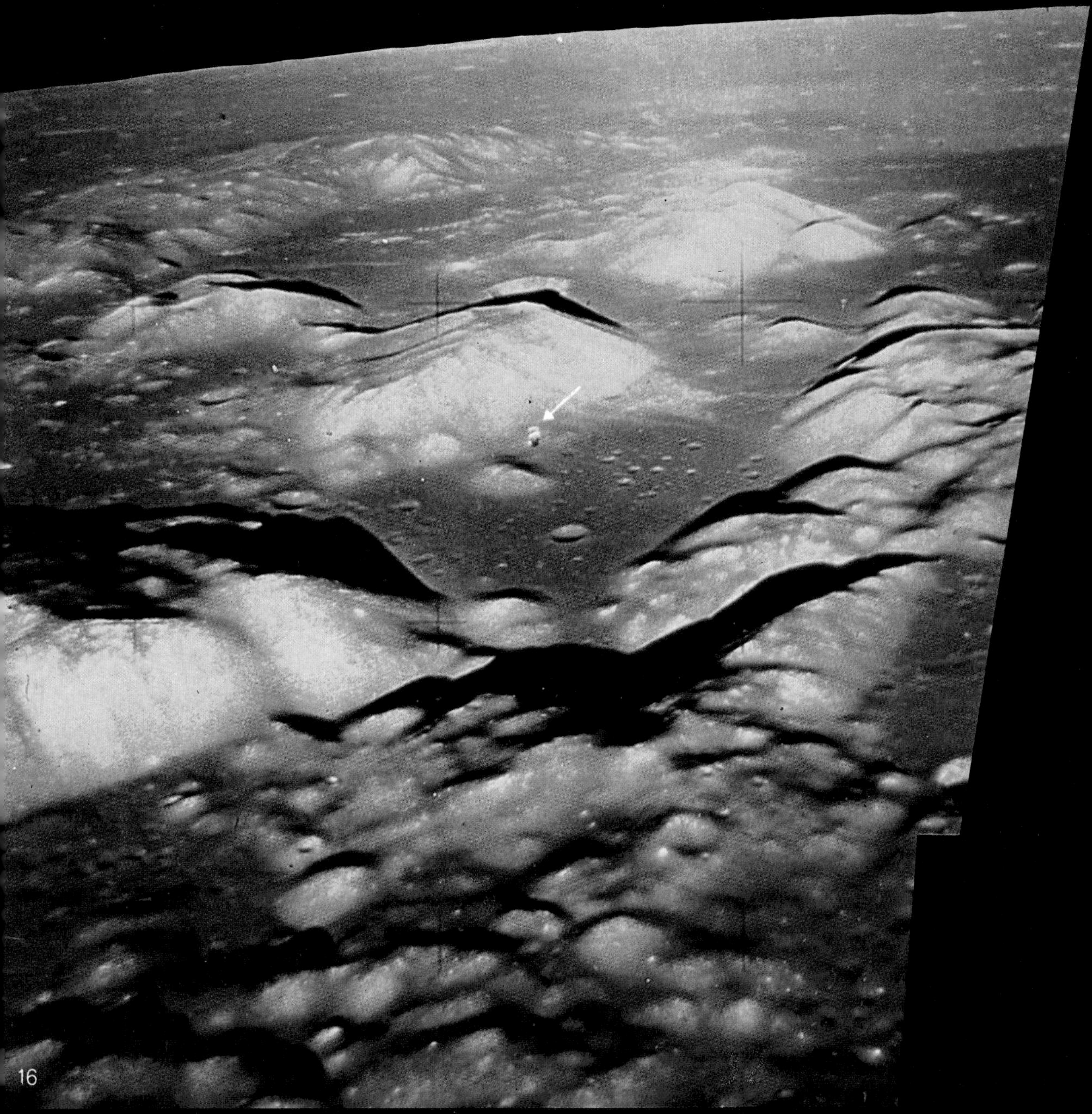

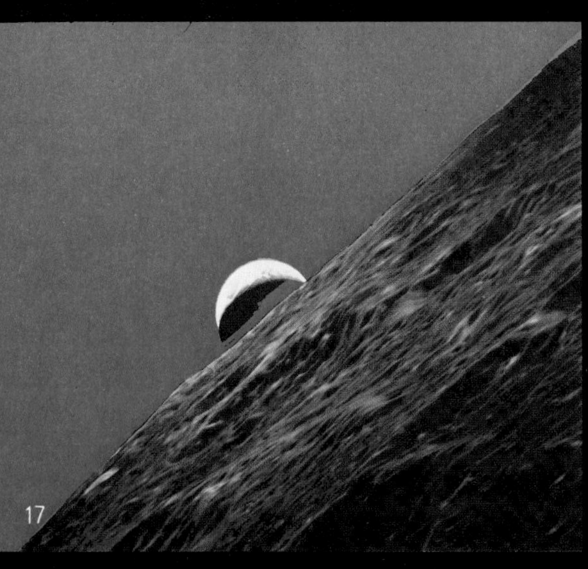

Fig. 1-17. The "blue planet" Earth is photographed from the Apollo 17 Command and Service Module just prior to the start of its journey back to Earth. Even at this resolution, Earth's atmospheric cloud patterns are discernible. The moonscape shown in this photograph is located on the back side of the Moon and is never observed directly from Earth.

AMONG THE MOON'S CRATERS IN A VEHICLE

Fig. 1-18. Apollo 17 Lunar Module pilot Eugene A. Cernan stands near the lunar roving vehicle at the start of a geological trip. Rock sample bags are attached to his chest. A checklist of tasks he will perform is attached to his left wrist. The astronauts performed a variety of scientific experiments while on the surface of the Moon, including deployment of instruments to collect data on geochemical, geophysical and atmospheric phenomena. Approximately 3 km in the background is the 2100-meter high North Massif. The umbrellalike equipment on the vehicle is an antenna used to communicate directly to Earth. The gold-colored television camera is located just to the right of the antenna. This photograph was taken by Geologist-Astronaut Harrison Schmitt. Astronaut Ronald E. Evans, piloted the Command and Service Module in lunar orbit while Schmitt and Cernan explored the surface of the Moon.

Fig. 1-19. This oblique view from the Lunar Module after it lifted off the lunar surface includes the craters Eratosthenes, 60 km in diameter, at left center, and Copernicus on the horizon at the right. To the left of Eratosthenes is the southern end of Montes Apenninus. Patterns of secondary craters are emphasized in this low-sun-angle photograph. The most conspicuous features seen are the central peak, concentric crater rims, and ejecta blanket. Eratosthenes is considered to be a classic example of impact crater morphology.

18

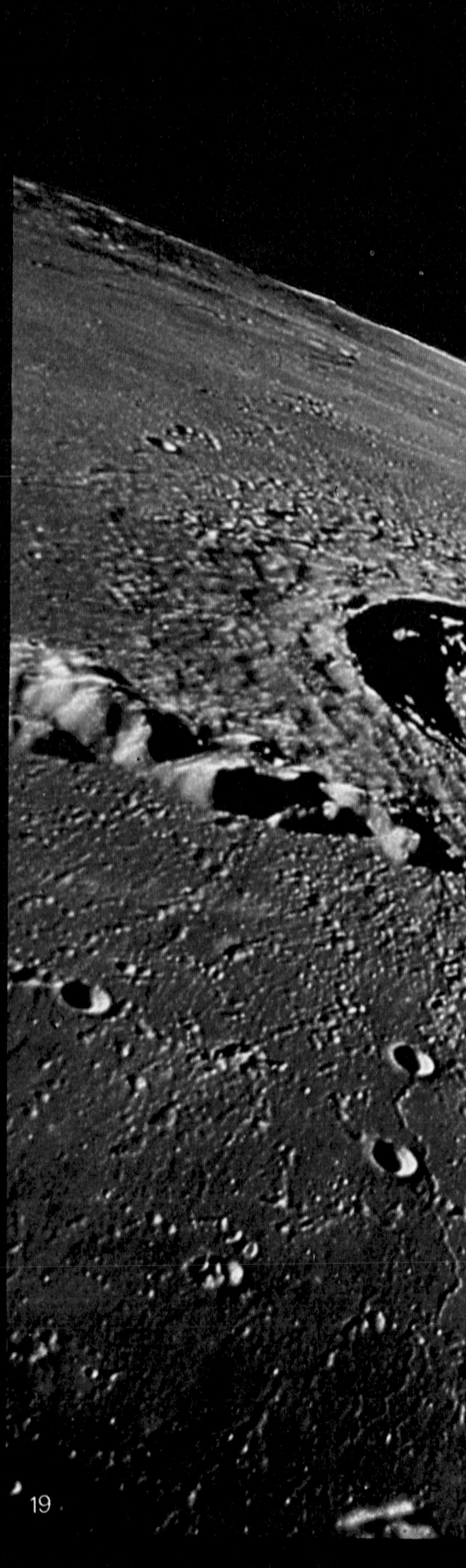

19

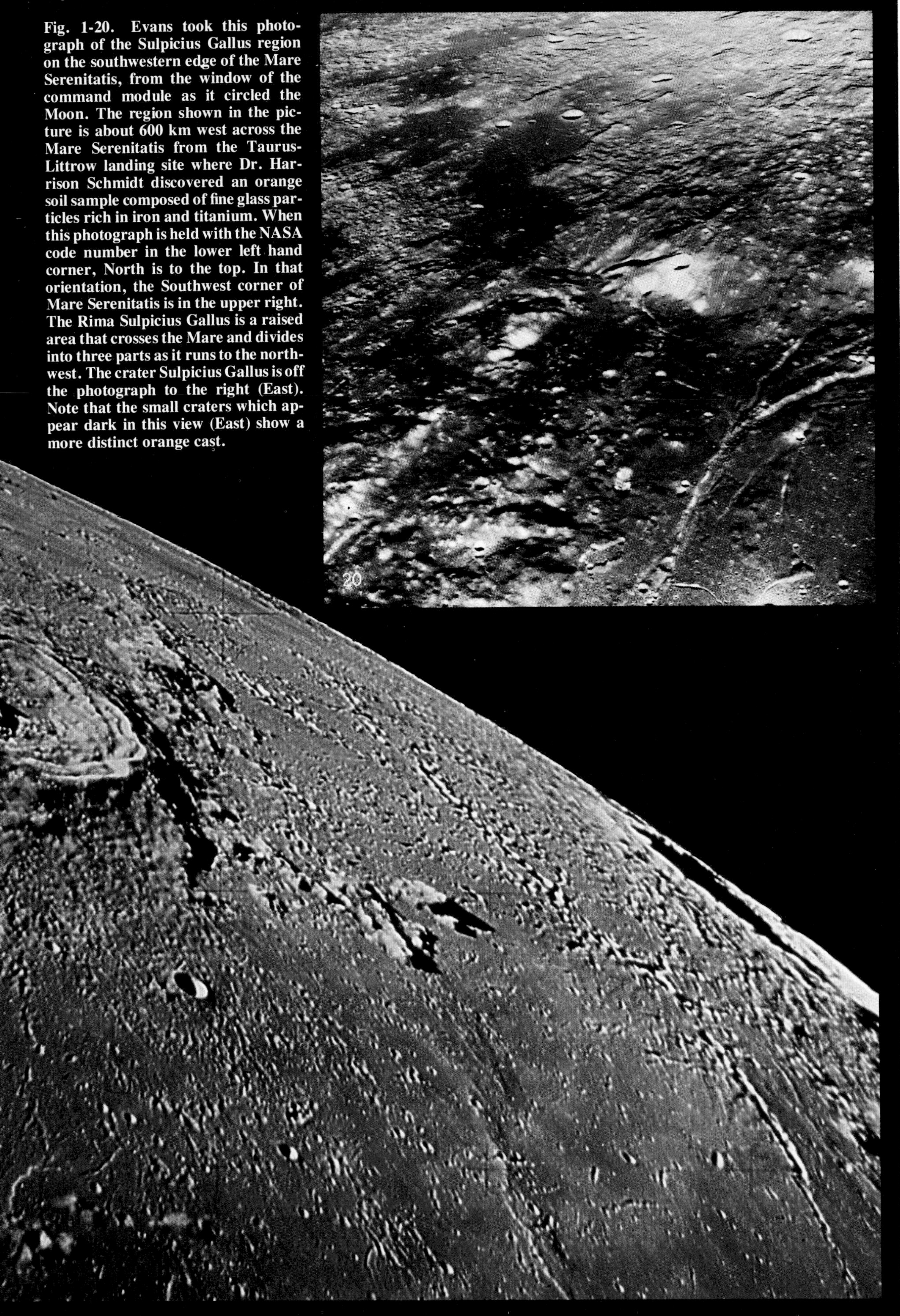

Fig. 1-20. Evans took this photograph of the Sulpicius Gallus region on the southwestern edge of the Mare Serenitatis, from the window of the command module as it circled the Moon. The region shown in the picture is about 600 km west across the Mare Serenitatis from the Taurus-Littrow landing site where Dr. Harrison Schmidt discovered an orange soil sample composed of fine glass particles rich in iron and titanium. When this photograph is held with the NASA code number in the lower left hand corner, North is to the top. In that orientation, the Southwest corner of Mare Serenitatis is in the upper right. The Rima Sulpicius Gallus is a raised area that crosses the Mare and divides into three parts as it runs to the northwest. The crater Sulpicius Gallus is off the photograph to the right (East). Note that the small craters which appear dark in this view (East) show a more distinct orange cast.

EARTH SEEN FROM ERTS 1

Fig. 1-21. The Earth Resources Technology Satellite is an unmanned spacecraft that is programmed to obtain information in the following scientific areas:

Agriculture. Information gathered will aid in land use planning, range management, identification and combatting of crop diseases and improved irrigation planning.

Geology. Information for use in the study of glaciers and volcanoes, earthquake fault systems, and in identifying terrain features associated with oil and mineral deposits.

Hydrology. Information for use in detecting water pollution trends; providing an inventory of surface water in lakes, reservoirs, and rivers; determining snow levels; and measurement of factors needed to predict the potential of floods and the location of water reserves.

Oceanography. Observation of environmental sea surface conditions which can be related to fish location, sources of pollution, behavior of major ocean currents, changes in shorelines due to storms. Maritime commerce can benefit from better charting of sea conditions, ice field observation and iceberg warnings.

Geography. ERTS data can be used to produce a constantly updated map showing the various changes in the Earth's surface, natural and manmade, of interest to such groups as urban planners and the transportation industry.

On the following nine ERTS photographs the true colors of surface features are not represented by the color appearing on the photograph. The purpose of the ERTS color photography is not to reproduce true colors, but rather to provide color photographs to help scientists interpret surface features or conditions.

Three colors, green, red and infrared, seen and recorded separately by the satellite, were combined at NASA's Goddard Space Flight Center, Greenbelt, Md. Healthy crops, trees and other green plants which are very bright in the infrared but invisible to the naked eye, are shown as bright red. Suburban areas with sparse vegetation appear as light pink and barren land as light gray. Cities and industrial areas show as green or dark gray and clear water is black/dark blue.

This scene of Monterey, Calif. provides an example for making accurate maps of geologic structure mainly on the basis of the numerous linear features displayed in the ERTS image. Such linear features often relate to geological faults, which reveal the dynamic processes taking place in the Earth's crust and are indicative of earthquake activity. The San Andreas fault is the major linear feature starting from the lower right-hand corner of the image and passing just below the city of San Jose which appears in the upper left. The Calaveras-Sunol fault, passes from northwest to southeast through the Calaveras Reservoir just east of San Jose. Several small east-west faults, previously unidentified, have been noted from this image in the area north of the San Luis Reservoir, the large water body near the center of the picture. One can see the California aqueduct which runs southwest from the San Luis Reservoir. Numerous canals and creeks which traverse the highly agricultural San Joachin valley are visible. The Salinas River, with its bordering croplands, enters the image at the bottom left, and runs northwest into the Monterey Bay.

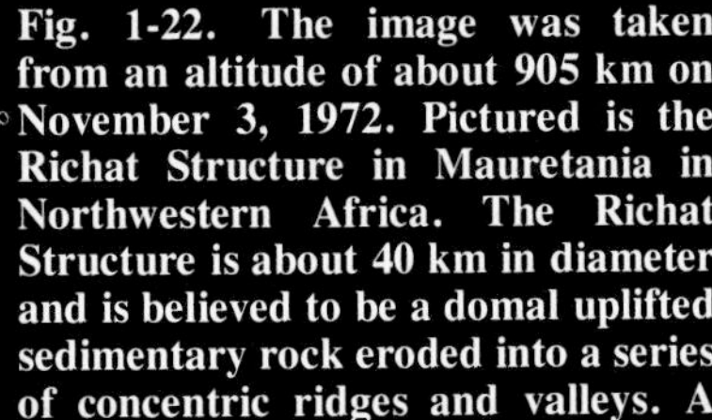

Fig. 1-22. The image was taken from an altitude of about 905 km on November 3, 1972. Pictured is the Richat Structure in Mauretania in Northwestern Africa. The Richat Structure is about 40 km in diameter and is believed to be a domal uplifted sedimentary rock eroded into a series of concentric ridges and valleys. A smaller, similar feature, the Semisyat structure, is about 30 km southwest of the Richat Structure and is about 8 km in diameter. Lava flow also is evident in the area.

Fig. 1-23. The prominent circular feature in the lower left is Lake Ponchartrain with the meandering Mississippi River below. The city of New Orleans south of Lake Ponchartrain is linked across the many bodies of water by numerous bridges, many visible here. Offshore islands produced by longshore currents are common. The Pascagoula River in the state of Mississippi is visible along the east side of the photograph as is the Pearl River in the center. The intense red indicates the high chlorophyl content of the surface vegetation in summer, which is primarily from the dense deciduous trees of the region. The Delta is a very spectacular example of a major river entering the ocean.

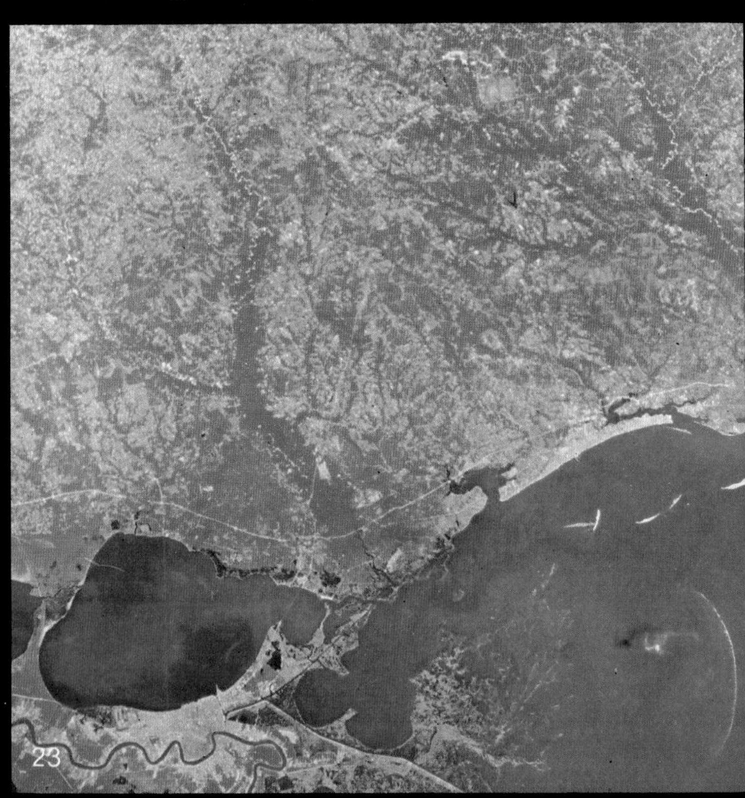

THE PLANET OF WATER

Fig. 1-24. The Bahama Islands of Grand Bahama, Great Abaco, and the Cay Berry Islands, together with their surrounding reefs and shallow waters create a scene of beautiful shades of blue. The Maris, situated within the crook of Great Abaco Island, clearly shows the areas of navigation hazards. Pelican and Marsh Harbors, behind their protecting reefs, are easily located as safe harbors. The Great Harbor Cay and Cay Berry Islands area will make a good study area for satellite bathymetry. Scattered over the entire area are fleecy white summer cumulus clouds.

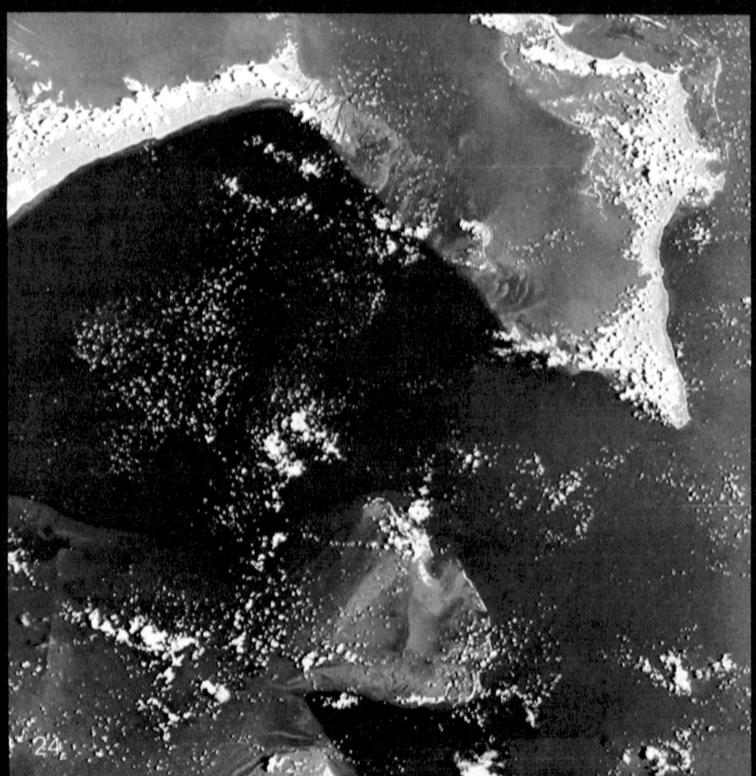

Fig. 1-25. A former outlet to the McKenzie River and a recently departed (geologically speaking) continental glacier is shown. Arctic ice is seen north of the cape with the largest being 8 km by 13 km in size. Kettles, too numerous to count, stud the tundra which is cut by meandering, braided rivers. The shallow depth along the coast shows as light blue in the water area. The parallel islands in the McKenzie River mouth at the lower left of the picture are the terminal moraines of the long-gone continental glacier.

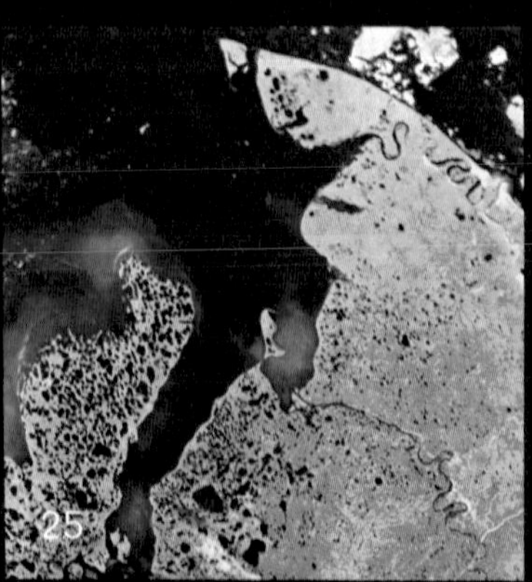

Fig. 1-26. The Gulf of Genoa and the city take up the lower portion of this ERTS-1 picture taken October 6, 1972 with the city of Genoa in the lower center. Pisa is further down the coast on the lower right, located on the Arno River. The Apennine Mountains stretch out across the center and the Po River is clearly visible, with its tributaries running all the way across the top from left to right into the clouds on the far right.

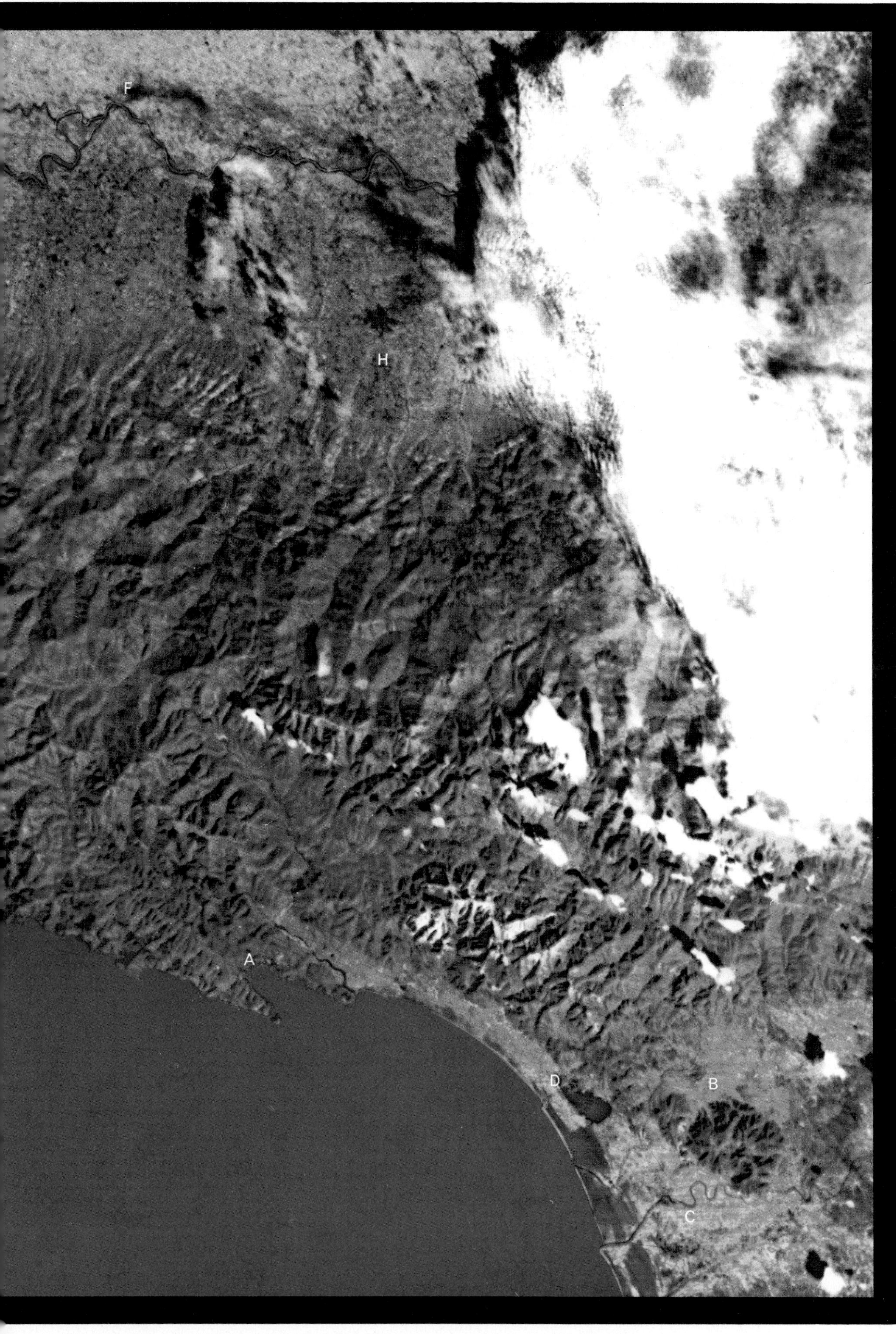
F
H
A
D
B
C

RIVERS AND MOUNTAINS OF THE EARTH

Fig. 1-27. This is a color composite photo of the Gulf of Suez, Suez Canal taken from the ERTS-1 at an altitude of 914 km.

Some of the notable geographical landmarks are:
Gulf of Suez (bottom right center)
Great Bitter Lake (center)
East Nile Delta (upper left)
Suez (lower center)
Suez Canal (top center to lower center)

Fig. 1-28. Since the ERTS cameras look straight down from very high up—about 914 km, or nearly 10 times higher than an airplane can fly—there is very little distortion in the pictures. Adjacent frames, each about 100 miles square, can be fitted together to form nearly true flat cartographic maps. This mosaic was made up from parts of 38 separate color pictures, each built up by photographing three black-and-white images of the area through color filters. This photomosaic includes the entire east coast of the United States, from Cape Cod at the north, Long Island, Delaware Bay, Chesapeake Bay, Cape Hatteras, and Cape Kennedy in Florida. A similar type of mosaic of the entire United States has been assembled. The value of this scale photograph lies in its ability to record meteorological and surface conditions at one time. In the past it was impossible to obtain an instantaneous view of an area as large as this.

Fig. 1-29. This clearly illustrates Earth's crustal folding in the Appalachian Mountains. When this mountain chain was formed, the crust of the Earth was compressed in a direction from the upper left to lower right. This caused a parallel orientation of alternating ridges and valleys. The Susquehanna (running downstream from the upper left to lower right) was able to erode the crust at a rate at least equal to the uplift of the crust and thus cut a river channel across the mountains. The Schuylkill River (running downstream from the upper right to right center and entering the Susquehanna River) was not able to downcut as fast, and as a result became a valley stream. Geologists can use this photograph to determine stratigraphic and internal structure relationships of rock formations.

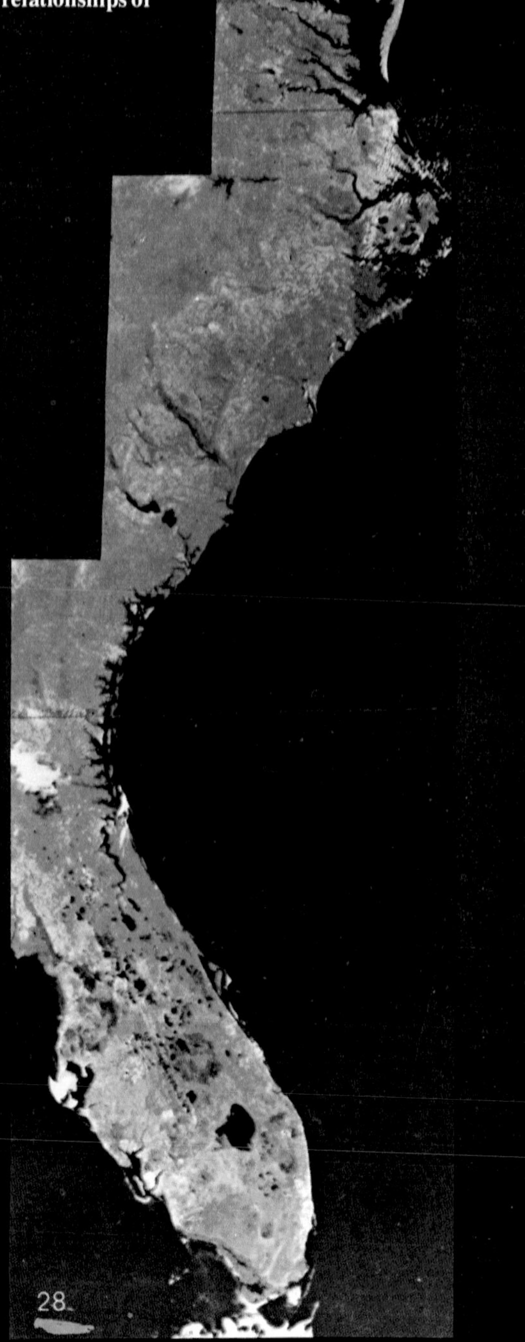

28

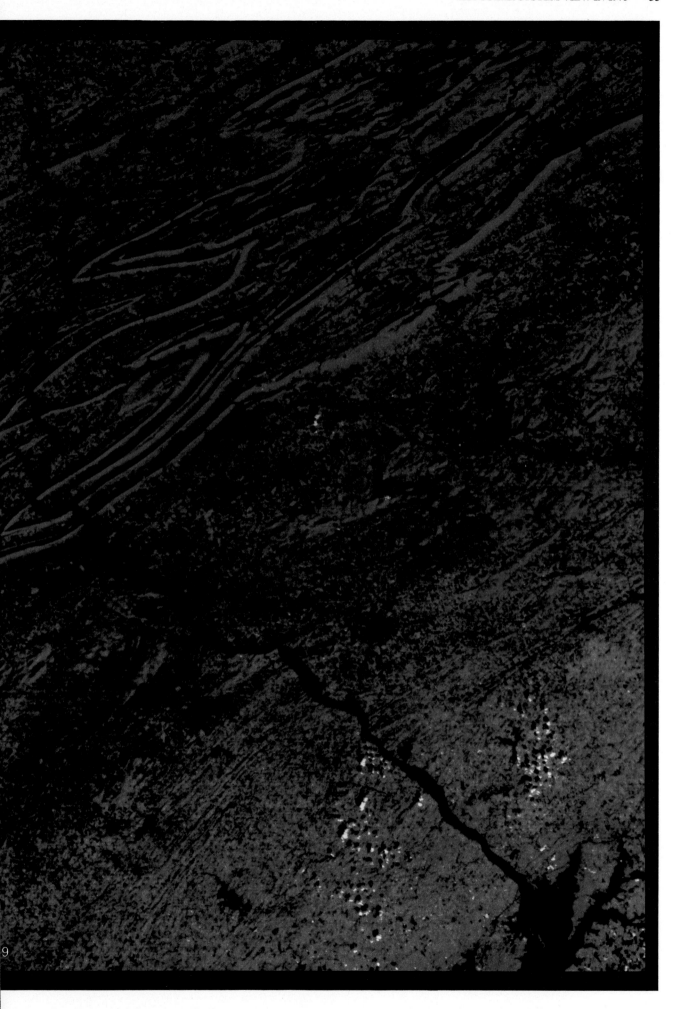
9

THE EARTH SEEN FROM SKYLAB

Fig. 1-30. A vertical view of the eastern coast of Sicily is seen in this Skylab 3 Earth Resources Experiments Package (5-inch Earth terrain camera) infrared photograph taken from the Skylab space station in Earth orbit. Mount Etna, the highest volcano (approximately 11,000 km) in Europe, is still active as evidenced by the thin plume of smoke emanating from its crest. Recent laval flows appear black in contrast to the older flows and volcanic debris that are red on the flanks of Etna. Numerous small circular cinder cones on the flanks represent sites of previous eruptions. Catania, on the Mediterranean coast south of Etna, is the largest of several cities and villages which appear as light gray patches on the lower slopes of the volcano. Piano di Catania, south of the city of Catania, is outlined by polygonal light and dark agricultural tracts. Several lakes, the largest of which is Lake Pozzilla, show up in dark blue in the photography. The unusual colors in the picture are due to the use of infrared film in which vegetation appears red. This is very evident on the slopes of Etna, in the Monti Nebredi area at upper left, and in the local areas in the lower part of the picture.

Fig. 1-31. A near vertical view of the area around Florence, Italy, as photographed from Earth orbit by the Multispectral Photographic Facility Experiment aboard the Skylab space station. The view extends from the Ligurian Sea, an extension of the Mediterranean Sea, across the Apennine Mountains to the Po River Valley. Florence is near the center of the land area. The mouth of the Arno River is at the center of the coastline. The city of Leghorn (Liverne) is on the coast just south of the Arno River. This picture was taken with an infrared film.

A

FLUVIAL EROSION AND MARINE CURRENTS ON EARTH

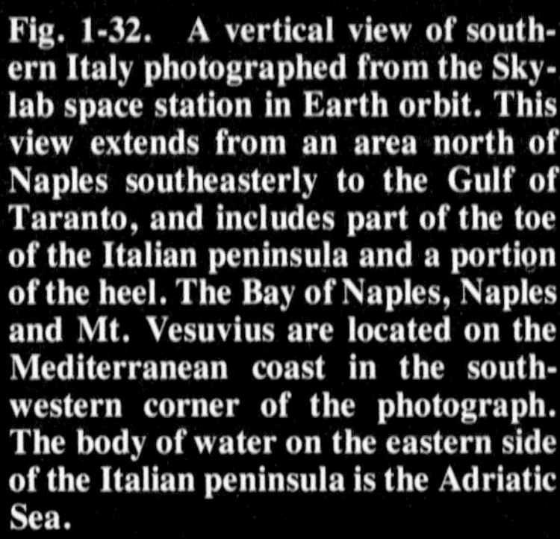

Fig. 1-32. A vertical view of southern Italy photographed from the Skylab space station in Earth orbit. This view extends from an area north of Naples southeasterly to the Gulf of Taranto, and includes part of the toe of the Italian peninsula and a portion of the heel. The Bay of Naples, Naples and Mt. Vesuvius are located on the Mediterranean coast in the southwestern corner of the photograph. The body of water on the eastern side of the Italian peninsula is the Adriatic Sea.

Fig. 1-33. This is a photograph of Denmark, southern Sweden and northern Germany, the Baltic Sea (A) and the North Sea (B). The cities of Kiel, Germany (C) and Copenhagen, Denmark are clearly shown.

Subsurface shallow ocean features can be detected at the mouth of the Elba River (E) as it exits into the North Sea. Sand bar deposits (F) indicative of ocean currents flowing parallel to the eastern shoreline of Denmark, appear as a narrow land mass.

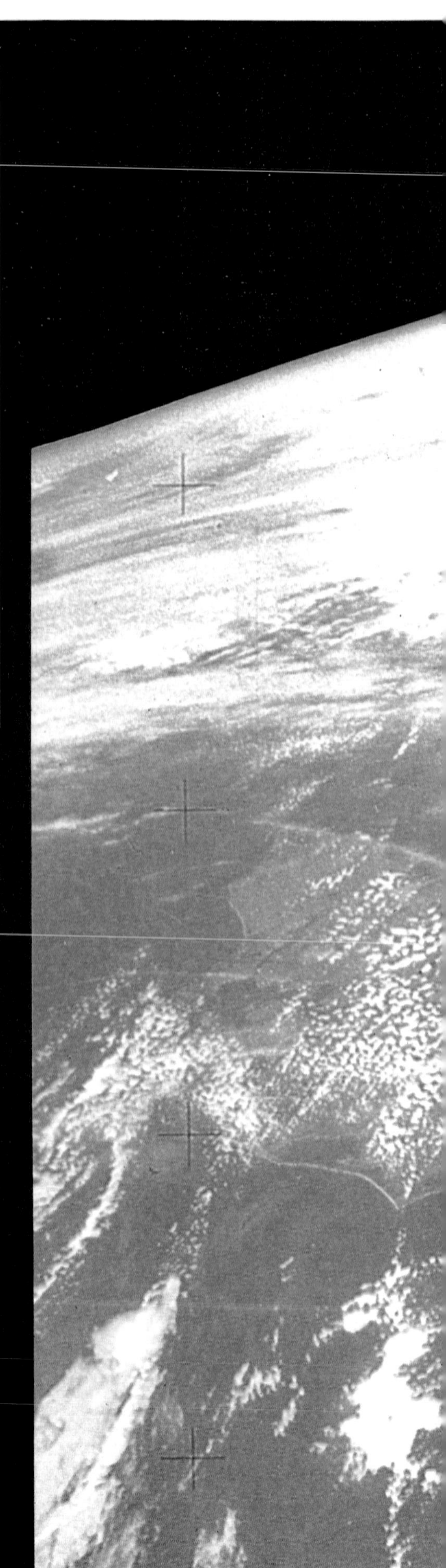

A
D
E
C

MARS GEOLOGIC RESULTS FROM MARINER 9

Fig. 1-34. Mariner 9 was launched from Cape Kennedy in May 1971 and arrived at Mars in November 1971. Reconnaissance pictures were taken of the dust-shrouded planet until January 1972 when the dust storms subsided sufficiently to begin mapping picture sequences. More than 7300 pictures were taken and about 1500 pictures of the planet were obtained by the wide-angle television camera with resolutions of 1 to 2 km. These pictures were made into a preliminary composite map of the entire surface of Mars at a scale of 1:24 million. A shaded relief map, the first detailed complete map of Mars, has been published at the same scale. The narrow-angle camera (500 m focal length) acquired high-resolution pictures (100 to 300 m) of about 2% of the surface.

Volcanic Features. The volcanic pile of Olympus Mons, which rises 24 km above the Amazonis basin floor is the largest known volcano of this type in the solar system. It is over twice as wide as the largest of the Hawaiian volcanic piles and is about equal in volume to the total extrusive mass of the Hawaiian Islands chain. The form of the flank flows and of the lava channels with natural levees is strikingly similar to those of Hawaii, suggesting that the flows may likewise be basaltic in composition.

Three other volcanoes, also of surprisingly large size, lie along the Tharsis ridge. The Amazonis basin floor is covered in many places by a succession of lobate-fronted flows that resemble terrestrial basalt flows and the basalt flows that fill the mare basin on the Moon. Widespread volcanism has occurred over Mars ranging from geologically youthful large volcanoes to much older, cratered and highly modified volcanic centers, indicating that Mars has been internally active over a long period of geologic time.

Tectonic Features. Sloping eastward from the Tharsis ridge crest is a large equatorial plateau or tableland. This plateau is broken by three sets of fractures running eastwest, northwest, and northeast. The lack of craters indicates that the rocks underlying the plateau are geologically young. East of the mosaic of fault blocks, the graben between blocks coalesce into a great equatorial canyon or rift valley system 6 km deep in places, that extends almost 5000 km to the east. This great equatorial chasm is comparable to the East African Rift Valley system and may represent the beginning of rifting on Mars. The steep lateral valleys in places appear to be debris channels for masses of wasted material.

Along the cliffs bounding the high plateau, layers and deposits are exposed which average about 100 m thick. The uppermost layer usually seems to be the most resistant and forms a rocky rim. In many areas the cliffs are bordered by masses of debris that apparently have slid into the adjacent lowlands. In other areas the material retains its cohesiveness and descends in a series of terraces.

Channels. Sinuous channels that have many tributaries cross the level high plateau surface. These channels descend to the east and north, becoming broader and more clearly defined as they do so. The tributaries and the form of the braided channels are unlike those of terrestrial and lunar volcanic sinuous rilles, and closely resemble terrestrial and stream channels. Their form and degree of freshness strongly suggest a flow of liquid water in the recent geologic past of Mars.

Polar Features. An ancient cratered terrain extends from north of the equator southward to the south polar region, where it is overlapped by younger units. The units of young layered deposits total 6 km in thickness, occur at a lower elevation than the surrounding ancient, cratered terrain and are covered by ice at the pole. Thus, these layered rocks occupy a saucer-shaped depression in both polar regions.

Eolian Features. In some areas, dark markings in wide-angle pictures have been resolved into sand-dune fields in the narrow-angle pictures. One dune field, about 50 km across, lies in the bottom of a crater. The spacing of crests of individual dunes is on the order of 1 to 2 km. The identification of dunes is significant because dunes indicate that saltation is operative on the surface despite the tenuous atmosphere. With saltation occurring, numerous eolian erosion and depositional features should be expected on high resolution pictures.

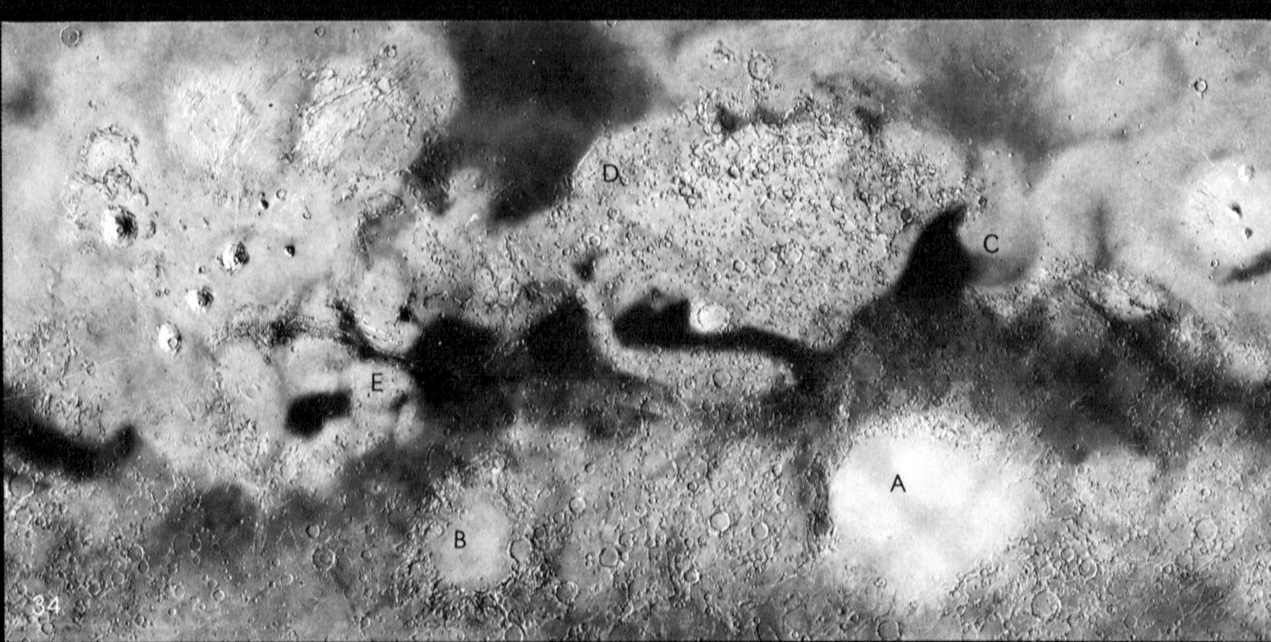

An equatorial 1:80,000 scale Mercator map of Mars. Topographic features have been derived from the U.S. Geological Survey and the albedo features are derived solely from International Planetary Patrol photographs obtained during 1969 and 1971, and collected by the Planetary Research Center, Lowell Observatory. Although this map does not reveal any consistent correlation between albedo features and topography over the entire planet, some relationship between the two is indicated in Hellas (A), Argyre I (B), Acidalium (C) and especially Syrtis Major (D). The most striking correlation is the coincidence of the canyon system, Valles Marineris (E), with classical albedo markings.

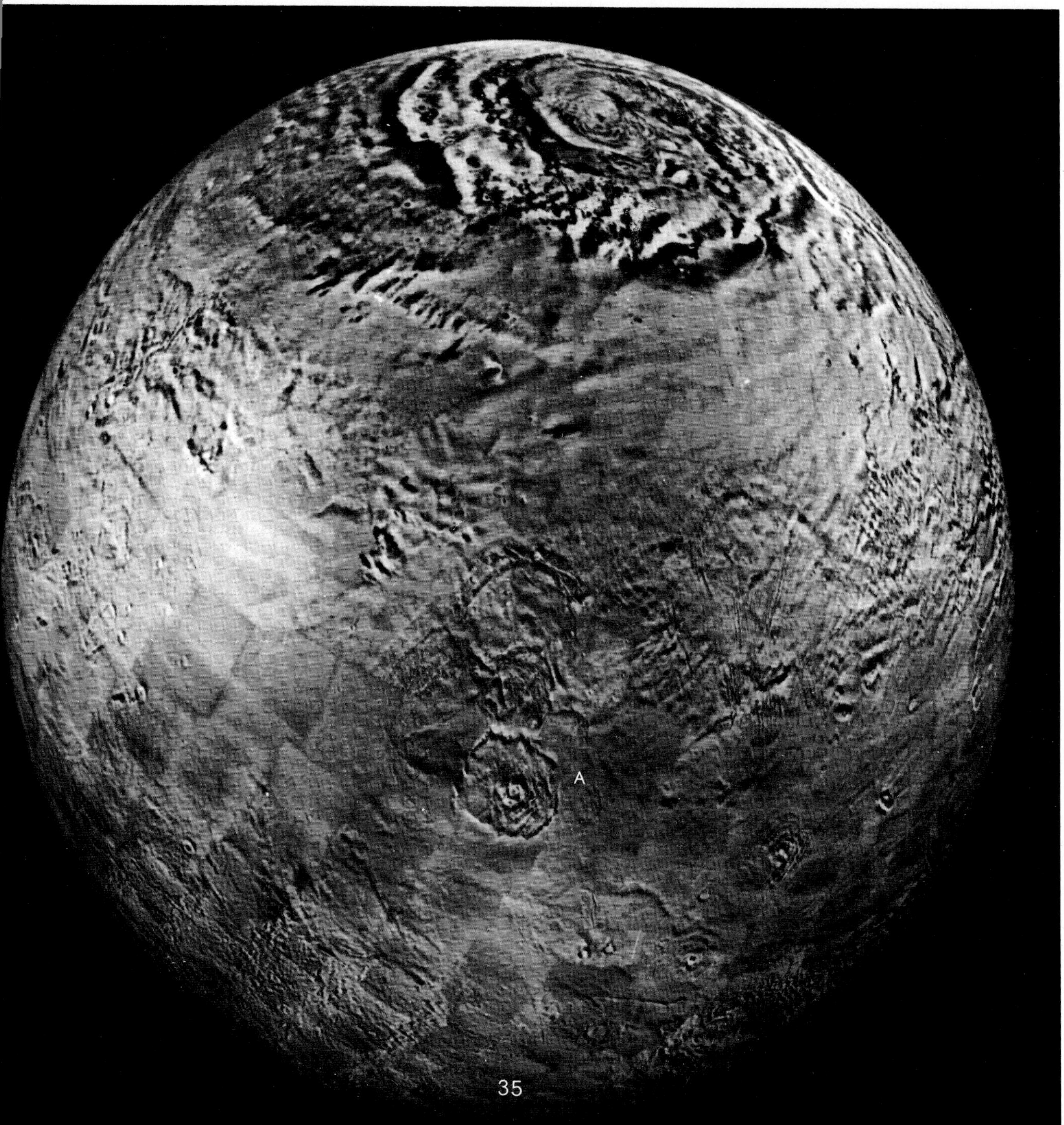

Comparison of these photographs with Fig. 1-1, clearly shows this advantage. A further comparison of these global views with small-scale higher resolution photographs shown in Figs. 1-4 through 1-6, illustrates the details that become discernable at these scales. The full study of any one planet in our solar system requires a variety of perspectives and scales.

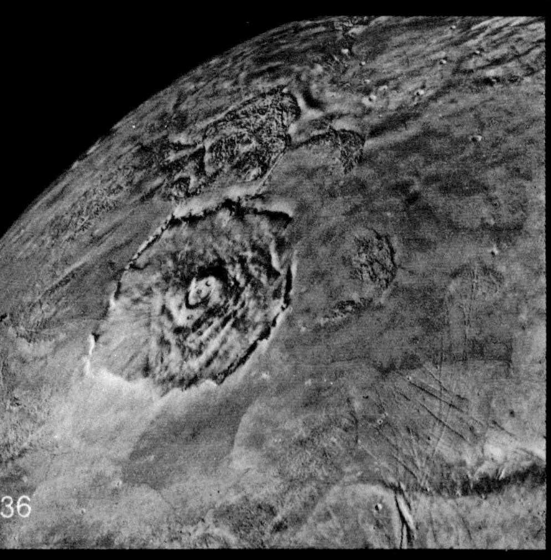

Fig. 1-35. Photographs of portions of a 4-foot globe that was assembled by the jet propulsion laboratory from photographs of Mars taken by Mariner 9. These photographs illustrate on a large scale features with a perspective that ordinarily cannot be achieved with flat projection mosaics.

Fig. 1-36. A section of the global mosaic photograph showing a portion of the north polar area and Olympus Mons. The distance from Olympus Mons to the center of the north polar area is about 4500 km. This synopic view shows only the extent of the polar topography in relation to the rest of the planet.

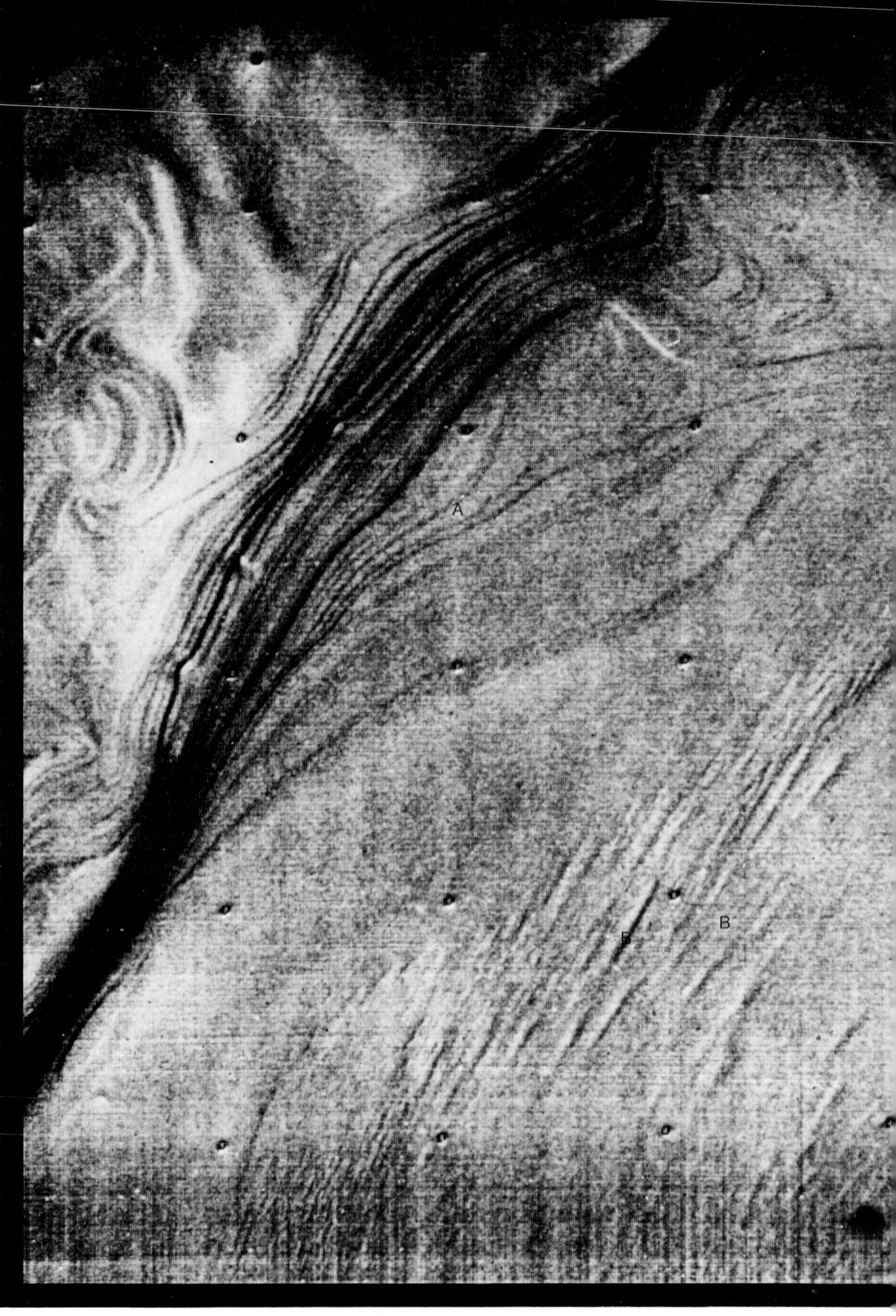
A
B
B

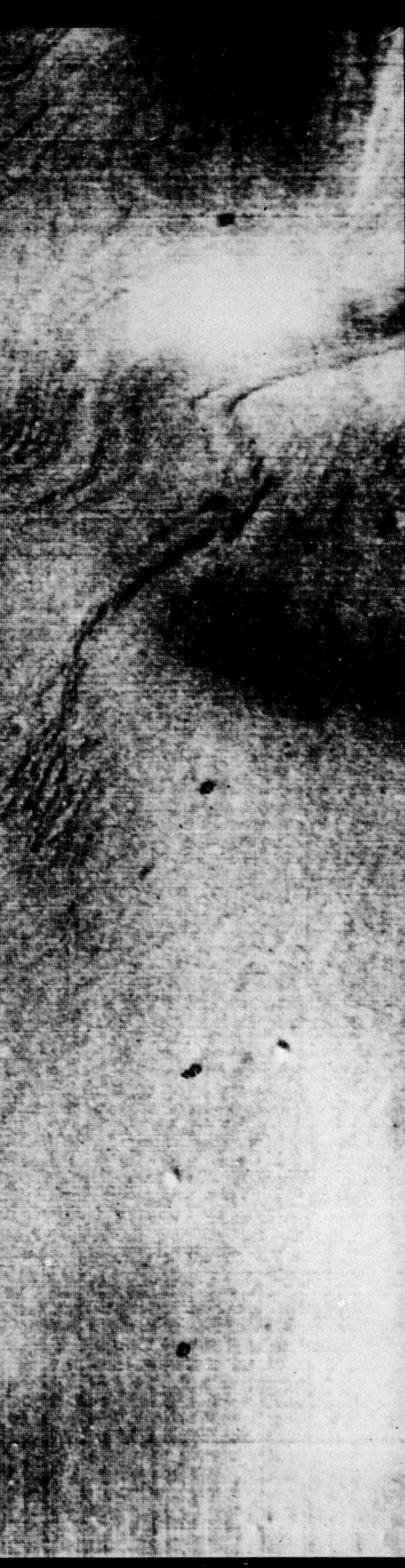

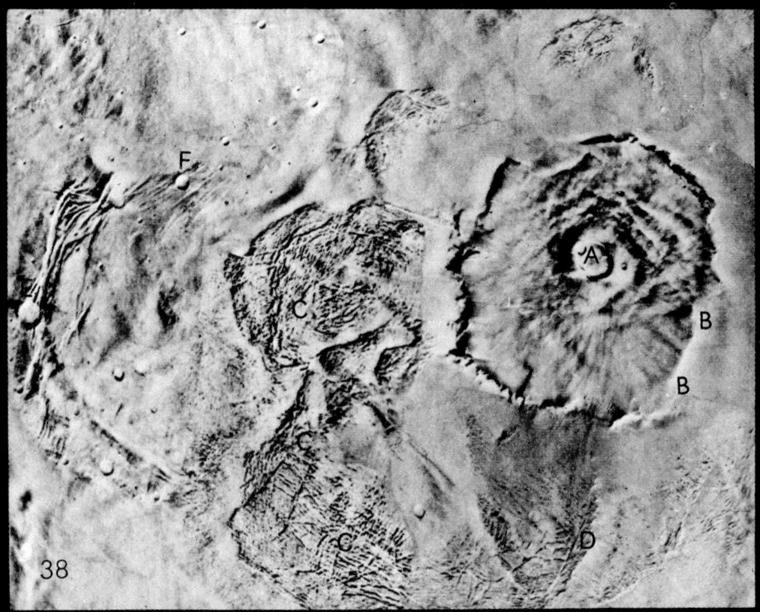

MARS: THE LARGEST VOLCANO

Fig. 1-37. Shows both laminated (A) and "etched" terrain near the south pole. The "etched" unit (B) is composed of alternating hard and soft rocks which when eroded by the wind, produce alternating topographic depressions and elevations. The effect of the wind upon Martian geomorphology is very evident not only in the polar area, but in all other parts of the planet. The scouring action of wind removed material in area "B" and deposited it in areas represented by "A." The uniformly bedded layers of the laminated terrain are estimated to range from 10 to 30 m in thickness. At least 20 of these layers have been counted, with at least 6 of them encircling the pole. Their edges are smooth, very different from the ridges in the etch pit terrain.

Fig. 1-38. Close-up view of Olympus Mons, the largest shield volcano yet observed on any planet in our solar system. It is about 600 km across its base and rises some 20 km above the surrounding plain. Steep cliffs drop off from the mountain flanks. The main crater at the summit, a complex of multiple volcanic vents (A) is 23 km in diameter. Olympus Mons is more than twice as broad as the biggest shield volcano on Earth. Mauna Loa in the Hawaiian Islands is 225 km in diameter and 4 km above the floor of the Pacific Ocean. The cliffs rising 5 km above the plain (B) to the south and southeast are not as sharp as those in other places because they are covered in part by wind-blown sands and silt. The grooved terrain (C) to the west is lava flows that have been modified by wind erosion and subsequently been partly covered with wind-blown material. The patterns are primarily caused by fracturing of the lava flows that originated from Olympus Mons. Close examination of this area reveals that the volcano has erupted at least 3 or 4 times and each time has left a lava flow that can be identified as a single rock unit. A distinctly different type of fracture pattern is noticeable in the area designated (D). This pattern is due to the occurrence of a different rock type that reflects a different mode or origin. The craters (E) and (F) are interesting because they were formed after the first major eruption. (The craters cut across the rock fractures.)

Fig. 1-39. In addition to the vast amount of data from Mars itself, Mariner 9 obtained close-up views of the Martian moons, Phobos and Diemos.

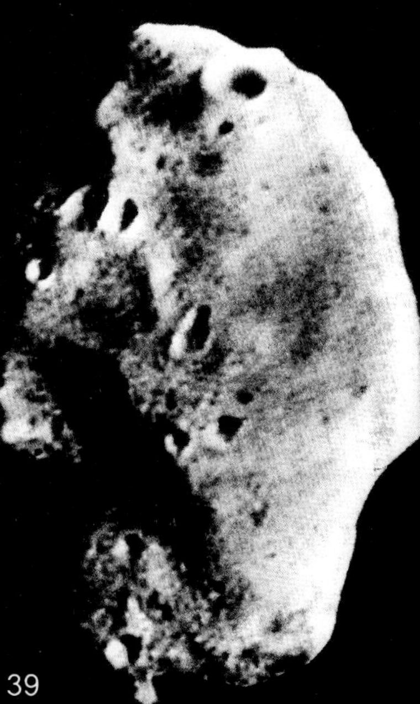

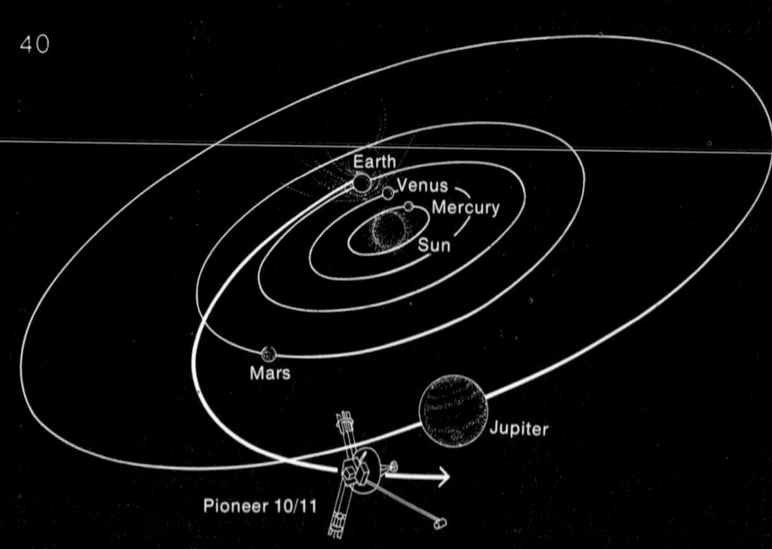

JUPITER SEEN FROM PIONEER 10

Fig. 1-40. In March 1971, the U.S. Pioneer 10 spacecraft was launched towards Jupiter. On December 4, 1973, it passed Jupiter at a distance of 130,000 km (diagram above). The objective of this mission was to investigate the interplanetary medium, to study the asteroid belt, and to make the first measurements in the environment of Jupiter. In addition to other specific questions to be answered, the mission attempted to resolve questions concerning the nature of the planet's atmosphere. Jupiter has been observed by astronomers for centuries, but an explanation of its thick cloud cover has never been made. Its gigantic red spot is a weird phenomenon that seemingly defies a logical explanation. It drifts along at a variable rate, in a rather well defined latitude range. Its position, relative to other features, is not constant. The characteristic is indicative that it is not tied to a permanent surface feature. Theories concerning its origin and behavior have all been qualified in nature and lack scientific proof. While Pioneer 10 was not able to obtain data into Jupiter's clouds, it was able to photograph the planet at a photographic scale that was sufficient to better understand its atmospheric characteristics.

Fig. 1-41. Shows Jupiter's clouds, Red Spot (A), and a shadow of Io, a satellite of Jupiter (B). This picture was taken on December 1, 1973, when the spacecraft was 2.5 million km from the planet. The photograph was made on tapes and processed by the University of Arizona.

The equatorial and polar regions appear to be more uniform in structure than the midlatitude region. The midlatitude region is highly structured and contains small identifiable units that do not appear to be associated with any global scale features. Four small, bright spots (C), each about 4000 km in diameter and surrounded by circular rings, appear in this region.

Striking wavelike patterns (D) are observed that could be caused by atmospheric drag on the margins of the less turbulent regions. The Red Spot appears as an oval with sharp tips. Its border is sharp down to the limit of resolution and is slightly darker than its interior.

A

D
B
D
C
C
C
C
41

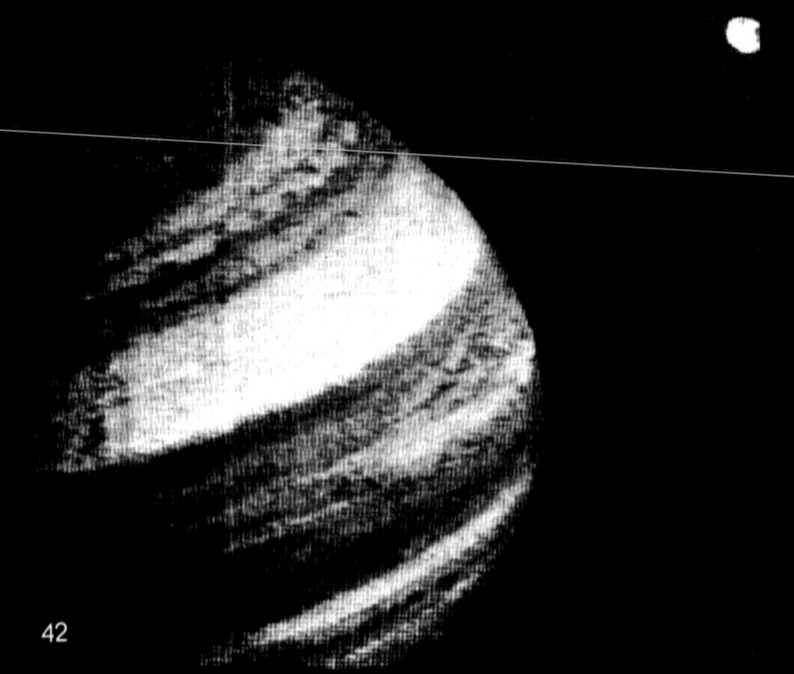

JUPITER AS SEEN BY PIONEER 11

Fig. 1-42. This illustration like the others on these two pages shows the images transmitted by the Pioneer 11 probe which approached Jupiter on December 3, 1974. The probe left Earth April 5, 1973. After a journey of more than a year and a half, having crossed the belt of asteroids without damage, it made a wide circle around Jupiter, and taking advantage of the sudden acceleration caused by the planet's gravitational field, set off for Saturn. In circling Jupiter, the probe reached the highest velocity ever reached by a man-made object: 171,000 km per hour. It reached Saturn in September of 1979. The illustration shows Jupiter and its largest moon, Ganymede.

Fig. 1-43. The Great Red Spot as it appeared to the sensitive instruments of the Pioneer 11 probe at a distance of 380,000 km. It has a diameter of about 40,000 km, about equal to the circumference of the Earth. Pioneer 11's investigations confirm that it is an enormous storm in a semistationary position. They also show another smaller one nearby. Pioneer 11 carried on board 12 scientific instruments weighing a total of 30 kg; the weight of the entire probe was 260 kg. The instruments were specifically positioned, like those aboard Pioneer 10, to analyze the gaseous covering of the great planet. Unlike Pioneer 10, however, Pioneer 11 passed close to Jupiter—41,000 km from the outer surface of the clouds—and thus passed inside the band of particularly intense radiation already noted by the Pioneer 10 probe. All the instruments survived undamaged, although some were saturated and were therefore unusable at the point of closest approach.

Fig. 1-44. This illustration shows the northern polar region of Jupiter as seen by Pioneer 11 eight hours into its trip towards Saturn. The ring which the probe made around Jupiter was oriented in such a way as to allow a view of the polar regions which Pioneer 10 was unable to show. The polar region is to the upper left of the illustration. The clouds around it show vortices and other structures indicative of internal turbulence.

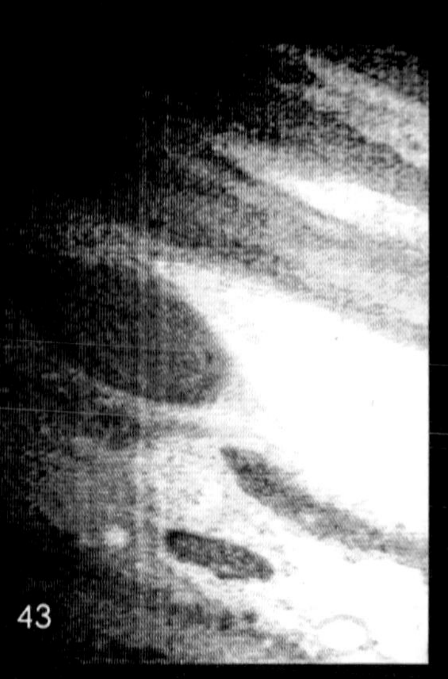

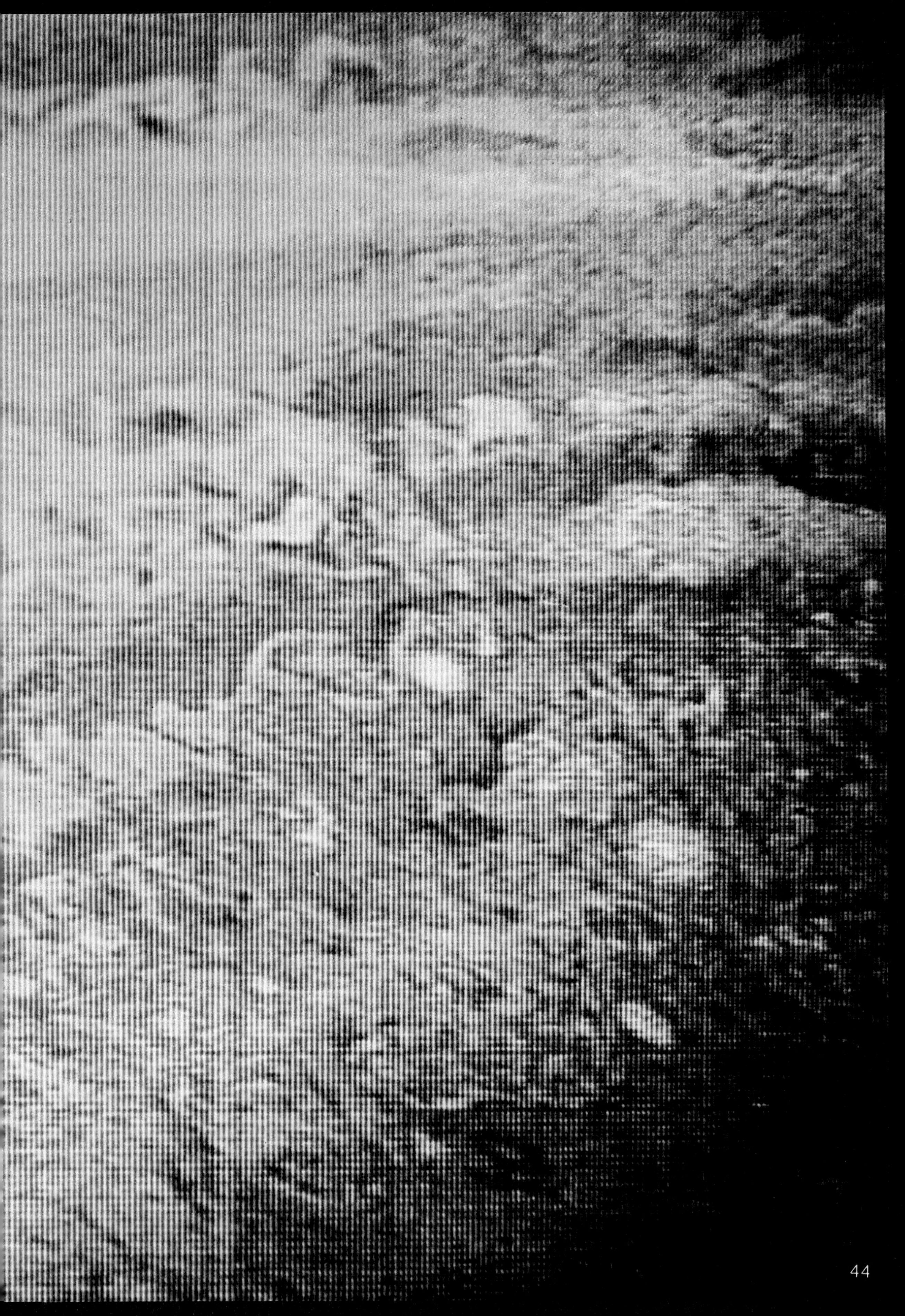
44

A magnetic analysis of the constituent particles of the soil shows that some of them are magnetized (ferro-magnetic particles), which can only be accounted for by the Martian magnetic field.

Although no organic compounds were found on the Martian surface, a remarkable kind of soil chemical activity was discovered, which does not prove the existence of life but is consistent with life processes. The addition of organic nutrient to the soil generates very rapid chemical activity which reaches a maximum very quickly and then subsides, corresponding to what appears to be rapid utilization of carbon compounds in the nutrient. This activity is stable up to a temperature of 18°C, but is greatly reduced by heat treatment for three hours at 50°C and disappears completely at 160°C.

The north polar region is covered with a perennial water-ice layer, devoid of craters, which implies either rapid rates of erosion or continuous deposition. The nonuniformity of the layered deposits strongly suggests a complex history of climatic changes while the layers were being formed. This permanent north polar ice cap is surrounded by a vast belt of dunes.

The abundance of water vapor in the Martian atmosphere varies with latitude, reaching a maximum in the 70° to 80° north latitude band during the northern midsummer season. The total column abundance of water vapor in the polar regions is consistent only with a surface temperature at the poles of 200°K and is incompatible with a frozen carbon dioxide cap at the pole. The north polar cap can therefore only be a water-ice cap whose thickness is no less than one meter and no greater than one kilometer. All in all, the atmosphere of the northern hemisphere contains a significant amount of water vapor.

The temperature of the Martian soil and surface varies drastically from night to day and from season to season. Lander 1 found small annual temperature variations, but lander 2 found large ones, showing that the latitude and the terrain play important roles in the variation of the climate of the soil. The daily temperature at the top of the soil varies from 183° to 263°K, but it is lower beneath the surface rocks. The peak temperature occurs about 2.2 hours after noon, which implies that significant amounts of sunlight are absorbed by dust grains in the lower Martian atmosphere.

The information communicated to the Earth from the Viking missions continued after the landers had ceased functioning, for the orbiters still operated, and Viking Orbiter 1 conducted an intensive investigation of the Martian satellite Phobos during the last 2 weeks of February 1977. VO-1 flew past Phobos at a distance from the surface of 80 km and sent back 125 pictures of that body. Its photographs show it to be an extremely cratered and pitted body, and quite irregular in shape, being only roughly spherical. Its value of GM (G = the Newtonian gravitational constant) was measured as $0.00066 \pm 0.00012\ \mathrm{km^3 sec^{-3}}$ and its mean density as $1.9 \pm 0.6\ \mathrm{gm\ cm^{-3}}$. It thus has the properties of a carbonaceous chondrite so that it is quite probable that Mars and its satellites had different origins. The most reasonable conclusion in view of these data is that both Martian satellites were captured by the planet, possibly from a group of asteroids that approached too close to the planet.

Certain conclusions that are significant for the Earth may be drawn from the Viking missions:

1. The relative abundances of the volatile elements on the surface of Mars and on the surface of the Earth are about the same and similar to those in ordinary chondrites.
2. The present Earth's atmosphere, and its oceans and the volatile components of Mars are the result of volcanism on both planets, but this degassification of the two planets was much more extensive on the Earth than on Mars.
3. Most of the degassification of Mars occurred much earlier in its history—during the first billion years of its existence—than it occurred in the history of the Earth.
4. Owing to the much smaller amount of degassification of Mars the Martian lithosphere must be much thicker than the Earth's lithosphere.
5. The absence of organic molecules on the Martian surface may be accounted for by the fact that the Martian surface is a highly oxidizing one as compared to the Earth's surface.

The Voyager Jupiter/Saturn Mission

The Voyager Jupiter/Saturn mission which was launched during the period 19 August/17 September 1977, and reached Jupiter in February 1979 will utilize the rare alignment of the outer planets during this epoch to complete its dual flyby. The placement of Jupiter relative to Saturn permits a spacecraft to fly by both planets with only the launch energy required for a Jupiter mission. The energy increase gained from the moving gravity field of Jupiter permits the spacecraft to reach Saturn in the relatively short time of 3–4 years, at a launch energy within the capability of the available launch vehicle.

1977 was by far the most desirable year for a launch of a Jupiter/Saturn mission. The relative position of Jupiter permitted a flyby geometry at that planet that avoided possibly intense radiation belts that might damage spacecraft equipment and yet was close enough (5 Jupiter radii) to permit good remote observation of Jupiter and its satellites.

The flyby conditions at Saturn are relatively unconstrained. Selected timing and flyby geometry will permit a very close approach to a satellite such as Titan, and a near approach to Saturn and its rings.

Then following the Saturn encounter, the spacecraft will leave the solar system and escape into interstellar space with a velocity of 3 astronomical units per year. It should be possible to acquire scientific information from the spacecraft out to a distance of at least 30 au from the sun, perhaps into interstellar space.

Jupiter as Seen by Voyager 1

Words can hardly describe the fantastic, well-nigh incredible, and completely unexpected features of Jupiter and its satellites that were revealed in the photographs transmitted to the Earth by Voyager 1 in its bypass of Jupiter. Jupiter's atmosphere and cloud cover, consisting of high velocity long-lived colored bands and long-lasting turbulent regions of apparent stability are dynamical marvels whose structure and energetics are still almost complete mysteries. The planet radiates back into space almost twice as much energy as it receives per second from the Sun, emits very strong radio waves that are in some unknown way related to its satellites, is surrounded by a very strong, rapidly varying magnetic field, and is a veritable riot of color. On March 4th 1979 Voyager 1 first passed Amalthea, the small innermost satellite, dipped under the south pole of Io, sped past Europa and Ganymede, and finally flew over Callisto's north pole in its approach to Jupiter.

The photographs of Io, looking like a huge pockmarked orange, are perhaps the most surprising information sent back by Voyager 1, since the surface of this small inner satellite looks like no other known body in the solar system. Completely devoid of craters and other evidence of meteorite encounters, and its color that of the poisonous mushroom, amanita muscaria, this satellite is almost frightening when first seen on these photographs. Shortly before Voyager 1 reached Io, or perhaps shortly after that, volcanoes began to erupt on its surface. Scientists had already identified at least one volcanic caldera on Io so that the discovery of the plumes, though exciting, was not unexpected. Io is the most active solid body thus far found in our solar system.

Shortly before Voyager began to photograph Jupiter directly, it sent back a picture of the Beehive cluster (a well-known loose cluster of stars in the Galaxy) which revealed a faint ring around Jupiter, which had never been seen before and whose existence had not even been suspected. This ring is, at most, some 30 km thick, at least 9000 km wide, and about 61,000 km above Jupiter's cloud tops.

The circulatory motion of the clouds around Jupiter is anything but orderly, consisting, as it does of very diverse, wildly turbulent patterns that move around Jupiter's surface like huge interlocking gears. These currents swirl around the Great Red Spot which stands out like a rock in roaring waters. The interior of the Red Spot is relatively calm, like the eye of a hurricane; it seems unaffected by the clouds flowing past it. A number of whitish ovals, which, like the Red Spot, are regions of anticyclonic, counter clockwise rotations, lie close to it.

STEPHEN E. DWORNIK

THE atlas of the solar system has presented us with a highly detailed view of the other planets which, like Earth, orbit around the Sun. We have obtained a concrete view of how these bodies, which objectively differ widely, have a whole series of characteristics in common: volcanoes, meteorite craters and valleys on Mars, Mercury and the Moon; clouds on Venus, hurricanes on Jupiter—all of which are also characteristic of Earth.

The author of this article, the astrophysicist A. G. W. Cameron, attacks this problem at its roots: How and when were the elements which compose the Earth, the Sun, and the other solar bodies formed? A distinction must be made between hydrogen, helium and all the other elements. Hydrogen and helium form almost the whole of the Sun's mass, while all the other elements together represent no more than 1.5%. The proportions of hydrogen and helium in the Sun are linked by a fixed ratio of about 3:1, which is to be found not only in almost all the stars of the Galaxy, but even outside the Galaxy in many of the hundred thousand million galaxies estimated to form our universe.

Thus it is clear that the formation of these two elements took place even before the primordial matter was divided up among the various galaxies. Hydrogen and helium are the most ancient elements of the universe. But what is their age?

It is difficult to find an answer to this question because so far we have not succeeded in defining accurately at what moment in the evolution of the universe the galaxies were formed. If we suppose that they were formed at a very early date, the age of hydrogen and helium must be close to that of the universe—between 12 and 16,000 million years. Those stars forming part of our galaxy which contain a proportion of heavy elements different from that of the Sun are very rare. Thus the formation of these elements must have taken place in the most primordial period of the formation of the galaxy, before most of the stars known to us were formed.

How were they formed? In the gigantic explosion (supernova) of a super massive star having a mass enormously greater than that of normal stars. When such stars exploded they lost into interstellar space much of their matter in the form of new elements formed as a result of the nuclear phenomena that took place before and during the explosion and because of it.

The Sun, the planetary system and the Earth itself were formed out of this interstellar material—gas, dust and granules containing the various elements. How did this take place?

Cameron has developed a model of his own. This model postulates a cloud of gas and dust containing a quantity of interstellar matter greater than would be necessary to form the bodies of the solar system as we know them. From this cloud of gas and dust, which he calls the massive solar nebula, the solar system was born. He uses this terminology to distinguish it from the minimum solar nebula described in other models. The minimum solar nebula allows for a smaller content of matter and hence does not permit the existence of phenomena which he considers of fundamental importance to an explanation of the present-day composition of the Sun and planets.

The massive solar nebula is subject to considerable gravitational forces that soon transform it into a rotating disk that is denser at the center and becomes progressively more rarified toward the periphery; its temperature is also highest at the center and lowest at the periphery. It is precisely because of this latter characteristic that the composition of each planet differs from that of the others: because in the hotter regions closest to the center only the substances with the highest melting points can solidify—for example silicates of aluminum and calcium and oxides of titanium—while in the coldest and outermost regions, even ammonia, water and methane reach their freezing points.

The rotation of the disk facilitates the formation of separate nuclei which, by collision with particles and by gravitational attraction, become larger and larger: these are the protoplanets. The greatest amount of mass, however, is building up at the center of the nebula which at a certain point reaches a density sufficient to give rise to thermonuclear reactions. Thus the Sun comes into existence, and initially it diffuses in all directions an impressive spray of particles (solar wind) which sweeps the residual gas and dust out of the solar system. By now—about 4.6 billion years ago—the solar system has been born, and one of the bodies that form part of it is the Earth.

ALASTAIR G. W. CAMERON

Director of the Planetary Science Division of the Astrophysics Center of Harvard University, where he also teaches astronomy. He was born in 1925 in Winnipeg, Canada. His interest in astrophysics goes back to the years 1952–1954 when, immediately after receiving his Ph.D. from the University of Saskatchewan (1952), he devoted himself, as a physics professor at Iowa State College, to the study of nuclear reactions responsible for the nucleosyntheses of heavy elements. In 1959–60 on visits to the California Institute of Technology and Mount Wilson and Palomar Observatories, he began to study the Sun and the planets and, above all the reconstruction of the primordial history of the solar system, based on the study of decayed matter found in meteorites. From 1961–1966, at the Goddard Institute for Space Studies, he began to devote himself to the study of the evolution of the Sun and stars, quasars and extragalactic radio-sources, neutron stars and X-ray sources, the structure and composition of the planetary atmosphere and finally, the origin of the solar system.

A. G. W. CAMERON

From Nucleosynthesis to the Birth of the Earth: Origin of the Planets and of the Earth

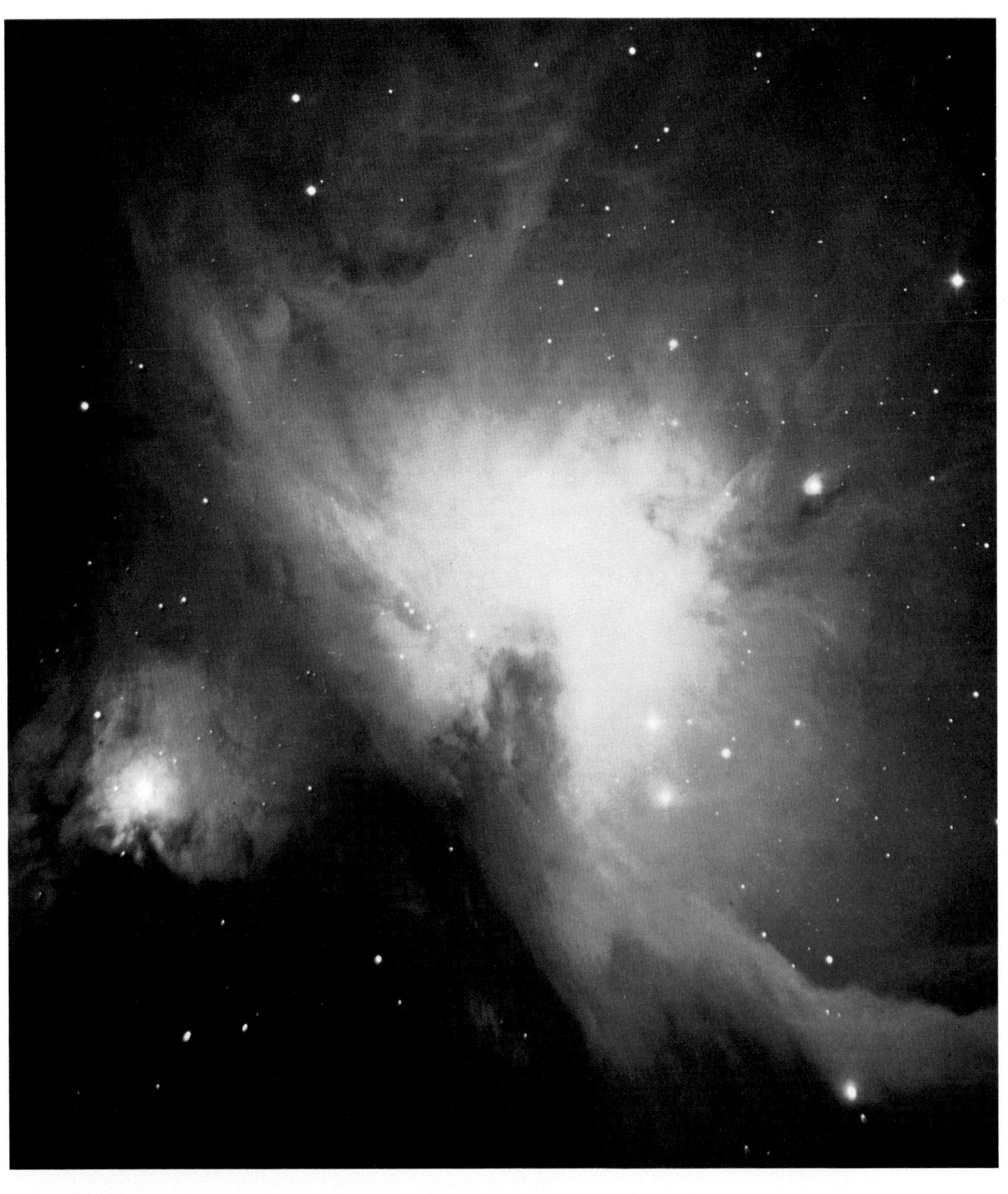

Overleaf: The Orion nebula, a gigantic cloud of gas and dust, possibly not very different from the solar nebula from which our solar system originated. Spectroscopic analysis has shown that it consists of the same elements as are plentiful on Earth. These large clouds of gas, several of which we know in our Galaxy, are not self-luminous, but they are bright because nearby stars either light them up or "excite" them, in the same way as the gas enclosed in a neon tube is excited. In the case of the Orion nebula both the brightness and colors depend on the excitation from very hot stars which light it up, and on the reflection caused by the characteristic fine dust. (Mount Wilson and Palomar Obs.).

Nucleosynthesis

Nearly two decades have passed since the relative abundance of the elements in the solar system were sufficiently well established that the specific nuclear processes responsible for the manufacture of the elements could be determined (see in particular Burbidge, Burbidge, Fowler and Hoyle, 1957). It was evident at that time that these nuclear processes must occur in stellar interiors, particularly in the late stages of the evolution of stars when the temperature becomes very high; and in some cases it was evident that the extreme conditions associated with supernova explosions would be needed to account for the manufacture of some of the elements. At that time it was natural to propose the theory that the galaxy initially started with either a pure hydrogen gas, or perhaps a mixture of hydrogen and helium, and that there was a gradual enrichment of the heavier element content in the interstellar gases as stars formed from these gases, evolved, and passed back to the interstellar medium the products of their internal nuclear reactions.

Fig. 2-1. This illustration shows a classic supernova photographed by astronomers between 1937 and 1942. In the top picture, the phenomenon at its greatest intensity in a photo taken on 23 August 1937 (exposure 20 minutes); the middle picture shows a clear drop in intensity in a photo taken on 24 November 1938 (exposure 45 minutes); below, there is no further trace of the phenomenon in this photo taken on 19 January 1942, despite the long exposure (85 minutes), which enables us to see other fainter stars *(Mount Wilson and Palomar)*. On the opposite page, a supernova observed in 1972; compare the photo on the left, taken in June 1959, with the one on the right, taken in May 1972, in the same area of the sky. The supernovas are now recognized as the structures in which the nucleosynthesis of the heavier elements occurs *(Hale Observatories)*.

In the intervening years a great deal has been learned about stellar evolution, particularly concerning the more advanced stages in the evolution of the stars, and the role of different stars in making contributions to nucleosynthesis has been somewhat clarified.

The composition of the Sun is approximately characteristic of that of normal stars. It is composed roughly three-fourths by mass of hydrogen, and one-fourth by mass of helium. All of the heavier elements put together yield a mass of only about 1.5% of the total (Cameron, 1973a). This ratio of hydrogen to helium has proven to be quite characteristic of other stars in the galaxy, including very old ones, and of the composition of interstellar gases in other galaxies, ranging over a wide variety of morphological types. Under these circumstances, it has become fairly clear that the helium content of our own galaxy, and presumably that of other galaxies as well, has resulted from some sort of nucleosynthesis taking place prior to the formation of the galaxies. It is not clear what this process is; it may be associated with the high temperature and density conditions at a much earlier stage in the expansion of the universe. Those stars which are strongly depleted in heavier elements relative to the Sun are very rare in our own galaxy, and hence it has become evident that most of the heavier element formation in our galaxy occurred very early in its history, before most of the stars presently observed were formed (Truran and Cameron, 1971).

With our recent improved understanding of stellar evolution, it has been clear that not only does the great majority of heavy element nucleosynthesis occur in supernova explosions, but that a small minority among these explosions is responsible for most of the heavy element formation. This small minority consists of the explosion of the more massive stars, of about 30 $M_\odot$ and greater. Relatively few of these massive stars are formed in space, but when they come to the catastrophic endpoint of their evolution, they eject a considerable amount of mass into the interstellar medium in the form of nuclear reaction products. The processes responsible for heavy element formation in these stars are known as explosive nucleosyntheses (Truran, 1973).

The Interstellar Medium

In the last few years, a great deal of fascinating new information has been obtained about the interstellar gases, partly as a result of the introduction of important new techniques into astronomy, such as radio telescopes and ultraviolet and infrared telescopes. About 10% of the mass of the galaxy is in the form of gas and condensed particles called dust or interstellar grains. These interstellar grains are responsible for the obscuration of starlight, forming black patches in the sky which obscure the light from the more distant stars. The gas is strongly concentrated toward the central plane of our galaxy in the form of a thin disk; the local density of gas near the solar system is about equal to the average density of the stars near the Sun.

The interstellar medium is neither quiescent nor uniform in density. The gas near the massive stars, which emit a great flood of ultraviolet light, tends to be ionized; this input of energy into the gas raises its temperature by perhaps a factor of 100 over that of much of the remainder of the gas, and hence the hot ionized gas tends to expand and form a relatively low-density medium. Since there is a tendency toward pressure equilibrium throughout the interstellar medium, the colder unionized gas tends to clump together in denser regions called interstellar clouds. These individual clouds may typically contain several hundred or several thousand solar masses of gas. The clouds are mainly concentrated in the regions of the spiral arms of our own galaxy; there are few if any recognizable clouds in the sparser regions between the spiral arms. Because new massive stars are continually appearing in the spiral regions, and because supernovae frequently occur in these regions, with a violent ejection of their constituent mass, the interstellar gases are being continually stirred into motion as a result of pressure inequalities and these violent events.

It appears probable that the spiral arms in our galaxy represent density waves propagating through the stellar distribution (Lin and Shu, 1964). There is a concentration of stars in the spiral arm regions, so that when a star in its motion around the center of the galaxy wanders into the vicinity of one of these arms, it feels an additional attraction due to the other stars which are clustered in the vicinity, and hence it spends an unusually long period of time in the vicinity of that clustering of stars as a result of the additional attraction which it feels from the other stars. While the star is in the vicinity of the arm, its own mass contributes to the attraction felt by other stars. This is a collective effect which causes a mutual star clustering, and the clustering tendency propagates throughout these stars in the plane of the galaxy.

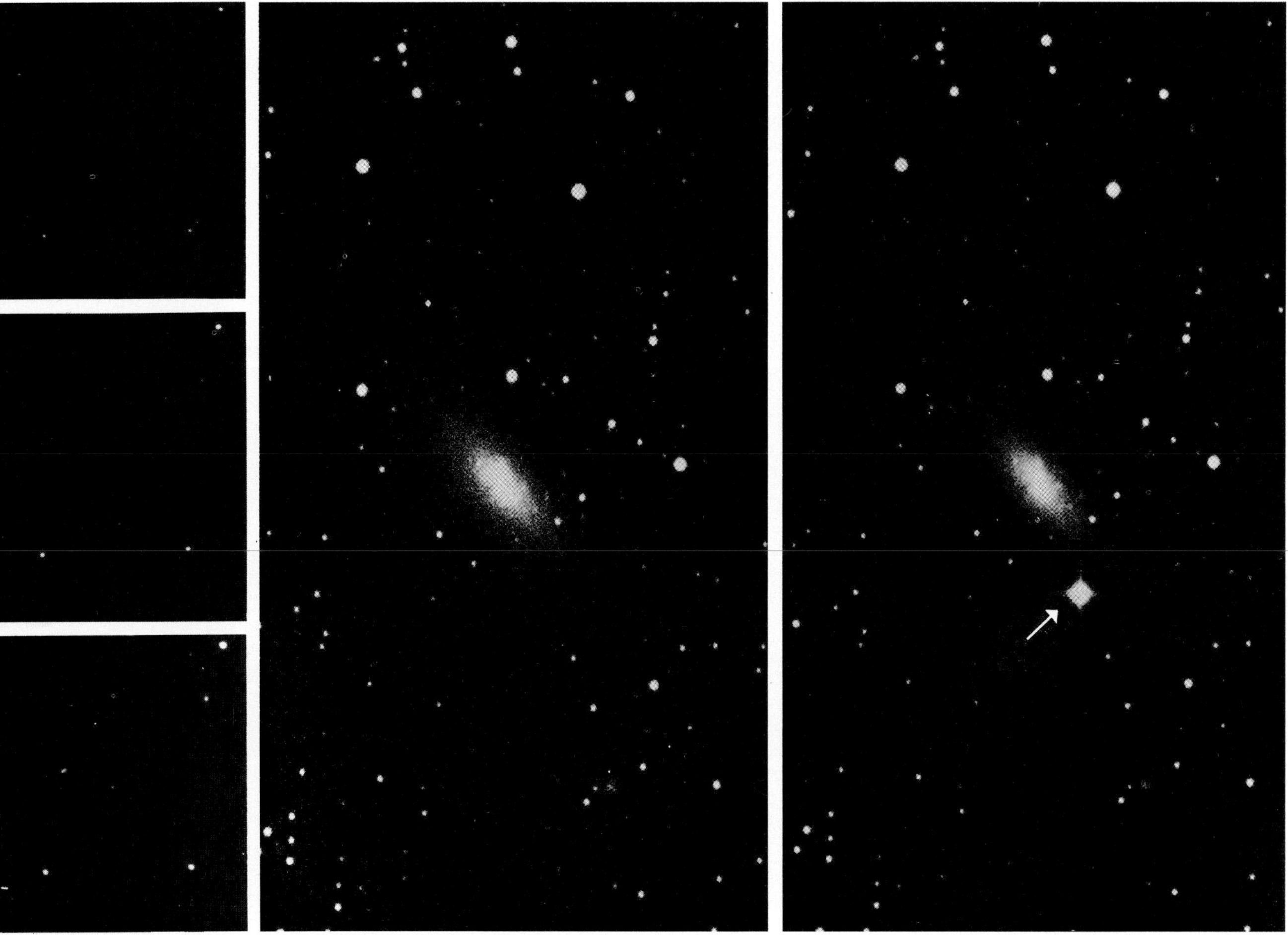

When the interstellar gas moves from an interarm region into one of the spiral arms, it also feels the extra attraction of the stars in the arm. The gas undergoes a shock deceleration, and it too spends an inordinately long time in the vicinity of the spiral arm (Roberts, 1970). Accompanying the shock deceleration is an enhancement of the gas density, which probably leads to various types of magnetic instability, involving the pressure of the magnetic field which is contained in the interstellar gas, and as a result, the denser clouds of neutral and colder gases are formed (Roberts and Yuan, 1970). Most of these clouds gradually heat up, expand, and dissolve into their surroundings as the gas leaves the vicinity of the spiral arm. However, some of the clouds which are formed appear to have an unusually high density. The strong clustering of the interstellar grains that is associated with them shields the interiors of these clouds from starlight, making it possible for very complex molecules to form and exist in their interiors. Radio astronomers have recently discovered a large variety of complex molecules in just such dense cloud configurations. Some of these molecules have a spectacularly large emission at certain characteristic radio frequencies as a result of masering action.

A concentration of gas in space in the form of an interstellar cloud has a tendency to pull together through its own gravitational attraction. However, ordinarily the internal pressure in the gas is more than adequate to overcome this tendency toward gravitational contraction, so that the typical interstellar cloud is quite safe from the likelihood of undergoing a gravitational collapse. In the case of the denser clouds, the issue is not so certain, and if very dense clouds are formed, it is likely that gravitational forces will overpower the internal pressure, and the gas will collapse toward a common gravitating center. The optimum conditions for such a collapse to occur are upon formation of a dense cloud when gas enters a spiral arm and undergoes shock deceleration and resulting magnetic instabilities. It is in just such spiral arm regions that the massive newly formed stars are found in the galaxy, and many of these young stars are observed to be formed in clusters or associations, and to be moving rapidly away from the centers of these associations, indicating a collective formation in a situation where the gravitational attraction which led to the collapse of a massive gas cloud became seriously weakened after formation of stars in the collapse process. It is probable that the appearance of hot massive stars in a newly formed association ionizes the gas which did not become incorporated into the stars, causing it to expand away from the association, and thus leaving the remaining stars in the association gravitationally unbound. It is probable that our Sun was formed as part of such an association.

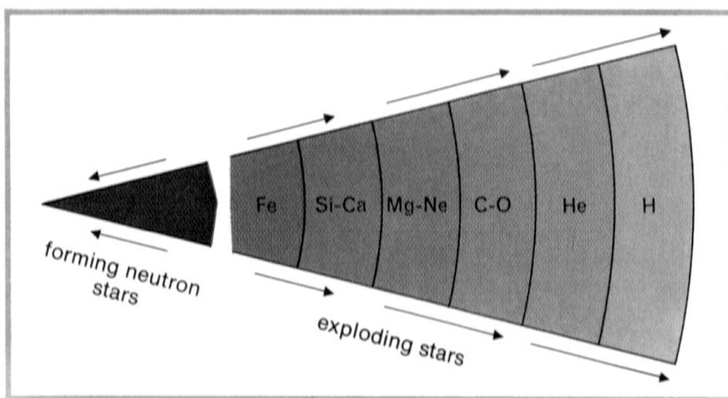

Fig. 2-2. A simplified diagram which shows the various strata of matter ejected during the explosion of a supernova. The innermost strata become hotter because of the shock and contain the heavier elements formed by explosive nucleosynthesis. The center implodes, tending to generate a neutron star.

Formation of the Solar System

Since the time of Descartes, scientists have been fascinated by the problem of accounting for the formation of the solar system in a scientific manner. We shall discuss here only the more recent approaches to the problem, most of which envisage some form of disk of gas and dust, probably formed in association with the formation of the Sun, from which the planets have evolved, mostly from chemically condensed materials.

There are two general approaches which can be taken in an attempt to deduce the character of this gaseous disk of gas and dust, which we shall call the primitive solar nebula. One approach is to try to argue backwards in time from the present distribution of mass in the planets, making the assumption that the general distribution of this mass was also characteristic of the distribution of the planetary constituents in the primitive solar nebula (Hoyle, 1960). It is an essential, but usually unstated assumption of this approach that the accumulation of the planets has been virtually 100% efficient in collecting these materials together from the primitive solar nebula.

We have mentioned that slightly more than 98% of the mass of material of solar composition consists of hydrogen, helium, and the rare gases. Most of the remaining material is composed of rather volatile "icy" substances: water, ammonia, and methane. The nonvolatile component, basically consisting of iron and rocky materials, constitutes only about 0.3% of the mass. With the assumption that the inner, terrestrial-type planets, Mercury, Venus, Earth, Mars, and the asteroids, consist principally of iron and rocky materials, one would have to multiply their combined mass by a factor of about 300 to obtain the total mass of gas from which it is assumed these planets would be derived in the inner portion of the primitive solar nebula. It is often assumed that the two major planets, Jupiter and Saturn, are essentially solar in composition (but see below), so that one would not add anything to their masses in order to determine the amount of material in the primitive solar nebula from which they were derived. The outermost major planets, Uranus and Neptune, have a characteristic density which indicates that they are principally composed of the "icy" constituents, presumably together with iron and rock, so that their masses must be multiplied by a factor of about 50 or 60 to obtain the total amount of mass in the outermost portion of the primitive solar nebula from which they were derived. Adding all this mass together, one obtains a mass for the primitive solar nebula which is about 1% of that of the Sun. We shall call this the minimum solar nebula.

It is evident that since the minimum solar nebula contains so much less mass than the Sun itself, one must assume that it was formed either because the Sun captured this gas from interstellar space, or because the Sun and the nebula were formed together, but the Sun formed directly to its relatively small size through the collapse of the interstellar gases. This poses a significant problem in that the angular momentum which one might conceivably associate with the primitive Sun, even if it were assumed to be spinning very much more rapidly than it is at the present time, is enormously less than the characteristic angular momentum that one would expect to reside in a solar mass of material collapsing from the interstellar medium. Few proponents of this type of theory have attempted to explain how this angular momentum was lost from the collapsing interstellar gas.

Another problem is associated with the recent discovery that the detailed composition of meteorites can be used as a form of cosmobarometer and cosmothermometer (Anders, 1972). Certain trace elements of minor abundance in the meteorites are observed to be strongly depleted in many of the classes of meteorites, but not others, and this gives clues both as to the temperature and the pressure in the primitive solar nebula at which the meteorite parent bodies accumulated. If the accumulation occurs at higher temperatures, then these trace elements are likely to be in gaseous form and strongly depleted in the chemically condensed solids; whereas at lower temperatures the trace elements will be condensed and hence quantitatively included in the solid materials which accumulate to form meteorite parent bodies. From these clues, we deduce that meteorite parent bodies typically were accumulated at a temperature of about 450°K and a pressure of about 10^{-5} atmospheres in the primitive solar nebula. While there appears to be no difficulty in obtaining a suitable temperature for this condensation in minimum solar nebula models, the pressure to be expected in the gas in such models is enormously less than the above indicated amount. Hence this severe deficiency in pressure is a problem which needs to be taken into account by the proponents of minimum solar nebula models.

The other method which may be used in an attempt to deduce the properties of the primitive solar nebula is to try to follow the behavior of a collapsing interstellar gas cloud. In our previous discussion, we followed a dense gas cloud as it formed and reached the point of gravitational collapse. Let us now try to trace what happens during that collapse.

As the gas is compressed toward higher density, the interior becomes shielded against heating by starlight, and cooling processes become very efficient, so that the temperature plummets to only about 10°K. The gas cloud will not be perfectly spherically symmetrical as it approaches the collapse condition, so that some parts of it will collapse more rapidly than other parts, and a fragmentation into progressively smaller in-

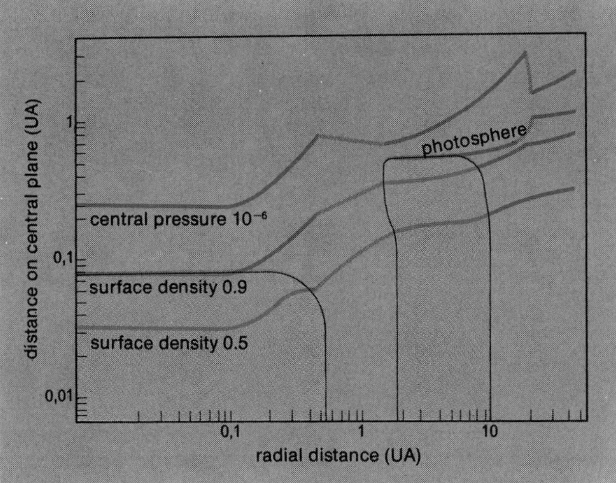

Fig. 2-3. The structure of the primitive solar nebula, based on the calculations of A. G. W. Cameron and M. R. Pine. The colored zones indicate the regions in which energy is carried from within towards the outer surface of the nebula, as a result of the convective movements which occur in the gas. In other regions the energy is moved by absorption and emission of radiation. From the level indicated in the diagram as the photosphere, energy is freely radiated into space. The photosphere coincides with the upper end of the innermost convective region and is not clearly defined between the two convective regions. Half the mass, in a column which is perpendicular to the plane of the nebula, is contained within the space delimited by the line of density on the surface equal to 0.5; the line of density 0.9, on the other hand, encloses 90% of the mass. The uppermost curve indicates where the pressure of the gas drops to values equal to 10^{-6} of those on the central plane of the nebula.

dependent pieces will occur within the gas. This fragmentation will only cease when gas masses of the order of that of the Sun have been reached. Furthermore, since the gas was brought to the verge of collapse by dynamically violent gas flows, it is likely that a great deal of internal turbulence will be induced within the collapsing cloud, and the strength of the turbulent motions will become enhanced as the collapse ensues. These internal gas motions will give a random contribution to the angular momentum of each of the fragments into which the gas breaks up; the total angular momentum that can be expected for the gas which forms the primitive solar nebula is sufficient to spread the gas out into a thin disk having dimensions comparable to that of the present solar system. This collapse process which forms a thin disk does not leave the Sun as a separate body at the center of the disk; the mass of the Sun is spread out through a large portion of the disk. Because the Sun loses a considerable amount of mass early in its history in the form of a vigorous stellar wind, called the T Tauri phase, and because not all the mass in the primitive solar nebula can become a part of the Sun, the total amount of mass probably associated with the formation of the solar system in such a scheme would be about two solar masses. Hence we may call such a model the massive solar nebula (Cameron, 1973b).

In the final stages of the collapse, the cooling efficiency of the gas becomes greatly reduced, since radiation can no longer readily escape from the gas, and hence the gas is heated toward higher temperatures. Both the temperature and pressure which are characteristic of the meteoritic cosmobarometers and cosmothermometers are readily accounted for within the massive solar nebula, since the pressure there is very much greater than in the minimum solar nebula models, because of the self-gravitational effects which tend to compress the gas toward the central plane of the nebula and to enhance the pressure there.

Some authors suggest that the formation and early history of the primitive solar nebula are dominated by effects associated with a strong magnetic field. However, because the collapsing interstellar cloud attains a very cold temperature relatively early in the collapse, it is very difficult for the collapsing gas to squeeze the originally associated interstellar magnetic field into the small dimensions, and this magnetic field thus readily expands out of the collapsing gas cloud. Hence it is unlikely that strong magnetic fields, except for those generated within planetary bodies and the Sun themselves, played a significant role in the early history of the solar system.

There have recently been some detailed calculations of the physical structure and properties to be expected of the massive solar nebula (Cameron and Pine, 1973). The basic structure of the nebula is shown in Fig. 2-3. In this figure both the radial distance and the height above the central plane of the nebula are plotted in logarithmic units. The nebula is very hot toward the center, well above 2000°K, and the temperature decreases radially outwards until it is 100°K or less near the outer boundary of the nebula. There are two regions in the figure which are shown in blue. These are regions in which the gas is stirred up by thermally driven convection. The central convective region results from the peculiar properties associated with the dissociation of hydrogen molecules. The outer convection region results from the very high opacity against radiative transfer of energy due to the presence of the condensed metallic phase of iron. In the regions shown in gray, radiative transfer can carry energy from the central portions of the nebula toward the upper and lower surfaces, from where it can be radiated away into space. We call these surfaces the photosphere, by analogy to the similar surface from which radiation escapes from the Sun.

Such a highly flattened system of hot gas is very unstable against dissipation processes. It is to be expected that circulation currents will exist in the gas which will transport angular momentum away from the center, concentrating it in the outer layers of the solar nebula, which will expand toward greater radial distance. However, the outward flow of angular momentum must be accompanied by an inward flow of mass throughout the majority of the solar nebula. This inward motion of the gas forms the Sun. All finely divided solids which are contained within the gas flow inwards into the Sun along with the gas. Since the amount of chemically condensible material that would be associated with the primitive solar nebula is about 100 times greater than that now present in the planets, at least in the inner terrestrial planetary region, it is evident that the accumulation processes leading to the planets were very inefficient in this model.

An estimate has been made of the rate at which chemically condensed solids may grow to form planetary bodies within this model (Cameron, 1973b). Let us return to the collapse phase of the interstellar cloud. As was remarked, there should be turbulent gas motions within this cloud, and these motions will stir up the interstellar grains which are collapsing with the gas, causing them to collide with one another. The interstellar grains are probably composed of a basic fluffy, icy structure, with cores of iron and rocky materials, and these therefore

TABLE 2-1. - CONDENSATION TEMPERATURES IN A GAS WITH A SOLAR COMPOSITION IN THE PROCESS OF COOLING

Temperature (K)	Substances
1800–1500	calcium, aluminum, titanium oxides, silicates
1450	ferrous metal
1400–1300	magnesium silicates
680	$Fe + H_2S \rightarrow FeS + H_2$
500–400	$3\ Fe + 4\ H_2O \rightarrow Fe_3O_4 + 4\ H_2$ water of crystallization
160	ice
110	$NH_3 \cdot H_2O$ in solid form
60	$CH_4 \cdot 8\ H_2O$ in solid form

probably adhere to one another upon collision. Mutual collisions among these collections of grains may build up snowballs having radii typically as large as 20 cm in the later stages of collapse, to form the solar nebula. The icy components of the solids which accompany the gas toward the inner part of the solar system, where the temperatures are higher, will evaporate away, leaving a fragile structure of the remaining iron and rocky particles.

After the solar nebula has formed, these condensed solid bodies will rapidly fall through the gas toward the midplane of the nebula as a result of the self-gravitational forces within the nebula. After collecting near midplane, it is probable, according to an analysis by Ward (1972) that gravitational instabilities set in within the relatively thin layer of solid material which has settled toward midplane, causing it to clump together to form sizable bodies. The gas in the solar nebula is partly supported by a pressure gradient in the radial direction, so that it rotates around the central axis more slowly than do massive solid bodies, which receive no such support. Hence the massive bodies will move fairly rapidly through the gas, picking up in the process those smaller bodies which are forced to flow more with the gas as a result of gas drag effects. In this way it is possible for bodies of at least asteroidal size, and perhaps considerably larger, to form within the primitive solar nebula on a time scale of the order of some thousands of years, which is also the probable time scale associated with the dissipation processes that formed the Sun out of the primitive solar nebula. Thus it is a feature of the massive solar nebula model that the early development of the Sun and planets should occur in a very short time scale indeed.

The behavior of the gas which flows together to form the Sun has been discussed by Perri and Cameron (1973). This gas remains at a relatively low temperature as it accumulates near the center of the solar nebula, too low for thermonuclear reactions to start at even the densities presently characteristic of the center of the Sun, 100 gm/cm^3. Only when the central density has become considerably greater than this can thermonuclear reactions begin, and then they are likely to release a great burst of energy in the interior, before the Sun can expand to a sufficiently large dimension to cut off the thermonuclear reactions by adiabatic expansion. This will cause the initial Sun to expand to a radius greater than the present one, and at that time the process which we previously called the T Tauri phase mass loss should begin. Several tenths of a solar mass may be lost during this process in the form of a strong stellar wind, about 10^6 times as intense as the present solar wind. This rapid outflow of gas from the Sun is likely to sweep up all remaining gases in the primitive solar nebula, carrying them into interstellar space. Following this loss of the solar nebula gases, the principal architecture of the solar system has been established, and it should contain several major solid bodies and vast swarms of smaller bodies which will suffer gravitational perturbations in their orbits and will be swept up in the further course of time by the major ones.

Structures of the Planets

In any model of the primitive solar nebula there is a decrease of temperature with increasing distance from the axis of rotation. Near the axis of rotation it is probably too hot for any condensed solids to exist, and at very great distance from the axis the temperature is probably low enough to allow the condensation of icy substances. An increasing number of chemical compounds will condense with increasing distance from the central axis. It might be expected that this order of volatility of the chemical compounds would give some clues as to the expected composition of the planets.

This problem has been addressed by Lewis (1972a,b). Nearest the spin axis only the least volatile compounds can be condensed; these are mostly oxides and silicates of aluminum, calcium, and titanium. The least volatile of all is corundum, an aluminum oxide. Only one body in the inner solar system appears to be composed largely of these substances; this is the Moon, and it has been suggested (Cameron, 1972) that the bulk of the Moon did in fact originate relatively far inwards toward the rotation axis in the primitive solar nebula, and that it was subsequently captured by the Earth.

Farther out, at temperatures between 1400 and 1500°K, first metallic iron condenses, and then the magnesium silicates. The abundances of these substances are considerably greater than those of the higher temperature condensates. The planet Mercury has an abnormally high mean density, after allowing for the effects of compression in the interior, and it is likely that Mercury has a great deal of metallic iron in the interior, and something of a deficit of magnesium silicates which have lower density. Venus, being formed at a greater distance, would have a full complement of the magnesium silicates, and thus one can understand its lower mean density after correcting for the effects of internal compression.

At a temperature of 680°K, metallic iron in contact with the gases of the primitive solar nebula would be converted to iron sulfide, the mineral troilite. Lewis proposes that one of the principal differences between the compositions of the Earth and Venus is that the Earth, being formed farther away, probably has a large internal content of troilite, whereas Venus probably does not.

At still lower temperatures, in the range 400 to 500°K, the remaining metallic iron will combine with oxygen if it is in contact with solar nebula gases, and several silicate minerals will acquire water of crystallization. Mars may have this composition.

No further major condensations occur until the temperature falls well below 200°K, when the ices, water, ammonia, and methane condense in that order with decreasing temperature. The two outermost giant planets, Uranus and Neptune, appear to be mostly composed of these ices, and thus it appears that they accumulated in a low temperature portion of the primitive solar nebula.

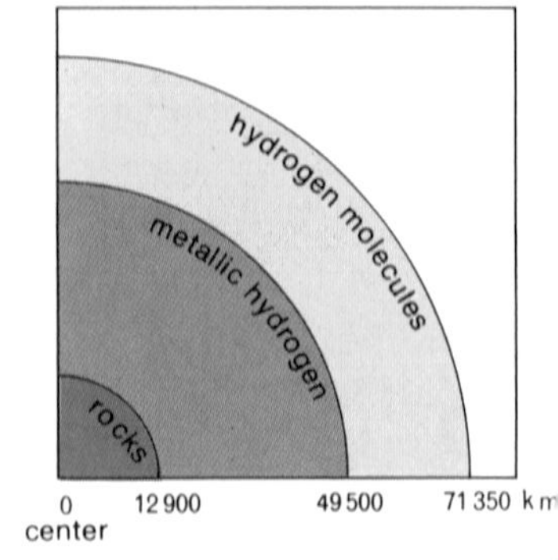

Fig. 2-4. Diagram of the inner structure, and photograph, of Jupiter. According to M. Podolak, this planet has a rocky core equal in size to 40 earth masses; the core is surrounded by a large amount of dust and gas caught from the primitive solar nebula from which the Earth too originated. On March 4th 1979 Jupiter was reached by the American space-probe Voyager 1, which passed it at a minimum distance of about 110,000 km, and substantially confirmed this model of the planet's internal structure *(Kitt Peak National Obs.).*

What of the intervening giant planets, Jupiter and Saturn? We have already mentioned that many people consider these planets to be essentially solar in composition. The mean densities are so low that the majority of the mass in the two planets must be in the form of hydrogen and helium. Many attempts have been made to build theoretical models of these planets entirely out of a composition of hydrogen and helium, by varying the ratio of the hydrogen to the helium. Fairly good models of Jupiter can be constructed in this way, in the sense that the radius, mass, and first gravitational moment can be fitted by such a model with a hydrogen-helium ratio comparable to that believed to exist in the Sun. However, the best of these models have nearly twice as much helium relative to hydrogen than the Sun is believed to contain. Models of Saturn have an even higher proportion of helium, and they do not fit the observable parameters of that planet nearly as well.

What kind of planetary structure would one obtain if one were to insist that the hydrogen-to-helium ratio in Jupiter and Saturn are truly solar? An attempt has been made to answer this question in recent calculations by M. Podolak (unpublished). Both Jupiter and Saturn must then have huge rocky cores at their centers, more than 40 earth masses in the case of Jupiter, and more than 10 earth masses in the case of Saturn. If the compositions of these planets (not only the hydrogen-helium ratio) were completely solar, then the amount of rocky material in Jupiter would be slightly less than one earth mass, and in Saturn the amount would be about one-quarter of an earth mass. Saturn must also contain an amount of water comparable in mass to that of the rocky core, more than 10 times the amount it would have for a completely solar composition.

If these results are correct, they suggest that Jupiter and Saturn may first have formed as large planetary bodies of chemically condensed material, the chemical condensation including some water ice in the case of Saturn. When these planetary bodies became sufficiently large, they captured very large amounts of gas from the primitive solar nebula.

That idea has gained additional credence from some theoretical calculations of F. Perri (unpublished) on the behavior of the gases in the solar nebula in the presence of massive planetary bodies. As the amount of mass in a planetary body increases, the gas in the solar nebula becomes gravitationally concentrated toward the planet, and an extensive atmosphere of gases is formed from the surface of the planetary body extending considerable distances into the solar nebula. There is a limiting condition in which the addition of further solid materials to the planetary core produces a dynamical collapse of the surrounding gas in the primitive solar nebula toward the planetary core. Even for the large planetary cores described above for Jupiter and Saturn, the condition under which gas collapse will occur to form the bulk of the planets, represents a considerably cooler state in the gas than that characterizing the accumulation of meteoritic parent bodies, according to the evidence of cosmothermometers and cosmobarometers described above. This is in accord with the idea that the accumulation of the planetary cores to the point where gas collapse can occur, took considerably longer than the accumulation of the meteorite parent bodies, so that the gas had more of a chance to cool by radiating its internal energy into space.

History of the Earth

Various theories of the origin of the solar system have important consequences for the thermal history of the Earth. There has been something of a debate in recent years as to whether the Earth accumulated in a hot or a cold form. It is sometimes argued that the presence of relatively volatile elements, present in the solid Earth interior, requires that the Earth accumulated at a cold temperature. Actually, this tends to say more about the conditions in which relatively small solid bodies must form in the primitive solar nebula, than the temperature conditions of the accumulated Earth itself. If the Earth itself were to form as a cool object, then ample opportunity must be given for the very large released gravitational potential energy of formation of the Earth to be radiated away during the formation process. Under these circumstances, the Earth would require hundreds of millions of years to be assembled. Such great lengths of time are inconsistent with the massive solar nebula described above, where the indicated dissipation time and planetary accumulation time are very much smaller, in the range of thousands of years.

If the Earth is formed in a relatively short period of time, it is impossible to avoid the consequence that it must have originally been very hot. If the Earth forms as a single main body with a large number of small bodies raining down upon the surface, then there is too little time for the bulk of the released gravitational potential energy to be radiated away during the

accumulation process, and the interior of the Earth must become extremely hot. At the opposite extreme, if several bodies are formed having a substantial fraction of the mass of the Earth, and these collide to form the final product, then the energy deposited in their interiors by the collisions also renders those interiors very hot. Temperatures of the order of 10,000°K or more can be expected in these cases. At the surface of such a large planetary body, rock is completely decomposed at such temperatures, and we are left with the intriguing possibility that the primitive Earth may have been formed with a primary atmosphere composed of the decomposition products of rocks! However, such a primitive atmosphere would cool very rapidly, and condense to a liquid phase in a time of about 10,000 years or so (J. Teller, unpublished).

If the Earth really formed as a cold body, then the presence of radioactive substances deep in the interior would cause a gradual heating until melting of iron sulfide (troilite) occurred. This mineral, having a higher than average density for the interior of the Earth, would settle toward the core of the Earth, releasing gravitational potential energy, and heating the interior enough to cause the melting and draining toward the center of the metallic iron.

On the other hand, if the Earth is formed initially as a hot body, then the iron sulfide and the metallic iron would promptly flow to the center of the Earth and provide a liquid core. At the high temperatures described, the overlying rocks would also be in liquid form, and energy would be rapidly transported toward the surface of the Earth by thermal convection in the liquid rocks.

In a massive solar nebula model, following the removal of the solar nebula gases by the T Tauri phase solar wind, there would be numerous smaller solid bodies in the general vicinity of the Earth, and the Earth would proceed to sweep these up by gravitational attraction in a period of some tens of millions of years or less. The Earth itself would not contain any significant abundance of the more volatile elements, having been formed from small bodies which reached chemical equilibrium with the solar nebula at a temperature perhaps in the vicinity of 600°K. However, the smaller bodies left over at the time of removal of the solar nebula gases would have acquired many of the more volatile elements during the later cooler phases of the solar nebula, and hence these would bring volatile elements into the Earth in the last stages of the gravitational accumulation of the Earth.

Among these volatile materials would be the water which is now in the terrestrial oceans. The atmospheric gases would also be brought in at that time. They probably included a large amount of organic compounds, but the carbon was later oxidized and further reactions with water laid this carbon down in the form of carbonate rocks. On Venus this deposition of carbonates has not occurred, so that Venus has an atmosphere with a tremendous abundance of carbon dioxide. However, Venus lacks the large amount of water which exists on the Earth, perhaps because the small bodies remaining in the vicinity of Venus had not picked up much water of crystallization.

The actual time required for the mantle of a hot initial Earth to solidify depends critically on the character of the secondary atmosphere of volatile materials introduced in this way. With molten rocks existing at the surface of the condensed part of the Earth, the water would not form oceans, but rather a massive steam atmosphere, highly opaque in the infrared parts of the spectrum. High up in this atmosphere the density would become small enough for radiation to escape freely into space, but at such a high altitude in the atmosphere the effective radiating temperature would be very much less than the temperature of the underlying molten rocks. Hence the atmosphere can greatly retard the cooling of the Earth compared to the situation where molten rocks would directly radiate into space. Many hundreds of millions of years may have been required for the solidification process. The oldest known rocks in the crust of the Earth are less than 4×10^9 years old, whereas it is known that the solar system and presumably the Earth also has an age of 4.6×10^9 years.

Following solidification and further cooling, the water vapor would gradually condense out of the atmosphere, forming the oceans. Many organic and nitrogenous chemicals would be dissolved in the primitive oceans, and various types of energy input, such as lightning, radioactivity, and heat, would transform some of these chemicals into very complex forms. It is generally believed that this is the type of process that led to the formation of the first living organisms in these primitive oceans, with the available chemicals being then rapidly consumed through feeding and replication of these first primitive living organisms. At this time the most violent events associated with the origin and early history of the Earth would have terminated, and the subsequent history of the Earth would be the more gradual but still dynamic affair which is described in our geological textbooks.

ALASTAIR G. W. CAMERON

Bibliography: Cameron A. G. W., *Abundances of the elements in the solar system* in Space Sci. Rev., October (1973a); Cameron A. G. W., *Accumulation processes in the primitive solar nebula*, in Icarus, XVIII, 407 (1973b); Cameron A. G. W., Pine M. R., *Numerical models of the primitive solar nebula*, in Icarus, XVIII, 377 (1973); Perri F., Cameron A. G. W., *Hydrogen flash in stars*, in Nature, CCXLII, 395 (1973); Truran J. W., *Theories of nucleosynthesis*, in Space Sci. Rev., October (1973); Anders E., *Conditions in the early solar system, as inferred from meteorites*, in: Elvius A. (ed.), *From plasma to planet*, Stockholm (1972); Cameron A. G. W., *Orbital eccentricity of Mercury and the origin of the Moon*, in Nature, CCXL, 299 (1972); Lewis J. S., *Metal silicate fractionation in the solar system*, in Earth Planetary Sci. Lett. XV, 286 (1972a); Lewis J. S., *Low temperature condensation from the solar nebula*, in Icarus, XVI,, 241 (1972b); Ward W. R., *1. The formation of planetesimals. 2. Tital friction and generalized Cassini's laws in the solar system* (tesi), California Institute of Technology (1972); Truran J. W., Cameron A. G. W., *Evolutionary models of nucleosynthesis in the Galaxy*, in Astrophys. Space Sci., XIV, 179 (1971); Roberts W. W., Yuan C., *Application of the density-wave theory to the spiral structure of the Milky Way system*, in Astrophys. J., CLXI, 887 (1970); Roberts W. W., *Large-scale shock formation in spiral galaxies and its implications on star formation*, in Astrophys. J., CLVIII, 123 (1969); Lin C. C., Shu F. H., *On the spiral structure of disk galaxies*, in Astrophys. J., CXL, 646 (1964); Hoyle F., *On the origin of the solar nebula*, in Quart. J. Roy. Astron. Soc., I, 28 (1960); Burbidge E. M., Burbidge G. R., Fowler W. A., Hoyle F., *Synthesis of the elements in stars*, in Rev. Mod. Phys., XXIX, 547 (1957).

THUS the Earth, according to astrophysicist Cameron, was born from a relatively cool solar nebula, but later, owing to aggregation processes and radioactive phenomena, it gradually warmed up. The geophysicist and seismologist Keith E. Bullen, author of the article which follows, comes to the same conclusions but by other paths, dealing with the inner nature of our planet. Seismology, in fact, leaves no doubt that the inside of the Earth is divided into layers which are progressively heavier towards the center. This stratification according to density can only have come about at a moment in which the whole planet was in a fluid state. As we shall see, seismological data also confirm the astrophysicist's deductions regarding the composition of our planet, and hence also a reliable model of the genesis of the Earth.

How can we know with such certainty the inner structure of the Earth, when the inside of the planet is unreachable and we can never hope to obtain samples of it for analysis?

If we have no direct vision of the inner structure of the Earth, this is so because it is opaque and light cannot pass through it. However, another form of energy which can be propagated inside the Earth does exist: the elastic energy produced by earthquakes. Seismographs are instruments that can "see" how mechanical vibrations are propagated inside the Earth, just as our eyes can see the way light is propagated.

This comparison between light and elastic energy is less qualitative and approximate than it might seem. Elastic energy, like light, is reflected and refracted precisely along the contact surfaces between the different media in which it is propagated at different velocities. And, like that of light, the speed of propagation of elastic energy is a function of the density of the medium, which in turn depends on the composition of the medium and the conditions within it. Thus an earthquake can tell us a great deal about the inside of our planet.

Let us examine in some detail the way this occurs. There are places on the Earth's surface which are exposed to continual stress. The rocks resist these stresses, accumulating ever greater quantities of elastic energy which is suddenly released when the rocks break up. This elastic energy spreads out in all directions, passing through all the rocks of the planet, passing both along the surface and through the Earth's interior. In this latter case, when the wave train meets a reflecting surface, part of the energy is reflected back to the surface.

When the seismic waves reach the seismological station, instruments record precisely both the time of arrival and the shape of the wave reaching the rocks at that point. These two data, picked up by a number of different stations, permit the seismologist to compute the path along which the wave has moved and the velocity at which it has been propagated in the various strata. Hence the seismologist can begin to build a model of the Earth's interior. To do this he must find a way to determine the physical characteristics (density, elastic properties) of the strata through which the wave has passed, to explain the mode and velocity of propagation. Hence the seismologist can deduce—partly on the basis of considerations lying outside the field of seismology—which substances might have the properties observed under the conditions existing at the depths in question.

Keith Bullen, the author of the next article, began in the forties to develop a series of models which he has updated over the years to take into account all the data regarding our planet's interior contributed by sciences other than seismology: by experimental petrology, by observation of the free oscillation of Earth, from the orbits of artificial satellites, by the propagation of vibrations due to nuclear explosions. The result is a fairly clear view of the Earth's interior which will be illustrated in the pages that follow: under a thin crust of light silicates, the mantle, consisting of magnesium and iron silicates, extends to a depth of about 2850 km. This borders on a liquid outer nucleus consisting of iron and nickel with some sulfur (or silicon) and about 2100 km thick. Inside this outer nucleus is a solid inner nucleus about 1300 km thick composed of nickel and iron.

KEITH E. BULLEN

Dr. Bullen formerly taught applied mathematics at the University of Sydney, Australia. He was born in 1906 in Auckland, New Zealand. His interest in the Earth's core started with the mathematical study of the behavior of seismic waves, which he began in the 1930s at Cambridge University under the guidance of the great English geophysicist, H. Jeffreys. In 1936, he was the first scientist to calculate the density of the Earth's core at varying levels to the very center. A sensational corroboration of Bullen's studies took place in the 1960s, following the great earthquake in Chile in 1960, when, for the first time, the free oscillation of the Earth was recorded and the results obtained, based on seismic waves, confirmed Bullen's calculations.

KEITH E. BULLEN

Investigating the Structure of the Earth

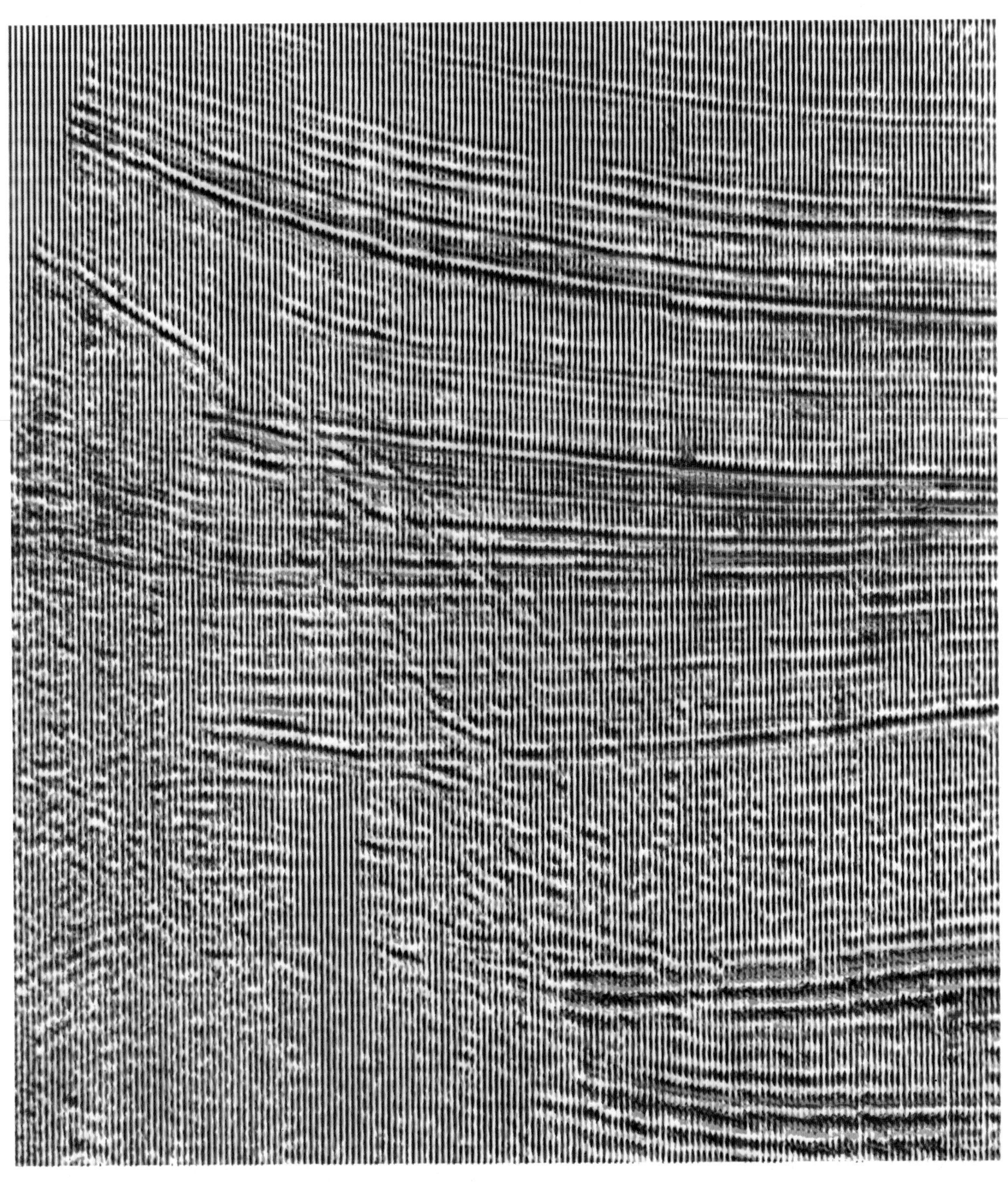

Overleaf: Although this odd image appears to be abstract in nature, it actually represents a new way of seeing the invisible. In fact it is the final result of a project aimed at a geophysical prospection of the subsoil and gives a very accurate picture of part of the inner composition of our planet. It is made by recording the behavior in the subsoil of artificially produced seismic waves, which are produced on the surface by explosions or with other techniques. The data are then passed to a computer specifically programmed to eliminate disturbances and pick up the most interesting signals (Prakla-Seismos).

Introduction

The present article is concerned with the study of the large-scale structure of the Earth, particularly the present internal structure. To a good approximation, the Earth can be treated as spherical with its various properties varying principally with the distance r from the center, and the article will for the most part be limited to this first approximation.

The study, as in all branches of natural science, is basically observational, with mathematical methods brought to bear to assist in co-ordinating the observations. An essential part of the procedure is to set up mathematical models of the Earth in which particular properties such as the density ρ, pressure p, gravitational intensity g, etc., are expressed as functions of r. The models are designed to represent sections of observational data, are continually tested against the data, and modified as new observations are gathered. With each fresh set of observations, the attempt is made to narrow the gap between Earth models and whatever may be that elusive thing called the "real" Earth.

A property common to all parts of the Earth, including the deep interior, the oceans and the atmosphere, is that they consist of deformable matter, solid for much of the interior, fluid for the oceans and atmosphere. An early task is therefore the formulation of suitable mathematical model representations of the deformable behavior of various regions of the Earth.

A representation commonly taken for solid regions is a generalization of Hooke's law, according to which each of the six components of stress is taken to be a linear function of the six components of strain. The relations expressing this dependence in its simplest form involve two coefficients (or parameters) which are suitably taken as the incompressibility (or bulk-modulus) k and the rigidity μ. The incompressibility measures the resistance of a material to stresses that are equivalent to a hydrostatic pressure, and the rigidity to stresses tending to distort the shape without change of volume. The Hooke's law model is justified only to the extent that it is found to be compatible with the relevant observations. The model has proved to be very useful in studying the behavior of the Earth under stresses of comparatively short periods such as occur in tidal movements and in the transmission of earthquake waves. It is less reliable when stresses with periods of the order of geological time are involved. Also, because of the huge stresses reached deep in the Earth, the linear theory of stress and strain is not adequate for some purposes. Except where stated, the generalized Hooke's law model is, however, suitable for the main purposes of the present article.

In these contexts where the model does serve, it is customary to call a material solid if μ is comparable with k, fluid if μ/k is small, say 1 per cent or less. Since the model includes cases in which the parameter μ is negligible, it can also serve for some purposes in respect of the oceans and atmosphere; but allowances have often to be made for additional parameters representing, for example, viscosity and thermodynamical effects. When the terms "solid" and "fluid" are used in connection with the Earth's interior, they are to be interpreted in the sense just defined and relate to behavior under stresses of limited periods. Later in the article, numerical values of k and μ will be given.

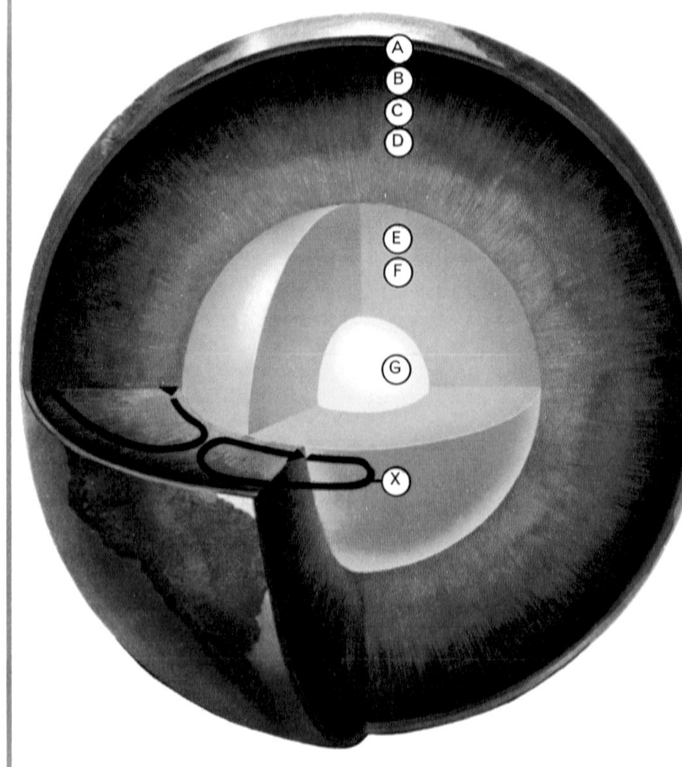

Although, as will be seen, five-sixths of the volume of the Earth below the oceans is solid, it is satisfactory for many purposes to represent the internal stresses in terms of a single parameter, which is equivalent to a hydrostatic pressure p, instead of the full six stress components; p corresponds to the mean of the three principal stresses. The selection of p as the dominant parameter is a consequence of further experimental evidence on the strengths of solid materials: the evidence indicates that deviatoric stresses, i.e. deviations from a pure hydrostatic pressure, become increasingly small compared with p as the depth z below the Earth's surface increases. At the same time, some of the effects of deviatoric stresses even where they are small compared with p, are significant at depths up to several hundred kilometers. (In the absence of deviatoric stress, there would be no mountains or separate continental and oceanic regions.)

The remainder of the article will be concerned principally with the processes whereby serviceable model representations are obtained for the internal distributions of ρ, p, g, k and μ in the Earth.

Internal Layers of the Earth

For purposes of exposition, it will be convenient to anticipate later detail by setting down now a nomenclature for the

Earth's internal layers, the main features of which were introduced by this author in 1940–42. The description here given is approximate only, but serves for the purposes of this article.

The region A is the so-called "crust" extending from the Earth's surface to the Mohorovičić discontinuity at depth about 35 km below continental shield areas; the depth is greater under some mountain ranges, but much less under the oceans. The crust, though very complex, is here treated as a single unit because of its comparatively small thickness. The Balkan seismologist A. Mohorovičić was the first to find evidence, from an earthquake in 1909, bearing on the existence of this important discontinuity.

Between the "crust" and the "central core" is the Earth's mantle which is solid throughout. The existence of the core was indicated in the course of earthquake studies by R. D. Oldham in 1906, and the depth of the mantle-core boundary was shown to be near 2900 km by B. Gutenberg in 1914. The mantle is conveniently divided into the regions B (with lower boundary at depth about 350 km), C (between 350 and 1000 km) and D (below 1000 km); D is subdivided into D' (1000-2700 km) and D'' (2700–2900 km).

The central core includes the so-called "outer core" or region E, extending from 2900 km to at least 4500 km depth, and the "inner core" or region G extending from about 5100 km depth to the center at depth 6370 km. Between E and G there may be one or more transition layers which constitute the region F. The existence of an inner core was indicated in work of the Danish seismologist Inge Lehmann in 1936, B. Gutenberg and C. P. Richter in 1938 and H. Jeffreys in 1939. There is overwhelming evidence that the outer core is fluid, and fairly strong evidence that the inner core is solid.

Fig. 3-1. The diagram opposite shows the principal "wrappings" which form the inner part of our planet. They are: the Earth's crust (A in the diagram), the mantle (B, C and D), the outer liquid core (E, F) and the inner solid core (G). To the left, the names of the various seismic wave-trains on the basis of the path taken: P, longitudinal waves; S, transversal; K, path through the outer core; I, through the inner core; c, reflection from discontinuity between mantle and core; i, between outer and inner cores; p, longitudinal waves emerging from the focus; s, similar transversal waves; X indicates the convection currents.

Sources of Evidence on the Earth's Internal Structure

Inferences on the internal constitution of the Earth became possible after 1798 when the Earth's volume V (1.083×10^{27} cm^3) and mass M (5.977×10^{27} g) had been determined. These determinations provide a first restriction which any acceptable Earth model must fit.

Intricate mathematical theory of the figure of the Earth, along with a body of astronomical evidence on such matters as the precession of the equinoxes and geophysical evidence on the Earth's rotation, etc., led in the late nineteenth century to an estimate of the coefficient y in the Earth's moment of inertia yMR^2, where R is Earth's radius. The value obtained for y before 1963 was 0.334. The early theory assumed model hydrostatic conditions throughout the Earth, but evidence in 1963 using analyses of the orbits of artificial satellites has reduced the estimate of y to 0.331. The value of y provides a second important restriction which Earth models must fit.

With knowledge available only on V, M and y, it was appropriate to consider Earth density models with only two parameters. Examples are the Legendre-Laplace model

$$\rho = Ar^{-1} \sin Br \tag{1}$$

and the Roche model

$$\rho = A - Br^2 \tag{2}$$

which were much used as early representations of the distribution of the density ρ and continued in use even as late as around 1930. When A and B are determined using the numerical assessments of y, V and M, these models give ρ as ranging from 2–3 g/cm^3 near the surface to 10–12 g/cm^3 at the center.

TABLE 3-1. VALUES OF THE PRINCIPAL PHYSICAL PARAMETERS INSIDE THE EARTH*

Zone	Depth z (km)	Density ρ (g/cm^3)		Pressure p (10^{12} dyn/cm^2)		Incompressibility k (10^{12} dyn /cm^2)
Region	z	ρ	p	k	μ	g
A	0–15	(2.8)		(0.6)	(0.4)	983
	15	3.3	0.004	1.0	0.7	983
B	350	3.6	0.18	1.8	0.7	994
C	1000	4.5	0.39	3.5	1.8	994
D	2900	5.6	1.37	6.5	3.0	1070
	2900	9.9	1.37	6.5	0	1070
E	4700	12.2	3.03	12.4	0	580
F	5200	12.7	3.35	13.6	1.9	420
G	6371	13.0	3.67	15.0	1.1	0

*The figures in parentheses represent the average values of these parameters which are assumed for the Earth's crust.

A big step forward became possible after seismology led in the course of the present century to fairly close estimates of k/ρ and μ/ρ at specific depths inside much of the Earth's interior. Details of this development will be considered in a separate section.

A central problem in constructing useful Earth models is to find means of deriving separate values of ρ, k and μ, given the seismic data on k/ρ and μ/ρ. The process involves appeal to much additional evidence, including laboratory experiments on rocks and other materials at high pressures and temperatures. Through these experiments and the seismic data, provisional identifications are made of materials below the Earth's surface and restrictions are set on the range of possible densities in some parts of the Earth. The investigations go hand in hand with various geochemical considerations.

Other evidence comes from the application of finite-strain theory (involving terms not considered in the simpler linear theory) to the study of various effects of high pressure in the Earth's deeper interior. In this way, Francis Birch has supplied important evidence, independent of seismology, on the dependence of k on p in the Earth, and especially on the gradient dk/dp. A variety of other theoretical considerations involving details on representative atomic numbers in particular internal regions are also brought to bear.

In the period since World War II, there have also been several further major advances contributing not only to improved assessments of k/ρ and μ/ρ but also, to some extent, to the assessment of ρ, k and μ separately. The advances include the use of nuclear explosions for seismological purposes, the erection of "array stations" which are large networks of stations with standardized equipment for recording nuclear explosions and earthquakes, the gathering of data on free Earth oscillations following some very large earthquakes, and the utilization of data from shock-wave experiments in which pressures up to those reached in the Earth's deepest interior are generated.

In another direction, still in its infancy, comparisons with other planets are starting to contribute evidence on the Earth's interior.

Evidence from Seismology

When an earthquake occurs, seismic waves issue from the source, or "focus" and, if the earthquake is strong enough, travel through all parts of the interior. The waves are associated with mechanical vibrations of the Earth's materials as the waves pass through. Bodily seismic waves travel in all direc-

Seismology and the Study of the Internal Structure of the Earth

The study of earthquakes is the main indirect method of gathering information about the Earth's interior. The indications supplied by seismograms (phases, waves and their parameters) are useful both for the study of the Earth's crust, with prospection by refraction and reflection and with the surface wave dispersion method (even if with considerable differences in the case of the continental plateaux*), and for the study of the mantle and the cores, using our knowledge of the propagation speed of waves, our knowledge of the presence and surface position of discontinuities affecting these speeds. But interpretation of the data always entails some uncertainty. It is thus necessary to correlate seismology with gravimetry, magnetism and the Earth's radioactivity in order to obtain other indirect possibilities of specific geological calculations or information, even though these may sometimes be localized.*

A basic factor is the exact calculation of the times of arrival of the P waves, the S waves and the surface waves. From the S—P difference it is possible to calculate the distance and, last of all, the propagation speeds of the seismic waves from the epicenter (i.e. from the barycenter of the zone which, on the surface, is on the vertical of the focus or hypocenter) to the observatories.

From a sufficiently complete series of propagation speeds it is possible to deduce the speed of the various waves at different depths.

It is evident that if the Earth's interior were isotropic, there would then be very clearly determinant hodochrones which were in a certain sense standardized. The analysis of hundreds of recordings from different stations around the world for the same earthquakes or for different earthquakes, and the consequent confirmation of the occurrence, at a certain point, of period variations, wave superpositions, new wave-trains of a particular dimension, times of arrival of particular phases being either late or early, and so on, has caused seismologists to formulate hypotheses in the first instance, and then theories and models corresponding to the nature and structure of the subsoil. Once this direction of research had been embarked upon, it developed to such an extent that today, with a 75% valid approximation, it is possible to think of seismograms as geological documents.

Lump together all the deductions which lead to a sufficiently acceptable picture of the Earth's crust, and you have something which is indicative rather than complex. Almost everything depends on the "travel-time" curves or hodochrones. The interpretation of these curves involves hypothetical presuppositions about the speed and depth of penetration of the seismic wave-trains. This leads to a conclusion which, in spite of everything, is not a negative one: the internal structure of the Earth is not univocally determined by seismology. In addition to a directly observable experimental model, the exact diagnosis (or the diagnosis nearest to the truth, at least)

A forshortened view of one of the most widely used seismographs, the Wiechert model. On paper blackened with lampblack the two needles trace the recordings corresponding to the horizontal components *(photo, G. Motto)*.

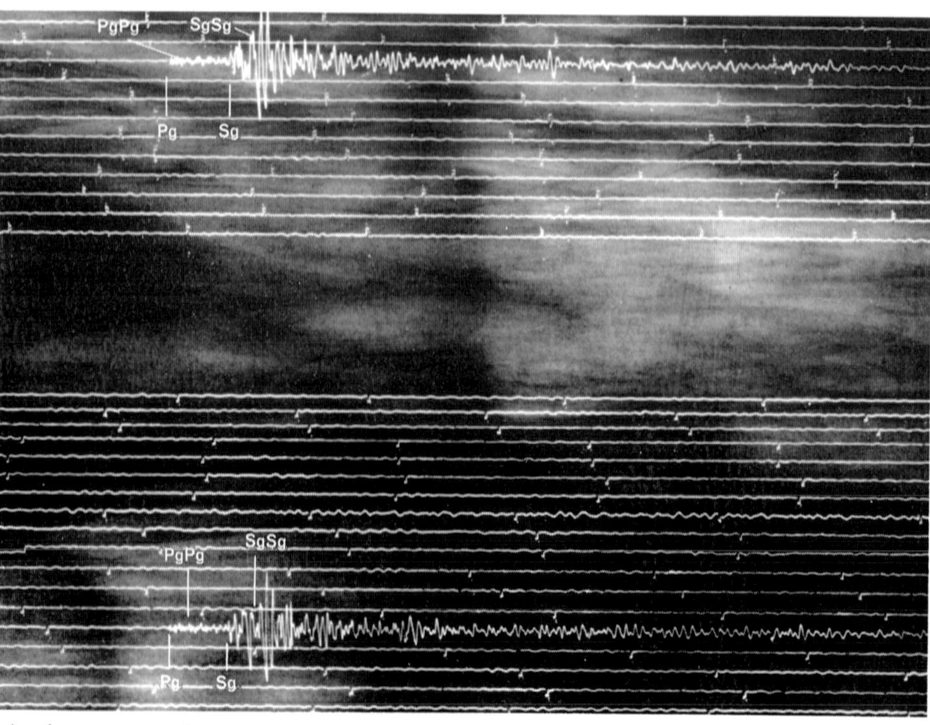

A seismogram made on 6 January 1969, corresponding to the two horizontal components of a mild earthquake (magnitude M = 4.25) with its epicenter in Tuscany about 180 km from the seismic station at Pavia. The seismogram shows the direct longitudinal waves P_g, P_gP_g, and the direct transversal waves S_g and S_gS_g.

tions through the Earth. Surface seismic waves travel over the Earth's surface, the amplitudes of the vibrations in this case diminishing with depth below the surface.

The bodily waves consist of two types: the primary or P waves in which particles vibrate in the direction of the advancing wave (like waves in sound); and the S or secondary waves in which the particles vibrate sideways. On the generalized Hooke's law model theory, the speeds α and β of P and S waves are given by

$$\alpha^2 = (k + 4\mu/3)/\rho, \qquad \beta^2 = \mu/\rho \tag{3}$$

In fluid regions of the Earth where μ is treated as negligible, β is effectively zero. (A fluid region is sometimes defined as one in which S waves cannot be detected.)

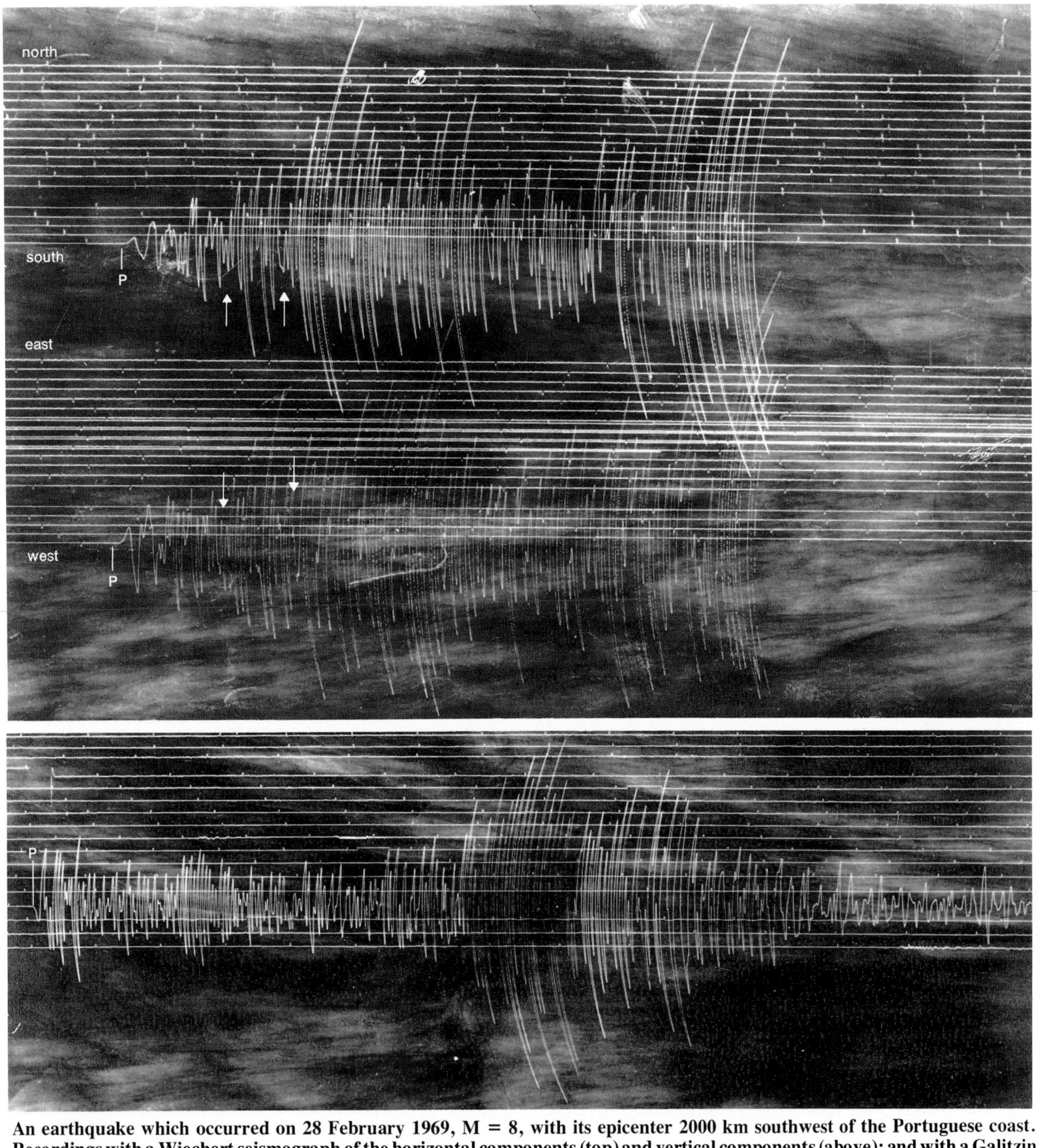

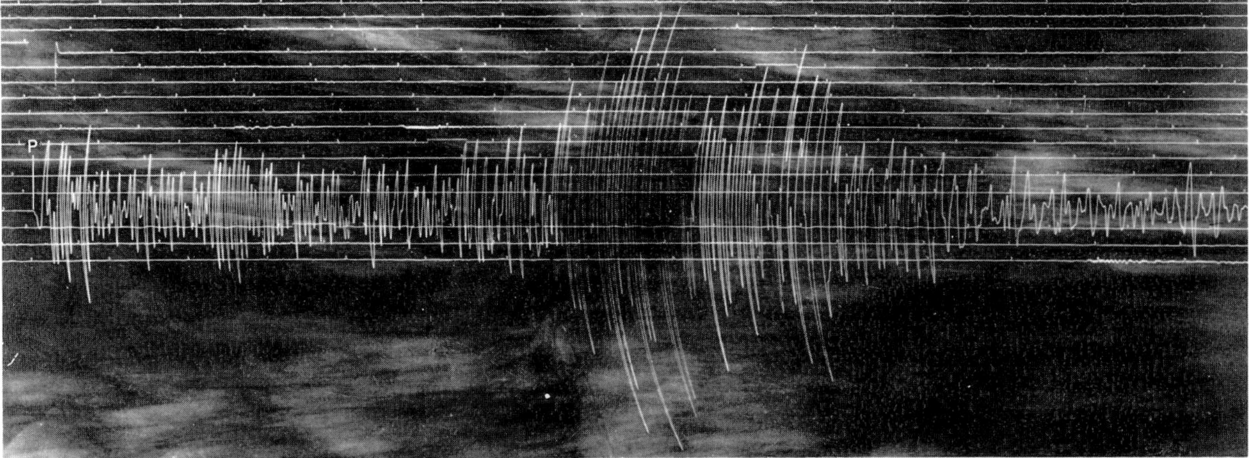

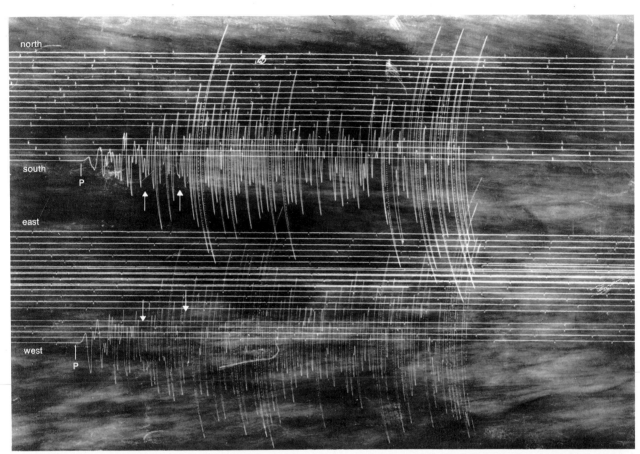

An earthquake which occurred on 28 February 1969, M = 8, with its epicenter 2000 km southwest of the Portuguese coast. Recordings with a Wiechert seismograph of the horizontal components (top) and vertical components (above); and with a Galitzin seismograph of the vertical component (below). From the Wiechert seismograms, which clearly show the arrival of the P waves, various phases (arrowed), and the size of the S waves, it is possible, with the help of other stations, to make a clear calculation of the epicentral coordinates. Based on the calculation of the time of arrival of the surface waves (visible on the Galitzin seismogram) which have covered the maximum circle several times, it is possible to obtain information about the medium crossed.

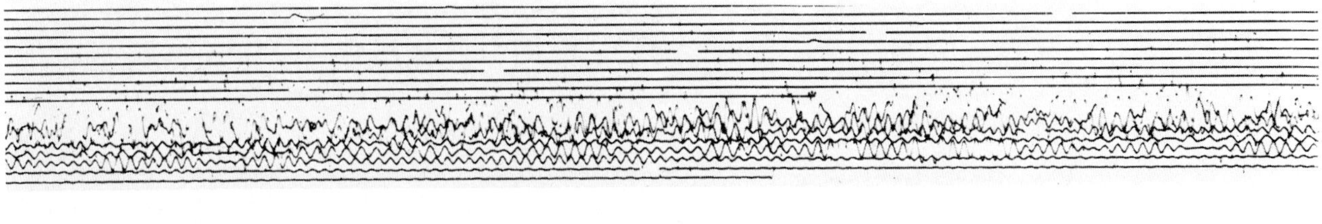

Seismological travel-time tables give the time t taken by a P or S bodily wave to travel from one point A of the Earth's surface down through the interior and up again to another point B of the surface in terms of the angle Δ subtended by AB at the Earth's center. The tables give t for all values of Δ up to 180° for P waves and to about 100° for S waves. The evolution of reliable tables was a slow process, partly because in the case of natural earthquakes there is no prior knowledge of the locations of the sources or the times of origin. There are also difficult problems in disentangling the records that are

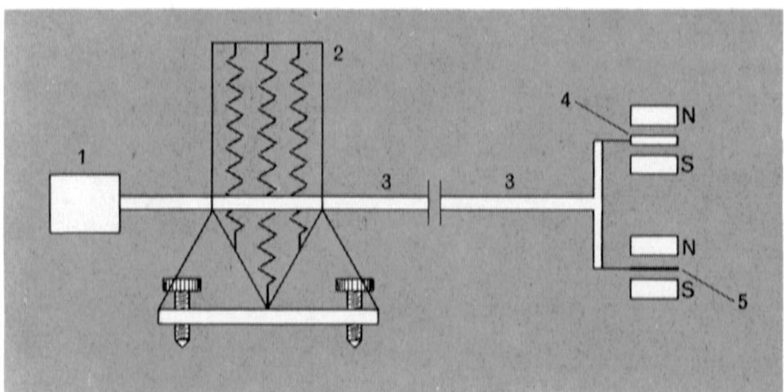

Top, an overall view of the vertical Galitzin-Pannocchia long-period seismograph, and photographic recording. On the left we can see the mass and the suspended elements; in the diagram below, these are respectively (1) and (2). In this diagram, (3) indicates the pendulum arm, to which a mobile coil (4) and a copper plate (5) with a shock-absorber function are attached. Nos. (4) and (5) can be seen in the photo on the right, which shows the inside of the protective housing on the right of the top photo. The movements of the coil generate a current which causes partial rotations of the mirror of a galvanometer on which a ray of light is reflected. As the angle of incidence varies, in relation to the rotations of the mirror, the angle of reflection also varies continuously, and thus so does the point at which the ray of light strikes a film wrapped around a rotating spool, producing seismograms like the one on the page opposite (*photo G. Motto*).

depends on a correlation between the speed of the seismic waves observed on Earth, and laboratory calculations corresponding to the parameters of elastic constants of mineralogically and chemically known samples. But in the laboratory one is still limited and conditioned by particular pressure values, anomalous states of aggregation, and many other technical factors.

Because every mineral-chemical type has its own particular propagation speed, which differs, even if only slightly, from the speed typical of other types, the interpretation of the hodochrones is all the more complicated: by local anomalies in earthquakes occurring nearby, and in distant earthquakes by the diversity of the means of propagation.

The hypothesis of the speed of seismic waves which are regularly and continuously on the increase makes it impossible to interpret the hodochrones. The hypothesis of discontinuously variable speeds (varying with the depth) leads to the interpretation of a stratified internal structure; this is borne out by many practical examples of recordings which have led to logically acceptable mathematical-physical models.

The analysis and interpretation of the hodochrones are basic to all research. They are the only point of departure for the solution of the problem of learning about the speed pattern inside the Earth.

Starting from Benndorf's relation ($v_0 = v \sin i_0$ in which v_0 is the real speed of the surface wave; v is the apparent speed of the ray along the surface; and i_0 the angle of incidence) and making use, among the various methods elaborated, of the Wiechert-Herglotz method, with the research repeated for various positions of the point of emergence, it is possible to calculate the propagation speed for different depths. This method is only valid up to a point if applied to the Earth's crust in which there are certain areas of discontinuity which the hodochrone itself can show up. The method becomes acceptable if the discontinuities are of a second type, i.e. if there are strong but constant speed variations. In practical terms the hodochrones are corrected for the 35–40 km of the Earth's crust.

The Earth's Crust

Disregarding the contribution of seismic prospection (applied geophysics), it is as well to remember that the phenomenon of the dispersion of surface waves (Rayleigh's theory) with variable speeds is very important in the calculation of the depths of the various strata: it increases with the wave-length and presumably increases likewise in oceanic routes as compared with purely continental routes.

Other speed variations are then connected to the means of propagation and to their dynamic-elastic characteristics. Because the maximum wave-length considered to date is of the order of about 2000 km, one can see that the movement of similar waves is affected right down to the core by the rest of the crust and by the mantle. A possible comparison between dispersion curves determined by seismograms and calculated theoretical curves makes it possible to reach fairly sound conclusions about the medium passed through (nature and thickness).

In general a network of seismic stations is used and the variations in speed are interpreted from place to place as based on the various thicknesses of the crust. Possible uncertainties and probable errors have been partly rectified with seismic prospection using large explosions.

With such considerations as these it has been possible to draw up the bases of methods for determining the thickness of the first stratum of the crust (H. Jeffreys has even been able to calculate a relation d-l, *where* d *is the thickness of the stratum and* l *the length of the* L *waves).*

These methods (developed by W. Rohrbach, von zur Mülhen, E. Tillotson and others) have produced good results in terms of estimating the thickness of the upper part of the Earth's crust, and these methods are very valid traced by seismographs at the world's seismological observatories. Good reliability had, however, been attained by 1940 in travel-time studies, and from that data tables compiled by H. Jeffreys and K. E. Bullen (the J.B. tables) have been used in compiling the International Seismological Summary. Since 1918, the ISS has given the origin time, epicenter (point of the Earth's surface vertically above the focus), focal depth and other details for the world's earthquakes.

By a sophisticated mathematical process, one can derive from the travel-time tables fairly reliable values of α through-

if applied to distances of at least 1000 km. Other methods are used for calculations made in narrower regions of lesser thicknesses (particularly artificial methods).

Today, nevertheless, tables have been published showing P and S speeds which are very close to the truth, and depth values worked out to the kilometer.

Mantle and Core

The Wiechert-Herglotz method works extremely well for the basic investigation. B. Gutenberg, followed by other seismologists, calculated the speed of the S waves down to a depth of 2900 km and singled out areas of discontinuity at several points: towards 80–150 km (the asthenosphere, i.e. the zone with slowed down elastic characteristics); towards 1000 km, in as much as there is another speed decrease, and lastly, at 2950 km with the fundamental (Gutenberg) surface which differentiates the mantle from the core. This latter area of discontinuity is of the first type: the drop in speed of the longitudinal waves, which are then the only ones to move into the (liquid?) core, is 45%.

*At a depth of about 5000 km the speed tends to increase once again (and now we are in the inner core), and at this point is is not certain whether the transversal waves affect at least the inner core (*kern *core).*

The nature of this innermost part of the Earth can be fairly well indicated whenever there are PKJKP waves (transversal waves passing through the inner core) in addition to the already known PKiKP waves (longitudinal waves reflected from the core). Coming back to the classification based on the speed of the longitudinal waves, we find that, after the moho discontinuity, this varies rapidly until the limit of the asthenosphere, undergoes a slight drop for about 100 km and then increases up to values of the order of 13.5 km/sec. up to the Gutenberg surface: the last variation occurs at a depth of about 5000 km.

A clear distinction between the core and the mantle is now confirmed. Other seismic evidence can be supplied by phenomena produced by the "shadow zone" which is a region with an epicentral distance of slightly more than 100° (about 11,000 km), about 40° wide (up to 15,500 km) and limited by the parallel which contains the points of emergence of the rays tangent to the core, and by the parallel formed by the emergent rays which, at a fairly wide angle, cut through the core itself. In the seismograms taken from the seismic stations in this zone, the P waves are missing, possibly because they have been diffracted from the tangency to the core, and we see the PKP waves which have touched the core. One can also calculate the dimensions of the core using as a basis (among the other methods) the times of the reflected (PcP) or refracted (PKP) waves.

But even mathematically speaking the method is somewhat laborious because of the difficulty of calculating (using Wiechert-Herglotz) the exact inclination of the direct waves issuing from the core.

FLORENZO CHIEPPI

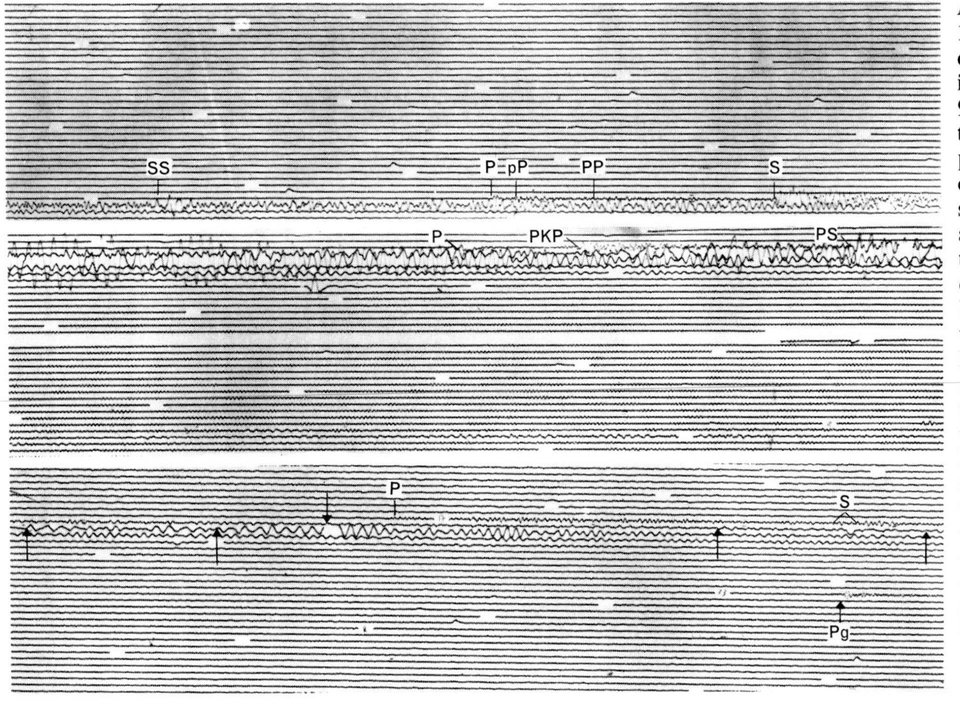

Above: a seismogram of 1.19.69 showing the earthquake (M = 6.25) with its epicenter north of Japan and 9200 km from the Pavia station; the beginnings of the phases P, S, PP and SS are clearly visible, whereas the surface phases are not evident at all: it can be deduced that the focus of the earthquake is deep down; this is confirmed by the appearance of the pP waves; in fact this reaches 210 km (130 miles). Center: a seismogram of 1.30.69 showing the earthquake (M = 7.25) with its epicenter SE of the Philippines, 11,700 km from the station; the P waves are weak, but the PKP waves, which have passed through the core, are evident; in relation to the earthquake's epicenter the observatory is near the seismic shadow area. Below: seismograms of 1.6.69 showing above, the earthquake (M = 6) which occurred to the east of the Santa Cruz islands, 15,400 km from Pavia; and below, the earthquake which occurred 180 km away in Tuscany; the beginnings of both the P and S waves are vague in the former, but other phases are visible (arrowed); in the latter the beginnings of the Pg waves are clearly visible.

Such an analysis of the speeds (especially of the longitudinals in km/sec (mi/sec.) carried out as mentioned, enables one to make a clear subclassification of the crust:

1. For speeds between 0.4 and 4 the strata are sedimentary, of variable depth (between 0 and 12 km), and the areas of discontinuity are uncertain and do not necessarily represent stratified horizons;

2. For speeds between 4 and 6.2 we find the granite stratum with the predominant Al silicates;

3. For speeds in excess of 6.2 (up to 8.3) we find basalts and possibly gabbros and immediately below the first major area of discontinuity: the Mohorovičić or moho discontinuity.

out practically the whole Earth and of β throughout the mantle. Values of α and β derived by Jeffreys in 1939 from the J.B. tables have been used in many Earth model studies. Thus through seismology good estimates of k/ρ and μ/ρ are available throughout the mantle, and of $(k + 4\,\mu/3)/\rho$ throughout the core.

Supplementary results include results derived from studies of the Earth's bodily tides; these are movements of the solid surface of the Earth similar to the oceanic tides but smaller in size. In 1862, Lord Kelvin made an estimate of the Earth's mean rigidity using data on bodily tides, and his approach combined with seismic results for the mantle enabled H. Jeffreys in 1926, H. Takeuchi of Japan in 1950, and M. S. Molodenski of the U.S.S.R. in 1955, to provide strong evidence that the outer core is fluid. Thus k/ρ and μ/ρ can be taken as well known in the Earth down to at least 4500 km depth.

Additional evidence on ρ, k and μ in the outer part of the Earth comes from analysis of seismic surface wave data. The foundations of surface wave theory were laid by Lord Rayleigh and A. E. H. Love using fairly simple mathematical models for the subsurface structure. The theory has since been greatly extended to allow for a variety of complexities in the structure, and observations of long-period surface waves now supply evidence on the structure down to about 400 km depth.

Studies of seismological records of nuclear explosions are enabling some of the travel times for bodily waves to be determined with sharply increased precision. This is because of an advantage over the case of natural earthquakes in that there can be prior knowledge of source conditions. Among other things, data from nuclear explosions are becoming particularly helpful in measuring effects on travel times of some of the deviations from spherical symmetry below the Earth's surface. Allowances due to the Earth's ellipticity of figure had already been calculated before 1940, but additional deviations from spherical symmetry are now known to be significant well below the surface. The most marked of these are differences between subcontinental and suboceanic structures which extend down to several hundred kilometers depth, but diminishing as the depth increases.

The largest seismological array station is in Montana, where 525 substations are spread over an area exceeding 10^4 km^2. Array stations are of special value in enabling "signals" of interest from particular earthquakes to be separated out from background "noise" on records, resulting in much sharper measurement of the onsets and other features of specific groups of arriving seismic waves. The mathematics of communication theory is used in the analysis. Array stations are currently adding important detail on $dt/d\Delta$ in P and S wave transmission through the Earth and thereby assisting in the finer determination of the variation of α and β with τ.

An important event occurred in May 1960 when a great earthquake in Chile set the Earth vibrating as a whole in its free oscillations, to a degree which resulted in these oscillations being indubitably recorded for the first time in history. The successful recording arose from pioneering work of Hugo Benioff who in 1952 had set up a special seismograph designed to record changes of strain during disturbances of the Earth. The usual seismograph records measure not strains but displacements of the ground. At the time of the 1960 earthquake, this instrument and several others recorded free oscillations of the Earth, both fundamental and overtone oscillations, with periods ranging up to nearly an hour. Similar oscillations have been recorded during several later earthquakes, notably the Alaskan earthquake of March 1964. There are now available more than 100 well-determined free Earth oscillation periods each of which sets a new restriction which an acceptable Earth model must satisfy.

The utilization of this new body of information involves prodigious mathematical calculations in which electronic computing is indispensable. The procedure is to start with an Earth model consisting of a set of distributions of ρ, k and μ, or of ρ, α and β, arrived at independently of the free oscillation data, compute the free oscillation periods for that model, and then test against the observed periods. Various devices, including the use of a Monte Carlo inversion scheme, are being used to arrive at improved models. The process is heavy because the determination of the periods for any one model is itself complex and lengthy, and many models have in practice to be rejected before one can be found which fits all the data within the observational uncertainties. Pioneering work on the determination of free oscillation periods for assigned Earth models was carried out by C. L. Pekeris. The first model to go close to fitting the currently available data was obtained by K. E. Bullen and R. A. Haddon in 1967, but studies still being carried out are likely to give further improvement in the fine detail.

Determination of the Earth's Density Distribution

As already stated, seismic data on the P and S bodily waves give much information on k/ρ and μ/ρ in the Earth. Seismic data also contribute in the following way to the direct determination of the density gradient $d\rho/dz$ at various levels below the Earth's surface.

Let ρ, p and g be the density, pressure and gravitational intensity at depth z below the Earth's surface or distance r from the center 0, and let m be the mass inside the sphere of radius r and center 0. It is sufficient for present purposes to ignore departures from hydrostatic conditions, so that we can to high accuracy write

$$dp/dz = g\rho. \tag{4}$$

The theory of gravitational attraction gives

$$g = Gmr^{-2}, \tag{5}$$

where G is the constant of gravitation.

Let $\phi = k/\rho$, where k is the adiabatic incompressibility. Then, by equations (3),

$$\phi = k/\rho = \alpha^2 - 4\beta^2/3. \tag{6}$$

For a chemically homogeneous material adiabatically compressed and free of phase changes, the usual definition of k gives

$$\phi = k/\eta = dp/d\rho. \tag{7}$$

We shall generalize (7) to the form

$$\phi = k/\rho = \eta\ dp/d\rho. \tag{8}$$

in which the coefficient η is an index of the degree of departure from the simple conditions pertaining to (7). (By change of phase is meant a change of property brought about by pressure, over and above the effects of simple compression. An example would be a change to diamond from some other form of carbon, the change being often associated with a rearrangement of crystal structure.)

By formal mathematics it can be deduced from equations (4), (5) and (8) that

$$d\rho/dz = \eta Gm\rho/(r^2\phi). \tag{9}$$

$$\eta = dk/dp - g^{-1}\ d\phi/dz. \tag{10}$$

The pair of relations (9) and (10) was derived by K. E. Bullen in 1963. In the particular case $\eta = 1$. the equation (9) reduces to

$$d\rho/dz = Gm\rho/(r^2\phi). \tag{11}$$

which had been used by E. D. Williamson and L. H. Adams in 1923 to estimate the density gradient $d\rho/dz$ under the simplified conditions of (7). An advantage of the more general relations (9) and (10) is that evidence on the values of dk/dp, g and $d\phi/dz$ can help in estimating η, thus enabling $d\rho/dz$ to be estimated even in some regions where the chemical composition and/or phase cannot be assumed uniform.

Through equation (6), values of ϕ are provided from seismic data. Then through equation (9) and the relation

$$dm/dr = 4\pi r^2\rho. \tag{12}$$

it is possible to arrive at fairly close estimates of the density distributions inside most regions of the Earth's interior. The needed information on η is derived either through equation (10) or, in the case of some regions, through independent evidence that η does not differ very significantly from unity. Since (9) and (12) are only differential relations, the process of applying them inside any region requires prior consideration of starting values of ρ and m to take at some level. Evidence from outside seismology is brought to bear on this point. At the top of the region B, ρ is commonly assumed to be of order 3.3 g/cm^3, and m is taken as the mass M of the Earth minus a conventional (small) allowance for the crust. (Evidence from geology, petrology and laboratory experiments are introduced in arriving at the value 3.3 g/cm^3.) In deeper parts of the Earth, other considerations are presented. The overall density distribution arrived at in this way has to fit the known values of M and y, has to fit the seismic data on k/ρ and μ/ρ, and has to be compatible with (9) and (10) throughout each internal region of the Earth.

Over the period 1936–42, a determination was made by K. E. Bullen, using a procedure broadly equivalent to that just described, of the Earth's density distribution down to the bottom of the outer core. It was possible at that time to indicate only lower bounds to the density at greater depths. A lower bound of 12.3 g/cm^3 was derived for the density at the Earth's center.

An important result of the calculations was the provision of strong evidence that the Earth's mantle contains a region in which there are significant departures from uniform chemical composition and phase. Having regard to details of the P and S velocity variations inside the mantle, it was inferred that the chief departures occur inside the region C. A plausible interpretation of the departures in terms of phase changes was proposed by H. Jeffreys and J. D. Bernal, and detailed later work of F. Birch and A. E. Ringwood indicates that changes of both phase and composition probably occur in this region. The favored view continues to be that the regions D and D' are close to chemical homogeneity. (The region D'' will be referred to later.)

A second important result is that there is a big jump in density from less than 6 g/cm^3 at the bottom of the mantle to nearly 10 g/cm^3 at the top of the core. Throughout the mantle (excluding the crust) the density increases steadily without any pronounced discontinuities.

The 1936–42 calculations were incorporated in a family of Earth models, referred to as A-type models. The particular model of the series, called Model A', which has the minimum value of 12.3 g/cm^3 for the central density has been used as a first approximation in later developments.

Observational evidence since 1942 has resulted in a number of amendments and refinements to the density distribution, though the amendments are quite small compared with the differences between the A-type models and models that had been constructed before 1936. The largest changes are a consequence of the revised estimate made in 1963 of the moment of inertia coefficient y. Various additional refinements have been made through a special consideration of incompressibility to be discussed in a later section. Studies by F. Birch in 1961–63 of evidence from shockwave experiments at pressures exceeding 10^{12} dyn/cm^2 (or 10^6 atm) have led to an estimate of about 13 g/cm^3 for the Earth's central density, enabling more definite models to be constructed for the part of the Earth below the outer core. Seismic evidence in 1962 and later on the P velocity variation also contributed to new detail in the lower core. Explicit use of the coefficient η enabled these new sources of observational evidence to be treated to good advantage. Free Earth oscillation studies made immediately after the 1960 Chilean earthquake showed that the density distributions in the 1942 A-type models were correct within less than 0.5 g/cm^3 at all depths down to about 5000 km. Later free Earth oscillation studies have contributed to a number of refinements in the delineation of density variation in the upper mantle. These studies have made it necessary to divide each of the regions B and C into at least two subregions. Array stations are adding to detail for the lower mantle.

At the present time, calculations designed to incorporate all this growing evidence are still in progress. Table 3-1 includes approximately the best current estimates of ρ at particular levels in the Earth. Also included are the approximate ranges of depths of the layers A, B, . . . , although a few simplifications have been made with a view to presenting broad results unencumbered by too much detail. The crustal region A is assigned a conventional thickness of 15 km, which is (very roughly) an average for continental and oceanic regions.

Pressure and Gravity in the Earth

The determination of the Earth's density distribution unlocks the door to the determination of many other internal properties of the Earth. By equation (12), m can be determined as a function of r, and the distributions of the pressure p and gravitational intensity g thence derived through (4) and (5). Values of p at various levels are included in Table 3-1.

An outstanding feature in the physics of the Earth's interior is the large increase in pressure between the surface and center, as compared with the increase in temperature (see Other Internal Properties, page 70). Thus the Earth can be looked upon in some ways as a special laboratory supplying evidence on the behavior of matter at pressures considerably beyond the ordinary laboratory range. Inside the core, the pressure ranges from about 1⅓ to 3⅔ million atmospheres.

At 35 km depth, the mean of the principal stresses has reached 10^{10} dyn/cm^2. Since all observed solids fracture or flow under deviatoric stresses less than or equal to this value, the pressure p becomes the dominant stress parameter below this depth.

Although the deviatoric stresses become increasingly small compared with p as z increases, they are significant below 35 km depth in several important problems. For example (mainly in the circum-Pacific region), a limited number of earthquakes have focal depths up to 700 km. The fact that S waves are generated as strongly as P waves in these earthquakes indicates that significant deviatoric stresses persist to these depths. The reduction in the estimate of the moment of inertia coefficient y from 0.334 to 0.331 provides further evidence on deviatoric stresses in the Earth. The older value 0.334 had been derived using purely hydrostatic theory for the Earth's interior, whereas the revised value 0.331 is derived independently of the hydrostatic theory, using observations of artificial satellites. Jeffreys has used the revision to provide new evidence on the magnitude of deviatoric stresses below the Earth's surface.

An interesting feature of the distribution of g is that, down to a depth of about 2400 km below the Earth's surface, g is nearly constant, being within about 1 per cent of 990 cm/sec^2. At the mantle-core boundary, g has its maximum value of about 1070 cm/sec^2. Throughout the core, g diminishes fairly steadily from this maximum to the value zero at the Earth's center.

Incompressibility and Rigidity in the Earth

With values of the P and S seismic velocities known in much of the Earth's interior, a simple calculation using the equations (3) yields values of the incompressibility k and the rigidity μ when the distribution of ρ has been found. Table 3-1 includes numerical results on k and μ.

The rigidity steadily increases throughout the mantle (with possible minor exceptions inside the outermost 200 km), and the value 3×10^{12} dyn/cm^2 at the base of the mantle is nearly four times that of ordinary steel. Calculations using the mantle distribution of μ combined with evidence from bodily tides show that the mean rigidity of the core is not detectably different from zero. The calculations imply that the outer core is effectively fluid, but do not lead to any inferences about the small inner core.

In striking contrast to ρ and μ, the incompressibility k is found to vary remarkably smoothly throughout most of the Earth. In particular, both k and its pressure gradient dk/dp appear to be nearly continuous at the mantle-core boundary. This finding stimulated special studies of equations of state (relations between k and p) for materials in general at high pressure. F. Birch brought finite strain theory and laboratory studies to bear on the problem, and the equation (10) has been used to throw light on both dk/dp and η inside the Earth.

The smooth behavior of k at the mantle-core boundary suggests that k is likely to vary fairly smoothly also across boundaries inside the core. Independent evidence on the dependence of k on p and the representative atomic number z of various materials has confirmed this suggestion to a good first approximation and set fairly narrow bounds to any departures there might be from smooth variation of k inside the core.

The evidence on k, along with seismic data on α, led this author in 1946 to infer that the inner core is probably solid in the sense defined. Stripped to essentials, the argument is as follows. Between the outer and inner core there is a fairly sharp increase in α; the detection of this feature led to the original discovery that an inner core exists. Inspection of the first of equations (3) shows that, if k is assumed to vary smoothly, the sharp increase in α can be accounted for only by a sharp increase in μ, since ρ is an essentially increasing function of depth. And a significant increase in μ implies the presence of solidity.

Since 1946, evidence pointing to the solidity of the inner core has been strengthened from a variety of sources. For example, it can be shown through (10), using F. Birch's evidence that the Earth's central density does not significantly exceed 13 g/cm^3, and seismic evidence on the variation of α inside the inner core, that $d\mu/dz$ has a significant (negative) value inside the inner core. The evidence thus suggests that, below the outer core, there is first a fairly rapid change from fluidity to solidity and then a mild trend back towards fluidity as the depth further increases. The best estimate of the mean rigidity of the inner core is 1-2 $\times$ 10^{12} dyn/cm^2, which is of the same order as for ordinary steel.

Properties of the region D'' (the lowest 200 km of the mantle) are also derived with the help of the theory on k and p. The region D'' appears to be one of continuously changing chemical composition or phase, the density gradient being about three times that inside the remainder D' of the lower mantle.

The various consequences of assuming that k is a smoothly varying function of p throughout the Earth's deeper interior (below, say, 1000 km depth) have been incorporated in a second series of Earth models, referred to as B-type models.

When k and μ are known, standard theory enables values of other parameters, for example, Young's modulus E and Poisson's ratio σ, to be derived for the Earth's interior through standard formulas such as

$$E = 9k\mu/(3k + 2\mu), \qquad (13)$$

$$2\sigma = (3k - 2\mu)/(3k + \mu). \qquad (14)$$

It is to be noted, however, that σ can be determined directly from α and β, since

$$2\sigma = (\alpha^2 - 2\beta^2)/(\alpha^2 - \beta^2). \qquad (15)$$

Young's modulus reaches 7.8 $\times$ 10^{12} dyn/cm^2 at the base of the mantle, is zero inside the outer core and is probably of order 4 $\times$ 10^{12} dyn/cm^2 in the inner core. Poisson's ratio is about 0.3 at the base of the mantle, is 0.5 inside the outer core, and probably averages about 0.45 in the inner core.

Other Internal Properties

The detailed study of the Earth's internal temperature T is a separate major task, but a few broad inferences can be readily drawn from results given above. The fact that throughout the mantle the Earth is solid, i.e. is below melting point, enables upper bounds to be set on the mantle temperatures. Although the temperature gradient substantially exceeds the order of 1°C/km close to the Earth's surface, it is thought to be appreciably less than 1°C/km throughout the whole of the lower mantle and core. Most estimates of T at the base of the mantle range from about 3000 to 5000°C. Because of convection inside the fluid outer core and the small size of the inner core, the value of T at the Earth's center is not likely to exceed the value at the base of the mantle by more than about 500°C.

Evidence from free Earth oscillations gives a preferred value for η significantly less than unity inside the outermost 300 km or so of the mantle. This result requires a temperature gradient (dT/dz) substantially greater than the adiabatic gradient and is compatible with direct findings from thermal studies that dT/dz is abnormally high in this part of the Earth.

By comparing the variations of p, ρ and k as found for the Earth with equations of state found experimentally and theoretically for a variety of rocks and metals, it is possible to set bounds to the representative atomic numbers Z for particular internal regions of the Earth and in this way to contribute evidence on chemical composition. The evidence shows the inner core to be almost certainly composed of iron and nickel with Z = 26–28. In the fluid outer core, Z is a few units less than the value 26 for pure iron. The preferred view is that the outer core consists of iron alloyed with some lighter material, which may be silicon or carbon. Another view is that the material in the outer core is a phase transformation of the material in the lower mantle. A third view is that the outer core consists of the iron oxide Fe_2O, which is unstable at ordinary pressures and cannot be produced in laboratories. Most of the mantle is thought to have a composition equivalent to that of ultrabasic rock, magnesium-iron silicate. Much additional evidence on composition and phase comes, of course, from detailed geochemical studies.

Bibliography: Bates D. R. (ed), *The planet Earth*, Pergamon Press, London (1964). Bullen K. E., 1957. In *The planet Earth* (Scientific American), Simon and Schuster, New York (1957). Bullen K. E., *Introduction to the theory of seismology*, Cambridge University Press, Cambridge (*3* 1965). Gaskell T. F. (ed), *The Earth's mantle*, Academic Press, London (1967). Jeffreys Sir H., *The Earth*, Cambridge University Press, Cambridge (1970). Richter C. F., *Elementary seismology*, Freeman, San Francisco (1958). Bullen K. E., *The Earth's density*, Chapman & Hall, London (1975).

THE history of Earth as a separate body began, as we have seen in the article by Cameron, about 4.6 billion years ago, in the great primordial nebula at a distance of 8 light minutes from the center at which the Sun was formed. Innumerable small fragments gathered together to form our planet. Examining its interior by means of seismology, we no longer find any trace of the chaotic structure it must have had at that time. Indeed, as we have seen in Bullen's article, the interior of our planet appears to have an extremely tidy structure, with a heavy nucleus, a lighter mantle and an even lighter outer crust.

How can this be explained? The only possible conclusion is that the planet warmed up to such an extent after its formation as to bring about a complete reorganization of its structure. When the planet was in a molten or semimolten state the force of gravity caused the heavier substances to accumulate at the center while the lighter ones rose to the surface. During this period Earth must have been an unapproachable ball of fire. Later the first slabs of solid crust must have appeared at various points on the surface; the earliest fragments of that thin planetary surface on which all the events in the evolution of life, including the recent appearance of the human race, were to take place.

The geological development of the planet, which has given its outer surface the aspect we observe today begins at this point.

When did this fundamental phase in Earth's evolution begin? What did it look like then? Answering these questions means that we shall have to determine the time-scale over which the planet, in its pregeological phase, set up yet another fixed point in its history.

The only way to attempt this is to find Earth's most ancient rock and try to discover what events led to its formation and what other events it later witnessed. The author of the article which follows, Stephen Moorbath, who has undertaken just this task is the man who discovered in southwestern Greenland, the oldest rocks yet found on our planet. Their approximate age is between 3.75 and 3.76 billion years. Are these, however, really fragments of the Earth's earliest crust? It appears that they are not: some of them are sedimentary rocks formed by fragments of earlier rocks. This means that other rocks must have existed before these sedimentary ones. These earlier rocks must have been destroyed by erosion and their debris transported by primordial rivers, and in all probability, deposited in a lake or sea, the earliest water-covered surfaces of which we have concrete proof.

When was the first terrestrial outer crust formed? The formation certainly took place more than 3.76 billion years ago, but the exact date is unknown. We do know, however, that at that date not only was there a solid crust, but this was being eroded by rain and wind in the same manner as today, and that the fragments detached by erosion ended up by accumulating in something similar to an ocean. The date at which the Earth's crust was born can be placed at between about 3.8 and 4 billion years ago. The period during which the Earth was a ball of fire thus lasted between 600 and 800 million years.

STEPHEN MOORBATH

Director of Oxford University's Laboratory of Geocronology and Isotopic Geochemistry. He was born in Magdeburg, Germany, in 1930. In 1959 he received his Ph.D. from Oxford University, where he has since worked in the study of the application of geochronological methods and with geochemical research on isotopes in the study of the Earth's evolution from the beginnings of its history. He integrated laboratory research with the geological survey of the terrain in Africa, the United States, Iceland, Norway, Scotland, and above all in Greenland, where he found what today are considered to be the oldest rocks on earth.

STEPHEN MOORBATH

Early Terrestrial Crust and Its Evolution

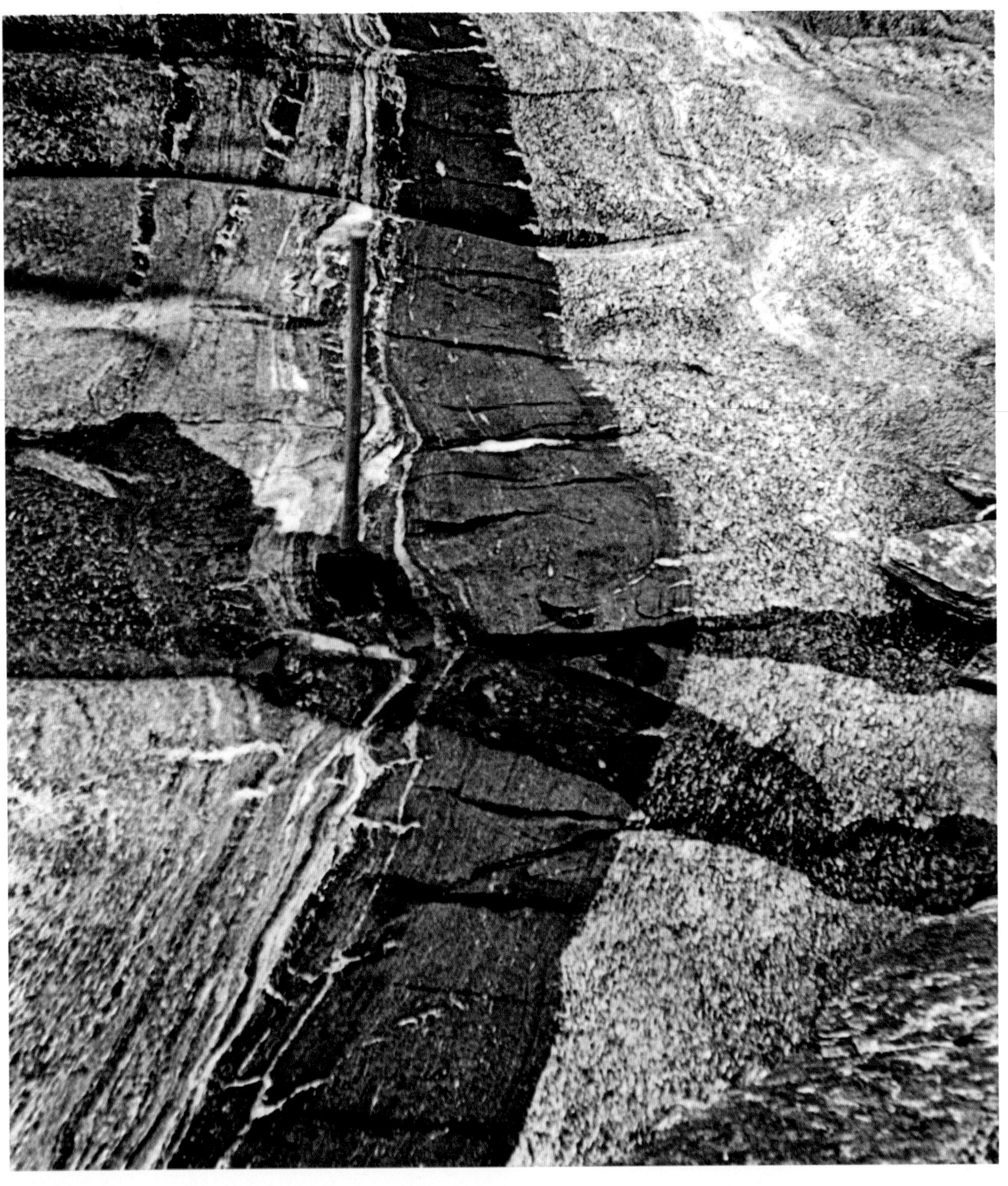

Overleaf: A view, molded by ice-flows, of the rocks which surface at the edges of Greenland's huge ice-cap. Research carried out during the 1960s has shown that some of these rocks, which can only be seen during Greenland's very brief summer, are extremely old: with an age of about 3.7 billion years, they are in fact the oldest rocks found to date on the Earth's surface. They may represent fragments of the Earth's first crust.

Such problems as the nature and composition of the primitive crust of the Earth, as well as the origin of continents and ocean basins, have been widely debated by geologists for many years. Many different hypotheses have been proposed, but it is only quite recently that sufficient geochemical, geophysical and other experimental data have accumulated regarding very ancient terrestrial rocks, to test some of these hypotheses, and to attempt to remove at least some of the conflicts which have arisen from the frequently ambiguous geological field relationships in early Precambrian (Archean) terrains.

Various hypotheses in the scientific literature propose as alternatives a primary acid ("sialic," "granitic," "continental") crust; a primary basic ("simatic," "basaltic," "oceanic") crust; a primary anorthositic (calcium aluminum silicate) crust; nucleation of granitic crust through geochemical differentiation of the mantle; development of sialic crust by sedimentary differentiation of a basic crust; primordial plate tectonics, early island-arc-like lineaments in either an original acid or basic crust; and so on. Of equal fundamental importance is whether the continental crust has grown continuously throughout geological time by addition of new material from the mantle, or whether the formation of continental crust

TABLE 4-1. RADIOACTIVE ISOTOPES USED FOR AGE DETERMINATION

Parent Isotopes and Type of Decay	Daughter Isotopes	Period of Reduction[1] (millions of years)	Zone of Determinable Geological Ages	Useful Geological Distributions and Materials
uranium-238 (α and β)	lead-206[2]	4510	rocks older than 10 million years	accessory minerals like zircon and blende in eruptive and metamorphic rocks
uranium-235 (α and β)	lead-207[2]	713		eruptive and metamorphic rocks, most of which have uranium and lead in sufficient amounts to be calculated
thorium-232 (α and β)	lead-208	13,900		uranium and thorium minerals; isotope limited to not very common geological environments; not used very much
potassium-40 (electron capture and β)	argon-40 calcium-40	11,850 1,470	rocks older than 100,000 years	many minerals which are commonly found in rocks such as biotite, muscovite, sanidine, hornblende and glauconite; basic and argillitic eruptive rocks
rubidium-87 (β)	strontium-87	50,000	rocks older than 10 million years	many potassium minerals contain rubidium in sufficient amounts for dating; for example, biotite, muscovite, potassic feldspar and glauconite; many common eruptive and metamorphic rocks some sedimentary rocks (for example the argillites) for which the method indicates the date of diagenesis

[1]In the literature one sometimes finds slightly different values, such as 4.7 billion years for ^{87}Rb.
[2]Given that two uranium isotopes decay and produce two lead isotopes, the ratio ^{207}Pb/^{206}Pb can itself be linked to the time and to the ratio U/Pb of a rock. This is often used for the determination of age instead of or as well as the U/Pb methods, and is normally called the Pb/Pb method.

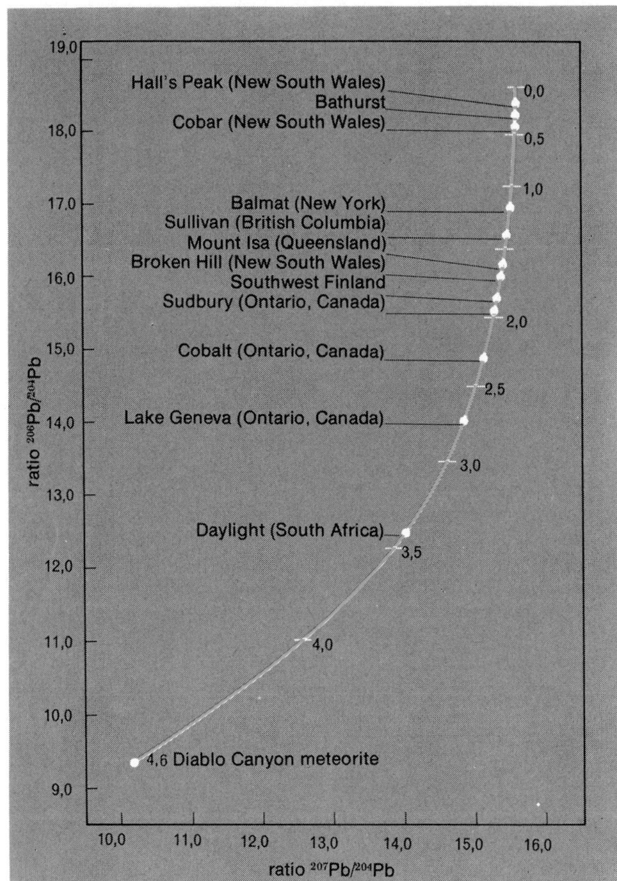

Fig. 4-1. Curve of the relation between ^{206}Pb and ^{204}Pb and of that between ^{207}Pb and ^{204}Pb, taken from some of the principal lead deposits from different geological periods. The curve, which is extrapolated into the past, passes through the values of these isotopic relations measured for the famous Diablo Canyon meteorite, which is 4.6×10^9 years old.

was largely completed early on in the history of the Earth, with younger crust representing repeated remobilization and regeneration of the initially formed crust. This is clearly relevant to the problem of whether a thin granitic crust formed very early in the Earth's history during rapid differentiation of the Earth into core, mantle and crust, or whether primordial differentiation was less complete and resulted in the gradual development of core, mantle and basaltic crust. No concensus of opinion has yet been reached on any of these fundamental problems, although certain possibilities are beginning to look more plausible than others.

The new global tectonics, described elsewhere in this volume, has provided a successful framework for diverse igneous, metamorphic and sedimentary phenomena in the past few hundred million years of earth history. Geologists are now beginning to test to what extent these principles can be extrapolated back to the early history of the Earth. Evidence is accumulating that some aspects of global mobility (plate tectonics?) had their early Precambrian counterparts, differing not so much in type as in degree. How far back in terrestrial time the principle of uniformitarianism can be extended will be one of the most stimulating geological and geophilosophical research problems in the earth sciences in the near future.

This article, in part, will deal with the recognition and characterization of early crustal remnants, using methods of isotopic dating based on long-lived radioactive nuclides. Some of these methods have now reached a high degree of analytical and interpretative refinement. A brief summary of the principal techniques is given in Table 4-1. Relevant aspects will be discussed as they arise in the text. Further details can be found elsewhere (Faure and Powell, 1972; York and Farquhar, 1972).

Geochronological Framework of the Earth

Age of the Earth. The Earth is considered to be about 4.6×10^9 years old, although no terrestrial rocks closely approaching this age have yet been found. Indeed, it is most unlikely that they ever will be. The evidence for the above estimate of the Earth's age is circumstantial and is based on the following analogies:

1. Isotopic age measurements by uranium-lead (U-Pb) and rubidium-strontium (Rb-Sr) methods on meteorites have yielded solidification ages close to 4.6×10^9 years (Tatsumoto, Knight and Allegre, 1973).

2. The oldest rocks and soils from the moon yield Rb-Sr and U-Pb ages close to 4.6×10^9 years, generally interpreted as the age of differentiation of the moon into its mantle and crust (Wetherill, 1971; Tera, Papanastassiou and Wasserburg, 1974).

3. The growth of terrestrial radiogenic lead isotopes approximates quite closely to a simple pattern which demonstrates that 4.6×10^9 years ago, the isotopic abundance ratios of lead in the Earth were identical to those in the parent body of the meteorites. The growth curve (Fig. 4-1) is based on several major lead ore deposits of different geological ages independently dated by other isotopic methods. When the growth curve is extrapolated backwards in time, using simple radioactive decay relationships, it passes through the measured lead isotope ratios in 4.6×10^9 year-old iron meteorites. The isotopic growth curve shows that, to a first approximation, the uranium/lead ratio of the source region of lead ore deposits is constant (with the exception of radioactive decay of uranium to lead) throughout geological time (Doe, 1971).

Thus although it is likely that the Earth attained something like its present form and consistency 4.6×10^9 years ago, the phrase "age of the Earth" is currently used to signify the time when the isotopic composition of lead in the Earth was the same as in the parent body of meteorites. In geological terms, however, this could possibly represent the time of accretion of the Earth and/or its first period of major chemical differentiation into core, mantle and, perhaps, crust.

Age and Nature of the Oldest Commonly Occurring Precambrian Rocks. There is a very marked concentration of isotopic dates in the general range 2.5 to 3.0×10^9 years from many continental areas on the Earth. There is little doubt that this was a very significant epoch in the development of our planet, perhaps marking the gradual beginning of a new tectonic and geochemical regime. It appears, furthermore, that a significant and sizable proportion of the continental crust already existed at that time. Detailed age work using Rb-Sr and U-Pb methods shows, for example, that at least 50–60% of the area of the North American continent already existed at about 2.5×10^9 years ago. Furthermore, it is clear that the younger orogenic belts near the presentday continental margins, such as the Appalachians in the east and the Cordilleras in the west, both of which can be interpreted in terms of the new global tectonics (continental drift, plate tectonics, subduction etc.), themselves lie in part on much more ancient rocks (Muehlberger, Denison and Lidiak 1967).

Rocks in the age range 2.5–3.0 × 10^9 years are found in many continental areas, including Scotland, Norway, Greenland, North and South America, Antarctica, various parts of Africa, Australia, India, Ceylon, Madagascar, Baltic Shield, Ukrainian Shield, Aldan Shield, and so on. They cover about 5% of the Earth's land surface.

Geological events within this age range which are being dated are of a surprisingly wide variety. Some of these 2.5–3.0 × 10^9 year dates obviously relate to high-grade (high pressure and temperature) metamorphism of preexisting rocks in pyrogene-granulite and amphibolite facies; other dates relate to massive igneous activity, both in the form of granitic and granodioritic masses of batholitic dimensions, or as extrusions of thick sequences of calc-alkaline (e.g. basalt, andesite, dacite) volcanics, as for example in the Yellowknife Province of Canada, where as much as 13,000 m of volcanic rocks were poured out onto what could have been oceanic-type crust approximately 2.7 × 10^9 years ago. There is little doubt that continents and oceans of a sort already existed at this period and, moreover, that the thickness of the continental crust was comparable, at least in some areas, with the thickness of modern continental crust, because otherwise it could hardly have supported granulite and amphibolite facies metamorphism in its lower part. On the basis of exposed stratigraphic thicknesses in Archean terrains, high pressure-temperature mineral assemblages in Archean granulite terrains, as well as trace element contents and ratios in Archean volcanic rocks, it appears that large portions of the Archean crust were 25 km in thickness, whilst corresponding depths to possible subduction zones were 85 km prior to the emplacement of extensive Archean granitic rocks between about 2.5 and 3.0 × 10^9 years ago. Although portions of the crust have been remobilized during subsequent orogenies and locally thickened to >40 km, the average thickness today of both remnant Archean crust and of younger continental crust is about 38 km (Condie 1973).

Many workers nowadays consider that the extensive granulite facies (charnockite) terrains exposed in ancient shield areas (cratons) represent the barren residue left behind after partial melting during deep-seated metamorphism removed anatectic melts containing geochemically incompatible and granitophile elements, as well as water from hydrated minerals, into upper parts of the crust. Such granulite (and some amphibolite) facies rocks are well known to be very highly depleted in elements like potassium, rubidium, uranium and thorium (to mention only the radioactive ones) and others, on account of ionic radius and calcium silicates under conditions of high pressure and temperatures at depth in the crust (Heier, 1973, 1974). At any rate during this great change in regime in the crust at around 2.5–3.0 × 10^9 years ago, large-scale production of perhaps the most permanent solid object in the crust, namely dry and barren granulite facies rock, was taking place deep down in the crust. At the same time, the raw materials of granite (in the widest sense of the word), including incompatible trace elements and water, were migrating upwards through the crust, resulting in the gradual formation of a compositionally layered crust, by some such method as shown in Fig. 4-2, taken from a paper by Fyfe (1970). The triggering mechanism for the creation of compositional layering could reside either in thermal and gravitational instabilities set up within several tens of millions of years within a geochemically fairly homogeneous, thick crust with an originally normal distribution of heat-producing elements (uranium, thorium, potassium, rubidium), or by tectonic processes beneath a thickened, continental crust involving continental rift, continental collision, continental or oceanic subduction, all mechanisms familiar in plate tectonic processes of more recent geological times. In Fig. 4-2, the rising granitic mass is shown as a bubble or "diapir," although it is open to question whether such masses are really rootless batholiths, or are underlain by a continuous zone of granitization and metasomatism down to lower crustal source regions.

Before about 3 × 10^9 years ago it seems, on the whole, unlikely that the continental crust was sufficiently thick or dry, or had low enough thermal gradients, to form much in the way of granulite facies rocks or, by corollary, much granitic material by partial melting at the base of an already thickened crust such as described above. Until now, no extensive granulite facies terrain exceeding about 2.9 × 10^9 years in age has been discovered. Older rocks than this, where still recognizably preserved, are generally in medium or low-grade metamorphic condition, tending to show effects of high temperature rather than high pressure.

As pointed out earlier, it is quite certain that many younger Precambrian (<2.5 × 10^9 years) and post-Precambrian orogenic belts have incorporated earlier material in various stages of reworking. For example, in the Lewisian basement complex of northwestern Scotland, so-called "Laxfordian" rocks yielding Rb-Sr and K-Ar mineral dates in the range 1.5–1.7 × 10^9 years can be clearly shown to be reworked, partly recrystallized "Scourian" rocks with an age of at least 2.7–2.8 × 10^9 years. Mineral dates (mica, hornblende etc.) usually record the time when the rocks were last subjected to a temperature of about 200–300°C., in this case during the Laxfordian metamorphism. However, when age measurements are carried out on such rocks by the whole-rock Rb-Sr (or Pb/Pb) methods, an age value of about 2.7–2.8 × 10^9 years is obtained. The possibility of getting two (or more) ages from a single rock depends upon the well-known fact that radi-

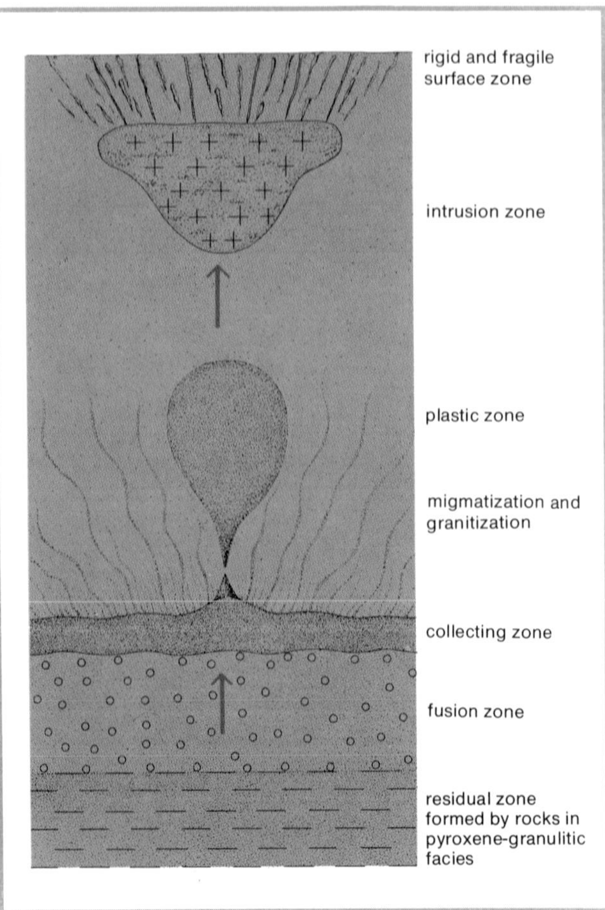

Fig. 4-2. A diagram which outlines the processes leading to the formation of a granitic pluton (area with crosses) and the creation of a clear compositional stratification in the actual body of the Earth's crust.

ogenic strontium – 87 readily diffuses out of a mineral like biotite mica down to temperatures of about 200–300°C., because it is not easily accommodated in the rubidium lattice site (equivalent to potassium lattice site) in which it is produced by radioactive decay. However, it does not as a rule move far, commonly being taken up by surrounding plagioclase feldspar ($NaAlSi_3O_8 - CaAl_2Si_2O_8$) grains, where strontium can substitute in lattice sites for the geochemically closely allied element calcium. In this way, quite small hand-specimens (ca. 2–3 kgs) of not too coarsely grained rock samples may remain closed systems with regard to strontium (as well as rubidium, uranium and lead) even through a postformational metamorphic episode of considerable intensity, such as the Laxfordian, to yield a typical Scourian age of about 2.7–2.8 × 10^9 years. Analogous examples have been reported from many other areas.

One of the main problems for future work is whether worldwide 2.5–3.0 × 10^9 years high-grade metamorphic rocks are themselves significantly older continental rocks, reworked, or whether their immediate precursors are new additions to the continental crust from upper mantle source regions and/or from something like basic oceanic lithosphere. It is usually extremely difficult to identify the original nature of high-grade metamorphic rocks because of the profound mineralogical and petrological change produced during metamorphism. This whole problem will be discussed again below.

The Age and Nature of the Oldest Precambrian Rocks, with Particular Reference to West Greenland. What happened on the Earth before about 3 × 10^9 years ago? Until quite recently, isotopic dates were few and sporadic, and in some cases untrustworthy. However, it is becoming clear that remnants of terrestrial crust older than about 3 × 10^9 years are much more widely distributed than was once thought, and that the Earth was extremely active early on in its history.

Some of the geologically most closely studied Archean rocks are found in South Africa and Rhodesia, (Anhaeusser 1973; Wilson 1973). These are the best areas for observing the spatially close, but genetically problematical, association of immense quantities of granites and granitic gneisses with so-called "greenstone" terrains, composed predominantly of volcanic and sedimentary rocks. A typical Archean greenstone belt, regardless of its geographical location, has the following principal characteristics: Early volcanic activity commonly ranges from ultrabasic to basic, implying a high and variable degree of partial melting in the underlying mantle. Successive sequences of volcanic rocks become progressively more silicic along calc-alkaline differentiation trends, to give basalts, andesites, dacites, and so on. Sedimentary rocks, including graywackes, shales, limestones, banded ironstones, cherts, are interbedded with the volcanic rocks, but predominate towards the top of the succession, which may reach a thickness of up to 30,000 m. Diapiric, rather homogeneous, granitic plutons commonly cut the greenstone belts, the latter being preserved as deep, variably deformed synclinal rafts between the plutons. Greenstone belts usually exhibit low-grade metamorphism, commonly at greenschist facies. Structures within the greenstone belts and the surrounding granites and granitic gneisses indicate deformation by vertically acting forces. Sediments from the best-known terrain of this type, the Barbeton Mountain Land of Swaziland, contain the oldest known organic remains (Nagy and Nagy, 1969). One group of these sediments (the Onverwacht Series), from fairly low down in the succession, has been dated at 3.35 ± 0.07 × 10^9 years by the Rb-Sr method, whilst a group of lavas has been dated at about 3.25 × 10^9 years by U-Pb methods (Hurley, Pinson, Nagy and Teska 1972; Sinha 1972). An ultrabasic rock ("basaltic komatiite") from near the base of the entire succession has yielded an Rb-Sr date of 3.5 ± 0.2 × 10^9 years (Jahn and Shih, 1974). Cross-cutting granitic plutons are probably in the range 2.8–3.2 × 10^9 years old (Allsopp, Ulrych, and Nicolaysen, 1968). Most of these dates are, for one reason or another, still preliminary and subject to confirmation.

The principal problem in granite-greenstone terrains is the relative age of the greenstones and the surrounding granitic gneisses. This is seldom clear from geological field evidence alone because of subsequent tectonic deformation of the contacts between the major units. In some cases, as in parts of Rhodesia (Wilson, 1973), it is probable that the gneisses form a true basement to the greenstone belts, but in other cases there is no evidence for a continental-type basement and the view is held that the basal rocks of the greenstone belts were deposited on a primordial oceanic-type crust. This is a vigorously debated problem and centers around the question of whether the earliest crust was of continental (granitic) or oceanic (basaltic) character. Detailed discussion of this complex and controversial topic is outside the scope of this article (for further reading, see Sutton and Windley, 1973). Systematic isotopic dating of granite-greenstone terrains, currently under way in several laboratories, should resolve these conflicts.

The oldest known rocks on the North American continent are the Morton and Montevideo gneisses of Minnesota, which could be as old as about 3.5 × 10^9 years, although the isotopic data is suggestive rather than confirmatory (Goldich, Hedge and Stern, 1970).

There have been frequent reports of rocks up to about 4 × 10^9 years old from the Soviet Union, although full details have not been made available in the scientific literature. However, recent work (Lobach-Zhuchenko and others, 1972) shows that former reports of rocks older than about 3 × 10^9 years from the eastern part of the Baltic Shield in northwest Russia are erroneous, owing to excess radiogenic argon in minerals analyzed by the K-Ar method. It is well known that radiogenic argon can be expelled from hot rocks at depth and absorbed by cooler minerals at a higher structural level under conditions where the pressure of argon outside the mineral exceeds that within. This has frequently been observed in both potassium-rich and potassium-poor minerals (in the latter it shows up more easily) and can give apparent ages greater than the age of the Earth! In the Baltic Shield, for example, some K-Ar dates on micas were found to range up to ca. 5.0 × 10^9 years. Application of the more appropriate Rb-Sr and U-Pb methods to gneisses of the Baltic Shield yielded characteristic and plausible ages of about 2.8 × 10^9 years. It will be of interest to see whether reported ages of 3.0–3.7 × 10^9 years from the Ukrainian and Siberian Shields will be substantiated.

By far the best-documented case for the existence and nature of really ancient rocks comes from the Godthaab and Isua areas within the heart of the Archean terrain of West Greenland (Fig. 4-2). The first detailed geological description of the area was given by McGregor (1973), and the complex sequence of geological events is presented in abbreviated form in Table 4-2. Rocks from this area yield the oldest known terrestrial dates at the time of writing (June 1974). Most of the isotopic dating has centered on the quartzo-feldspathic (granitic, granodioritic, tonalitic, dioritic) Amîtsoq gneisses from around Godthaab and Isua, which yield an Rb-Sr whole rock age (Fig. 4-3) of about 3.75 × 10^9 years (Moorbath, O'Nions, Pankhurst, Gale and McGregor 1972), as well as almost identical Pb/Pb whole-rock ages (Black, Gale, Moorbath, Pankhurst and McGregor 1971, and unpublished data) and U-Pb zircon ages (Baadsgaard, 1973). It is not yet quite clear whether the measured age represents the emplacement of the presumed igneous precursors of the Amîtsoq gneisses, or their regional metamorphism in the amphibolite facies. The Amîtsoq gneisses were definitely metamorphosed and deformed prior to the intrusion of the Ameralik dikes (Table 4-2) and it is also certain from several lines of evidence based on the detailed study of strontium and lead isotopes, that the emplacement of the precursors of the Amîtsoq gneisses did not precede their metamorphism by more than about 0.1 × 10^9 years. There is, furthermore, much evidence that the Amîtsoq gneisses (and subsequent rock units) in the Godthaab area underwent further intense metamorphism and deformation about 2.8–2.9 × 10^9 years ago (Black, Moorbath, Pankhurst and Windley, 1973; Pankhurst, Moorbath, Rex and Turner, 1973), but it appears that these later tectonothermal events were weak or absent in the Isua area, where cross-cutting, undeformed Ameralik dikes preserve igneous texture and mineralogy.

After the discovery of the great age of the Amîtsoq gneisses, it was thought that they could represent a remnant of the earliest crust of the Earth. However, this cannot be the case, because there is clear evidence for the existence of even earlier

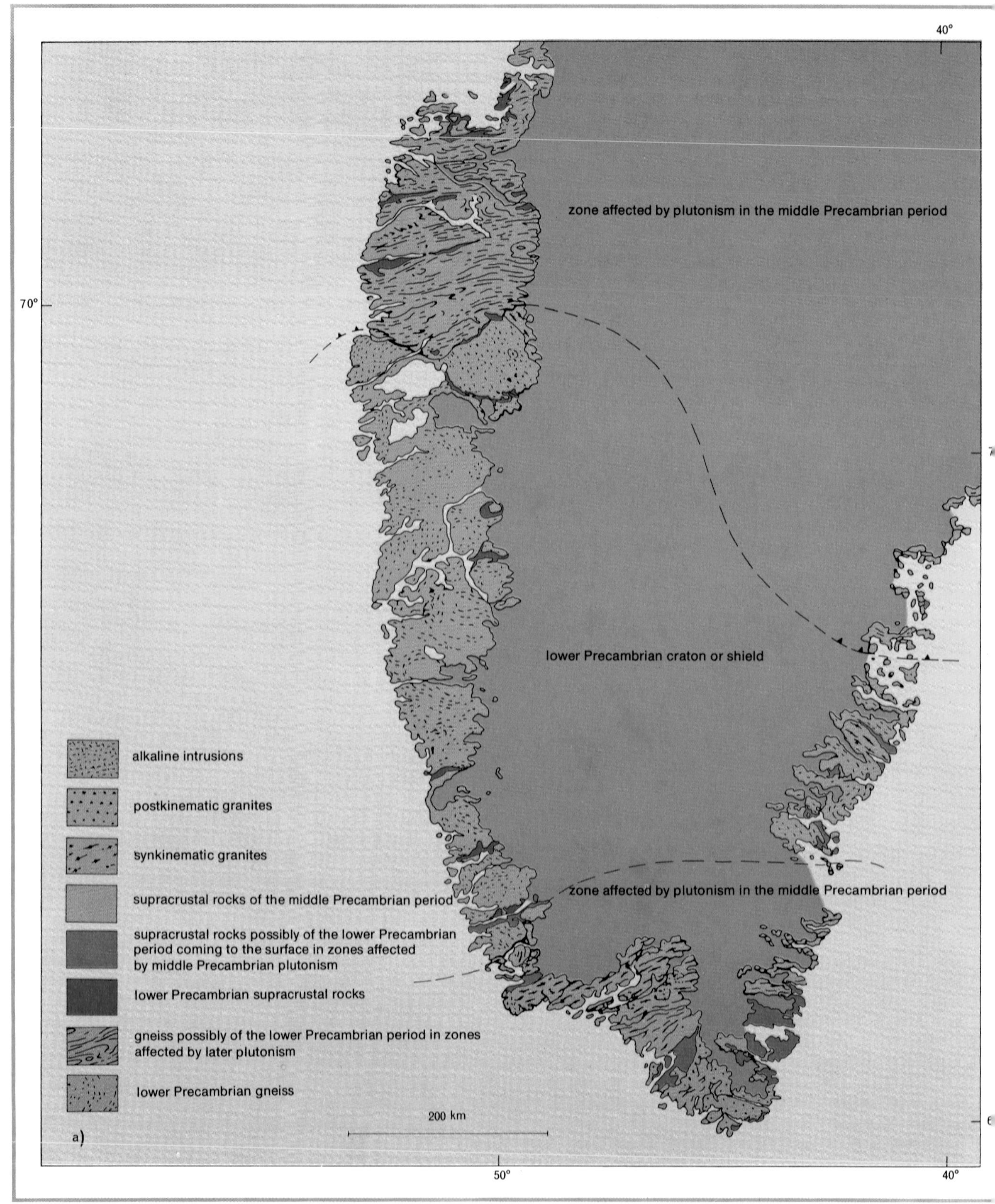

rocks. McGregor has discovered inclusions in the 3.75×10^9-year-old Amîtsoq gneisses in the Godthaab area which are sufficiently abundant to be separable as a distinct lithostratigraphic unit (McGregor and Bridgwater, 1973). There, so-called "pre-Amîtsoqs" form rafts and enclaves up to several hundred meters in length in the gneisses. The rock types include:

1. Layered amphibolites.
2. Striped green and black rocks with magnetite, orthopyroxene and hornblende.
3. Quartz-rich rocks with magnetite, orthopyroxene, grunerite and garnet.
4. Basic rocks rich in hornblende and biotite.
5. Pods of enstatite-olivine ultrabasic rocks.
6. Quartz-rich biotite-gneiss with garnet, plagioclase, grunerite and graphite.

Fig. 4-3. On the opposite page, in the diagram (a), we see the geology of southern Greenland. On this page, in (b) a view of Isua, where the highest point (about 1500 m) is formed by a ferriferous banded formation which is 3.76×10^9 years old: these are the oldest sediments known on Earth, being even older than the South African series. In (c) we see the Amîtsoq gneisses crossed by the Amerilak seams which, despite their age, still retain the original eruptive structure and texture. In (d) an Amerilak seam cuts through a virtually undeformed section of the Amîtsoq gneisses which are more than 3.7×10^9 years old. In (e) a large and undeformed Amerilak seam cuts through the Amîtsoq gneisses which are 3.7×10^9 years old; both the rocks are crossed by a pegmatite which is 2.6×10^9 years old. In (f) an undeformed Amerilak seam, the dark area in the foreground, crosses a typical section of the Amîtsoq gneisses.

All these inclusions are metamorphically concordant with the Amîtsoq gneisses themselves, so that their original nature has to some extent been obscured. Nonetheless, it is probable that some of them are remnants of basic lavas, layered basic igneous rocks, ultrabasic rocks, clastic sediments and, in particular, sedimentary banded ironstones. Indeed, most of the mafic minerals in all these rocks are iron-rich. The rocks are almost certainly broken up remnants of a supracrustal volcanic-sedimentary succession, with the characteristics of a typical greenstone belt, into which the precursors of the Amîtsoq gneiss were intruded. Because of the intense heating and metamorphism to which the inclusions were subjected, with resulting migration and homogenization of radiogenic isotopes between inclusions and country rock, it is most unlikely that they will preserve isotopic ages older than those given by the Amîtsoq gneisses.

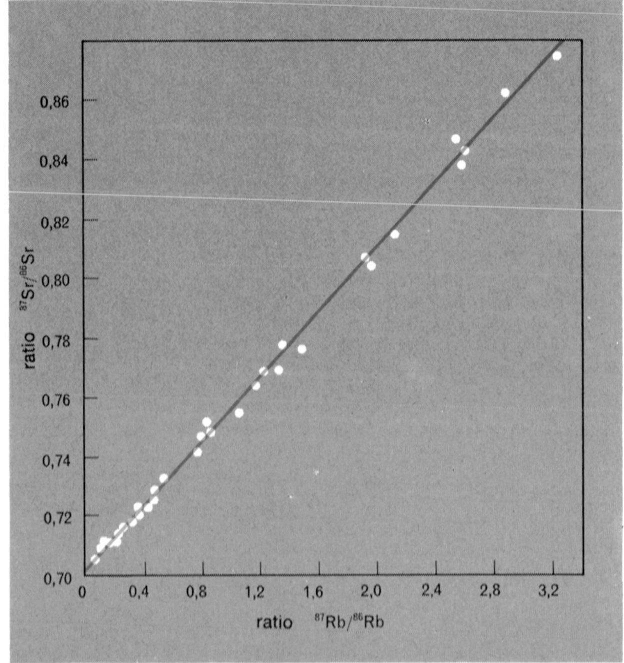

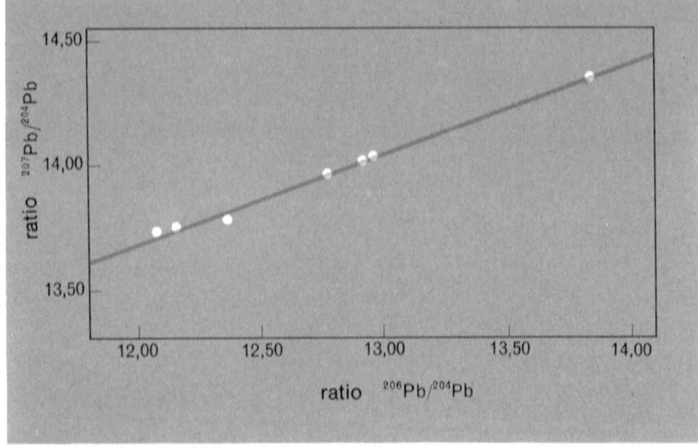

Fig. 4-4. The series of photos above and on the opposite page shows a group of samples collected in the Godthaab and Isua region. The first four from the left are Amîtsoq gneisses coming from the Godthaab region. The fifth sample was collected in the banded ferriferous formation which comes to the surface in the Isua region and forms the oldest known sedimentary series, 3.76 × 10^9 years old. The dark bands are formed predominantly by magnetite, while the light areas are formed mainly by quartz, with small amounts of other silicates and carbonates. The diagram at the top of the page indicates the results of whole rock age calculations using the rubidium-strontium method, carried out on 32 samples (the white dots) of Godthaab gneisses. The slope of the curve is proportional to the age which turns out to be 3.75 × 10^9 years. To the right, the results obtained at Isua with the lead-lead method. The age is 3.76 × 10^9 years. The white dots represent the individual calculations.

Some 150 km northeast of Godthaab, on the edge of the inland ice at Isua (Fig. 4-2), the geological situation provides an interesting comparison and contrast with Godthaab. (The geology of the intervening region is hardly known as yet). At Isua, granitic gneisses regarded as equivalents of the Amîtsoq gneisses of the Godthaab area, yield a Rb-Sr whole-rock age of 3.70 ± 0.14 × 10^9 years (Moorbath, O'Nions, Pankhurst, Gale and McGregor 1972) and an almost identical Pb/Pb whole-rock age (unpublished data). The gneisses at Isua enclose a 2–3 km thick synclinal succession of supracrustal rocks, not unlike a smaller and more highly metamorphosed (greenschist to low-amphibolite grade) version of a greenstone belt from southern Africa. The rock types are very varied and include graywackes, quartzites, slates, calcareous rocks, conglomerates, basic volcanics, ultramafic pods and lenses, garnet-chlorite schists, and so on. Of particular interest is a thick unit of banded iron-formation, consisting of alternating bands of magnetite and quartz-grunerite-chlorite, and including some carbonate lenses. Rocks of this type occur in many Archean greenstone belts older than about 2.6 × 10^9 years (Goldich, 1973) and are regarded as chemically precipitated, water-deposited sediments (Eugster and Chou, 1973), for which some authors favor a biogenic origin (La Berge, 1973).

A successful attempt to date the Isua banded iron-formation by the Pb/Pb method was reported by Moorbath, O'Nions and Pankhurst (1973). This is the first time such rocks have been dated directly. A Pb/Pb isochron age of 3.76 ± 0.07 × 10^9 years (Fig. 4-4) was obtained, which is the oldest age ever reported for a sediment. The age was interpreted as the time of metamorphism of the banded iron-formation, although a depositional age cannot be ruled out. Nevertheless, it is a reliable minimum age for these sediments. Although this age is within analytical error of the nearby Amîtsoq gneisses of the Isua (and Godthaab) areas, field evidence (Bridgewater and McGregor, 1974) suggests, though not completely unequivocally, that the igneous precursors of the gneisses intruded the Isua volcanic-sedimentary supracrustal succession. It is highly likely that the Isua supracrustal succession is the in situ stratigraphical equivalent of the numerous inclusions of earlier rocks in the Amîtsoq gneisses of the Godthaab area, referred to above.

What about the basement on which the Isua supracrustals were deposited? This is no longer discernible in the field, but in the summer of 1973, Bridgwater and McGregor (1974) discovered a thick conglomeratic unit within the succession, containing numerous cobs and boulders of granitic composition ranging from a few centimeters to 2 m in diameter, set in a fine-grained carbonate-bearing matrix. This easily identifiable unit could be traced for at least 14 km along strike. Occasionally, the boulders are almost undeformed. They are metamorphically concordant with the enclosing sediments, but look as if they were originally acid volcanic rocks, or fine-grained, intrusive granitic rocks. They are certainly completely recrystallized. The boulders yield a Rb-Sr whole-rock age in close agreement with the nearby iron-formation, probably a metamorphic age (Oxford unpublished data). At any rate, if the geological correlation is correct, these boulders, with their typical granitic chemical composition, represent the oldest discernible rocks in the Archean of West Greenland, and may have been derived from a terrain which formed a granitic basement, or at least a nearby continental-type crust, to the basin of deposition of the Isua volcanic-sedimentary assemblage.

Geological work at Isua has only recently commenced in earnest, but the high degree of exposure, and freshness of the rocks (at an altitude of 1000–1500 m in an Arctic desert climate), the great variety of rock types exposed, the absence of any major tectonothermal events later than about 3.7 × 10^9 years, all combine to make Isua one of the most interesting and critical areas in the world for future study of the geological relationships between contrasted rock formations of early Precambrian age.

TABLE 4-2. THE OCCURRENCE OF GEOLOGICAL EVENTS DURING THE ARCHEAN PERIOD IN GREENLAND[1]

1. Formation of a very old granitic crust (gneiss included in the supracrustal Isua formations); more than 3.75 × 10^9 years old.
2. Deposit of Isua supracrustal formations (basic and ultrabasic lava, quartzites, stratified ferriferous formations, calcareous rocks); some parts of the series may be as old as (1).
3. Deformation of Isua supracrustal formations.
4. Intrusion of syntectonic and late-tectonic granites, forerunners of the Amîtsoq gneisses.
5. Deformation and metamorphism of the Amîtsoq gneisses and of the Isua supracrustal formations; about 3.75 × 10^9 years ago.
6. Intrusion of a large number of basaltic seams (Amerilak seams) inside the Amîtsoq gneisses, and the Isua supracrustal formations, possibly in conditions of regional metamorphism.
7. Extrusion of basic lava (locally "pillow" lava), intrusion of ultrabasic masses, deposit of sediments including pelites, alluminiferous quartzites, and smaller calcareous elements (Malene supracrustal formations).
8. Main deformation (?) between the Amîtsoq gneisses and the Malene supracrustal formations with the consequent formation of a stratified regional complex.
9. Placement of ultrabasic elements, especially along the separation surface between Malene supracrustal formations and Amitsoq gneiss; placement of stratiform and gabbroid anorthosites.
10. Intrusion of the principal series of calc-alkaline, syntectonic and late-tectonic rocks (Nûk gneiss) in the form of subconcordant levels; age, partly 3.04 × 10^9 years and partly younger.
11. Intense deformation with the formation of principal folds and intense folding followed by a less intense deformation which has produced a widespread dome and basin structure; age between 3.0 and 2.8 × 10^9 years.
12. Further placement of syntectonic and late-tectonic granites; later placement of norites.
13. High degree of metamorphism which has interfered with (11) and (12) (culminating with the widespread crystallization of minerals in granulitic facies north and south of the Godthaab region) in late or post-tectonic conditions; age between 2.7 and 2.8 × 10^9 years.
14. Regional deformation in the northern part of the Archean complex and formation of friction zones.
15. Placement of potassic granites in the Godthaabsfjord region and of fairly widespread pegmatites; age about 2.6 × 10^9 years.
16. Placement of various groups of basic seams; age between 2.6 and 2.0 × 10^9 years.
17. Slight regional metamorphism and consequent zero-setting of the Rb/Sr ages throughout the Archean block; age about 1.6 × 10^9 years.

[1]The events are arranged from the oldest to the most recent; the table is taken from D. Bridgwater, V. R. McGregor, J. S. Myers (1974) and based mainly on the chronology of the Godthaab region indicated by McGregor (1973) but modified to include information coming from other areas and recent isotopic data.

Geological Significance of the Amîtsoq Gneisses

The petrological, mineralogical and geochemical (major element) characters of the Amîtsoq gneisses (and of their igneous precursors) are extraordinarily similar to those of rocks of "calc-alkaline" type (rich in calcium, with variable but substantial amounts of alkali elements and silica) formed throughout subsequent geological time. Rocks of the calc-alkaline series commonly occur in orogenic belts of any geological age in which continental masses have collided, or oceanic crust has been subducted beneath continental or other oceanic crust. The most plausible petrogenetic processes that have been suggested are:

1. Chemical and gravitational differentiation of igneous magmas derived directly from upper mantle source regions.
2. Partial melting of basic oceanic crust.
3. Partial melting of the base of the continental crust itself, or
4. A combination of the foregoing.

The Amîtsoq gneisses exhibit certain trace element peculiarities. Their incredible depletion in uranium (already referred to earlier in connection with pyroxene-granulite metamorphic rocks in the 2.5–3.0 × 10^9 year age range) compared to average crustal rocks is well shown up by the unradiogenic character of the lead isotope composition in many samples, with lower $^{206}Pb/^{204}Pb$ and $^{207}Pb/^{204}Pb$ ratios than any other known terrestrial rock. This uranium-depleting event occurred at about 3.7 × 10^9 years ago and could be interpreted as due to a combination of amphibolite-facies metamorphism accompanied by rapid dehydration of originally rather wet rocks, containing a lot of hydrous minerals. It is known that hexavalent uranium can easily move out of rocks under such conditions. Dehydration and degassing of such early terrestrial crustal (and mantle) rocks contributed greatly to the Earth's oceans and atmosphere.

Another interesting geochemical feature is the high degree of relative and absolute depletion in the heavy rare-earth elements shown by several samples of Amîtsoq gneiss (O'Nions and Pankhurst 1974). This has also been found in ancient gneisses from Minnesota and Ontario (Arth and Hanson 1972). A typical rare-earth distribution pattern is shown in Fig. 4-5. This depleted heavy rare-earth content, which contrasts strongly with many other igneous rock series, suggests a magmatic source in which the residual phase had a high proportion of garnet, since garnet is the only known rock-forming mineral which preferentially retains the heavy rare earths (Schnetzler and Philpotts, 1970). This evidence suggests that eclogite could be a possible parent (Hanson and Goldich, 1972). Green and Ringwood (1968) suggested that quartz-diorite (tonalite) and granodioritic magmas are the low temperature melting fractions obtained by partial melting of slightly wet eclogite or amphibolite of basaltic composition at mantle depths. Hence, some of the precursors to these ancient gneisses, including the Amîtsoq gneiss, could have been derived by partial melting of either eclogite at a depth of about 100 km, leaving a residue of garnet and clinopyroxene with or without quartz, or by the partial melting of amphibolite at somewhat shallower depth, again leaving a residue of garnet and clinopyroxene. The mechanisms of metamorphism and transport of the parent basalt or gabbro to mantle depth are not known, but the following possibilities exist: (a) basalt or gabbro may transform to amphibolite or eclogite along with a subduction zone, with subsequent partial melting at mantle depths; or (b) the magmas may be produced by partial melting of amphibolites at the base of a thick pile, or by conversion of basalt or gabbro to eclogite at the base of a thick pile, followed by sinking of the dense eclogite into the mantle and partial melting at about 100 km depth.

If the interpretation of this important preliminary rare-earth element result is correct, it would appear that these early calc-alkaline igneous rocks—the precursors of the ancient gneisses—were formed from a previously existing crust in a petrogenetic two-stage (at least) process. Direct differentiation of calc-alkaline rocks from the upper mantle in a one-stage process hardly seems feasible, since it is well known from a study of geologically young oceanic volcanic rocks that such a process would not produce heavy rare-earth element fractionation of the type observed in the ancient gneisses. Furthermore, it is also known that metamorphic processes themselves have little, if any, effect on rare-earth element patterns in a rock undergoing metamorphism.

The unavoidable conclusion from all this is that the igneous precursors of the Amîtsoq gneisses are themselves the result of complex (multistage) and fairly time-consuming crustal processes.

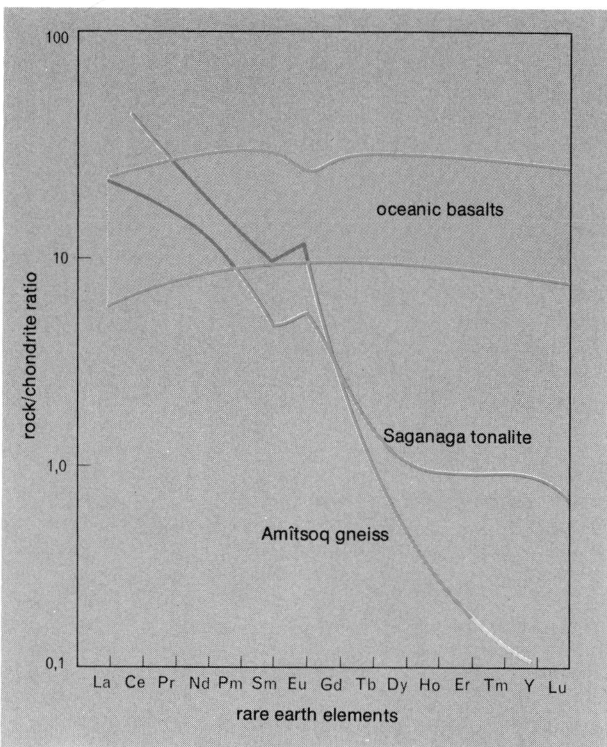

Fig. 4-5. Concentration of rare earth chemicals in relation to the medium of chondritic material. The basalt field on the oceanic ridges contrasts clearly with the Saganaga tonalite field and the Amîtsoq gneiss field, here represented by a model.

Have the Continents Grown Through Geological Time?

There has been much debate on this problem in the scientific literature. Contrasting hypotheses are that:

1. The continental crust was differentiated very early in the history of the Earth and has been reworked during successive tectonothermal episodes ever since, few areas having escaped later reworking—the scarcity of very ancient rocks being indicative of the mean rate at which normal geological processes erase the record of older events;
2. The continental crust has grown in a more or less regular manner from the earliest beginning until the present time;
3. The rate of growth of continental crust has either accelerated or decelerated in the course of geological time.

Much of the evidence is ambiguous, and no final solution to this fundamental problem is yet possible. The present author favors a decelerating rate of growth of continental crust when averaged through geological time. It was pointed out earlier that perhaps as much as 50–60% of the total volume of continental crust was in existence by about 2.5 × 10^9 years ago, and detailed isotopic age work is in progress on many ancient shield areas to test this. Growth of continental crust since that time, and perhaps before then, has probably proceeded by accretion processes at continental margins above subducted oceanic crust in an island-arc environment. A detailed description of the petrogenetic processes in such a setting has recently been given by Ringwood (1974). The main consequence is irreversible chemical differentiation of the mantle. The whole process commences in the gravitationally unstable low-velocity zone beneath midoceanic ridges. Diapirs of molten mantle source rock ("pyrolite") rise upwards and undergo partial melting leading to the generation of basaltic magma together with residual unmelted peridotite. These components ultimately form the basic oceanic crust, underlain by refractory peridotitic lithosphere. This composite lithospheric plate then moves outwards from the ridge, sliding over the weak, incipiently molten low-velocity zone, ultimately reaching a trench and undergoing subduction into the mantle. Here, further differentiation occurs, involving partial melting of oceanic crust as well as mantle above the subducted plate, leading to production of increasingly silica-rich orogenic-type magmas which rise upwards and contribute to the growth of new continental crust. The latter forms either as lateral addition to an existing continental margin, or as an intra-oceanic island arc system which is subsequently accreted to the continental margin. Eventually, the region beneath the growing continental crust becomes depleted in low-melting point components, and becomes a refractory plate of peridotite and eclogite which sinks deep into the mantle, perhaps down to at least 700 km.

Most of the continental crust may have formed by such irreversible mantle differentiation processes over the whole of geological time (Hurley, Highes, Faure, Fairbairn, and Pinson, 1962; Engel, 1963; Green, 1972a). Current rates of orogenic volcanism are sufficient, when extrapolated to past eras, to have formed the entire continental crust in about 3–4 $\times$ 10^9 years, whilst the mean composition of the continental crust appears to be similar to the composition of average andesites erupted in island arcs (Taylor, 1967). If the present rate of generation of new lithosphere at midoceanic ridges is assumed to have applied over the past 3.5 $\times$ 10^9 years, then about 30–60% of the mantle must have differentiated to yield the continental crust (Dickinson and Luth, 1971; Ringwood, 1972). Calculations based on various geochemical parameters have yielded essentially similar results.

On the above plate-tectonic model, refractory eclogite and peridotite in the lithospheric plates sink deep into the mantle (>150 km), having become irreversibly differentiated and incapable of ever participating again in the formation of basaltic magmas at midoceanic ridges. The complementary differentiate ultimately forms the continental crust, which grows through geological time by accretion of island arcs and by addition of the calc-alkaline suite, either as volcanic eruptions, or their plutonic-intrusive counterparts.

A method which may be applied to the problem of deciding whether a given metamorphic basement complex represents either a much older, reworked continental crust, or a relatively juvenile addition to the continental crust from upper mantle source regions, is based on the growth of the strontium isotope ratio $^{87}Sr/^{86}Sr$ with geological time. The principles are described briefly below. For further reading, reference should be made to the recent book by Faure and Powell (1972).

In any rock or mineral, the ratio $^{87}Sr/^{86}Sr$ increases with time because of the radioactive decay of ^{87}Rb to ^{87}Sr, with a half-life of 5.0 $\times$ 10^{10} years (Table 4-1). The growth rate of $^{87}Sr/^{86}Sr$ over a given time is proportional to the Rb/Sr ratio in a given sample. In the upper mantle, the $^{87}Sr/^{86}Sr$ ratio has changed from a value of 0.699 at 4.6 $\times$ 10^9 year ago (by analogy with the known values in meteorites and on the Moon) to a value of about 0.703, which is the average for the immediate source of modern oceanic ridge basalts. (It is not yet known with any certainty whether the growth rate from 0.699 to 0.703 was linear or not, but see Faure and Powell, 1972, page 132). This corresponds to an Rb/Sr value in the upper mantle source region of about 0.02. However, since the Rb/Sr ratio of average continental crust is about 0.25–0.30 (although it varies greatly between different rock types), it follows that the $^{87}Sr/^{86}Sr$ ratio increases at a much faster rate in the continental crust than in the upper mantle. The principle of applying this method to the problem of "reworking" versus "accretion" is shown in Fig. 4-6, taken from the classic work in this field by Faure and Hurley (1963).

An example of the application of the method is provided by rocks from West Greenland. A hotly debated problem is whether the voluminous, younger Nûk gneisses (Table 4-2) of the Godthaab area, as well as the adjacent extensive gneiss areas to the north and south of the Godthaab area, are reworked 3.75 $\times$ 10^9-year-old Amîtsoq gneisses, or whether they represent massive new additions to the continental crust not long before ca. 2.8–3.0 $\times$ 10^9 years ago, which is the measured date of these younger gneisses (Black, Moorbath, Pankhurst and Windley, 1973; Pankhurst, Moorbath, and McGregor, 1973). The controversy has been discussed recently in a paper on the tectonic regime of the Godthaab area Archean rocks (Bridgwater, McGregor and Myers, 1974), in which the individual authors themselves hold conflicting views.

The initial $^{87}Sr/^{86}Sr$ ratio of the Amîtsoq gneisses 3.75 $\times$ 10^9 years ago was very close to 0.701 (Fig. 4-6). The average Rb/Sr ratio of the Amîtsoq gneisses is approximately 0.3 (the mean of many measured samples) and quite close to the average for continental crust. It can easily be calculated that by about 2.8–3.0 $\times$ 10^9 years ago the average $^{87}Sr/^{86}Sr$ ratio of the Amîtsoq gneisses, of the type exposed in the Godthaab and Isua areas, was about 0.715. Remobilization of large volumes of representative Amîtsoq gneisses of about 2.8–3.0 $\times$ 10^8 years ago would have yielded rocks with this initial $^{87}Sr/^{86}Sr$ value. But what do we actually find? It turns out that the initial $^{87}Sr/^{86}Sr$ ratio of the younger gneisses is approximately 0.702 (Pankhurst, Moorbath, and McGregor, 1973; also, unpublished data). Clearly, the younger gneisses so far investigated are not remobilized or reworked Amîtsoq gneisses (Fig. 4-6).

The explanation in closest accord with the respective initial $^{87}Sr/^{86}Sr$ ratios of 0.701 and 0.702 for the Amîtsoq and Nûk gneisses (and their broad equivalents) is that their respective plutonic igneous precursors were produced from upper mantle source regions, through the intermediate step of some Archean equivalent of continental accretion, not more than 0.1 $\times$ 10^9 years before their measured isotopic age, and possibly a good deal less.

It thus appears that the emplacement of calc-alkaline rocks of Nûk type may represent a considerable thickening of this part of the crust at about 2.8–3.0 $\times$ 10^9 years ago. However, this is subject to confirmation from many further strontium isotope measurements in this complex area.

A similar situation exists in the extensive "gray-gneiss" complex of the Outer Hebrides of northwest Scotland, which had yielded Rb-Sr and Pb/Pb whole-rock ages of about 2.7 $\times$ 10^9 years, with an initial $^{87}Sr/^{86}Sr$ ratio at that time of 0.701 (Moorbath, Powell, and Taylor, 1974). Contrary to the views of some workers (Bridgwater, Watson, and Windley, 1973), this precludes derivation of the gneisses by remobilization of a much older continental crust with anything like an average crustal Rb/Sr ratio. The result is much more in line with derivation of the precursors of the grey gneisses, which are known to have a typical igneous calc-alkaline geochemistry, from upper mantle source regions (probably via a continental accretion mechanism) not more than about 0.1–0.2 $\times$ 10^9 years before the measured isotopic date.

From this, and other recently published work, it is evident that just because younger continental crust can often be shown to be reworked ca. 2.5–3.0 $\times$ 10^9 years-old continental crust, it cannot be automatically assumed that 2.5–3.0 $\times$ 10^9 years-old continental crust represents reworked continental crust of significantly greater age. Certainly, some such reworked older crust must exist, and will undoubtedly turn up with further search. For example, Taylor (1974) has recently discovered gneisses in northern Norway which yield a Rb-Sr metamorphic age of about 2.3 $\times$ 10^9 years, but with an initial $^{87}Sr/^{86}Sr$ ratio of 0.713. Detailed study by the Pb/Pb age method conclusively demonstrates that these rocks were already in existence about 3.5 $\times$ 10^9 years ago.

Detailed age and isotope studies on individual segments of Archean crust (and on crust of *any* age) will perhaps provide a better answer to the question of continental reworking versus continental growth.

General Discussion

The fact that chemically highly differentiated granitic and other calc-alkaline rocks existed as early as 3.7–3.8 × 10^9 years ago, gives substantial support for the Earth being hot during the first billion years of its history. Most models of the Earth's thermal history nowadays suggest a very hot initial stage (Hanks and Anderson, 1969). The general geochemical equilibrium of the Earth could hardly have been achieved without a very large degree of melting at an early stage. Early, rapid heating has been attributed to rapid release of gravitational accretion energy, together with energy from short-lived radionuclides. Subsequently, thermal energy produced by long-lived radionuclides (uranium, thorium, potassium) became the principal driving mechanism for terrestrial process.

Complex convective structures would have been present at an early stage. Crust will appear when surface temperatures approach 800°C., which could be very early on indeed (Fyfe, 1973a). After the appearance of liquid water at 100°C. or less, thicker crust would rapidly form. Some models (Fanale, 1971) suggest that the formation of a hydrosphere might be an early event which would itself accelerate initial cooling. It was seen earlier that water-deposited sediments certainly existed on the Earth by 3.76 ± 0.07 × 10^9 years ago (Moorbath, O'Nions, and Pankhurst, 1973).

The very earliest crust was probably basic and ultrabasic in character, but subject to extreme mobility on account of postulated, complex convective cells with a short wavelength pattern of less than about 100 km in extent (Fyfe, 1973b; 1974), and forming a great contrast to the comparatively simple and large-scale pattern of global convection cells thought to exist at the present day. The earliest semistable crustal fragments could have been granitic, formed by some primitive precursor of subduction and continental accretion? Such a crust would concentrate radioactive elements (which are geochemically incompatible with calcium-iron-magnesium silicates at deeper levels) and would be capable of floating on basaltic liquids below. Gradual production of granitic crust and hydrosphere would result, the commencement of the great process of formation of continental crust by irreversible differentiation of the mantle. Furthermore, continental crust, once formed, is probably permanent throughout geological time because of its comparatively low density. Continental buoyancy inhibits extensive destruction of continental lithosphere by subduction. It is most unlikely from first principles that substantial amounts of continental crust could ever be recycled through the mantle.

The initial radioactive crust could have been associated with thermal gradients of up to about 100°C. per kilometer and would not need to be thicker than about 10 km before melting at its base could occur giving rise to igneous overturn of the material. Convective flow of basic liquids under the original basic crust would tend to produce granitic material in regions of descending currents, so that these proto-subduction zones would be zones of intense crust and mantle mixing in which the first calcalkaline igneous assemblages could appear (Fyfe, 1973a).

The early Earth, 4.0–4.5 × 10^9 years ago, can thus be visualized as partially molten, with widespread, massive basic volcanism at positions of rising convection cells, and intermediate to acid volcanism in down-flow regions, leading to the gradual accumulation of granitic and sedimentary rocks at the surface.

The greenstone belts might be visualized as mini-oceans, the primitive ancestors of present-day oceans, but basically different only in degree and not in kind, and subject to the same type of mechanisms. Perhaps they are the remnants of small, thin and unstable lithospheric plates, which were especially vulnerable to rifting destruction as a consequence of being carried down into the region of melting by subduction. A plate tectonic interpretation for Archean greenstone belts has been advocated by Hart, Brooks, Krogh, Davis, and Nava (1970) and Jahn, Shih, and Murthy (1974). Greenstone belts have been variously compared to proto-oceanic ridges and to proto-island-arcs, whilst the petrology and geochemistry of greenstone belt volcanic rocks show features comparable with both modern oceanic ridges and island arcs, which nowadays form highly contrasted geological, petrological and geochemical environments. Since these early "oceans" were probably so numerous and small compared to modern ones, (ca. 50–100 km dimension), it is hardly surprising that oceanic and continental characteristics exist in close juxtaposition and that regional relationships between greenstone belts and surrounding granites and granitic gneisses are not always clear-cut. It seems likely to the present author that greenstone belts may form on both oceanic and continental crust and that different parts of a single greenstone belt may be underlain by either. It may be that the lower ultrabasic and basic assemblages in some greenstone belts (e.g. Swaziland) are more akin to a proto-oceanic ridge environment, whilst the stratigraphically higher calc-alkaline volcanic assemblages are more akin to a proto-island-arc setting (Jahn and Shih, 1974).

Green (1972b) has suggested that Archean greenstone belts are the terrestrial analogues of lunar Maria. He interprets some greenstone belts as very large impact scars, initially filled with impact-triggered melts of ultramafic to mafic composition, and thereafter cooling with further magmatism, deformation and metamorphism to the presently observed greenstone belts. The hypothesis originated from the observation that the lower parts of the volcanic sequence of some greenstone belts include members which crystallized from ultramafic liquids extruded at the Earth's surface at 1600–1650°C. Such liquids may be products of 60–80% melting of their mantle source, implying much more catastrophic conditions of mantle melting than observed in more recent geological times. However, it appears that the terminal impact cataclysm on the Moon (and by analogy on Earth?) may have occurred at about 3.9 × 10^9 years ago, causing widespread

Fig. 4-6. The radioactive decay of the ^{87}Rb has, in the 4.6 × 10^9 years of the Earth's existence, caused an increase in the upper mantle of the $^{87}Sr/^{86}Sr$ ratio from 0.699 tc 0.703. From the values measured on crustal rocks of the ratios $^{87}Sr/^{86}Sr$ and $^{87}Rb/^{86}Sr$, one can draw the curve of development of the ratio $^{87}Sr/^{86}Sr$ for each rock, and extend it back in time. The intersection gives both the time in which the materials had a common origin and the initial $^{87}Sr/^{86}Sr$ ratio. If this latter coincides with the line *A*, this indicates that the material derives from the upper mantle and not from older, re-melted crust. Lines *B* and *C* show this occurred 3 × 10^9 and 1 × 10^9 years ago.

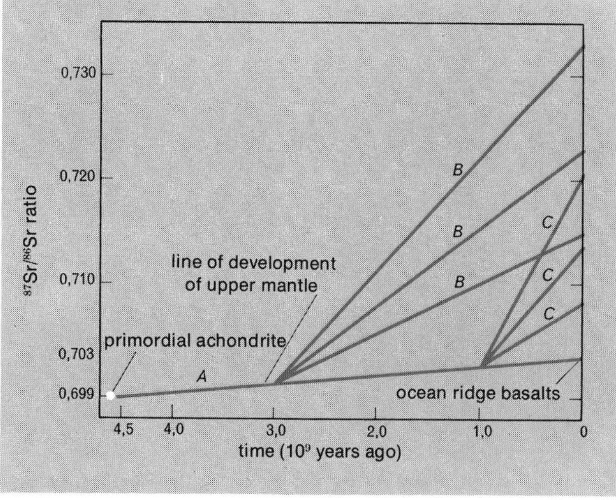

metamorphism and element redistribution, and probably associated with the formation of several major basins by late impacts which occurred within a relatively short time interval of less than 0.2 × 10^9 years (Tera, Papanastassiou, and Wasserburg, 1974). In contrast, it is now known that the formation of Archean greenstone belts on Earth was occurring at least within the interval ca. 3.8–2.7 × 10^9 years ago, and therefore does not correlate with lunar events.

It is more likely, therefore, that the Archean greenstone terrains were formed by endogenic processes. If one postulates a plate-tectonic interpretation for early crustal development and assumes:

1. Much thinner and smaller plate configuration (prototype plates and low velocity zone at higher level).
2. A higher thermal gradient due to higher radioactive heat production (about three to five times at about 3.5 × 10^9 years ago).
3. Higher water content in the upper mantle due to a lesser degree of mantle outgassing in the Archean, then it is possible to generate magmas by large degrees of partial melting at shallower depths (Jahn and Shih, 1974).

The formation and destruction of thin lithospheric plates, within the general framework of a high-frequency conveyor-belt system, probably characterized the early Earth and provided a continental accretion-type mechanism for producing new sialic crust, and also for thickening any preexisting sialic crust. Crustal thickening may also result from the collision and welding-together of continental fragments (Dewey and Burke 1973). At any rate, as the primitive oceans grew larger, the primitive sialic crust gradually coalesced into thick segments, leading to the gradual formation of new continental crust. It is at the base of these thick partly older and partly juvenile continental segments, that long-lived radioactivity eventually produced enough heat by some 2.5–3.0 × 10^9 years ago to produce granulite facies metamorphism and partial melting at depth, resulting in the production of granites, granodiorites etc., by expulsion of granitophile and geochemically incompatible elements as well as water upwards, to yield a compositionally strongly layered crust as described previously (Fig. 4-2), and which exhibits an exponential decrease of rate of heat production with depth. Thermal gradients 2.5–3.5 × 10^9 years ago were about two to five times the present-day value, and formation of granulite facies rocks as well as partial melting could have occurred at the bottom of a crust about 15–25 km thick. The evolution of the Archean of the Godthaab area of West Greenland has recently been discussed within the general framework of the above model, involving horizontal tectonics and crustal thickening (Bridgwater, McGregor and Myers, 1974).

After about 2.5 × 10^9 years ago, the sizes of plates, convection cells, oceans and continents become progressively more like those observed at the present day, so that geological processes become more comparable in scale to relatively recent ones. Mobility and irreversible chemical differentiation of the outer parts of the Earth have clearly been going on from the time at which the earliest known rocks were produced. It is the author's belief that the principle of uniformitarianism, namely that the present is the key to the past, has applied to the Earth for at least 3.8 × 10^9 years. The incredibly complex study of the early evolution of the Earth will surely provide an exciting field for scientific research for many years to come.

STEPHEN MOORBATH

Fig. 4-7. An alternative situation to that outlined in the previous diagram is that the initial $^{87}Sr/^{86}Sr$ ratio of a given crustal rock is significantly higher than the typical ratio of the mantle (curve *A*). The diagram shows in this case that the remelting which occurred 1 × 10^9 years ago of a continental crust aged 3 × 10^9 years (see lines *B* and *C*) is, conversely, the curve of development of the ratio $^{87}Sr/^{86}Sr$ for an average Amîtsoq gneiss of 3.75 × 10^9 years ago to today and the initial value of the $^{87}Sr/^{86}Sr$ ratio (point E) of a Nûk gneiss: these Nûk gneisses cannot be dislocated or remelted Amîtsoq gneisses.

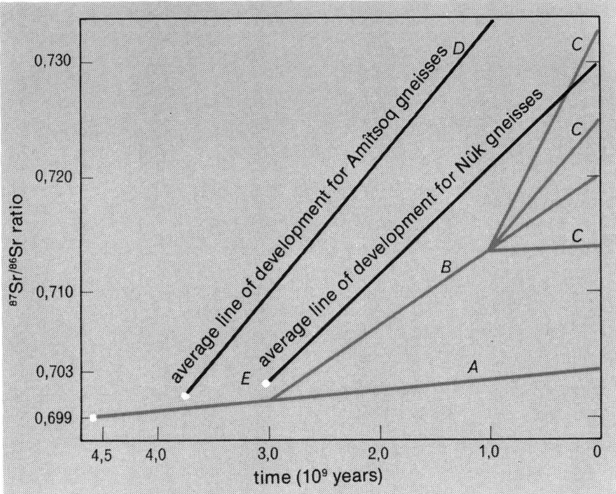

Bibliography: Bridgwater D., Escher A., Jackson G. D., Taylor F. C., Windley B. F., *Memoir 19*, in: Pitcher M. G. (ed.), *Geology of the Arctic*, Tulsa (1973); Bridgwater D., McGregor V. R., Myers J. S., *A horizontal tectonic regime in the Archaean of Greenland and its implications for early crustal thickening*, in Precambrian research, **1**, 179 (1974); Moorbath S., Powell J. L., Taylor P. N., *Isotopic evidence for the age and origin of the Grey-gneiss complex of the southern Outer Hebrides, Northwest Scotland*, in Journ. geol. Soc. Lond. O'Nions R. K., Pankhurst R. J., *Rare earth element distribution in Archaean gneisses and anorthosites, Godthaab area, West Greenland*, in Earth Planet. Sci. Lett. **22**, 328 (1974); Taylor P. N., *An early Precambrian age for granulite gneisses from Western Langøy, Vesteralen, North Norway*, Earth Planet, Sci. Lett. Bridgwater D., McGregor V. R., *Field work on the very early Precambrian rocks of the Isua area, southern West Greenland*, in Geol. Survey of Greenland, **65**, 49 (1974); Fyfe W. S., *Archaean tectonics*, in Nature, CCXLIX, 338 (1974); Heier K. S., *A model for the composition of the deep continental crust*, in Fortschr. Miner., L, 174 (1974); Jahn B. M., Shih C. Y., *On the age of the Onverwacht Group, Swaziland Sequence, South Africa*, in Geochim. Cosmochim. Acta, XXXVIII, 873 (1974); Jahn B. M., Shih C. V., Murthy V. R., *Trace element geochemistry of Archaean volcanic rocks*, in Geochim. Cosmochim. Acta, XXXVIII, 611 (1974); Ringwood A. E., *The petrological evolution of island arc systems*. in Journ. geol. Soc. Lond., CXXX, 183 (1974); Anhaeusser C. R., *The evolution of the early Precambrian crust of Southern Africa*, in Phil. Trans. Roy. Soc. Lond., A CCLXXIII, 359 (1973); Baadsgaard H., *U-Th-Pb dates on zircons from the early Precambrian Amîtsoq gneisses, Godthaab district, West Greenland*, in Earth Planet. Sci. Letters, XIX, 22 (1973); Black L. P., Moorbath S., Pankhurst R. J., Windley B. F., *$^{207}Pb/^{206}Pb$ whole rock age of the Archaean granulite facies metamorphic event in West Greenland*, in Nature Phys. Sci., CCXLIV, 50 (1973); Bridgwater D., Watson J., Windley B. F., *The Archaean craton of the North Atlantic region*, in Phil. Trans. Roy. Lond., A CCLXXIII, 493 (1973); Condie K. C., *Archaean magmatism and crustal thickening*, in Bull. geol. Soc. Amer., LXXXIV, 2981 (1973); Dewey J. F., Burke K. C. A., *Tibetan, Variscan and Precambrian basement reactivation; products of continental collision*, in Journ. Geol., LXXXI, 683 (1973); Eugster H. P., Chou I. M., *The depositional environments of Precambrian banded Iron-Formations*, in Econ. Geol., LXVIII, 1144 (1973); Fyfe W. S., *The granulite facies, partial melting and the Archaean crust*, in Phil. Trans. Roy. Soc. Lond., A CCLXXIII, 457 (1973); Fyfe W. S., *The generation of Batholiths*, in Tectonophys, XVII, 273 (1973); Goldich S. S., *Ages of Precambrian banded Iron-Formations*, in Econ. Geol., LXVIII, 1126 (1973); Heier K. S., *Geochemistry of granulite facies rocks and problems of their origin*, in Phil. Trans. Roy. Soc. Lond., A CCLXXIII, 429 (1973); La Berge G. L., *Possible biological origin of Precambrian Iron-Formations*, in Econ. Geol., LXVIII, 1098 (1973); McGregor V. R., *The early Precambrian gneisses of the Godthaab district, West Greenland*, in Phil. Trans. Roy. Soc. Lond., CCLXXIII, 343 (1973); McGregor V. R., Bridgwater D., *Field mapping of the Precambrian basement in the Godthaabsfjord district, southern West Greenland*, in Geol. Survey of Greenland, *55, 29 (1973);* Moorbath S., O'Nions R. K., Pankhurst R. J., *Early Archaean age for the Isua Iron-Formation, West Greenland*, in Nature, CCXLV, 138 (1973); Pankhurst R. J., Moorbath S., McGregor V. R., *Late event in the geological evolution of the Godthaab district West Greenland*, in Nature Phys. Sci., CCXLIII, 24 (1973); Pankhurst R. J., Moorbath S., Rex D. C., Turner G., *Mineral age patterns in ca. 3700 m.y. old rocks from West Greenland*, in Earth Planet Sci. Letters, XX, 157 (1973); Sutton J., Windley B. F. (ed.), *A discussion on the evolution of the Precambrian crust*. in Phil. Trans. Roy Soc. Lond., CCLXXIII, 315 (1973); Tatsumoto M., Knight R. J., Allegre C. J., *Time differences in the formation of meteorites as determined from the ratio of lead-207 to lead-206,* in Science, CLXXX, 1279 (1973); Wilson J. F., *The Rhodesian craton - an essay in cratonic evolution,* in Phil. Trans. Roy. Soc. Lond., A CCLXXIII, 389 (1973); Arth J. G., Hanson G. N., *Quartz diorites derived by partial melting of eclogite or amphibolite at mantle depths,* in Contr. Mineral. and Petrol, XXXVII, 161 (1972); Faure G., Powell J. L., *Strontium isotope geology,* Berlin (1972); Green D. H., *Magmatic activity as the major process in the chemical evolution of the Earth's crust and mantle,* in: Ritsema A. R. (ed.), *The upper mantle,* in Tectonophys, XIII, 147 (1972); Green D. H., *Archaean Greenstone belts may include terrestrial equivalents of lunar maria?*, in Earth Planet, Sci. Letters, XV, 263 (1972); Hanson G. N., Goldich, S. S., *Early Precambrian rocks in the Saganaga Lake-Northern Light Lake area, Minnesota-Ontario,* II, *Petrogenesis,* in Geol. Soc. Amer. Memoir, CXXXV, 179 (1972); Hurley P. M., Pinson W. H., Nagy B., Teska T. M., *Ancient age of the Middle Marker Horizon, Onverwacht Group, Swaziland Sequence, South Africa,* in Earth Planet. Sci. Lett., XIV, 360 (1972); Lobach-Zhuchenko S, *Geochronological constraints for the geological evolution of the Baltic Shield,* Leningrado (1972); Moorbath S., O'Nions R. K., Pankhurst R. J., Gale N. H., McGregor V. R., *Further rubidiumstrontium determinations on the very early Precambrian rocks of the Godthaab district, West Greenland,* in Nature Phys. Sci., CCXL, 78 (1972); Ringwood A. E., *Phase transformations and mantle dynamics,* in Earth Planet. Sci. Letters, XIV, 233 (1972); Sinha A. K., *U-Th-Pb systematics and the age of the Onverwacht Series, South Africa,* in Earth Planet. Sci. Lett., XVI, 219 (1972); York D., Farquhar R. M., *The Earth's age and geochronology,* Oxford (1972); Black L. P., Gale N. H., Moorbath S., Pankhurst R. J., McGregor V. R., *Isotopic dating of very early Precambrian amphibolite facies gneisses from the Godthaab district, West Greenland,* in Earth Planet. Sci. Letters, XII, 245 (1971); Dickinson W. R., Luth W. C., *A model for plate tectonic evolution of mantle layers,* in Science, CLXXIV, 400 (1971); Doe B. R., *Lead isotopes,* Berlin (1971); Fanale F. P., *A case for catastrophic early degassing of the Earth,* in Chem. Geol., VIII, 79 (1971); Wetherill G. W., *Of time and the moon,* in Science, CLXXIII, 383 (1971); Fyfe W. S., *Some thoughts on granitic magmas,* in: Newall G., Rast N. (ed.), *Mechansim of igneous intrusion,* London (1970); Goldich S. S., Hedge C. E., Stern T. W., *Age of the Morton and Montevideo Gneisses and related rocks, South Western Minnesota,* in Bull. geol. Soc. Amer., LXXXI, 3671 (1970); Hart S. R., Brooks C., Krogh T. E., Davis G. L., Nava D., *Ancient and modern volcanic rocks, a trace element model,* in Earth Planet. Sci. Letters, X, 17 (1970); Schnetzler C. C., Philpotts J. A., *Partition coefficients of rare earth elements between igneous matrix material and rock-forming phenocrysts,* in Geochim. Cosmochim. Acta, XXXIV, 331 (1970); Hanks T. C., Anderson D. L., *The early thermal history of the Earth,* in Phys. Earth Planet. Interiors, II, 19 (1969); Nagy B., Nagy A. L., *The early Precambrian Onverwacht microstructures; possibly the oldest fossils on Earth?,* in Nature, CCXXIII, 1226 (1969); Allsopp H. L., Ulrych T. J., Nicolaysen L. O., *Dating some significant events in the history of the Swaziland System by the Rb-Sr isochron method,* in Canad. Journ. Earth Sci., V, 605 (1968); Green T. H., Ringwood A. E., *Genesis of the calc-alkaline igneous rock suite,* in Contr. Mineral. and Petrol., XVIII, 105 (1968); Meuhlberger W. R., Denison R. E., Lidiak E. G., *Basement rocks in continental interior of United States,* in Bull. Amer. Assoc. Petrol. Geol., LI, 2351 (1967); Taylor S. R., *The origin and growth of continents,* in Tectonophys., IV, 17 (1967); Engel A. E. J., *Geologic evolution of North America,* in Science, CXL, 143 (1963); Faure G., Hurley P. M., *The isotopic composition of strontium in oceanic and continental basalt: application to the origin of igneous rocks,* in Journ. Petrol., IV, 31 (1963); Hurley P. M., Hughes H., Faure G., Fairbairn H. W., Pinson W. H., *Radiogenic strontium-87 model of continental formation,* in Journ. geophys. Res., LXVII, 5315 (1962).

As we have seen in the article by Moorbath, the Earth acquired a solid crust and something similar to the present-day oceans between 3.8 and 4.0 billion years ago. This is a date of fundamental importance for our history, since from then on there existed the basic conditions capable of permitting the survival of the earliest life forms. As we shall see in a later article by Nagy and Nagy, what are probably, if not certainly, the most ancient traces of life, appeared between 3.4 and 3.2 billion years ago. During the 600 million years between the formation of the primordial crust and the development of these life forms, those as yet unknown phenomena occurred which led from inorganic to living matter.

These phenomena are the subject of the next article, which deals with all those lines of research which are pursued today to increase our understanding of how life emerged on the Earth.

Ponnamperuma first considers the possibility that these phenomena may have taken place outside our planet and that living matter or some immediate precursor of living matter reached Earth at a later time. Indeed the study of interstellar gases has revealed the existence of a whole series of organic substances. Though similar to those fundamental to living matter, these substances are only to be considered as very distant relations to them. Ponnamperuma has also found traces of organic substances of this type in meteorites and lunar samples, but no evidence of compounds truly characteristic of living matter.

The presence of organic matter outside Earth is in any case highly interesting because it suggests that, if in some other part of the solar system, or further afield, suitable conditions have existed, some form of life, based on a chemical system analogous to that of terrestrial life may have developed there too. In the solar system the localities of greatest interest are Mars and Jupiter. In August 1975 two Viking probes were launched toward Mars. These automatic vehicles achieved soft landings and made soil analyses in the search for living organisms. The results which indicated great chemical activity in the Martian soil were not conclusive. Jupiter is, however, of special interest since it is a sort of gigantic laboratory in which phenomena which occurred during the early history of Earth are taking place today.

It is believed by some that the color of the Great Red Spot is due to the presence in that region of great quantities of organic matter. If it were proved that elementary life forms have developed there, the answer to the question of how life on Earth began might also be resolved in a short time by direct observation. The photographs of the vast turbulences in the Jovian clouds sent back by Voyager 1 cast serious doubt on the existence of life there.

Much more realistic observations are also taking place on a much smaller scale in Earth-based laboratories. Ponnamperuma describes the apparatus with which he simulates Earth's atmosphere and oceans as they were billions of years ago, and also the present-day conditions of other planets. With this equipment, he, like many other research workers all around the world, is attempting to reproduce the reactions through which organic matter became alive. The fundamental difficulty lies in the fact that everything is hypothetical in this kind of research—the nature of the materials involved, the quantities and relative proportions, the source of energy. The variables are of such a kind and so numerous that it is almost impossible to repeat the fundamental reaction under laboratory conditions. To date the results have all been negative, but the experiments continue.

CYRIL PONNAMPERUMA

Director of the Exobiology Laboratory of the Department of Chemistry at the University of Maryland. He was born in Galle, Sri Lanka, in 1923. His interest in the origins of life, both on Earth and outside its atmosphere, goes back to his meeting with J. D. Bernal, one of the pioneers in this field, which took place in London while Ponnamperuma was studying for a degree in chemistry after having received a degree in philosophy in India. He continued his work at Berkeley under the guidance of Nobel Prize winner, Melvin Calvin. In 1962, he began to work with the exobiology program that NASA started at the Ames Research Center where, in 1965, he became Director of the Chemical Evolution Branch.

Ponnamperuma obtained important research results discovering amino acids in meteorites, demonstrating in the laboratory the possibility of the synthesis of many important molecules on the biological level, and simulating in the laboratory the environmental conditions of the planets and of primordial earth.

The co-author of the article is Peter M. Molton, English organic chemist, born in Wolverhampton, England, in 1943, who has, for many years, collaborated with Ponnamperuma chiefly in laboratory research on the possibility of life on Jupiter.

CYRIL PONNAMPERUMA
AND PETER MOLTON

Recent Advances in Our Search for Life Beyond the Earth

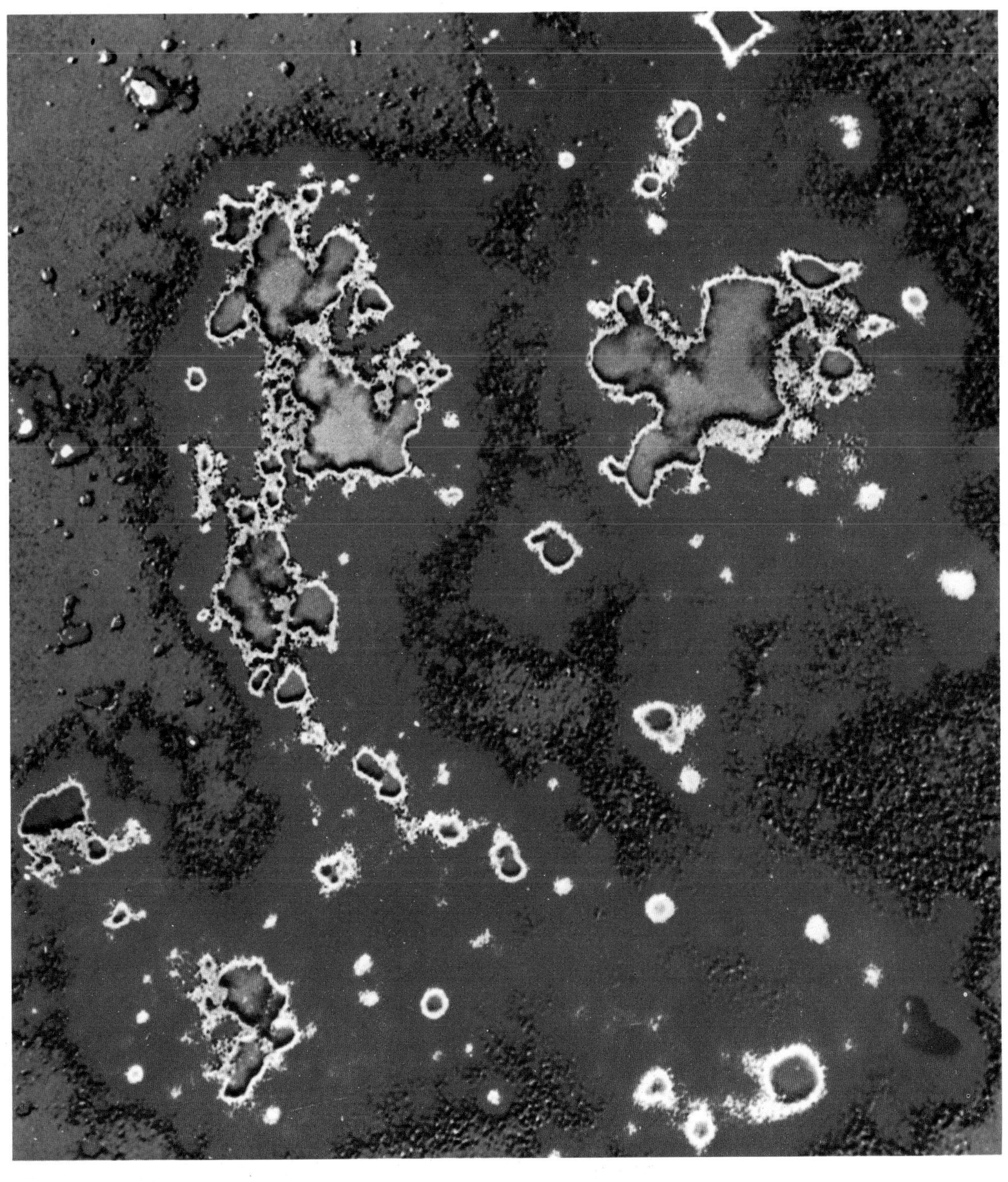

Overleaf: the Great Magellanic Cloud photographed in ultraviolet light by the astronauts on the Apollo 16 mission. The original photo was in black and white; the colors were introduced at a later stage by means of computer analysis of the various levels of emission. The overall effect is typical of a young galaxy like ours, containing a hundred billion stars or so and a large quantity of interstellar gases. It is generally thought that the conditions which brought about the birth of life on Earth also occurred in other heavenly bodies, in our galaxy and in others. Early in 1975 two techniques of direct research into technologically advanced forms of life were introduced: systematic investigation with radiotelescopes to pick up possible extraterrestrial messages carried out with the Green Bank radiotelescope in the U.S. and the transmission of a terrestrial message into space via the Arecibo radiotelescope in Puerto Rico (NASA).

The attempt to discover our own origins and to find our place in the universe is being made along several different routes. There is the archaeological and geochemical method, that of digging up fossils and examining them for clues to the process of evolution. In the laboratory we can simulate assumed primitive Earth conditions and outline the first chemical steps towards our own living system. We can extend this method to investigate the possible evolution of life on other planets by using our flasks to simulate conditions there, or we can examine the adaptability of terrestrial life to alien conditions. Looking further out, other stars are being scanned for the planetary systems that we believe must be orbiting them, and the interstellar medium itself is being analyzed for organic chemicals. The meteorites that fall upon the Earth have been analyzed and found to contain familiar compounds, and active volcanoes have been examined for clues to their part in the original synthesis.

Fig. 5-1. Here and on the opposite page, two impressions of the Viking mission, the first specifically aimed at research into forms of life on another planet. Setting off a month apart in the summer of 1975, the two space-probes landed on Mars in the summer of 1976. Once in the Martian orbit about 1500 km from the surface, each probe dispatched its landing module to the planet's surface. This module (opposite page) has two stereoscopic television cameras, a mechanical arm for gathering samples and a small biological laboratory with three pieces of analytic equipment. Pictures and data were sent back to the probe which was still in orbit, and from there they were transmitted back to Earth *(NASA)*.

All of these approaches have been successful to some extent. There is a large and rapidly growing effort on the part of humanity to solve the question of its origin and that of the origin of other species on other planets. During the last two years exobiology has taken great strides forward and has given us a considerable amount of data to digest.

The Interstellar Medium

The question of the origin of life on the Earth has taken on a new depth of meaning as a result of the discovery that organic molecules exist in interstellar space. The existence of ammonia and water vapor has been demonstrated, and other simple molecules such as hydrogen cyanide and the cyanide radical have been found. The radio source in Sagittarius (Sgr. B2) is a rich source of such molecules, and an intensive search has yielded evidence for the existence of methanol, formaldehyde, formic acid and cyanoacetylene. The concentrations of these molecules are in the range 10^{15}–10^{16} per cm^2. No doubt other molecules exist in this and other sources. There is even a suggestion that porphyrins and polyaromatic hydrocarbons may exist in interstellar space. A highly significant recent observation of the presence of hydroxyl radical in two extragalactic sources—NGC 253 and M 82—indicates that organic synthesis in space is not restricted to our own galaxy, but that there may be other more complex molecules beyond the sensitivity ranges of our instruments.

These observations lead us to believe that the mechanism of organic synthesis proposed for the primitive Earth may be valid throughout space, and offer hope that life may indeed exist elsewhere than on our own planet. Although interstellar space is an extremely dilute medium for any kind of organic reaction, it has been shown that the interstellar gas is highly ionized by ultraviolet radiation produced by the inverse Compton effect, but it is not appreciably heated by this radiation. Thus, the probability of reaction between any two colliding species is high.

Another aspect of interstellar synthesis is that the compounds formed are familiar to us, having been proposed as precursor materials in the evolution of our own biochemical system. Perhaps it would be stretching the evidence too far, but this would suggest that life on other planets may be based on the same chemistry as our own.

Meteorite Studies

Turning to our own solar system, to the comets and meteorites, we find evidence for the existence of organic compounds in them. Since these objects are supposed to be the residue from the material which formed the planets, the original solar nebula would appear to have contained organic compounds. Much of the recent work has been done on meteorites, since these can be examined directly in the laboratory. In particular, the Murchison, Murray and Orgeuil meteorites have received attention.

In the past there has been a good deal of controversy over the nature of the organic compounds in meteorites. There is general agreement that polymeric organic material is present, but many of the results concerning the presence of other simple compounds are inconclusive, since meteorites are subject to terrestrial contamination and early experiments were performed with inadequate controls. The Orgeuil meteorite, for example, has been stored in the presence of paraffin wax for a century.

The fall of the Murchison meteorite on September 28, 1969, at latitude 36°36′ and longitude 145°12′ gave an opportunity to apply modern techniques of analysis to minimally contaminated material. Fragments of this meteorite—a type II carbonaceous chondrite—were chosen with the fewest cracks and least possible obvious contamination, and extracted, hydrolyzed and examined for the presence of amino acids by the same rigorous methods used for the analysis of lunar samples. This included conventional ion-exchange chromatography, gas chromatography of the amino acid N-trifluoro-acetyl beta-butyl ester derivatives, and mass spectrometry. In this case, however, adequate controls and standards were used throughout the analysis, and maximum precautions were taken to avoid accidental contamination by contact with human skin, etc.

Amino acids were found in the samples, both D and L enantiomers being present in approximately equal quantities. There were both "normal" and nonprotein amino acids in the extracts. This in itself tends to rule out terrestrial contamination. Further evidence for the extraterrestrial origin of these amino acids is the existence of both enantiomers, since a terrestrial amino acid would have been in the L form. If the amino acids are extraterrestrial in origin, the problem arises of how they were formed. The two possibilities are that there was an abiogenic synthesis, or that after initial formation by an extraterrestrial organism, racemization occurred. The second suggestion cannot be ruled out, but it is unlikely since there are a number of nonprotein amino acids present in the extracts, and by analogy with terrestrial biochemistry these are difficult to explain. Even so, there is no necessity for an extraterrestrial organism to utilize exactly the same amino acids as a terrestrial one. In the absence of other evidence, the abiogenic hypothesis seems the most likely, since in laboratory experiments five of the nonprotein amino acids identified in the meteorite have also been identified in extracts from the products of primitive Earth or planetary atmosphere experiments. In addition the presence of all of the isomers of simple amino acids suggests a random synthesis.

We can discount one further possibility for the origin of these amino acids. During its passage through the Earth's atmosphere, the outer crust of the meteorite was melted. Abiogenic synthesis from methane and ammonia or other simple compounds could have occurred within the meteorite if the temperatures were high enough. The evidence suggests that nowhere was there a sufficiently high temperature inside the meteorite for this to have occurred—and the samples were taken from inside the meteorite. It is significant that these same amino acids have also been identified in extracts from the Murray and Orgeuil meteorites. It is very unlikely that bodies falling upon the Earth at such widely different places and times would absorb the same contamination to the extent of containing the same 12 nonprotein amino acids—but it is highly likely that the same products would be obtained from an abiogenic synthesis wherever it occurs.

These same meteorites and nine others have recently been examined for the presence of hydrocarbons, also using rigorously clean techniques. The aliphatic hydrocarbons of the Murchison meteorite are largely saturated alkanes. The gas chromatographic trace from these hydrocarbons shows a resemblance to the trace obtained from the products obtained from passing a spark through methane. The mass spectral data showed that the resemblance was maintained to the extent of both samples giving the same fragmentation pattern, suggesting that the Murchison aliphatic hydrocarbons may have arisen abiogenically.

From the other meteorites examined, Belsky and Kaplan found aliphatic hydrocarbons from methane to *n*-heptane (excepting *n*-hexane) together with unsaturated and branched-chain hydrocarbons in extracts from the Orgeuil, Murray, Cold Bokkeveld, Mokoia, Mighei, Erakot, Karoonda, Bjurbole and Abee meteorites, using gas chromatographic and mass spectrometric techniques. A wide range of isoprenoids, including members of the farnesane and phytane series, and cyclo-pentyl-, cyclohexyl-, normal, mono- and dimethylalkane series were found in extracts of the Essebi, Grosnaja, Mokoia, Murray, Orgeuil and Vigarano meteorites. The identification of isoprenoids associated with such a wide range of other hydrocarbons is significant, pointing towards an abiogenic origin for the isoprenoids.

An investigation of the aromatic hydrocarbons in the Murchison meteorite led to the identification of naphthalene, and 1-, 2-, 1n3-, 2,6-, 1,4-, and 2,3-mono- and di-methylnaphthalenes, one C-3 and two C-4- naphthalenes, biphenyl, diphenylmethane, acenaphthene, fluorene, fluoranthrene, pyrene, anthracene, phenanthrene and methylphenanthrenes. An indication of the concentrations of these hydrocarbons is given by the naphthalene concentration of 6.4 μg/g. of meteorite.

Yet further evidence for the authenticity and abiogenic origin of these compounds was obtained by isotope fractionation of various carbon-containing fractions from the Murchison meteorite, compared to the PeeDee Belemnite standard. The meteorite was enriched in C^{13} in every fraction except carbonate. The δC^{13} values of +4.43 to +5.93 for the extractable organic material of the meteorite are in a range very different from terrestrial organic matter.

This evidence suggests very strongly that organic synthesis occurred outside the Earth, in or on meteorites or the material which formed them, by an abiogenic process.

A most important discovery in this investigation into the organic components of meteorites has been the identification by gas chromatography and high and low-resolution mass spectrometry of pyrimidines. Once again, complete analytical controls were applied. The compounds thus far identified were 4-hydroxypyrimidine, 4-hydroxy-2 or 6-methylpyrimidine, a hydroxymethylpyrimidine, and homologous keto-N,N-dimethyltetrahydromethyl, ethyl and propyl-pyrimidines. The analytical procedure used in isolating these compounds was to extract the crushed meteorite and adsorb organic material in the extract onto a celite-charcoal mixture which had been extensively washed. After elution with water, formic acid or pyridine, the product was trimethylsilykated and analyzed by gas chromatography/mass spectrometry. There was no detectable trace of any naturally occurring purine or pyrimidine in these meteorite extracts, which in itself argues for the authenticity of these compounds, since they could not have arisen from terrestrial contamination.

This is the first evidence for the abiogenic synthesis of nucleic acid precursors outside the Earth, and the first direct evidence that this may have occurred on the primitive Earth—laboratory simulations tell us if a given process *can* occur, but give no indication as to whether it *did* occur.

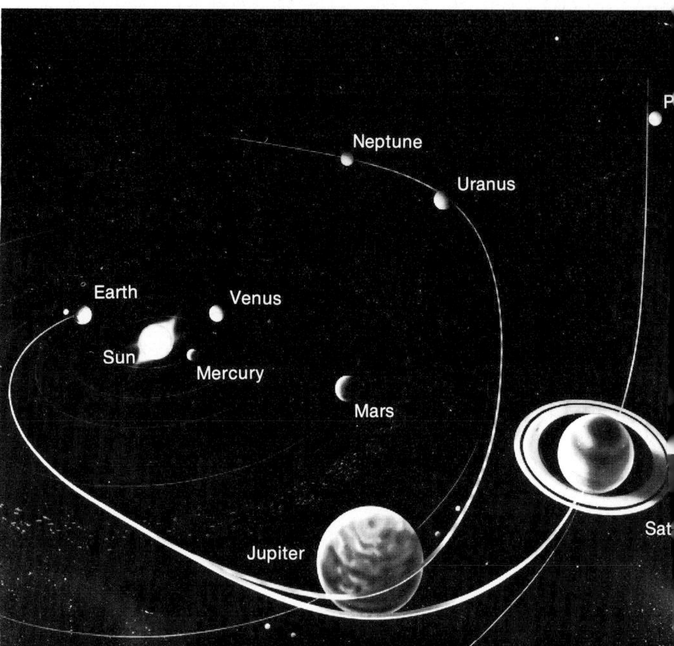

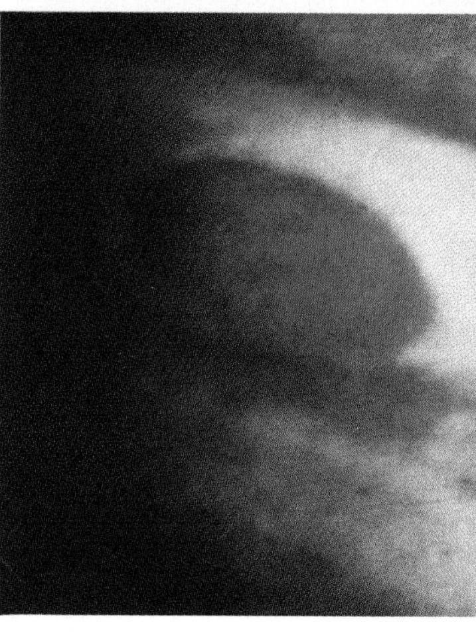

Fig. 5-2. Jupiter, Saturn's rings and the satellites around these planets are all places in which various scholars consider the presence of some form of life to be possible. The diagram above shows the missions of Pioneer 10 and Pioneer 11. A year apart, in December 1973 and December 1974, both probes traveled close to Jupiter, and confirmed that the Great Red Spot (photo on the right below) is a huge semipermanent hurricane, the color of which is due to the fact that it encloses large amounts of organic substances. In 1979 Pioneer 11 reached Saturn (opposite page), and supplied us with the first direct information about the nature of this planet. More recently, in 1979, the photographs of Voyager 1 confirmed these facts. *(NASA)*

Analysis of Lunar Material

We are fortunate in recent years to have had the opportunity to analyze lunar samples. Previously, meteorites were the only source of nonterrestrial material, and much of that available was badly contaminated. From the outset the lunar rock has been kept away from any source of contamination, with the possible exception of the rocket exhaust during landing. However, the samples were taken from points distant from the landing, and therefore uncontaminated by exhaust gases. They were kept in vacuum-tight boxes during the return to Earth, and were broken open and dispatched under sterile conditions. The analyses were carried out in clean laboratories with filtered air. The entire sequence of extractions of the lunar dust was carried out in a single vessel, further reducing possibilities for contamination.

The total carbon content of the samples was determined by measuring the volume of CO_2 evolved when a 1 gm sample was outgassed at 150°C. at a pressure of below one micron, and burned at 1050°C. The most consistent values obtained were between 140–160 μg/g. for the Apollo 11 material, and 110 μg/g. for the Apollo 12 sample.

The amount of carbon which could be converted into volatile carbon-hydrogen compounds was determined by pyrolyzing about 30 mg of the lunar dust at 800°C. in an atmosphere of hydrogen and helium. The resulting volatile compounds were estimated by a flame ionization detector. The average value for the Apollo 11 sample was 40 μg/g. For the Apollo 12 sample, hydrolysis of the fines with DC1 yielded CH_4 along with deuterated hydrocarbons, confirming the presence of 7-21 μg/g. of carbon as carbide and about 2 μg/g. as indigenous methane. By vacuum pyrolysis to 1100°C., gases from this sample were detected in the relative abundance $CO >> CO_2 > CH_4$.

Isotope measurements on the total sample from Apollo 11 resulted in a $\delta^{13}C$ value of +20 relative to the PDB standard, and a $\delta^{34}S$ value of +8.2 relative to the Canyon Diablo meteorite; the Apollo 12 $\delta^{13}C$ value was +12. These figures are much higher than those reported for intact meteorites, i.e. $\delta^{13}C$ of −4 to −20 and $\delta^{34}S$ of −2 to +2.

There was no evidence of indigenous biological structures obtained by examination of the dust, microbreccia surfaces, or of thin sections of the microbreccias made by light and electron microscopy.

To detect extractable organic compounds in the lunar samples, a complex analytical scheme was followed. Some of the lunar dust was treated with a 9:1 mixture of benzene and methanol. The extracts were examined by fluorescence excitation and emission spectroscopy, and evidence for the presence of porphyrins was found. Since similar spectral responses were obtained from extracts from the exhaust products after tests of the lunar descent rocket engines, it is possible that these pigments were synthesized in the exhaust. The level of porphyrins detected was very low—10^{-4} μg/g.

Capillary gas-liquid chromatography showed that any single n-alkane from C_{12} to C_{32} was not present at concentrations of 2×10^{-5} μg/g. The absorption spectrum of the benzene eluate showed bands at 224, 274, and 280 nm. Since these absorption bands were also present in the sand and solvent banks, the presence of aromatic compounds in the lunar sample cannot be inferred.

After benzene-methanol treatment the bulk fines were dried in a rotary evaporator and extracted with water. This extract was desalted and analyzed for the presence of free amino acids and carbohydrates by gas-liquid chromatography, using N-trifluoroacetyl-n-butyl esters of amino acids and trimethylsilyl derivatives of sugars. Amino acids were not present at a concentration of 10^{-5} μg/g. of sample, and sugars at 6×10^{-4} μg/g.

The residue from the water extraction was hydrolyzed with 1 N HCl. Hydrogen sulphide evolved during the hydrolysis was collected. Its concentration was about 700 μg/g., with an average δ^{34}D value of +8.0. The hydrolysate was treated with charcoal to absorb any bases which may have been present, and the charcoal was extracted with formic acid. The extract was derivatized with BSA [bis(trimethylsilyl)trifluoroacetamide] and examined for the presence of trimethylsilyl derivatives of purine and pyrimidine bases by gas-liquid chromatography. None were found at levels up to 4×10^{-3} μg/g. (The technique used here was the same as that used for the identification of pyrimidines in meteorite extracts).

Finally, the 6 N HCl hydrolysate of the sample obtained after refluxing with acid for 19 hr. at 125°C. was desalted and examined for amino acids by ion-exchange chromatography and gas-liquid chromatography of the trifluoroacetyl-n-butyl ester derivatives. Although the limit of detection was 2×10^{-3} $\mu g/g.$, none of the amino acids commonly found in protein were present.

Although the limits of detectability of the various types of compounds determined in the lunar material was in the nanogram range, there was no detectable trace of normal alkanes, isoprenoids, hydrocarbons, fatty acids, amino acids, sugars, nucleic acid bases, or of polymeric material such as proteins or nucleic acids. Porphyrins detected were probably derived from the rocket exhaust. Although this result is disappointing, it was not entirely unexpected since the surface of the Moon has been exposed to irradiation from various sources for millions of years, and any compounds initially present would have been destroyed.

Planetary Exobiology

The question of whether or not there is life other than our own in the solar system is of great importance to our view of the probability of there being other life in the universe as a whole—if life can evolve separately twice in one solar system,

TABLE 5-1. METEORITE CLASSIFICATION

Lithoid meteorites with or without small metal inclusions
CHONDRITES (generally, when the nickel increases in the metal inclusions, the bivalent iron increases in the magnesium silicate; distinction is made between "gray," "white," and "veined" varieties)
1. enstatite chondrite
2. bronzite chondrite
3. hypersthene chondrite
ACHONDRITES (categorized on the basis of calcium content);
1. achondrites with little calcium aubrites (enstatite) urelites (clinobronzite and olivine) amphoterites (bronzite and olivine) rhodites (broken bronzite and olivine) diogenites (hypersthene and olivine) chassignites (mainly olivine)
2. achondrites with abundant calcium angrites (mainly augite) nakhlites (diopside, olivine) eucrites (clinohypersthene and anorthite) shergottites (clinohypersthene and maskelinite) howardites (hypersthene and anorthite)

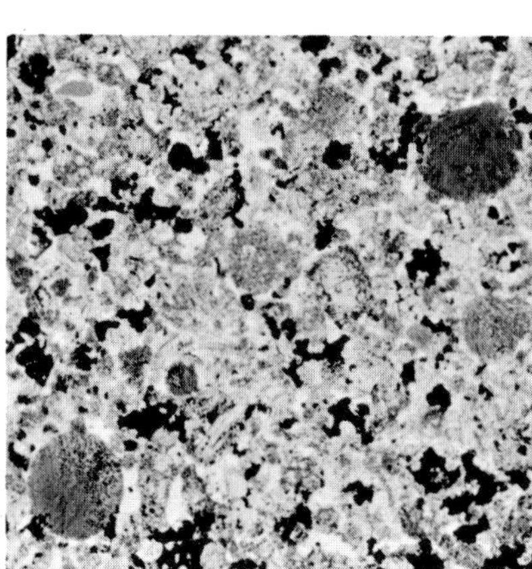

Fig. 5-3. Left, the lithoid meteorite weighing 843 g which fell near Bruderheim in Canada, covered with melted crust. Right, a bronzitic chondrite: a microphotograph of a thin cross-section of Miller's meteorite (Arkansas) which shows 1 mm chondrules formed by olivine or pyroxene or both; the matrix consists of olivine, pyroxenes and plagioclases, nickel-iron and troilite *(NASA, R. A. Oriti, Griffith Obs.; B. Mason).*

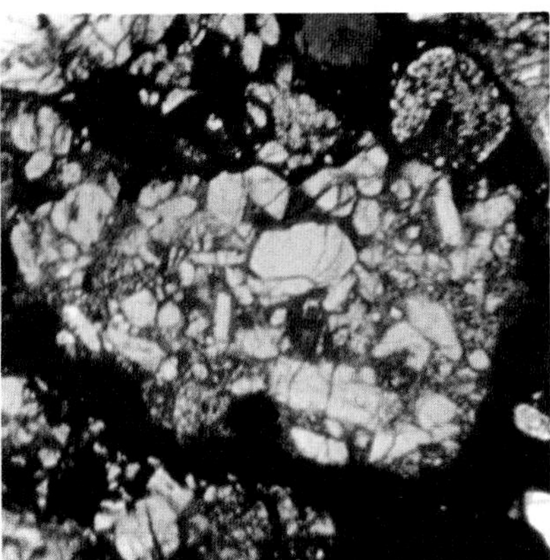

Fig. 5-4. A lithoid chondrule, in a microphotograph of Sharps's chondrite; it has an irregular contour and is made up of olivine crystals embedded in a basically microcrystalline mass *(R. W. van Schmus).*

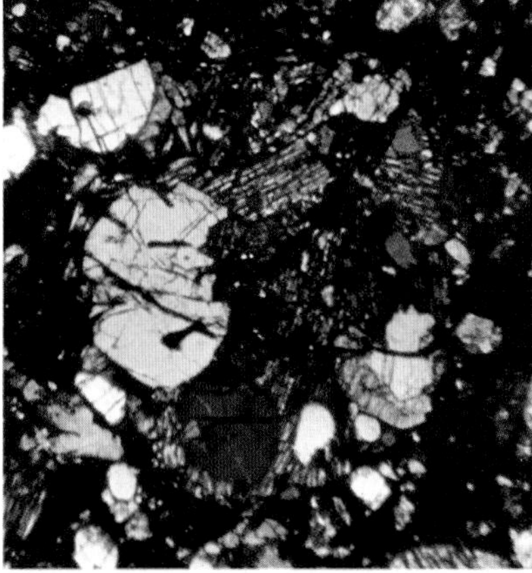
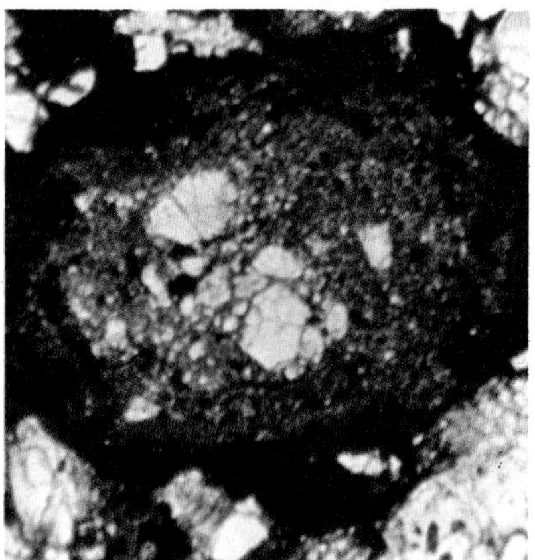

Fig. 5-5. Left, a porphyritic chondrule, in a microphotograph of Sharps's chondrite using the Nicol prism method. The chondrule consists mainly of polysynthetically twinned clinobronzite and olivine. Right, a zoned chondrule, again in a microphotograph of Sharps's chondrite, in which the dark bands which can be seen on the edge derive from the accumulation of metal granules and sulphides *(R. W. van Schmus).*

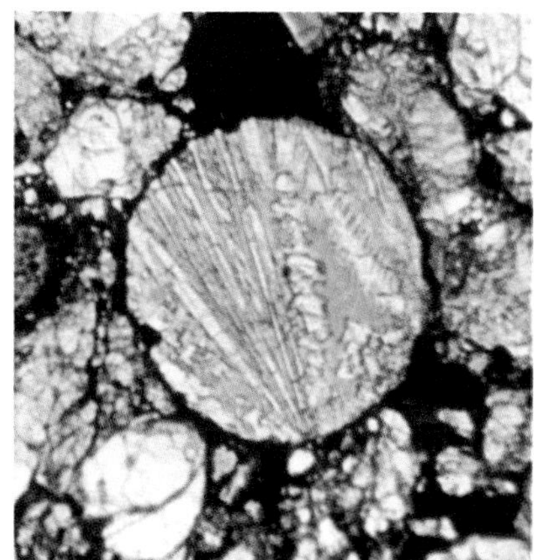

Fig. 5-6. A glassy chondrule, in a microphotograph of Sharps's chondrite; it consists of clearly radiate aggregates of olivine crystals embedded in a basically glassy isotropic mass *(R. W. van Schmus).*

then it may be assumed that it will arise wherever there are correct conditions. Apart from this question is the ancillary one of chemical evolution. We can only surmise about the chemical steps which led to the evolution of life on the Earth, but in the giant planets there is a present-day example of chemical evolution which is accessible to our telescopes. It is for this latter reason that the planet Jupiter, the nearest giant planet to us, is of great importance. The terrestrial-type planets Mars and Venus have received their share of attention in recent years, but it is Jupiter that has received the most concentrated interest. Although it is the nearest of the giant planets, and being the largest is relatively easy to observe, data on the structure of its atmosphere and the processes occurring therein is limited since the clouds covering the planet prevent direct observation of lower levels of the atmosphere. The cloud-top temperature of about 150°K is generally accepted, and it is believed that the clouds themselves are made of ammonia ice, but there is no agreement on the cloud-top pressure. Values ranging from 0.1 to about 10 atmospheres have been suggested. This variation is a consequence of the difficulty of determining the H_2/He ratio by direct observation—the values quoted are obtained from the pressure-broadening

TABLE 5-2. METEORITE CLASSIFICATION

Stony-iron Meteorites (intermediate type)
pallasite (olivine in iron) sorotites (troilite in iron) siderophyres (bronzite, tridimite and iron) lodranites (bronzite, olivine and iron) mesosiderites (hypersthene, anorthite and iron)
Metal Meteorites (masses formed essentially by nickel-iron with occasional silicic inclusions)
HEXAHEDRITES AND ATAXITES: with little nickel (less than 6% nickel content)
OCTAHEDRITES: 6–14% nickel content; grouped on the basis of the widths of the kamacite bands into very coarse, coarse, medium, fine, and very fine octahedrites (the nickel content increases in the same order). When polished and cleaned, octahedrites show Widmanstätten structures; taenites form thin lamellae at the edges of the kamacite bands with high nickel-iron content (γ iron); plessite, a mixture of kamacite and taenite, is sometimes found between bands of kamacite.
ATAXITES: with abundant nickel (normal nickel content of 14–30%).

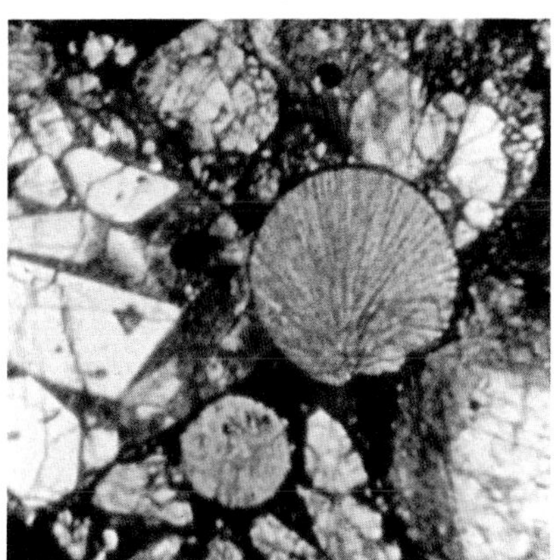

Fig. 5-7. Radiate pyroxene chondrule, in a microphotograph of Sharps's chondrite; it consists of a radiate aggregate of elongated clinobronzite crystals, a mineral in the pyroxene group *(R. W. van Schmus).*

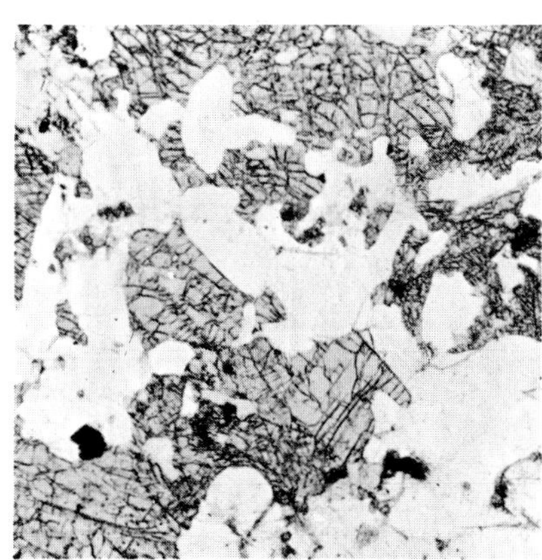

Fig. 5-8. Left, a eucrite, an achondrite formed by pyroxene (gray) and plagioclase (light), in a microphotograph of a thin cross-section of the Moore County meteorite. Right, a polished cross-section of the pallasite which fell near Albin, Wyoming; it consists of olivine crystals, which look like colored inclusions compared with the basic mass which is in turn formed of a nickel-iron alloy *(NASA, R. A. Oriti, Griffith Observ.).*

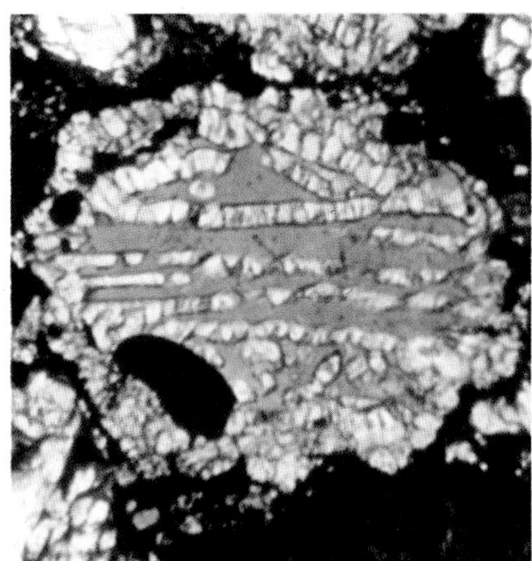

Fig. 5-9. An olivine chondrule, in a microphotograph of the Mezö-Madaras chondrite; it is formed by an olivine crystal with an interstitial isotropic glass (black with Nicol prism method) rich in Na and Al *(R. W. van Schmus).*

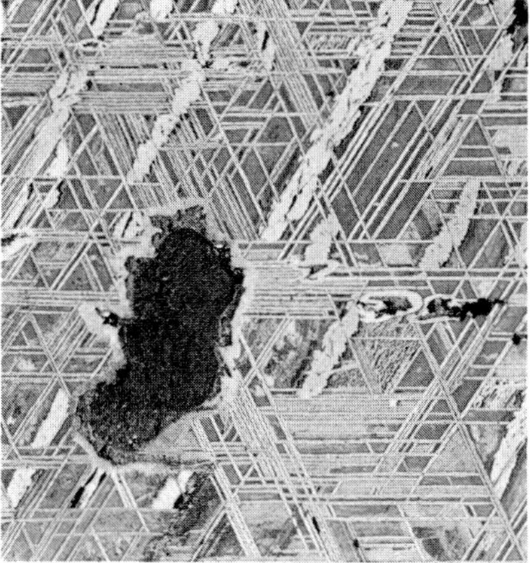

Fig. 5-10. Left: a polished cross-section of the octahedrite which fell at Edmonton, Kentucky; in it we clearly see the typical Widmanstätten figures and the presence of irregular troilite inclusions which appear as gray "splashes" *(Smithsonian Inst. Museum of Nat. History).* Right: a polished section of the hexahedrite which fell near Calico Rock, Arkansas, characterized by clear Neuman figures, typical of hexahedrites *(R. A. Oriti, Griffith Observ.).*

of lines from minor atmospheric constituents, e.g. methane and ammonia. Since the true composition of the atmosphere is of the greatest consequence to any discussion of atmospheric structure, a number of attempts have been made to determine the H_2/He ratio more accurately. One means by which this may be achieved directly is by occultation of a bright star by the planet. Such an occultation occurred recently, so it should not be long before the composition of the atmosphere is known somewhat more accurately than previously.

Based on assumptions of atmospheric composition, a number of models have been proposed for the structure of the lower atmosphere. One such is the Gallet-Peebles model, which suggests the presence of liquid water and higher temperatures beneath the outer layer of ammonia ice clouds. The pressure at this level has been variously suggested as between 1–100 atmospheres. This model, if true, is of great importance to exobiology since there are conditions within the Jovian atmosphere which would not be inimical to terrestrial life. Several cloud layers are suggested in the model, including ice, ammonium sulphide, and aqueous ammonia solution.

Fig. 5-11. Direct research into forms of life or traces of life has a precedent which dates back to the earliest days of space exploration and the studies carried out with lunar rocks, three microphotographs of which are shown here. They were taken using polarized light, with a technique, capable of showing up the nature of the component minerals. No trace of any organic substances was found which might have in some way linked up with the activities of organisms. The minerals present are: pyroxene (brown, green, orange); plagioclase (white, gray); ilmenite (black); olivine (middle photo, purple, blue) *(NASA)*.

The question of abiogenic synthesis of organic compounds from the methane and ammonia which are known to be present in Jupiter's atmosphere has been examined in the laboratory. In a number of experiments, methane-ammonia mixtures (1:1) were subjected to semicorona or electric spark discharges, in the absence of water. While the methane and ammonia were consumed, hydrogen and nitrogen were produced, together with a small amount of hydrogen cyanide. On examination of the volatile products from the reaction, using gas-liquid chromatography, acetonitrile, propionitrile, aminoacetonitrile and its C- and N-methyl homogues were identified. It was suggested that the hydrogen cyanide polymerized to the red tar which was also produced in this reaction. Acid hydrolysis of this nonvolatile material and subsequent ion-exchange and gas chromatography showed that glycine, alanine, sarcosine, iminodiacetic acid, iminoacetic-alpha-propionic acid and iminoacetic-beta-propionic acid were formed.

Since the red tar material that was the major product of this reaction appeared to be similar to the material forming the Great Red Spot on Jupiter, it was examined further by gel permeation chromatography on Sephadex LH 20, ultracentrifugation and infrared spectroscopy. The substance was separated into several bands, and had a maximum molecular weight of a few thousands. Its properties did indeed appear similar to those of the material of the Great Red Spot, suggesting that abiogenic synthesis on Jupiter may have reached an advanced stage—that of polymer formation. The fact that some common amino acids are produced by hydrolysis of this polymer is of great significance.

A theoretical examination of the possibility of Jovian abiogenic synthesis has been carried out, using a computer to calculate equilibrium concentrations of given components and assuming a complete thermodynamic equilibrium. The high temperatures required could easily be attained in lightning. It is interesting that colored compounds such as azulene and polynuclear aromatics tend to form under these conditions. Some of the product concentrations thus obtained are indicated in Table 5-3:

TABLE 5-3. PREDICTED EQUILIBRIUM IN THE JOVIAN ATMOSPHERE AT HIGH TEMPERATURES AND MODERATE TO LOW PRESSURES.

Pressure Temperature	1 atm. 1500°K	10^{-6} atm. 1000°K
Noble gases	0.4	0.4
Hydrogen	0.6	0.6
Methane	4×10^{-3}	6×10^{-5}
Acetylene	2×10^{-4}	2×10^{-3}
Ethylene	3×10^{-5}	7×10^{-7}
Ethane	2×10^{-7}	1×10^{-2}
Benzene	2×10^{-9}	1×10^{-7}
Naphthalene	1×10^{-12}	4×10^{-9}
Ashphalt (Yellow)	6×10^{-25}	1×10^{-8}
Nitrogen	7×10^{-5}	6×10^{-5}
Hydrogen cyanide	6×10^{-5}	9×10^{-5}
Ammonia	2×10^{-7}	2×10^{-13}
Acetonitrile	6×10^{-8}	2×10^{-10}
Azulene (Blue)	5×10^{-14}	2×10^{-9}
Aniline	3×10^{-19}	2×10^{-23}
Azobenzene (Red)	3×10^{-28}	3×10^{-33}

Applying the Gallet-Peebles model to Jovian atmosphere reactions, Sagan and Khare irradiated a mixture of methane, ammonia, water vapor, ethane and hydrogen sulphide with ultraviolet light of 2537 and 1849 A° wavelength at temperatures of 215–400°C. for up to 25 days. Various combinations of conditions were used. The products were analyzed by ion-exchange chromatography and by paper autoradiographic chromatography. Cysteic acid, cystine and serine were found, as well as H_2S_2, dimethyl and diethyl disulphides and diethyl sulphide.

Assuming that the Gallet-Peebles model is correct, and that there is some organic synthesis occurring within the Jovian atmosphere, there would be a level about 200 km below the

visible clouds where one would find liquid water, in which was dissolved ammonia, hydrogen sulphide, organic materials, and possibly traces of minerals from meteoritic bombardment of the planet. The temperature would be 25°C. and the pressure between 1–100 atmospheres. These conditions are not very different from those found on the Earth. Using this assumption further, P. Molton and C. Ponnamperuma tested several common microorganisms to see if they could tolerate these conditions. The conditions used were that a 1 ml. suspension of *E. Coli* B, *B. subtilis, Serratia marcescens,* or *A. aerogenes* was placed inside a steel bomb with a gas mixture containing 0.5% each of methane and ammonia, 43% of helium, and the remainder hydrogen, for 24 hr., at 25°C. and 100 atm. pressure. After this treatment the organisms were recultured and their viability determined, by plating. All of the organisms tested survived this treatment, there being little mortality beyond that shown by nongassed controls after standing in air for 24 hr. The conclusion to be drawn from this simple experiment is that given the conditions suggested by Gallet and Peebles, terrestrial organisms could find a home on Jupiter. This in turn raises the interesting possibility that since both abiogenic synthesis and survival of terrestrial organisms have been shown possible in the case of Jupiter, indigenous Jovian life may not be an impossibility. This is the reason that the upsurge of interest in Jupiter is likely to be continued. The evidence obtained from the photographs of Jupiter transmitted by Voyager 1 in 1979 indicates that there is no liquid or solid surface of water below the clouds. Instead, Jupiter appears to be gaseous into its deep interior.

In previous years the planet Mars was the focus of attention from the standpoint of exobiology. Speculation as to the existence of life on this planet has decreased as the results of the Mariner 6 and 7 missions became available. With the Mars lander mission (Viking) well under way, and with a factual solution to the question possibly obtainable by 1975, there has been little new work done on simulation experiments. In fact, the results of the Mariner experiments indicated that there was a polar ice cap temperature as low as 150°K, which would mean that solid CO_2 would be a major component of the cap. This is a blow for those would expect water ice on Mars. The other conditions (atmospheric pressure, etc.) tend to preclude the existence of liquid water anywhere on Mars. Solar ultraviolet radiation at wavelengths down to 1970 A° may reach the surface, and thus cause the decomposition of any protein, nucleic acid, etc. that may be present. There is one indication that things may not be as bad as suggested here—the "blue haze" of Mars appears to be an aerosol, which would absorb much of the short-wavelength ultraviolet before it reaches the surface. One significant laboratory result was the discovery that irradiation of powdered Vycor with simulated Martian sunlight in the presence of 0.06% of CO in CO_2 saturated with water vapor at 1 atm. pressure led to the formation of glycolic acid, formaldehyde and acetaldehyde, and possibly other organic products. This is the first case of abiogenic organic synthesis being demonstrated in a relatively oxidizing atmosphere. If such a synthesis occurs on Mars, there may be quantities of ultraviolet-absorbing material in the atmosphere, and the surface conditions may not be as hostile to life as would be expected from the Mariner results.

Venus has been something of a disappointment from the exobiological standpoint. With an atmosphere of 97% CO_2, below 2% of nitrogen and 0.1% of free oxygen, there is little prospect for any abiogenic synthesis. However, there may be from 6–11 mg/1. of water vapor present at the 0.6–2 atm. pressure level, if a result from the Venera probes is valid. Various suggestions for the composition of the clouds have been made, some more plausible than others. Among those suggested have been ferrous chloride dihydrate, ammonium chloride, mercury, and carbon suboxide. No suggestion was given for a mechanism to keep these components airborne.

One recent theoretical report showed that liquid water could exist on Venus' surface under the known temperature and pressure conditions (400–700°K and 40–80 atm.). If oceans exist, they would be extremely acid from the carbon dioxide dissolved in them, and would in consequence be extremely saline also.

Despite the inhospitable conditions of the Venerian surface, the possibility of life cannot be completely ruled out. J. Seckbach and W. Libby have demonstrated that the alga *Cyanidium calidarium* will grow at 50°C. at 50 atm. pressure of CO_2

over a 15-day period. Although these conditions do not approach true Venerian conditions in their severity, they do exceed those under which it had been previously assumed that this organism could survive. It thus becomes profitable to speculate on possible life forms indigenous to Venus.

Prebiological Evolution on the Earth

There has been considerable discussion on the nature of the Earth's original atmosphere, it being generally accepted that our present oxygen/nitrogen mixture cannot have been the original atmosphere. It is very difficult to obtain even circumstantial evidence on the actual nature of the original atmosphere, but it is generally assumed that it must have approximated the solar nebula in composition, and must have been rich in hydrogen. By taking into account the amount of hydrogen necessary for the reduction of iron, carbon, nitrogen, oxygen and sulphur, a concentration of hydrogen of 1.5×10^{-3} atm. has been calculated under equilibrium conditions. The residual ammonia and methane would have constituted the substrate for the simulation experiments of Miller, from which the oceanic concentrations of amino acids, etc. would have been formed. However, the ammonia itself would have rapidly been removed by solution in the sea, reaction with clay, and photolysis by solar ultraviolet radiation. Perhaps 40,000 years would have been sufficient for the removal of the ammonia, according to recent calculations. The nature of the processes by which this occurred and the compounds formed have been investigated in detail by various workers over the past 20 years and will not be repeated here. However, some recent work has been done on the methane which would have been left as a major component of the atmosphere after removal of hydrogen and ammonia. Photolysis of methane leads to the formation of progressively more complex hydrocarbons, with elimination of hydrogen gas. This would have escaped from the Earth in a geologically short time. It has been calculated that the result of this process of gradual removal of hydrogen would have been the formation of an oil slick over the oceans. This slick could have been tens of feet thick—although this is unlikely since further reaction of the oil with sea water with lightning as a stimulus would have generated fatty and hydroxy-acids from the oil. An experiment has been performed in which methane was subjected to a semicorona discharge in the presence of water. A number of common fatty acids were identified through their methyl ester derivatives by gas chromatography. This may have been the method by which fatty acids and sugars (from condensation of the intermediate formaldehyde) were formed.

Over a period of time, the methane in its turn would have become depleted in the atmosphere. The gradual replacement of methane and ammonia by other gases, presumably from volcanic action, is a part of the Earth's early history which seems to have received scant attention. Photolysis of water vapor in the upper atmosphere would have generated small concentrations of free oxygen, but amounts larger than fractions of a percent would have had to come from photosynthesis. It has been estimated that the primeval atmosphere containing methane and ammonia was superseded by one containing some free oxygen about 1.45 billion years ago. The generation of an ozone layer would have removed much of the solar ultraviolet radiation that would otherwise have reached the surface. From this point, living organisms would have contributed to the formation of our present-day atmosphere.

While the atmosphere was evolving in the manner assumed, life was evolving from inanimate chemicals in the oceans. The gap between the formation of a dilute solution of organic chemicals in the ocean and the first living cell is a large one. Intermediate between the two must have come a method of polymerizing amino acids and purine and pyrimidine bases to proteins and nucleic acids respectively.

The formation of proteins has been studied recently. The obvious way in which they could have formed under primitive Earth conditions is by hydrolysis of nitriles and amidines formed in an electric spark. If a polymer was initially formed, it could hydrolyze to a protein or smaller peptide. Peptides have in fact been identified in the reaction products of methane, ammonia and water, in one case having five amino acids joined in a chain. An alternative way of forming peptides which may have occurred on the primitive Earth is by the action of heat on ammonium cyanide—and again peptides have been identified in the products from this reaction.

The intermediacy of carbodiimides in aqueous condensations is well known in organic chemistry. The simplest carbodiimide exists as its tautomer, cyanamide. This could easily have been formed in a spark discharge. When an aqueous solution of glycine and leucine was exposed to ultraviolet light in the presence of cyanamide the dipeptides gly-gly, gly-leu, leu-gly, leu-leu and the tripeptides gly-gly-gly and leu-gly-gly were identified in the products.

It has been suggested that amino acids may have condensed to form peptides when small pools dried up in the heat of the sun. Pursuing this idea of a "hypohydrous" condensation further, the condensation of glycine with itself in the presence of orthophosphates was investigated. Isocyanite was found to be an effective catalyst in the reaction. The results of this reaction are listed in Table 5-4.

Fig. 5-12. Two experimental pieces of apparatus for laboratory examination of the problem of the origins of life. This equipment usually consists of special glass receptacles in which the gases are introduced which are considered to have been present at the dawn of the Earth's history, or which are present today on a planet where there is a desire to study the possibility of the development of a life form. This mixture is then charged up with energy in the form of electric discharges or ultraviolet rays. On this page: a simulation of the atmosphere on Jupiter; opposite: simulation of the original atmosphere of Earth *(NASA)*.

While these experiments show that amino acids may have condensed to form peptides under primitive Earth conditions, they give no indication as to how a *nonrandom* amino acid sequence could have arisen. It has been argued that the amino

TABLE 5-4. CONDENSATION OF GLYCINE IN THE PRESENCE OF ORTHOPHOSPHATES.

Orthophosphate	Diglycine yield,* percent A	Diglycine yield,* percent B
Na_2HPO_4	0	1.85
$Na_3PO_4 \cdot 12H_2O$	6.4	0
$NaH_2PO_4 \cdot H_2O$	0.5	10.2
$NH_4H_2PO_4$	0	2.6
$(NH_4)_2HPO_4$	0	0
K_2HPO_4	14.4	0.2
$CaHPO_4$	0	0.14
$Ca(H_2PO_4)_2 \cdot H_2O$	7.0	1.9

Composition: Column A orthophosphate (1 mM), glycine (1 mM)
Column B orthophosphate (1 mM), glycine (1 mM), KNCO (1 mM)
Conditions: 20 days at 95°C.
*Includes a small percentage of triglycine.

acids themselves have properties which tend to define the sequence of units in a peptide, but on the other hand the amino acids do not contribute appreciably to sequence determination during cellular protein synthesis. One attempt to produce nonrandom sequencing from a complex mixture has been made. The adenylates of sixteen free amino acids were reacted in alkaline buffer and directly analyzed on sieve columns. The products were predominantly monomeric, however, with no large peptides being obtained. There were many small peptides produced. One significant result was the formation of a mixture of oligomers whose amino acid composition did not reflect that of the starting mixture, suggesting that this could be a useful line of approach to the problem of nonrandom sequencing of amino acids.

A similar problem exists in consideration of the way in which purine and pyrimidine bases were phosphorylated under aqueous conditions, and the condensed in a nonrandom manner to form nucleic acids. The concept of hypohydrous condensation has proven useful in this case also. When the nucleosides adenosine, guanosine, cyridine, uridine and thymidine were heated with sodium dihydrogen orthophosphate, NaH_2PO_4, the nucleoside monophosphates were formed. At a temperature of 160°C., the highest yield of monophosphate formed was about 18%. The reaction could take place at much lower temperatures, a small yield of monophosphate being obtained even after heating at 50°C. In this case, three days heating was required.

In these experiments, the different monophosphate isomers were separated by autoradiography, paper chromatography, ion exchange and electrophoresis. The use of C^{14}-labeled nucleosides and P^{32}-labeled phosphate clearly showed that other compounds were formed besides the monophosphates. The electrophoretic mobility of one of these compounds corresponded to that of the dinucleoside U-P-U. This identity was confirmed by the techniques listed above. The dinucleotide U-P-U-P was similarly identified, and there was evidence for the formation of tri- and tetra-nucleotides also. These results indicate that thermal phosphorylation of nucleosides can lead to the formation of small polynucleotides. Linking together of several small chains formed in this way would give the first small "nucleic acids."

Recent studies on the pathway of phosphorylation have shown that condensed phosphates are formed when orthophosphates are heated to 160°C. At lower temperatures, partial transformation to linear phosphates was observed. Because of its simplicity to low-temperature condensation of inorganic phosphates it is very attractive as a source of condensed phosphate on the primitive Earth. Polyphosphates produced by heating may have been a major source of nucleotide formation, even in aqueous solution. Simple heating of an aqueous solution containing adenosine and linear polyphosphate has been shown to lead to the formation of a mixture of monophosphate isomers. In a typical experiment, an amount of tri- and higher polyphosphate salt containing the equivalent of 20 millimoles of phosphorus was mixed with 2 mmoles of adenosine in a 50 ml. flask, dissolved in 20 ml. of water, and refluxed for 4–6 hr. The hot solution was then diluted with water, cooled, and run directly onto a column of Dowex 1 × 2 formate ion-exchange resin. The identities of the isomers were established by comparison with standards, by thin-layer chromatography, by degradation with *E. coli* alkaline phosphatase, and by periodate oxidation. Pyrophosphate was ineffective in causing this condensation.

The conditions of these experiments may be considered to be genuinely prebiotic. The temperatures used are within reasonable limits, and the reactions are not inhibited by water unless it is present in large excess. Hence, micromolecules and macromolecules of biological significance could have been formed on the primitive Earth, giving rise to the proteins and nucleic acids which are the forerunners of those present in living systems today.

The interactions between nucleic acids and amino acids are of great importance in the development of a living system. Such intersections depend on the size, composition and conformation of the interacting species. A simplified model of such complex systems was tested by immobilizing 10 representative amino acids individually on a prepared chromatographic support by the formation of an amide linkage. Ion-exchange equilibria were determined by equilibrating the amino-acyl resins with a solution containing five competing 5′-ribonucleotides and analyzing the ionic composition of the amino resin after separation from the solution. It was found that binding behavior is indeed dependent on the nature of the base and of the amino acid, indicating that interactions take place at the monomeric level which would be of interest in elucidating the interaction mechanisms between polymeric nucleic acids and amino acids.

While these experiments are of great assistance in elucidating the mechanism by which a living system could first have formed, the picture is still very incomplete. One major question which remains unanswered and almost uninvestigated is the method by which a purine or pyrimidine base became attached to a ribose or other sugar residue. This is just one example of a major problem which will have to be solved before even the mode of formation of nucleic acids can be understood. Replication of nucleic acids inside cells is a source of controversy still. While we do not know the method of reproduction of a cell at the present day in all its molecular complexity, we cannot hope to understand the same problem removed in time by several billion years. The gap between our knowledge of primitive chemical synthesis and the formation of the first cell is not entirely accidental—much of our knowledge of the evolutionary process comes from a study of the fossil record. This can be examined back in time only as far as there are recognizable fossils, and understandably this does not include a record of the first forms of life. In order to reach further backwards in time it is necessary to investigate the chemistry of primitive organisms by analyzing the traces of recognizable biochemicals which they left behind them even after all traces of their skeletons had disappeared. This rewarding and important field provides another arm with which the problem of the origin of life on Earth can be studied.

Organic Geochemistry

Evidence in support of the presence of life on Earth as far back as 3.4 billion years has been presented. The Swaziland formation in S. Africa (about 3×10^9 years), the Gunflint chert in Ontario (dated 2×10^9 years), and the Bitter Springs formation in Australia (1×10^9 years) have all yielded microfossil evidence of unicellular life.

Amino acids, the n-alkanes, isoprenoids pristane and phytane, and porphyrins have been identified in these rocks and sediments. However, there is no unambiguous method of dating the organic matter in the rocks. In the absence of such criteria the molecular evidence appears to be on weak ground.

One example of the pitfalls to which these determinations are subject was the discovery that the porosity and permeability of the Precambrian Onverwacht chert is such that over a period of 3×10^9 years, 0.35 g. of a C_9-C_{30} suspension of *N*-alkanes in water can flow through 1 m^3 of the rock. Biological markers, such as the isoprenoids mentioned above, could have been brought into the rock at a later age and may not have been indigenous to the early Precambrian environment.

The eruption of Surtsey in 1963 gave a considerable boost to organic geochemistry in that for the first time it was possible to observe presumed primeval processes at first hand, obtain fresh and uncontaminated material, monitor the ways in which this material was contaminated by wind and sea, and look for traces of organic chemicals that must have been formed by volcanic processes.

Examination of the volcanic ash freshly fallen from Surtsey showed that traces of glycine, serine, alanine and aspartic acid were present, but no hydrocarbons. Methane and possibly isobutane, however, were detected by mass spectrometry of a sample of the emitted volcanic gases. Of great importance to prebiological chemistry is the way in which phosphates were made available to the environment instead of being tied up in insoluble rocks. Data concerning the kinetics of weathering and the amount of phosphorus produced in the volcanic gases was available from this eruption.

On the biological side, an organism identified as a fossil in the Gunflint chert of S. Ontario (dated 2×10^9 years) and recently found in a restricted environment as a living organism, *Kakabekia barghoorniana,* was isolated from several sources in Iceland. The fauna that might have been confused with this organism were not present in the severe environment of the Iceland terrain.

Continued investigation of the chemical prebiological evolution that may have occurred on the primitive Earth, together with fossil and geochemical evidence, should allow us to narrow the gap between our knowledge of the living and the nonliving states of matter from which eventually we were formed. Extension of this knowledge into space should allow a deduction of the probability of life elsewhere in the universe.

CYRIL PONNAMPERUMA AND PETER MOLTON

Bibliography: Chadha M. S., Flores J. J., Lawless J. G., Ponnamperuma C., *Organic synthesis in a simulated Jovian atmosphere,* in Icarus, XV, 39 (1971); Khare B. N., Sagan C., *Synthesis of cystine in simulated primitive conditions,* in Nature, CCXXXII, **5312**, 577 (1971); Kvenvolden K. A., Lawless J. G., Ponnamperuma C., *Non-protein amino acids in the Murchison meteorite,* in Proc. Nat. Acad. Sci., LXVIII, **2**, 486 (1971); Lasaga A. C., Holland H. D., Dwyer M. J., *The primordial oil slick,* in Science, CLXXIV, **4404**, 53 (1971); Ponnamperuma C., Skehan J. W., *Boston college environmental center summer institute on Surtsey and Iceland,* NASA TM X-62,009 (1971); Rabinowitz I., Chang S., Ponnamperuma C., *Possible mechanisms for prebiotic phosphorylation,* in Kimball A. P., Oro J. (ed.), *Prebiotic and biochemical evolution,* Amsterdam (1971); Sagan C., *The solar system beyond Mars: an exobiological survey,* in Space Sci. Revs., XI, **6**, 827 (1971); Schwartz A., Ponnamperuma C., *Phosphorylation of nucleosides by condensed phosphates in aqueous systems,* in Kimball A. P., Oro J. (ed.), *Prebiotic and biochemical evolution,* Amsterdam (1971); Turner B. E., *Detection of interstellar cyanoacetylene,* in Astrophys. J., CLXIII, **1**, L35 (1971); Ball J. A., Gottlieb C. A., Lilley A. E., *Detection of methyl alcohol in Sagittarius,* in Astrophys. J., CLXII, 3, L203 (1970); Belsky T., Kaplan I. R., *Light hydrocarbon gases, ^{13}C, and origin of organic matter in carbonaceous chondrites,* in Geochimica et cosmochimica acta, XXXIV, **3**, 257 (1970): Chang S., Williams J., Ponnamperuma C., Rabinowitz J., *Phosphorylation of uridine with inorganic phosphates,* in Space Life Sci., II. **2**, 144 (1970); Kvenvolden K. A., Ponnamperuma C., *A search for carbone and its compounds in lunar samples from Mare Tranquillitatis,* NASA SP-257 (1970); Ponnamperuma C., *Chemical evolution and the origin of life,* in N. Y. State J. Med., LXXI, **10**, 1169 (1970); Ponnamperuma C., Klein H. P., *The coming search for life on Mars,* in Quart. Rev. Biol., XLV, **3**, 235 (1970); Seckbach J., Libby W. F., *Vegetative life on Venus? Or investigations with Algae which grow under pure CO_2 in hot acid media at elevated pressures,* in Space Life Sci., II, **2**, 121 (1970); Calvin M., *Chemical evolution. Molecular evolution towards the origin of living systems on the Earth and elsewhere,* New York (1969); Feldman P. A., Rees M. J., Werner M. W., *Infrared and microwave astronomy,* in Nature, CCXXIV, **5221**, 752 (1969); Woeller F., Ponnamperuma C., *Organic synthesis in a simulated Jovian atmosphere,* in Icarus, X, **3**, 386 (1969); Hodgson G. W., Ponnamperuma, C., *Prebiotic porphyrin genesis: porphyrins from electric discharge in methane, ammonia, and water vapor,* in Proc. Nat. Acad. Sci., LIX, **1**, 22 (1968); Hayes M. J., *Organic constituents of meteorites – a review,* in Geochimica et cosmochimica acta, XXXI, **9**, 1395 (1967); Ponnamperuma C., Pering K., *Possible abiogenic origin of some naturally occurring hydrocarbons,* in Nature, CCIX, **5027**, 979 (1966); Young R. S., Ponnamperuma C., McGaw B. K., *Abiogenic synthesis on Mars,* in Florkin M. (ed.), *Life science and space research,* III, Amsterdam (1965).

WHAT was the appearance of the first forms of life that emerged on our planet: what was the appearance of the Earth itself? And in what environment did this most remote ancestor which links us in a relationship to the whole of the animal and vegetable world appear? An answer to these questions comes from a huge rock formation more than 15 km deep, the "Swaziland sequence," in South-East Africa, and is the main subject of the following article. It is, however, necessary to begin with some prior considerations.

Scientists who deal with life in the past study fossils, animal or vegetable remains which have come to form parts of rocks. Life today has invaded all available environments—waters, dry land, air. However, if we compile a backward-looking ecological history, we soon realize that this is a relatively recent state of affairs: life left the water to conquer land and air only during the last 450 million years—that is, during the last tenth of the history of the planet. All earlier remains are of organisms that lived in water. The reason probably is that the atmosphere did not contain enough oxygen to form the ozone layer which now prevents ultraviolet radiation from reaching the Earth's surface in a quantity sufficient to cause irreparable harm to living organisms: until this atmospheric layer was formed only the waters provided sufficient protection. Anyone searching for the earliest traces of life must thus examine the sedimentary rocks which were mainly deposited in submarine environments. Furthermore, for these traces to be preserved, the rocks must not have undergone excessive change (metamorphism) after their formation. In an absolute sense the most ancient sedimentary rocks are those discovered by Moorbath in Greenland. These however have undergone extremely radical metamorphic action which makes them useless for the above purpose. The rocks belonging to the Swaziland sequence are slightly less ancient: they were laid down about 3.4 billion years ago. Professor Nagy and his wife, who are the authors of the next article, have analyzed these rocks meter by meter in their search for something similar to the remains of a living organism, and have found simple roundish structures, a few thousandths of a millimeter in diameter. It is extremely likely that these are unicellular blue algae. The chance remains, however, that they might be physical structures in no way connected with living matter.

The first absolutely certain traces of the activity of algae and bacteria are found at Bulawayo in South Rhodesia. Here, in rocks between 2.7 and 3 billion years old, we found stromatolites, structures whose formation can still be observed today, linked with the activity of colonies of blue algae. It thus appears that these, together with anaerobic bacteria, traces of which are much more difficult to find, were the first inhabitants of the planet, appearing about 3.2 billion years ago. The most ancient fossil whose biogenic origin is now universally accepted is from the Transvaal, where the first undisputable filamentary algae appears in a rock dating from 2.2 billion years ago. This is no primitive form; it already has a billion-year evolution behind it. At least another billion were to pass before the first animal life appeared on the face of the Earth.

BARTHOLOMEW NAGY

Director of the Organic Geochemistry Laboratory of the University of Arizona at Tucson. Born in 1927 in Budapest, Hungary, he moved to the United States in 1948, becoming a U.S. citizen in 1955. He began studying the earliest forms of life while working on carbonaceous meteorites with Nobel Prize winner, Harold C. Urey, at the University of California at San Diego. Working closely with his wife, Lois Anne Nagy, co-author of the following article, he analyzed the oldest sedimentary rocks on Earth in a search for life forms. He has followed, and still follows, current research on extraterrestrial life, having been part of the group of consultants at NASA (Houston) working on the landing of the Viking probe to Mars in 1976.

BARTHOLOMEW NAGY
AND LOIS ANNE NAGY

The Oldest Known Sediments and the Early Forms of Life

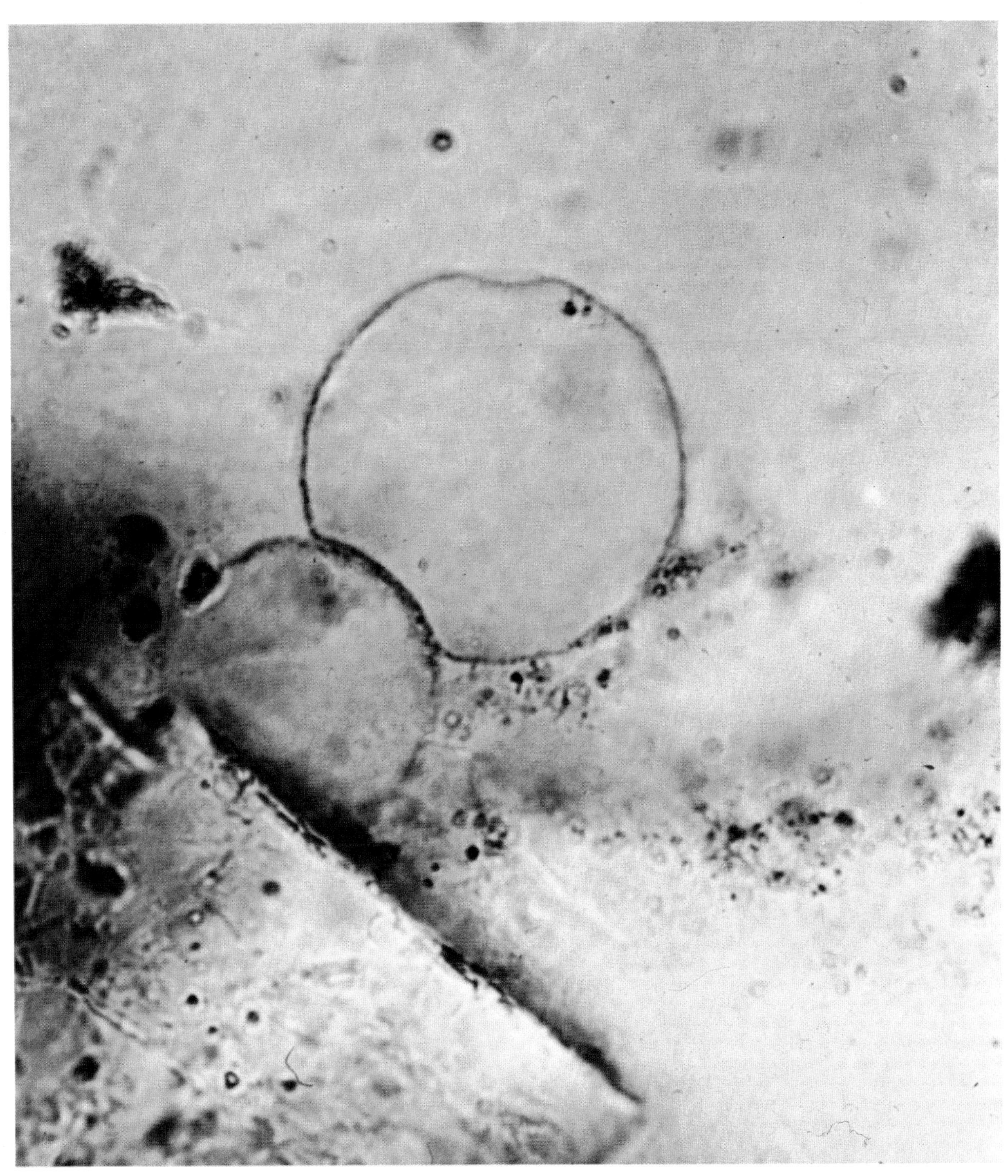

Overleaf: A thin cross-section of a rock from the Canadian Gunflint Formation, more than 2 billion years old, showing two clearly identifiable cells in contact with each other. These and other forms found in the same formation are among the oldest microfossils about which there is no longer any uncertainty. The first probable traces of life, however, date back to about one billion years before this (photo S. Barghoorn).

When the first forms of life appeared, what was our planet like? This is one of the questions that has puzzled geologists, astronomers, chemists, and philosophers for centuries. Much of this question is still to be answered, but with modern analytical techniques some of the pieces of the puzzle are beginning to fall into place.

The Earth is believed to be 4.5×10^9 years old. The oldest known terrestrial rocks are granitic-gneisses which have been found in the Godthaab District of Western Greenland. These rocks have been dated by S. Moorbath and his co-investigators at Oxford University using the Rb-Sr dating method. The determined age of these rocks is 3.75 billion years old. Unfortunately, there do not seem to be any known unmetamorphosed rocks in this assemblage, so little is known about this early part of geological history. In trying to understand primitive life, or life precursor processes in the ancient geological record, unmetamorphosed sediments must be studied. These are rocks which have remained basically unaffected by heat and pressure during geological time. The oldest known, basically unmetamorphosed sedimentary rocks are 3.4 billion years old and they occur in the Eastern Transvaal of South Africa.

The Barbeton Mountain Land

These sedimentary rocks are located in an area known as the Barbeton Mountain Land in South Africa. This area has been known since the latter part of the 19th century for gold mining and its colorful history of pioneering and gold rush days. We are also told that in these years for every mile of railroad track built a man was killed by a lion, and that areas of lower elevation were infected with malaria and sleeping sickness. Boulders imbedded in what was once a road still bear the scar marks made long ago by wagon wheels carrying the malaria victims to higher, uninfected elevations. This region today is still sparsely populated, Fig. 6-1.

Geological Setting

The Swaziland Sequence is geologically complex that is unique in the excellent state of preservation of its ancient unmetamorphosed sedimentary rocks. This sequence is situated in the stable Kaapvaal Craton and consists mainly of primitive ultramafic volcanic rocks, mafic and acid volcanics, and a minor amount of sediments. D. R. Hunter has suggested that in this area during the earliest Precambrian age (prior to 3.4 billion years ago) deposition of the sediments and volcanic rocks occurred on the primeval crust. This earliest depositional environment is difficult to assess. No remnants have been found to date to reveal the nature of the surface on which these first rocks were deposited. It is possible that the original sedimentation and granitization considerably predates the Swaziland Sequence. Tonalitic diapir domes which outcrop at the western and southern flanks of the Swaziland Sequence suggest that the ancient gneiss basement complex was reactivated at a later geological time. One of the intrusive granite domes is seen in Fig. 6-1. This idea, i.e., the later disturbance of the basement could be one way to explain why the base of the Onverwacht (the oldest member of this Sequence) is underlain by younger intrusive granites. Another mechanism which might have played a role in the preservation of this area is connected with some recent information about continental drift and plate-tectonics. It was proposed by K. Burke and J. T. Wilson that the African plate has moved relatively little during geological time. They suggest that the first shifting of the continents took place 100 to 200 million years ago with the break-up of Gondwanaland. The African plate then remained quiet until the present phase of uplift; volcanism and rifting

Fig. 6-1. The Onverwacht series which contains the oldest known sedimentary rocks. On this page, a view of the outcrop, the Barbeton Mountains in South Africa; on the opposite page, the Komati river valley with the rocks of the Theespruit Formation, the part of the Onverwacht Series with the oldest sedimentary rocks of all, 3.4 billion years old.

began only about 25 million years ago. It is also thought that one of the most important factors governing plate movement are plumes, which are "hot spots" some 400 km deep within the Earth. A plume has an average life of about 100 million years and competition between old plumes and new ones causes plate motion.

The Swaziland Sequence is stratigraphically more than 15,000 m thick and is made up of three rock Series. The youngest member is the Moodies Series, the middle one is called Fig Tree (age 2.9×10^9 years old), and the oldest is called the Onverwacht which is 3.4×10^9 years old. The word Onverwacht is an Africaans word meaning "unexpected"; this is the Series which is most pertinent to this discussion. Some of the Onverwacht is composed of pillow lavas (these are lavas which have been quenched in water thus forming large "pillow-like" structures). The sedimentary rocks which are in the vast minority are mainly cherts, siderites, and some shales. Siderite is iron carbonate ($FeCO_3$) and it occurs in some places as distinct bands next to black chert. Hand specimens of these rocks are seen in Figs. 6-2 and 6-3, and despite the large accumulation of igneous and metamorphic rocks many of the sediments within the Onverwacht are relatively unmetamorphosed. In hand specimens this can be seen by the presence of ripple marks, (Fig. 6-3) cross-bedding, and lamination. These and other sedimentary structures would have been obliterated had there been any widespread and intense metamorphic activity affecting the sediments. It has been suggested by C. Anhaeusser that some of the volcanic activity occurred in slow systematic cycles and that the Swaziland area may have been covered until relatively recently by younger Precambrian rocks which were removed by erosion. This, plus the relatively stable nature of the African plate might help to explain the excellent state of preservation of these ancient sediments. The base of the Onverwacht (Sandspruit formation) contains no known sediments and the first sedimentary horizon occurs in the Theespruit formation which is 2100 m stratigraphically above the younger intrusive basement granite. The oldest unmetamorphosed sedimentary rocks are basically cherts. Fig. 6-4 shows an outcrop of the Theespruit formation and a freshly broken surface of one of the chert layers. Fig. 6-3 shows a hand specimen from the same formation. Isotope age determinations have been made on an argillaceous chert from the middle of the Onverwacht. This work was performed by P. Hurley and his colleagues using the Rb-Sr dating method. The age obtained on this, what is called the Middle Marker horizon, is 3355 ± 70 million years. Fig. 6-4 shows one of the Middle Marker outcrops. There are other sedimentary horizons which are still undated below the Middle Marker; how much older they may be is not known. Other scientists have dated some of the granites and gneisses in the area and they have reported ages which are younger but comparable to that of the age compared on the Middle Marker sample.

Organic Geochemistry

A substantial amount of work has been done on the chemistry of these old rocks by scientists in South Africa, Europe, and the United States. In this article some of the highlights of this research will be discussed. We are using the word "organic" as it is used by chemists to refer to carbon compounds and not necessarily to molecules which have something to do with life processes. One important point is to keep in mind that almost all sedimentary rocks are porous to some degree. This includes rocks which appear dense and have no pore channels visible when viewed in petrographic thin sections under the microscope at magnifications of 1000 times. A few years ago B. Nagy calculated the volume of liquid that could percolate through a cubic meter of dense Onverwacht chert during a period of 3 billion years assuming average sedimentary basin pressure gradients. The chert had 0.5 percent porosity and 5.7×10^{-7} millidarcy permeability. It was calculated that during a period of 3×10^9 years one cubic meter of water containing 0.35 grams of dissolved saturated hydrocarbons (C_9-C_{30} n-alkanes) could flow through a cubic meter volume of this rock. This porosity and permeability makes it difficult to interpret correctly the origin of trace amounts of "biological type" hydrocarbons and other compounds reported in the solvent extracts of these old rocks. The major problem is that there is no way of telling if these soluble hydrocarbons are as old as the rock or if they were brought in more recently by percolating solutions.

Another, but more difficult, way of studying the organic chemistry of these rocks is to analyze the insoluble polymer-like organic matter known as "Kerogen." There are several methods for doing this, but the way in which it is done in the authors' laboratory is by ozonolysis or pyrolysis in vacuum or in an inert gas atmosphere. The oxidizing agent ozone is used to cleave double bonds in molecules, then the products are made volatile by making methyl esters which can be analyzed by combined gas chromatography-mass spectrometry. The mixture of esters is injected into the gas chromatograph which separates the various compounds which are then directly introduced into the mass spectrometer where they are bombarded with electrons and further broken down. The fragmentation pattern is recorded on a chart; this spectrum enables one to discover the original compound. Fig. 6-5 shows a gas chromatogram of methyl esters prepared from the ozonolysis products of the originally insoluble organic matter in an Onverwacht sedimentary rock. The numbers on the chromatogram represent peaks from which mass spectra were obtained. The mass spectra showed kerogen fragments which are usually obtained from aromatic compounds. Because of the predominance of such compounds it seems that the Onverwacht polymer is basically aromatic in nature, consisting of condensed aromatic nuclei which are connected by short aliphatic chains. Aromatic substances, such as coal, occur also in rocks of younger geological ages. (Interestingly, the complete chemical composition of coal is still not understood.) This aromatic composition of the Onverwacht kerogen is noteworthy because samples from the younger Fig Tree Series (2.9×10^9 years old) which were studied in the authors' laboratory yielded an abundance of aliphatic compounds of the types which are usually thought to be biological in origin. Fig. 6-6 shows capillary column gas chromatographs of pyrolized Onverwacht and Fig Tree kerogen. The younger Fig Tree sample is mainly aliphatic in nature while the older Onverwacht samples are aromatic.

Considerable research has also been devoted to the chemical composition of the igneous and metamorphic members of the Onverwacht. One of the interesting and important discoveries made by geologists of the Economic Geology Research Unit, University of Witwatersrand, Johannesburg, South Af-

Fig. 6-2. Here and on the page opposite, samples of the Swaziland Sequence which, in addition to the Onverwacht Series (the oldest) also includes the Fig Tree Series (intermediate) and the Moodies Series (the most recent). Above, left, thin laminations in an Onverwacht flint; right, a yellowish and fairly broad band of siderite—a thin layer of flint can be seen at the bottom.

rica was that the alkali (mainly potassium) content is low in the igneous rocks below the Middle Marker horizon in the Onverwacht. In the horizons above the Middle Marker the alkali content is within the normal range for rocks.

Carbon isotope studies of the organic matter in the Onverwacht and other younger Precambrian rocks have given additional information. Recent work by D. Z. Oehler and her coworkers on the reduced carbon in the lower and middle Onverwacht and Fig Tree rocks, yielded values comparable to the carbon isotope rations obtained in much younger Precambrian and younger rocks. It was found that the $o^{13}C_{PDB}$* value was −28.7 per mil for the middle and upper Onverwacht. This falls in the range of photosynthetically produced biochemicals. By comparison the reduced carbon in the Onverwacht (Theespruit) gave $\delta^{13}C_{PDB}$ value of −16.5 per mil. This value would be unusual for most biochemical compounds. From this evidence, if it can be confirmed, it would seem that prior to 3.4×10^9 years ago life, at least in this area, had not yet fully evolved or that life was scarce. There are indications that at the time of the Middle Marker sedimentations or shortly thereafter there was an evolutionary and/or geological change. This is shown by the normalization of the alkali content in the igneous rocks and the change in the $\delta^{13}C$ values.

Microstructures and Microfossils.

The Onverwacht rocks have been studied extensively for any trace of biological remnants or microfossils. There are several ways in which rocks can be studied under the microscope. The surface is cleaned, then ground into a fine powder and treated with various chemicals before preparing glass microscope slides. Another way is to look at the rock directly and this is done by preparing petrographic thin sections. In this procedure a rock is cut, polished and cemented to a glass microscope slide with an appropriate adhesive. The other side of the rock is then polished down to between 15 μm to 25 μm in thickness and then a cover slip attached to it. Using transmitted light microscopy and magnifications of 1000 microstructures in the rock can be seen and photographed. Fig. 6-5 shows a petrographic thin section of one of the Onverwacht rocks. In this thin section one can see a good example of the sedimentary structure of cross-bedding which is common in these rocks.

The authors' laboratory has been studying the micropaleontology of the Onverwacht sedimentary rocks since 1967. All the known sedimentary horizons have been studied within the Onverwacht and they yielded simple round and "cup-shaped" microstructures. We still wish to call these structures "microstructures" and not *microfossils* because of their very simple morphology and because their origin is still not known. Fig. 6-5 is shown to illustrate how careful one must be in evaluating the origin of a microstructure based on morphological criteria alone. This microstructure which looks biological comes from a petrographic thin section of an Onverwacht pillow lava. This rock crystallized out from a silicate melt and the microstructure had nothing to do with a life or biological system. Fig. 6-5 shows typical Onverwacht microstructures which are found most abundantly in the cherts. Fig. 6-5 is a scanning electron micrograph of one of these forms. These structures occur in a wide size range and may vary from 3 μm to 200 μm. They are very simple in morphology and heavily mineralized. Treatment with hot 6N hydrochloric acid, which would remove most carbonates and oxides, followed by 48 percent hot hydrofluoric acid, which dissolves silicates, and the oxidizing agent ozone, had no affect on these structures. Many of the larger structures have one or more holes in them. A double structure, Fig. 6-6, was studied in detail under the microscope by means of optical cross-sectioning. The way this is done under the microscope is by taking the particle completely out of focus and bringing it back into focus slowly; the particle is then photographed at every 0.5 μm intervals. When these photographs are mounted side by side the whole can be studied in great detail. In one photograph a certain portion of the particle may be very sharply in focus while this same portion of the particle may be out of focus four photographic frames away, and a new area will be in focus. Since the photographs were taken every 0.5 μm this would mean that the distance between the sharply focused portion and the one which is now out of focus would correspond to 2 μm. These hollow microstructures can be studied and measured very accurately by this method. Fig. 6-6 shows photomicrographs of one microstructure which was photographed using the optical cross-sectioning technique. The outer surface of the structure is shown, then 6 μm inside the particle. Other types of microstructures which are seen in Fig. 6-7 fall in a much narrower size range. They are found most abundantly in the siderite bands but have also been found occasionally in cherts. The same simple morphology is seen throughout the Onverwacht. These microstructures may be remnants of single celled blue-green algae, or precursors to the first living cell which was to evolve at a later time, or they could be organic nonbiological particles which had nothing to do with life or prelife.

Stromatolites

The younger Precambrian Transvaal Sequence (1.9 to 2.3 billion years old) is located to the north of the Onverwacht type area. These sedimentary rocks are of particular interest in

*PDB. The $^{13}C_{PDB}$ (per mil) value is defined as

$$13\delta = \frac{(^{13}C/^{12}C)\text{ sample} - (^{13}C/^{12}C)\text{PDB}}{(^{13}C/^{12}C)\text{PDB}} \times 1000$$

where PDB refers to the Peedee belemnite standard from the Peedee formation, Upper Cretaceous of South Carolina.

Fig. 6-3. Above left, thin bands of siderite (iron carbonate) in black flint, in an Onverwacht sample; the arrows indicate small ripples in the siderite band which may have been caused by wave movement; right, structures which clearly betray the activity of slight wave motion of the type existing in lagoons, in a sample taken from the Moodies Series.

studying early life forms because recently nine new stromatolite horizons have been found in them.

Stromatolites are laminated sedimentary rock structures which are formed by bacterial filaments or algal mats trapping and binding mineral particles. There are no known inorganic processes which could produce these structures. The oldest known stromatolite in the geological record occurs in a limestone at Huntsman's Quarry in Bulawayo, Rhodesia. The estimated age of this stromatolite is between 2.7 and 3 billion years old. Unfortunately, this rock has been mildly metamorphosed and no microstructures have been found in it to date. It had been assumed until recently that the Bulawayan stromatolite was formed by algal activity. However, current studies of stromatolites still in the process of formation in hot pools at Yellowstone National Park, have revealed that photosynthetic flexibacteria together with algae can build stromatolites. One may ask, in light of these findings, if the Bulawayan stromatolite could basically be a bacterial stromatolite rather than an algal one.

In studying a petrographic thin section of one of the Transvaal stromatolites, well-preserved remnants of what appear to be blue-green algae have been found. This stromatolite is basically a dolomitic-limestone; it is finely laminated and according to Logan's classification it is of the laterally linked type. The base of the Transvaal Formation has been dated at 2.3×10^9 years old. Our sample was located approximately 700 m stratigraphically above the base, and therefore estimated to be 2.2×10^9 years old. Carbon isotopic studies on the carbonate minerals of this rock have yielded a $\delta^{13}C$ value close to 0 which is in the marine carbonate range.

Numerous tapered and segmented algal filaments have been found (Fig. 6-8). Some of the filaments contained enlarged and thick-walled cells which may represent heterocysts or akinetes. Some species of modern blue-green algae contain these enlarged specialized cells called akinetes or heterocysts (the morphology and function of these cells will be discussed below).

One of these forms from the Transvaal is morphologically similar to the modern blue-green algal family Nostocaceae. This new microfossil which is shown in Fig. 6-10 has been named *Petraphera* vivescenticula. Fig. 6-10 shows a drawing for morphological comparison with the modern blue-green alga *Raphidiopsis* mediterranea Skuja (after K. V. N. Rao). Some features which both the modern form and the fossil have in common are: comparable (but not identical) size, terminal cell narrowed to a sharp point, and enlarged thick-walled cell near the middle of the filament. In the case of the modern algal form this enlarged cell would be an akinete or resting cell, and

Fig. 6-4. The oldest sedimentary rock and the marker horizon. Left, an outcrop in the Theespruit Formation of the Onverwacht Series; this is the oldest known sedimentary rock; its age has not been calculated directly but we know that it is older than the Middle Marker (right), which has been dated at 3355 ± 70 million years with radioactive isotopes.

in the fossil form it could possibly have served the same type of function. Also, numerous broken filaments have been found, as well as single double-walled cells in the Transvaal thin sections. Finding these morphologically complex fossils of 2.2×10^9 years old would suggest that by this time in geological history algal plant life was becoming abundant and had already begun to diversify.

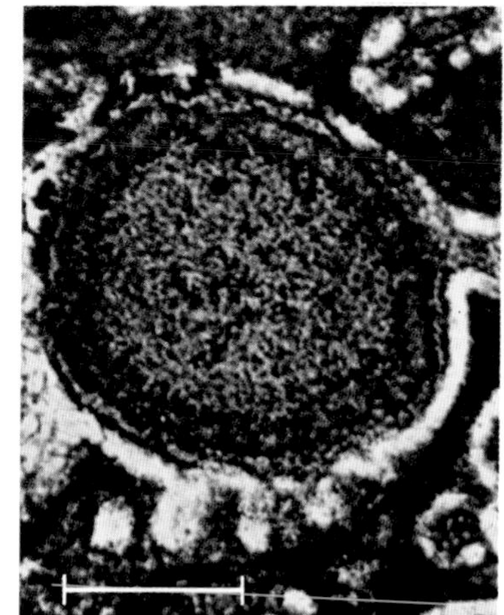

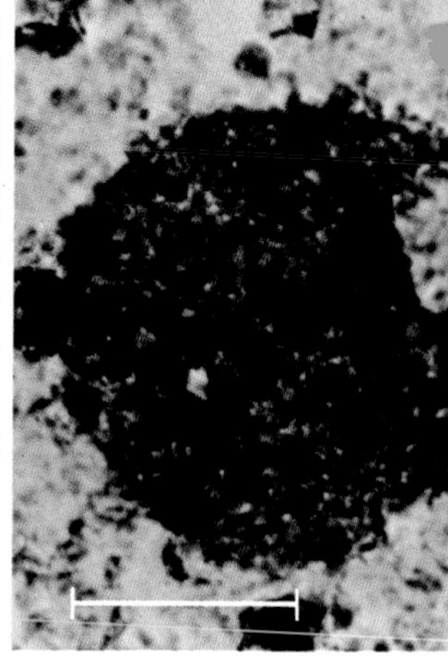

Blue-green Algae—What are They and What can They Tell Us?

There is still considerable controversy among scientists as to the nature of Earth's atmosphere 3 billion years ago. The ozone layer in the upper atmosphere may not have developed yet, so more ultraviolet radiation could have penetrated to the surface of the Earth. Also, it is thought that the carbon dioxide concentration in the atmosphere was considerably higher and that there was very little free oxygen. The oxygen content of our present atmosphere is due mainly to the release of oxygen by plants. In early Precambrian times however an abundant plant population may not have yet evolved. We know from the geological record that there was liquid water at this time. How much water there may have been is not known, but it is thought that there were at least shallow seas. All bodies of water have a zone called the thermocline. This is a division (or boundary layer) between the warmer upper layer and the cooler bottom layer, which do not mix. Below the thermocline early forms were certainly protected from the harmful effects of ultraviolet radiation.

It is generally assumed that the first life on Earth consisted of an aerobic bacteria, and that bacteria and blue-green algae, the simplest of all the algae, perhaps evolved from the same ancestor. Blue-green algae are to be considered the first true photosynthetic plants, that is, they can utilize carbon dioxide from the atmosphere, use sunlight as an energy source, and respire oxygen. There are photosynthetic bacteria but they differ from blue-green algae in several ways and they cannot respire oxygen. Modern blue-green algae apparently have evolved little from their ancestors. They remain simple non-nucleated cells which reproduce by division. One of the remarkable features of blue-green algae is their ability to adapt

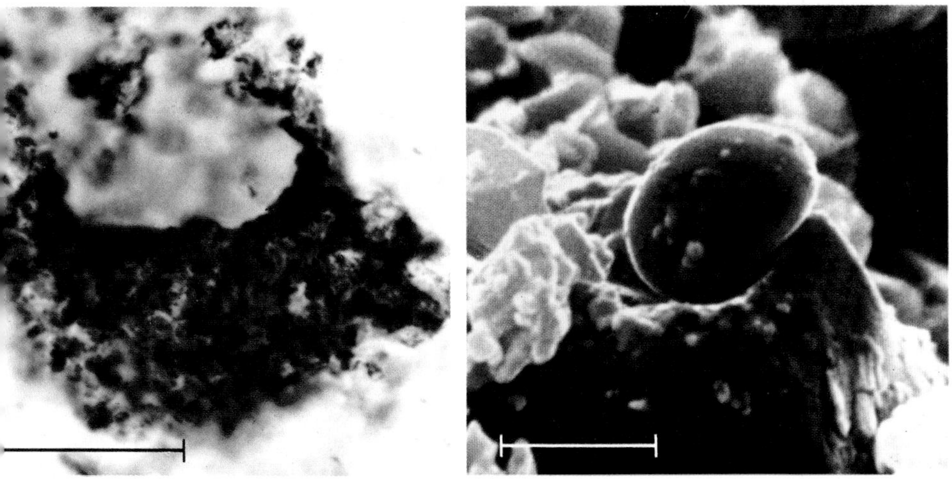

Fig. 6-5. Microstructures and microfossils. On the opposite page: left, a microstructure photographed in a thin cross-section of an Onverwacht pillow lava; despite its appearance, which might be misleading, this structure which comes from a lava has nothing to do with life or prebiotic processes; right, a possibly biogenic microstructure found in an Onverwacht flint; the bar represents 20 μm. It is evident how the purely morphological criterion can lead to errors in the identification of biogenic structures. On this page: a dimensional comparison between a frequent Onverwacht microstructure (left) and the smallest structure (right) detected to date: the bar in the first represents 20 μm, and in the second only 3 μm.

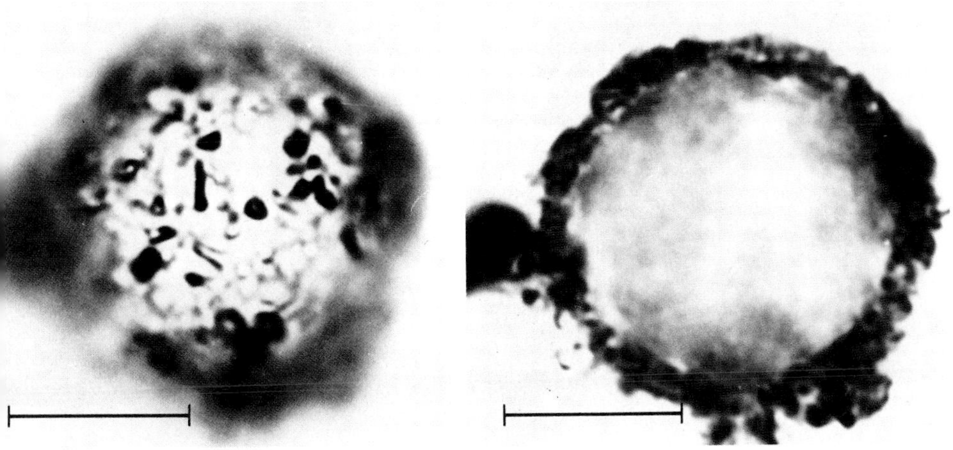

Fig. 6-6. Microstructural study method. On the opposite page, an overall view which shows the double microstructure with deep cavities and perforations in the outer wall; these perforations are common in many of the larger forms; the bar represents a length of 30 μm. Given the height of the object, note how only part of it is in focus: this observation lies at the basis of the technique of later optical sections. On this page: two pictures of the same object taken by focusing first the interior of the cavity (left) and then the outside (right). By taking a series of photos and moving the focus by a given amount, it is possible to make a detailed reconstruction of the form and characteristics of the object. The bar represents 10 μm.

Fig. 6-7. Left two pictures taken at Shark Bay in Australia. The rock fungi, shown in an overall view on the opposite page, and in greater detail on this page, are currently forming stromatolites. Stromatolites are curved, superposed laminae forming curious groups of varying shapes and sizes, found in calcareous rocks ranging from the Archeozoic period right up to the present day. The oldest formations of this type are the Bulawayo stromatolites, 2.9 billion years old. With regard to these, and other very old stromatolites present in many rock formations, there has been a great deal of discussion about whether they might be the product of the activity of organisms or of some type of physical or chemical phenomena. The Shark Bay stromatolites have clarified the mechanism whereby this structure is formed. A thin covering of blue algae, extending over part of the seashore which is only submerged at high tide, retains billions of tiny grains of sand in a sort of web when the tide ebbs. A new thin layer is added at each tide. The result is the lamina structure typical of the stromatolites which are closely connected with the activity of living organisms *(photo S. Barghoorn.)*

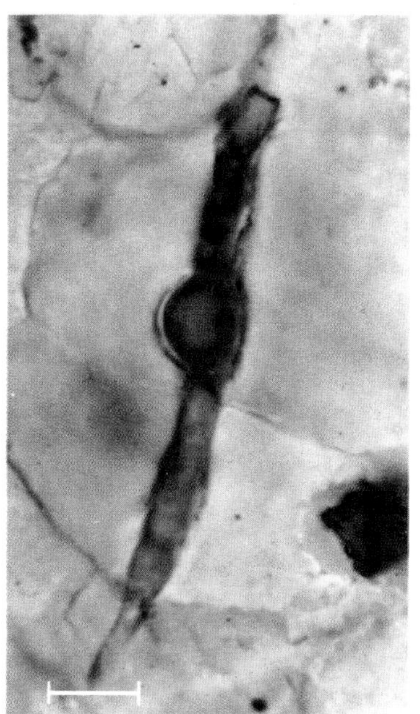

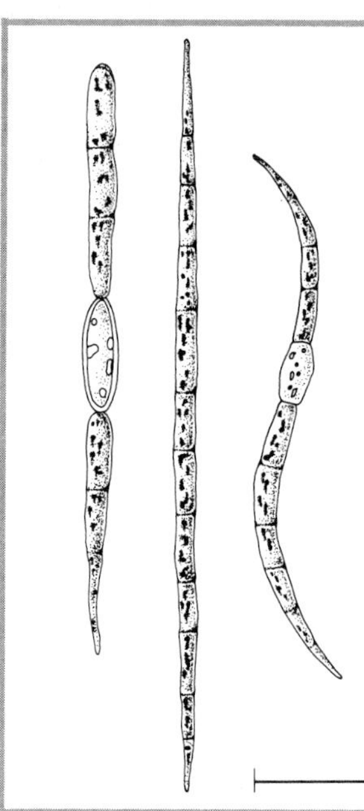

and survive under a wide range of adverse environments. Possibly this property made blue-green algae suitable candidates for early life on Earth.

The first group of blue-green algae to be recognized as a special class were blue-green in color, in fact they can be found in all the colors of the visible spectrum. Their differences in color are thought to be largely controlled by environmental factors. There are more than 2000 known species of blue-green algae which are to be found all over the world in both fresh and salt water. They are found as single cells or in random colonies held together by a gelatin-like mass; some occur in filaments with one cell following others like a string of beads. Others which are more developed are found with segmented filaments and with specialized cells. Two types of specialized cells which are comparable to those found in the Transvaal rocks are the akinetes and the heterocysts. Both are enlarged thick-walled cells but they differ in shape and function. Heterocysts are usually colorless or pale yellow (although in some tropical species they remain green) thick-walled cells arranged at regular intervals along the filaments. They may be found in the middle or at the ends of the filament and they are associated with nitrogen fixation. Akinetes, which are usually larger and longer than heterocysts, are pigmented germination cells which may often occur next to heterocysts; however, not all

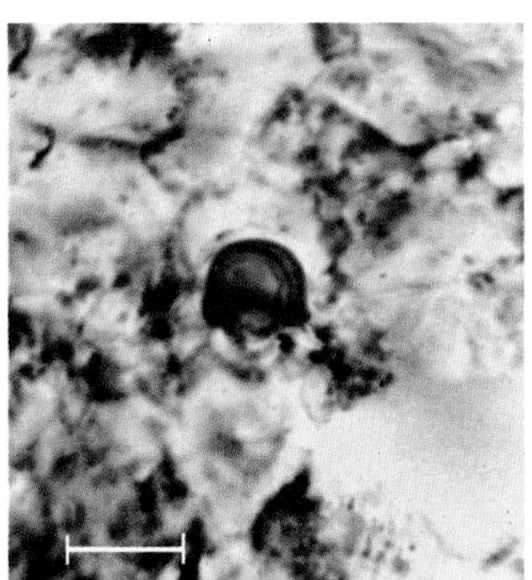

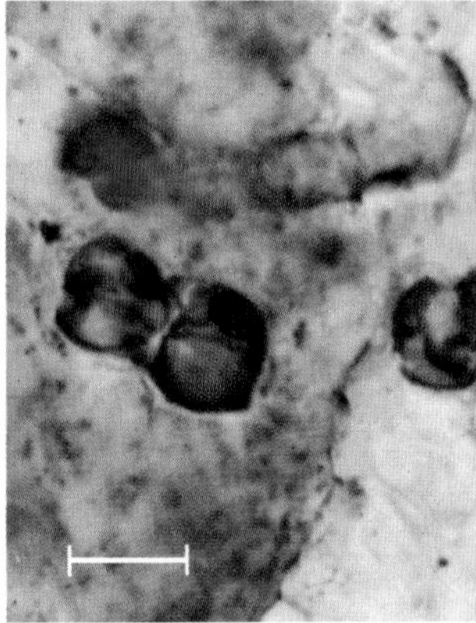

Fig. 6-8. Algal filaments on a thin cross-section of a sample of a stromatolitic sedimentary rock from the Transvaal Sequence (2.3 billion years old), which was taken 700 m above the bottom of the Transvaal Sequence (2.3 billion y.o.) is considered to be 2.2 billion years old. The bar represents 10 μm.

Fig. 6-9. Microfossils of isolated (left) or double (right) cells with double walls, present in thin sections of samples coming from sedimentary rocks of the Transvaal Sequence; their biogenic origin is now undisputed. The age of the samples is about 2.2 billion years. In both cases the bar represents 10 μm.

akinetes form next to heterocysts. There are some species of blue-green algae that form only akinetes, and have the ability to remain viable for long periods of time even under desiccated and somewhat elevated temperatures. Some species can live in Antarctic climates while others live in thermal springs at temperatures near or above 70°C; still others can be found in caves having only minimal lighting. Laboratory experiments using simulated primitive atmospheric conditions have shown that eight strains of blue-green algae can live under mildly reducing conditions. In some of these experiments the partial pressure of CO_2 was 1700 times greater than in the atmosphere today. Blue-green algae are relatively sensitive to the pH of the aqueous solutions in which they live. Various experiments have shown that blue-green algae can live in aqueous solutions having pH values between 4 and 9. There are other organisms which can tolerate even wider variations in pH. On the extreme acid side, *Sulfolobus* bacteria which are found in sulfur-rich hot springs live under pH conditions of 1.5; but on the other hand, the organism *Kakabekia* has been found living at a pH of 12.

Unfortunately, not a great deal is known about the process of fossilization, although a few laboratories have performed experiments in simulating "instant fossils." One such experiment was performed in our laboratory using the blue-green alga *Anabaenopsis*. A saturated solution of ferric chloride was allowed to penetrate among the cells and then the microscope slide was sealed. Fig. 6-11 shows photomicrographs of these algal cells 9 months later. Continued laboratory experiments may shed more light on the complicated process of fossilization.

Fig. 6-10. On the opposite page, the microfossil *Petraphera vivescenticula* (left) photographed in thin cross-section in a sample coming from the stromatolites of the Transvaal Sequence, with an attributed age of 2.3–2.2 billion years. This is one of the oldest microfossils recognized as such with any certainty. It is an algal filament and shows, among other things, akinetic-type specialized cells. It is clearly comparable, morphologically speaking, with the present-day blue alga *Raphidiopsis mediterranea* (right) in which there is a swollen cell which, as it happens, is an acineta. In both cases the bar represents a total length of 10 μm.

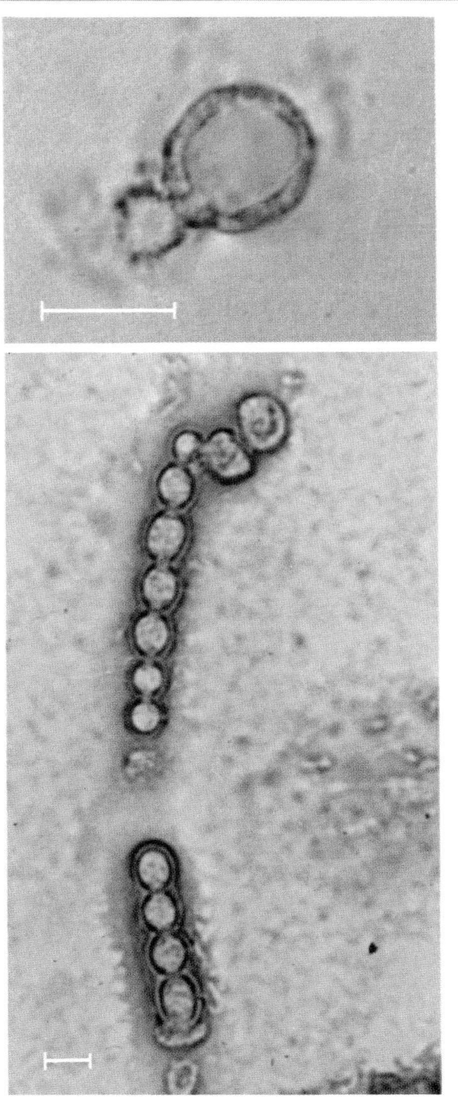

Fig. 6-11. Laboratory fossilization experiments, carried out to clarify the mechanisms of this phenomenon. The photos show two aspects of a present-day blue green alga, *Anabaenopsis*, nine months after being fossilized by cellular absorption of a ferric chloride solution. The bar represents 10 μm.

TABLE 6-1. BASIC STAGES OF BIOLOGICAL EVOLUTION IN THE PRECAMBRIAN PERIOD

Age (billions of years)	Event or Structure	Place of Origin
~3.7	oldest dated terrestrial rocks	western Greenland
~3.4	simple, round microstructures of uncertain origin	Onverwacht Series, eastern Transvaal (South Africa)
~3.4	oldest dated sedimentary rocks	Onverwacht Series, eastern Transvaal (South Africa)
3.0–2.7	oldest known stromatolite	Bulawayo Formation (southern Rhodesia)
~2.2	oldest diversified algal filaments	Transvaal Sequence (South Africa)
~1.9	presence of fossils considered beyond dispute	Gunflint Formation (Canada)
~1.2	cells with a nucleus	Beck Spring Formation (California)
0.9	first large association of fossil eukaryotic cells with indications of sexual reproduction	Bitter Spring Formation (Australia)

Summary

From the available geological, chemical and paleontological information it would seem there is some evidence, but no conclusive proof, that life existed on Earth as long as 3.4 billion years ago.

By 2.2 billion years ago blue-green algae were abundant and had already begun to diversify in their morphology. During later Precambrian periods the process of evolution can be well seen and has been documented in the fossil record.

The Precambrian period, which encompasses approximately two-thirds of the history of the Earth is a truly remarkable age of evolutionary progress. With further research one hopes to be able to define exactly when the first life on Earth arose.

BARTHOLOMEW NAGY AND LOIS ANNE NAGY

Bibliography: Mosser A. G., Brock T. D., *Bacterial origin of sulfuric acid in geothermal habitats,* in Science, CLXXIX, 480 (1973); Anhaeusser C. R., *The evolution of the early Precambrian crust of Southern Africa,* in Econ. Geol. Res. Unit, Univ. Witwatersrand, Info. Circ., 70 (1972); Burke K., Wilson J. T., *Is the African plate stationary?,* in Nature, CCXXXIX, **5372**, 387 (1972); Button A., *Algal stromatolites of the early proterozoic Wolkberg group, Transvaal Sequence,* in Econ. Geol. Res. Unit, Univ. Witwatersrand Info. Circ., **69**, 1 (1972); Hurley P. M., Pinson W. H., Nagy B., Teska T. M., *Ancient age of the Middle Marker horizon, Onverwacht group, Swaziland Sequence, South Africa,* in Earth Planet. Sci. Letters, XIV, 360 (1972); Moorbath S., O'Nions R. K., Pankhurst R. J., Gale N. H., *Further ribidium-strontium age determinations on the very early Precambrian rocks of the Gothaab District, W. Greenland,* in Nature, Physical Sci., CCXL, 78 (1972); Oehler D. Z., Schopf J. W., Kvenvolden K. A., *Carbon isotopic studies of organic matter in Precambrian rocks,* in Science, CLXXV, **4027**, 1246 (1972); Walter M. R., Bauld J., Brock T. D., *Siliceous algal bacterial stromatolites in hot springs and geyser effluents of Yellowstone National Park,* in Science, CLXXVIII, 402 (1972); Hunter D. R., *The granitic rocks of the Precambrian in Swaziland,* in Econ. Geol. Res. Unit. Univ. Witwatersrand Info. Circ., **63**, 1 (1971); Nagy L. A., *Ellipsoidal microstructures of narrow size range in the oldest known sediments on Earth,* in Grana, XI, **2**, 91 (1971); Schopf J. W., Oehler D. Z., Horodyski R. J., Kvenvolden K., *Biogenicity and significance of the oldest known stromatolites,* in J. Paleo., XLV, **3**, 477 (1971); Stoecker R. R., *Survival of blue-green Algae under primitive atmospheric conditions,* in Space Life Sci., III, 42 (1971); Nagy B., *Porosity and permeability of the early Precambrian Onverwacht chert; origin of hydrocarbon content,* in Geochim. et Cosmochim. Acta, XXXIV, 525 (1970); Schopf J. W., *Precambrian micro-organisms and evolutionary events prior to the origin of vascular plants,* in Biol. Rev., XLV, 319 (1970); Scott W. M., Modzeleski V. E., Nagy B., *Pyrolysis of early Precambrian Onverwacht organic matter ($> 3 \times 10^9$ yr. old),* in Nature, CCXXV, **5238**, 1129 (1970); Nagy B., Nagy L. A., *Early Precambrian Onverwacht microstructures: possibility the oldest fossil on Earth?,* in Nature, CCXXIII, **5212**, 1226 (1969); Echlin P., *Primitive photosynthetic organisms,* in: Hobson G. D. (ed.), *Advancements in organic geochemistry. Proceedings of 3rd International conference,* Oxford (1966); Desikachary T. V., *Cyanophyta,* New Delhi (1959).

IN southern Australia a rock formation is exposed which is of paramount importance in the reconstruction of the history of our planet, and of the evolutionary stages leading from the first simple life forms which, as we have seen in the article by the Nagys, appeared about 3.2 billion years ago, to those of the present day. It consists of common quartzitic sandstones, ancient sands which geological events have transformed into extremely hard rock.

Like so many other similar rock formations, this appeared to be of exclusively local interest until R. C. Spring led the whole world of geology to discuss these "Ediacara sandstones." In 1947 he was carrying out geological studies in the region when he observed a series of traces which, though of an unusual aspect, were unmistakeably fossils. The Austrian-born paleontologist Martin F. Glaessner, was working at the University of Adelaide, and he took an immediate interest in these strange fossil forms. In the late fifties Glaessner revisited the region several times and, together with his colleagues, succeeded in finding an incredible variety of such remains. It was impossible to assign a precise age to these rocks and hence to the fossils, but there was every reason to believe them to be more ancient than the paleozoic rocks in the surrounding region. Confirmation of this hypothesis soon followed since isolated fragments of similar fossils had been found in other parts of the world, and in all cases, these were in rocks whose age was around 700 million years. The difference lay in the fact that at Ediacara the fossil remains were not fragments but a very rich assortment, so varied in character as to permit far-raching studies. There was no doubt that these were animal remains, or rather impressions left in the mud of the sea bed by animals having neither a shell nor a skeleton—the most ancient animals known to us.

These were between 10,000 and 100,000 times larger than those first life forms that had appeared in southern Africa 2.5 billion years before, and they consisted not of a single cell but of millions of cells. However, to gain a complete understanding of the great qualitative leap forward made by evolution, we shall have to look more deeply into the body of the animals and blue-green algae which had preceded these later life forms. The later forms have highly complex cells, enclosed in a cell membrane, in which specific functions are entrusted to specialized organelles; the name given by biologists to this kind of cell is eukaryote. The cells of blue-green algae are on the other hand much simpler and much more primitive: they have no cell wall, the substance bearing hereditary characteristics is directly immersed in the cytoplasm, and they have no specialized organelles designed to serve as separate functional units. These are the so-called prokaryote cells.

Today this type of cell is still to be found in blue-green algae, whilst the basis of the whole of the rest of the plant and animal kingdom is the eukaryote cell. The passage from the former to the latter was thus a fundamental turning-point in evolution, a turning-point that was only reached after more than 2 billion years, half the life of the Earth.

How this passage took place is still no more than an hypothesis, because we have only the two extremes as evidence of it, with no intermediate stage to provide a link. The most interesting hypothesis seems to be that of the biologist Lynn Margulis, according to whom the prokaryote cell turned into a eukaryote cell by progressive association with other living structures which then became the organelles. To prove this, however, we would have to find the first eukaryote cell; and so far this has not happened.

MARTIN F. GLAESSNER

Martin F. Glaessner taught paleontology at the University of Adelaide, Australia where, until 1971, he was the Director of the Center for Precambrian Research. He was born in 1906 in Aussing (today known as Usti nad Labem, Czechoslovakia). He first specialized in paleontology in 1924, working in Vienna under the guidance of the great Austrian paleontologist, O. Abel, and in 1931 he received his Ph.D. His specific interest in the first animal forms goes back to the 1950s when, as a paleontologist at the University of Adelaide, he found the first fragments of the oldest association of animal fossils, the fauna of Ediacara.

MARTIN F. GLAESSNER

Precambrian Paleobiology: Middle and Late Precambrian Life and Environment

Overleaf: The fossil imprint of a jellyfish (Mawsonites) *12 cm in diameter, found in Upper Precambrian quartzites in southern Australia; the Ediacara fauna, which dates back 6–700 m.y., is the first extensive grouping of fossil animals known.*

For an understanding of the history of life during the time from the appearance of its earliest forms to the diversification which characterizes the Phanerozoic Eon, an evaluation of four distinct sets of facts is necessary:

1. The framework of measured geological time in which this sequence of events occurred.
2. The traces of Precambrian life which are recorded as fossils.
3. The distinct steps in evolution observed in the fossil record or required by theory.
4. The major changes in the environment of life which could have influenced its evolution during Precambrian time.

Time Scales

The events in the history of life which concern us here took place in the 2 billion years from about 2.5 billion years to about 570 million years. The time before 2.5 billion years is designated by geologists as the Early Precambrian (often also as Archean), the time from 2.5 to 1.8 or 1.6 billion years as the Middle Precambrian, from about 1.6 to about 570 million years as the Late Precambrian. The Cambrian Period is the time from about 570 to 500 million years. Periods in history, human as well as Earth history, are "real" in the sense that events during each of these time divisions had something in common that differed from the characteristics of the preceding and succeeding periods. Just as the concept of the Middle Ages, for example, is a realistic, sensible and useful one, so is the concept of the Middle Precambrian. It does not follow that the boundaries between periods can be determined by observation to infinitely more precise limits. Nobody is interested in the precise month, week, day or hour when the Middle Ages ended. No one should be unduly concerned about an uncertainty in the boundary between the Middle and Late Precambrian of some millions, or tens of millions of years. Admittedly, it would be better for mutual understanding of what we are talking about if all terms were more closely defined, if in this instance the uncertainty were less than 200 million years, and in the case of the beginning of the Cambrian less than 20 million years. Progress towards better measurements and more uniformity in the use of geological time terms is being made. We should not expect, however, a precise definition of each geological time boundary based on observation or measurements, because such a precise moment in the Earth's history could only be defined by an instantaneous, global catastrophe. Unlike the fathers of geological science 200 years ago we do not consider such catastrophes, if they did occur, as recurrent, standard time markers. Geological time is marked by a sequence of events in the evolution of the Earth and of life on Earth, as represented by fossil remains in rocks. A serious problem arises from the fact that during the Precambrian evolution did not produce the great diversity of life which only in later times left great numbers of different fossils in rocks of different ages. As long as it was not known that the Precambrian represented about 85% of documented geological time this did not seem to matter. The great length of Precambrian time during which much of the Earth's crust and its mineral resources were formed, requires us to answer this question: Can the standard method of dividing geological time by biological events (e.g. Paleozoic = Age of trilobites, Mesozoic = Age of dinosaurs, Cainozoic = Age of mammals) be continued back into Precambrian times? If so, how far back? If not, why not? These questions are of interest not only for the geologist who needs a comprehensive, reliable and practical time scale but also for the biologist who wants to trace the evolution of life as far back and as continuously as it can be documented.

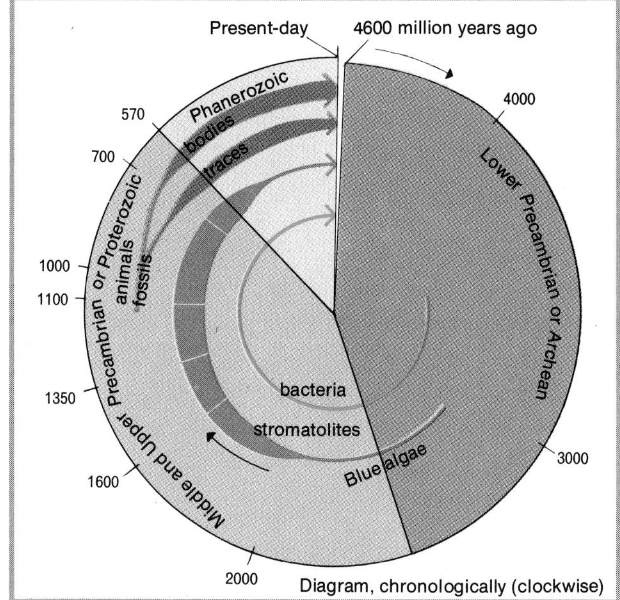

Fig. 7-1. Geological evolutionary period: animal life only appeared in the last quarter of the Earth's history, and only became abundant in the Phanerozoic period which covers the last 570 million years; but bacteria and blue algae are among the earliest forms of life and traces of them are to be found in the oldest sedimentary rocks; there are also traces of stromatolites, which were certainly biogenic.

The Fossil Record of the Middle and Late Precambrian

Fossils are the only documents which can prove the reality of life history during distant periods of time. Only a small proportion of living organisms has a significant chance of being fossilized. The bias operates against organisms living far from areas of sedimentation, against animals which are entirely soft-bodied rather than shell-bearing or provided with skeletons, and against organisms embedded in older rocks which were affected more frequently, or for longer periods, by destructive forces of heat and pressure from recurrent Earth movements. We know now that these forces have not entirely destroyed the fossil record of Precambrian time. It is in fact surprisingly rich in remains of primitive aquatic plants. What is surprising, after the discovery of organic remains in rocks of Early and Middle Precambrian age is the fact that they all represent the most primitive stage of the organization of the living cell, the prokaryote stage. At this stage there is no nuclear membrane, the hereditary material of the cell (DNA) is spread through the cytoplasm, and its other functional units are not structurally separated as they are in the form of organelles (chloroplasts, mitochondria and complex flagella) in the eukaryotic cells. All bacteria and the blue-green algae (Cyanophyta) are prokaryotes. They are the only organisms represented by fossil remains in the oldest sedimentary rocks and through probably 2 billion years of Early and Middle Precambrian time. This does not mean that during that time there was no evolutionary progress. We see mostly the external shape of these organisms. Hidden behind the simple exterior are probably quite considerable internal and functional changes. There is clear evidence, however, that evolutionary progress in the direction of diversification was slow during a period almost four times as long as Phanerozoic time. If the earliest phase of evolution during the Precambrian was distinctly slow, the rates of evolution at the close of this long time span were remarkably fast, if judged primarily from the fossil record known at present. Before we can examine the relation between changes in the rates of evolution and changes in the environment of life which could have caused this speeding up, we must briefly survey the fossil record itself.

The first abundant assemblage of fossils occurs in the Gunflint Iron Formation of North America. It was found by S. A. Tyler and described by Elso Barghoorn, Preston Cloud, and others. It is between 1.6 and 2 billion years old. According to William Schopf it comprises predominantly prokaryotic microorganisms including blue-green algae (Cyanophyta) some of which belong to genera which are still living, and also chemosynthetic bacteria. No eukaryotic organisms have been identified with certainty in this assemblage or in other similar ones of similar age from the Biwabik Iron Formation of Minnesota, the Belcher Group of the Hudson Bay area of Canada, or the Pretoria Group of South Africa. This does not mean that the nucleated cells could not have existed at that time. It is not yet technically possible to identify with certainty dark bodies observed in some fossilized cells, as remains of the nucleus and to distinguish them from similar looking condensed accumulations of decayed protoplasm. It is clear, however, that a comparison of these early assemblages of fossils with later ones, of Late Precambrian age, shows that at this time a greater diversity of aquatic plant life existed, including undoubted evidence of eukaryote organization of plant cells. The first of these diversified assemblages was discovered by Elso Barghoorn and William Schopf in cherts in the Bitter Springs limestone of Central Australia. Other discoveries were made in South Australia and Queensland, in California, Montana, Michigan, Finland, Sweden, Scotland, northern France, and elsewhere, in sediments 1.3 billion to about 700 million years old. The 30 species of microscopic plants described from the Bitter Springs cherts include not only blue-green algae but also green and possibly red algae, dinophyceans and other microorganisms. A number of sections of these cells could be arranged in a series which strongly suggests successive stages of mitotic cell division. This can occur only in eukaryotes. It seems likely that tetrads of spores suggesting meiosis and sexual reproduction are also present in these fossil microfloras. These discoveries seem to prove that eukaryote cell organization and sexual reproduction originated more than 1 and less than 2 billion years ago. Before considering the importance of these evolutionary steps for the evaluation of evolutionary rates we have to refer to the geological importance of the dominance of blue-green algae and associated bacteria during the Middle and much of Late Precambrian time. Studies of calcareous sedimentation in some shallow areas of the present

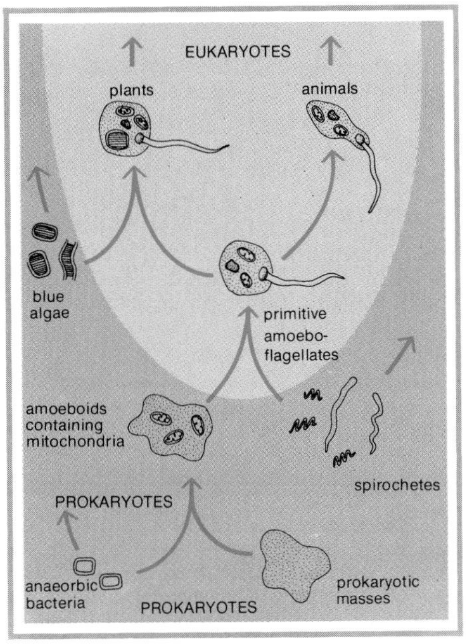

EUKARYOTES
plants
animals
blue algae
primitive amoebo-flagellates
amoeboids containing mitochondria
spirochetes
PROKARYOTES
anaeorbic bacteria
PROKARYOTES
prokaryotic masses

Fig. 7-2. Evolution of prokaryotic cells in eukaryotic cells in the symbiotic theory proposed by L. Margulis; in fact the only organisms conserved in Lower Precambrian rocks and Middle Precambrian rocks are prokaryotes: the emergence of eukaryotes which appeared about 1 billion years ago, i.e. 2 billion years after the prokaryotes, was probably connected with phenomena of symbiosis ***(from L. Margulis).***

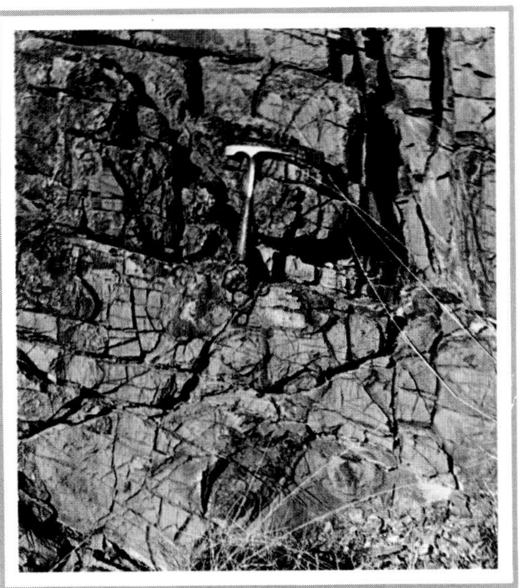

Fig. 7-3. The surface of a stratum of the Bitter Springs Limestone Formation in central Australia. The darker concretions are small masses of flint which contain fossil algae remains belonging to some thirty species, dating back about one billion years.

Fig. 7-4. Stromatolites: specific structures with a definite biogenic origin, the formation of which is today linked with the daily and seasonal variations in the activities of algae and bacteria. Left, Bulman Precambrian stromatolites in northern Australia, 1.25 billion years old *(photo Haddon King)*. Right: stromatolites in the Bitter Springs Limestone Formation which are about 1 billion years old. *(photo M. R. Walter)*.

seas and their margins have shown the importance of living blue-green algae and bacteria for the formation of calcium carbonate layers. This occurs partly through precipitation of this substance in the process of respiration which removes carbon dioxide from the water, and partly through entrapment of calcareous particles in the gelatinous sheath surrounding the algal filaments. Depending on the composition of the organic community and on local conditions, the calcareous layers produced by the daily and seasonally varying life activities of the algae take various shapes: separate ovoid nodules of various sizes, continuous wrinkled layers, or cylindrical, conical or branching structures. These layered structures in sedimentary rocks were named stromatolites before their origin was understood. As the study of Precambrian and early Paleozoic lime-

Fig. 7-5. A false fossil which, after in-depth analysis, turns out to be of physical origin: these are the imprints of dried mud found on the stratifications of a quartzite in central Australia, probably 1 billion or so years old. It is obvious that a great deal of caution is needed when analyzing the oldest fossil forms.

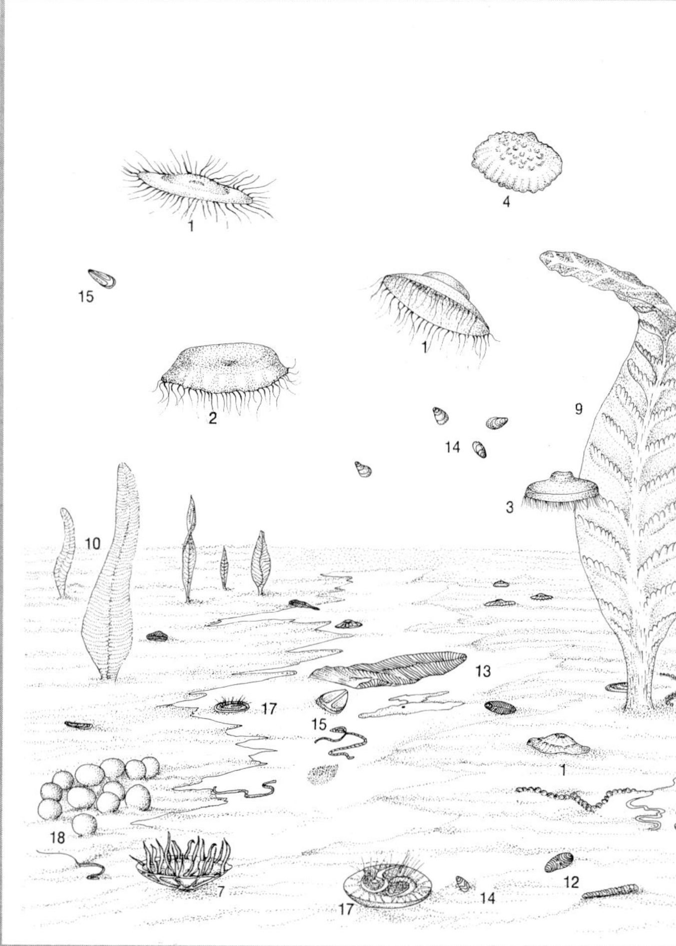

Fig. 7-6. On this page and opposite: the Ediacara fauna of southern Australia. This page, an attempt to reconstruct the way the environment must have looked as it developed. The diagram shows in detail: 1–6, various forms of jellyfish; 7, *Conomedusites;* 8–10, *Pennatulacea;* 11, *Spriggina;* 12–14, *Dickinsonia (Annelida);* 15, *Parvancorina (Crustacea);* 16, *Precambridium (Trilobitomorpha* or *Chelicerata);* 17, *Tribrachidium;* 18, *Algae* (?). Opposite: top left: *Cyclomedusa* which has been found not only at Ediacara but also in strata of the same age in the southern U.S.S.R. (diameter, 8 cm); top right: the imprints of two organisms, *Dickinsonia,* a primitive worm, on the left of the photo, and *Parvancorina,* probably a primitive crustacean, in the box at bottom right; below left; *Spriggina,* an extinct worm 5 cm long; below right: *Tribrachidium,* an extinct organism with a diameter of about 2.5 cm.

stones progressed it became clear that many of these rocks were in fact stromatolitic. We know now that this is a geological expression of the biological fact that blue-green algae and bacteria were the dominant forms of life during Precambrian time. Their decline in early Paleozoic time was puzzling until it was realized that growing Algal mats, wherever they can still be found, provide food for grazing molluscs which evolved in Paleozoic time. Conditions of high salt concentration discourage the molluscs, but are tolerable for algae. In hypersaline lagoons on the west coast of Australia they survived to form structures which are very similar to the ancient columnar stromatolites. In the Gunflint and Bitter Springs Formations the algae described by E. Baghoorn and J. W. Schopf were found in cherts which are closely associated with stromatolites. The algae probably participated in their construction.

No animal remains have been found associated with stromatolites. No fossil animals occur in any rocks as old as the Gunflint or the Bitter Springs Formations. We shall see that this does not mean that animals could not have existed 2 or 1 billion years ago. Many specimens of what were claimed to be animal fossils in rocks of such great age have been shown, particularly through the critical efforts of Preston Cloud, to be either younger or not of animal origin. What remains are possibly fossil sponge spicules in rocks from northern Australia (about 1.5 billion years old) and a few possible traces of animals burrowing in sediments from the Grand Canyon of Arizona, the Beltian rocks in Montana, and the Vindhyans of India. All these possible traces of animal life in rocks 1–1.2 billion years old still need careful verification.

The first abundant and unquestionable assemblage of Precambrian animals was found in quartzitic sandstones at Ediacara in South Australia by R. C. Sprigg in 1947 and later detailed studies were made by the author and his colleagues at the University of Adelaide from 1957 to the present time. The precise age of the rocks containing this fauna cannot be determined directly. Various components of this assemblage have been found in other countries where the rocks containing them could be dated approximately as ranging from about 600 to 700 million years old. The composition of this fossil fauna is of very great interest. It is dominated by cnidarian coelenterates representing medusae, hydrozoans, and probably soft corals resembling living pennatulids. These were probably sessile and anchored to the sea floor, while the hydrozoans and scyphozoans were floating. There were also at least two kinds of segmented worms which could crawl and swim, and traces of grazing and burrowing wormlike animals, probably sediment feeders, are common. The most advanced and complex metazoans were primitive arthropods, one resembling primitive trilobites or merostomes remotely like the living horseshoe crab, the other a primitive crustacean remotely like the living *Triops*. Some of these animals—medusae, soft corals, worms—were large, up to 30 or 40 cm in diameter or length. None seemed to be predators adapted to feeding on large particles of food. Most significantly, none had mineralized tissues such as the calcareous skeletons of later corals or the calcified external shells of crustaceans or molluscs. As long as there was only one known occurrence of this peculiar fauna it could be claimed that it was the result of accidental preservation of soft-bodied animals due to local conditions. Elsewhere very different animals, possibly shell-bearing, might have existed. However, other discoveries have disproved this idea. One of the distinctive pennatulid-like forms was found in very different fine-grained rocks (hornfels) at Charnwood Forest in Leicestershire, England, by Trevor Ford, together with medusa-like impressions, and named "*Charnia*." A very similar fossil was found in dolomitic rocks in northern Siberia. Medusae specifically identical with those from Ediacara were discovered in argillaceous strata in southern Russia and a trilobite—or merostome-like organism, *Vendia* is preserved in a core of siltstone from northern Russia. Hydrozoa, medusae and *Charnia*-like imprints occur in large numbers in tuffaceous rocks, probably deposited in deep

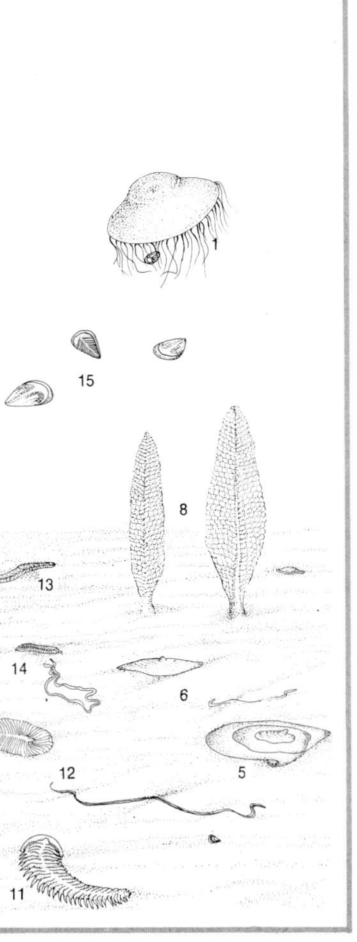

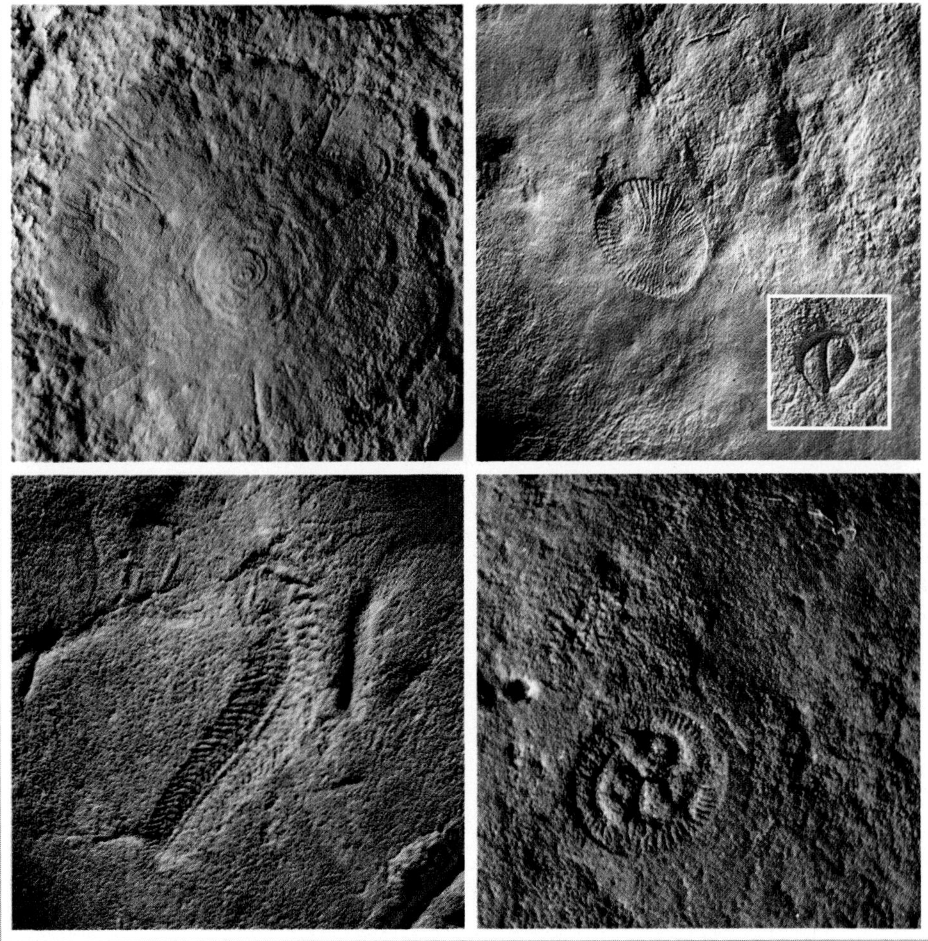

water, in eastern Newfoundland. Obviously, the conditions under which these animals lived were very different at these localities, and yet they display the most significant features of the Ediacaran fauna. Therefore this fauna must be considered as representative of the world of animals of Late Precambrian time and for the stage of evolution of animals before the beginning of the Cambrian Period with which the Paleozoic Era commences in the traditional geological time-scale.

The richest and first described Late Precambrian fauna is that of the Nama Group of southwest Africa. It was originally described by G. Gürich in 1930, later by Rudolf Richter in Frankfurt, and recently by H. D. Pflug in Giessen and Gerald Germs in Cape Town. Most of these fossils were found in massive quartzites. They resemble the pennatulid-like forms of the Ediacara fauna where the African genus *Pteridinium* also occurs. The work of Pflug and Germs has shown that other forms, while generally similar, are of much greater complexity than can be observed, because of the three-dimensional preservation of the African fossils which were embedded in soft and partly silicified rock, while most of the Ediacara fossils are two-dimensional impressions on bedding planes. On the other hand the three-dimensional preservation is connected with severe alteration of the rocks and the interpretation of original biological structures is difficult and contentious. Pflug believes these organisms to be so complex as to be beyond the coelenterate grade, foreshadowing most of the structural plans of higher metazoans, but he also finds microscopic structures resembling plant cells in their surface layers. He is therefore inclined to accept the dubious hypothesis of Sir Alistair Hardy who once expressed the view that metazoans could have evolved from higher plants. Obviously, the compound leaf-shaped "Petalonamae" of southwest Africa are a special element of the Nama and Ediacara faunas which require much further study. Medusae and, according to a personal communication from G. Germs, segmented worms, also occur in the Nama fauna. Limestones found by Germs in the Nama Group contain what may be the oldest fossils with shells. They are small, organic and calcareous, originally flexible tubes, which he correctly ascribed to worms and which he named *Cloudina*.

Steps in Precambrian Evolution

These new observations on the distinctive Late Precambrian level of animal evolution have to be evaluated from the viewpoint of the general evolution of the metazoans. How do the new data fit the picture of subsequent evolution based on the

TABLE 7-1. COMPOSITION OF THE EDIACARA FAUNA

Type and Relative Percentual Incidence	Orders	No. of Genera	No. of Species
Coelenterata (67%)	Medusae	6	8
	Hydrozoa chondrophora	3	3
	Conulata	1	1
	Scyphozoa	3	4
	Anthozoa pennatulacea	3	4
Annelida polychaeta (25%)		2	7
Arthropoda (5%)	Trilobitomorpha (or Chelicerata)	1	1
	Crustacea branchiopoda	1	1
Tribrachidium (3%)		1	1
traces			8
Total		21	38

abundance of shelly fossils in Cambrian and younger Phamerozoic rocks? How do they fit phylogenetic hypotheses based on the study of the living fauna which, although far from metazoan origins, can provide much more detailed morphological, embryological and genetic information on its evolution than the fossil record?

It is known that most major elements of the present marine fauna existed in Ordovician time, about 4.5 billion years ago. Since then there have been some major extinctions and some substitutions such as the replacement of most brachipods by molluscs, of trilobites by crabs and lobsters, and of shell-bearing cephalopods by naked squids and octopus, but few substantial additions. This means that an essential modernization of the marine fauna was achieved in about 100 million years after the beginning of the Cambrian when the marine fauna

Fig. 7-7. Imprint of *Charnia*, possibly a colonial coelenterate with no skeleton, found at Charnwood Forest in Great Britain; similar fossils have been found in Precambrian rocks in Newfoundland, northern Siberia and Australia; some of these rocks have been dated and found to be about 680 million years old *(from material supplied by T. D. Ford, Univ. of Leicester)*.

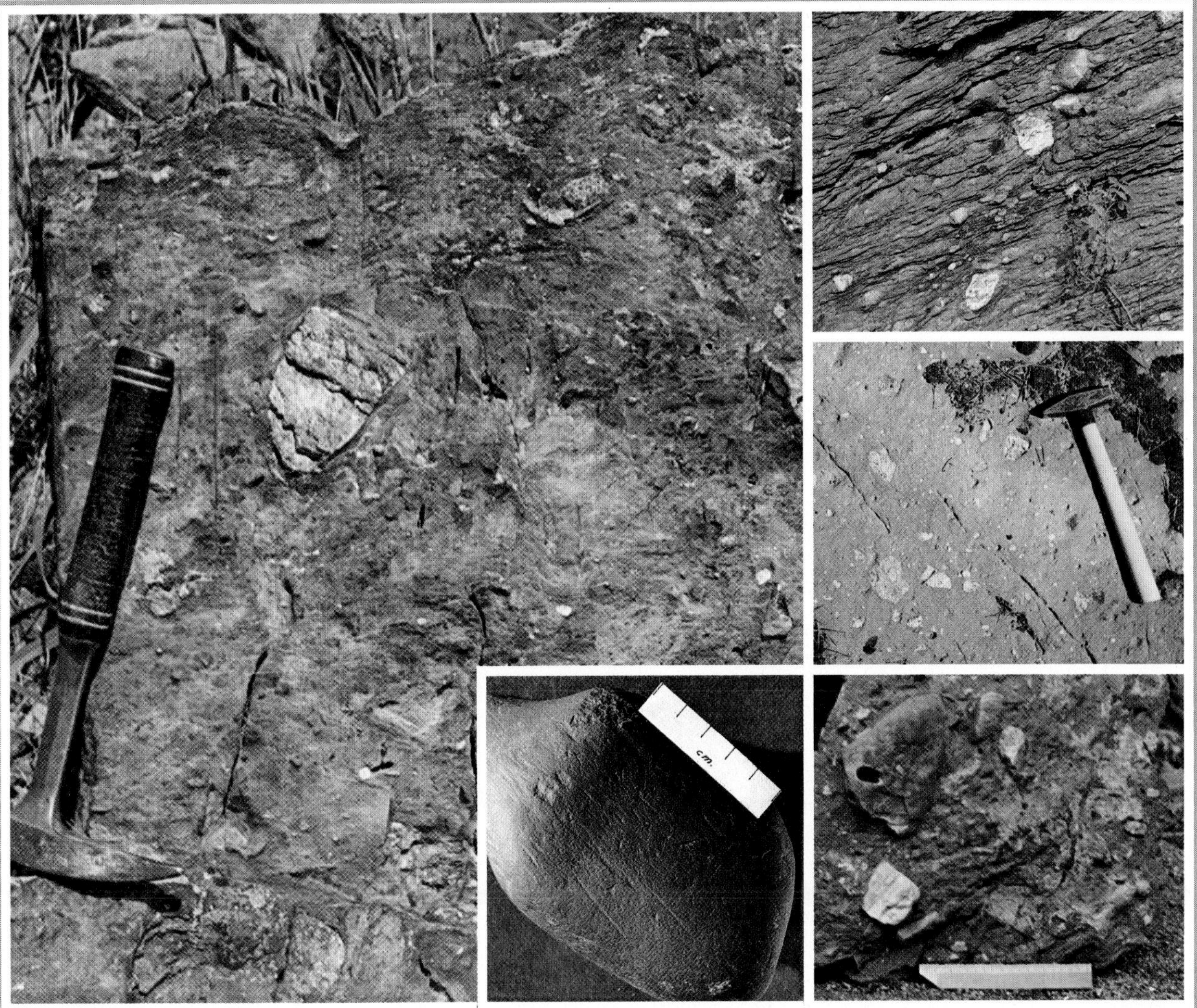

Fig. 7-8. Upper Precambrian glacial deposits. In the large photo, the Precambrian tillite of Sturt Gorge near Adelaide in southern Australia and in the inset a typically glacial striated pebble found in it. Right: top, the Precambrian tillite of the Windermere Group in the Selkirk Mountains in Canada; center, the Precambrian tillite of Lake Mjosa in southern Norway; bottom, the alta tillite from northern Norway. The Upper Precambrian glacial deposits are found all over the world and for this reason suggest glacial expansions somewhat larger in size and longer-lasting than those occurring in the Quaternary period.

was still very different from all subsequent faunas: Archeocyatha, which soon became extinct, brachiopods of which only very few had calcareous shells, worms with and without hard tubes, and molluscs to which probably the inhabitants of the conical hyolithid shells were related. Early crustaceans are known and soon trilobites with thin phosphatic shells appeared, but their great morphological variety suggest strongly that they must have existed earlier, possibly without hardened and therefore preservable exoskeletons. This assumption is supported by traces of burrowing in many basal Cambrian sediments resembling the later traces of activities of trilobites. Brachiopods with thin chitinous coverings could have existed earlier. Mollusca without shells still exist today and complex trails in basal Cambrian strata were probably made by naked molluscs rather than by worms. In the first few millions or tens of millions of years of Cambrian time the ability of metazoans to produce mineralized tissues evolved. Their biological and evolutionary significance varied considerably. It would be anthropocentric thinking if we assumed that all shells were simply there for protection, responses to the sudden appearance of hypothetical, threatening predators. It would also be incorrect to assume that the first shells appeared in many unrelated groups at one moment of time which could be conveniently designated the beginning of the Cambrian. There are good reasons for assuming that some elements of the Ediacara fauna had chitinous "shells," while others had calcareous or horny spicules. These could and did become fully mineralized later. Russian workers have documented the gradual appearance of very rare tubular hard shells in the youngest Precambrian strata, followed by rich "shelly" faunas in the basal Cambrian and by faunas with trilobites later. The abundant worm tubes in some beds in the Nama Group are probably the oldest known "shells." Similar but smaller ones of the same type, once wrongly thought to be aberrant Archeocyatha, are equally common in the Lower Cambrian of Siberia.

The evolutionary conclusions to be drawn from the composition of the Late Precambrian fauna, compared with what came before and after, are clear enough. Segmented worms were already highly differentiated, some were able to build tubular shells. They must have had a long history. Similarly, cnidarian coelenterates were of many kinds. some still living, others extinct at various times and probably primitive. This indicates a long history of this phylum as well. On the other hand, the Arthropoda of the Ediacara fauna are few, not highly differentiated, resembling primitive or larval living forms. Hence they are thought not to have originated much earlier. Mollusca are represented in the Late Precambrian only by trails, apart from a few specimens of hyolithids in the transitional beds to the Cambrian. This suggests that their most primitive forms had no shells. No more can be said about them.

All students of the evolution of the Metazoa agree that the family tree of the invertebrates, above the level of coelenterates and sponges, has two main branches. The phyla just mentioned are on one branch, together with the Platyhelminthes, Nemertini and other wormlike forms, many of which are parasitic and totally unknown as fossils. The other branch which leads to the vertebrates and man includes the echinoderms and the peculiar Pogonophora, the tubes of which occur in the Late Precambrian and Cambrian. The echinoderms are considerably differentiated when they first appear in the Lower Cambrian. Specialists in the study of this group agree that they must have had direct ancestors in the Precambrian but they don't make guesses about their possible morphology and appearance. It has been suggested that the unique *Tribrachidium* could be related to primitive echinoderms but they all possessed calcareous plates and *Tribrachidium* had none. There is no firm evidence either for or against this suggested relationship. It would be wiser to wait for further discoveries than to speculate on phylogeny while our knowledge remains obviously very incomplete. This does not mean that phylogenetic speculation on the evolution of the lower Metazoa is generally and totally useless and valueless. Some of the data on which it is based come from comparative morphology, others from studies of embryology and ontogenetic development of living animals. These data, summarized in 1968 by the Russian zoologist Ivanov, indicate clearly that the earliest Metazoa would have been small, without resistant cell walls or skeletons, and therefore incapable of being preserved as fossils in any kind of sediment. Should we then believe that the definite paleontological record which goes back about 6.5 billion years is true and complete and that the animals originated about that time? Or should we conclude that for valid zoological reasons we cannot expect their early history to be recorded, and accept the hypotheses, based on the high degree of differentiation of coelenterates and annelid worms in the Ediacara fauna (and on the possibility of earlier existence of sponges) that animals originated much earlier in the geological history of the Earth, at least 1 and possibly 2 billion years ago? The choice between these hypothesis must be influenced by studies of the environment of life during Middle and Late Precambrian time, its changes as far as they are actually recorded in the rocks, and the environmental requirements for animal life.

Main Changes in the Environment

Three major factors of the environment are known whose changes during the relevant interval of time could have made it suitable for animal life. We need only consider the sea as a habitat, since we know that the major expansion of the biosphere to the land took place not earlier than in mid-Paleozoic time. In Precambrian time only some algae and possibly mosses are believed to have existed on land. The uniformity of sedimentary rocks suggests that the composition of sea water is not likely to have changed fundamentally since the end of the early Precambrian. The main changes affecting life in Precambrian time concerned, first, the composition of the atmosphere which exchanges its components freely with the ocean; second, the climate which at certain times became glacial over large areas; and third, the configuration of continents and ocean basins. It is generally accepted that the oxygen content of the atmosphere increased during Precambrian time. A theory proposed by the physicists Berkner and Marshall explains the process but does not determine the rate of oxygen accession. The authors of the theory turned therefore to the history of life as it was known in the 1950s to give fixed points in the curve of oxygen accession. One was the time of the appearance of life on land, which depended on the fact that at an oxygen level of 10% of the present level most of the dangerous ultraviolet radiation is absorbed in the atmosphere before reaching sea level. This makes life on land possible. The other point was chosen on grounds which are no longer valid. It was thought that animal life on a significant scale started at the beginning of Cambrian time which was then dated at about 600 million years. Animals are characterized by their dependence on respiration. Microorganisms change from fermentation to oxydative respiration when the oxygen level reaches 1% of the present atmospheric level (Pasteur point). At that level, dangerous ultraviolet radiation penetrates sea water only to a depth of about 10 m, making life in shallow water possible. As we now know, differentiated metazoans existed before the Cambrian and could live in shallow water, hence the Pasteur point must have been reached before the beginning of the Cambrian.

The source of oxygen accession was the photosynthesis in plants. Blue-green algae which, incidentally, are very resistant to ultraviolet radiation, and chemical compounds probably related to the existence of chlorophyll, are found in rocks of Early Precambrian age, hence photosynthesis is likely to have commenced yielding oxygen at least 3 billion years ago. The biologist Lynn Margulis has developed a theory according to which the origin of the eukaryote cell with its nucleic membrane and organelles was not by evolutionary differentiation of the prokaryote cell but partly by symbiotic incorporation into it of other microorganisms such as aerobic bacteria (to form mitochondria), spirochetes (to form flagella) and blue-green algae (to form chloroplasts): With the formation of the eukaryote cells, oxygenic respiration became possible. From the viewpoint of biochemistry there need not have been a long interval between the Bitter Springs flora and the first air-breathing organisms. Metazoan animals such as coelenterates could have appeared within the next 100–200 million years, i.e. 800 million years or more before the present. The development of miotic cell division and sexual reproduction with its spreading of genetic variability and rapid response to selection pressures, seems to have been the main stimulus to the increase in rates of evolution, which appear to have been slow before sex was "invented." Alfred G. Fischer has presented in a diagram three modes of oxygen accession. Model 1, assuming gradual accession from the first appearance of photosynthetic plants conflicts with geological observations on the state of oxydation of detrital minerals in ancient rocks. Model 3 is the classical and probably faulty Berkner-Marshall view. Model 2 seems to be in best agreement with the present data.

The second environmental factor is that of widespread glaciation in Precambrian time. To obtain a picture of its possible influence on life we must distinguish between Middle Precambrian (2.3–1.7 billion years) and Late Precambrian (780–660 million years) glacial epochs. They are documented by sedimentary rocks closely resembling those produced by the influence of glaciers or icebergs in a continental or, more frequently, in a marine environment. The most distinctive glacigene rocks are often (but somewhat imprecisely) called tillites. There is no clear global picture yet of the Middle Precambrian glaciation, though tillites about 2 billion years old from North America and South Africa have been known for almost 70 years. The Late Precambrian glaciation, named Eocambrian, Varangian or Laplandian, is well documented by characteristic tillite occurrences and other geological signs on all continents except Antarctica. It is clear from its time span that it was not an event comparable to the Pleistocene glaciation of the last million years.

If the position of the poles for Late Precambrian time as determined by paleo-magnetic methods is accepted as correct, the most spectacular occurrence of tillites and glacial land forms in Scandinavia and Australia would have been close to the equator at that time, as Brian Harland pointed out in 1964. This would, in his opinion, indicate a major, worldwide, glacial epoch. During its time span of over 100 million years there were several glaciations resulting in the deposition of two or three separate tillites, hundreds of meters thick and separated by thousands of meters of other, interglacial sediments. Harland and Rudwick have pointed out far-reaching effects that could be expected from such a period of extreme climatic fluc-

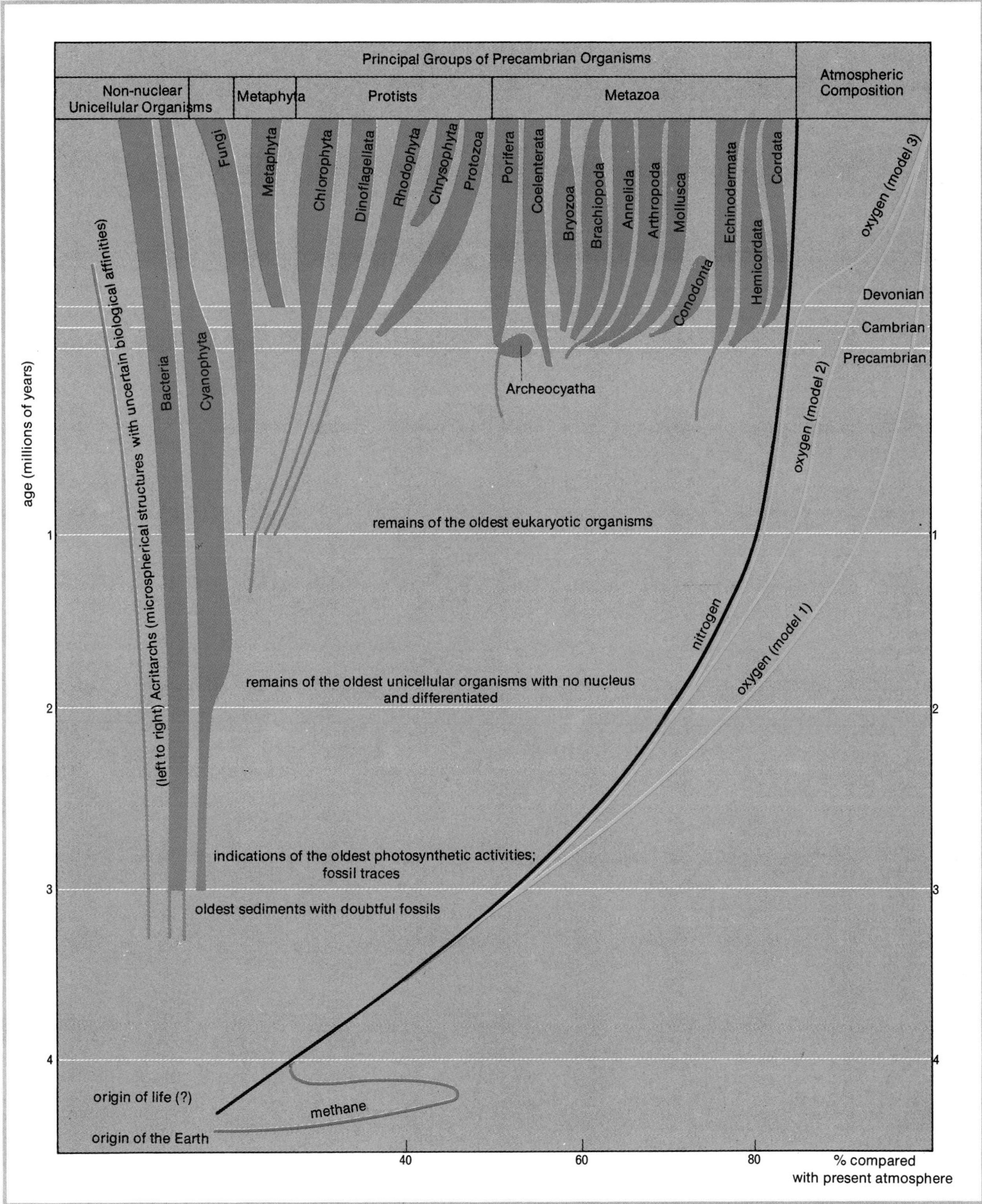

Fig. 7-9. Progressive evolution in the composition of the Precambrian atmosphere towards compositions similar to those of today, and stages in the evolution of organisms. The increase of atmospheric oxygen corresponds to three different models, the most likely of which is model 2. In horizontal coordinates (abscissae) the percentage of atmospheric gases; in ordinates, the times; the last 1 billion years has been doubled in extension *(from A. G. Fisher, 1972)*.

tuation. They believe that the end of this period provided a "trigger mechanism" which set in motion an expansion of the habitable marine realm over shelf areas, through the rise of sea level caused by the melting of vast glaciers and though an improvement of the world climate. This triggered off assumed rapid evolution leading to the supposedly sudden appearance of the Cambrian fauna. Such arguments, though they are attractive at first glance and possibly contain an element of truth, fail to account for the 50–100 million years time lag between the last Precambrian glaciation and the beginning of the Cambrian, and for the fact that subsequent glaciations did not produce similar effects. The glaciation was followed by and locally contemporaneous with the deposition of sediments containing the Ediacara fauna which is now known to have been widespread. This faunal expansion could have benefited from an amelioration of the world climate. Much more has to be learned about the effect of glacial climate on life in the ocean before the hypothesis of its "trigger-effect" can be fully evaluated.

Other geographical factors must also be considered as possibly influencing the diversification and spread of marine faunas. Admittedly, we have no clear picture of the distribution and size of continents and oceans in Late Precambrian time. Many geologists believe that this was one of the periods in Earth history when only one supercontinent existed. This subsequently broke up into several pieces and the transgressions of the Cambrian seas followed. The consequences of these changes were considered by two American investigators, the biologist Valentine and the geologist Moore, from the viewpoint of the influence of global tectonic events on marine ecology, food resources and evolutionary strategy in the development of adaptations enabling the fauna to cope with such changes.

The picture drawn by them, of fluctuating resources under conditions of supercontinent, and stabilization of resources concomitant with transgressions such as that at the beginning of the Cambrian could, in their opinion, account for the differences between the fauna of the Late Precambrian and that of the Early Cambrian.

Clearly, there is still much to be learned, by experiment, observation and exploration, before the complex interactions of a changing environment and changing forms of life in the distant past can be understood. Much progress has been made. As usual, new discoveries have led to new questions. More questions have been asked than have been answered but they will lead forward to new discoveries.

MARTIN F. GLAESSNER

Bibliography: Schopf J. W., *On the development of Metaphytes and Metazoans*, in J. Paleont., **47** (1973); Fischer A. G., *Atmosphere and the evolution of life. Main currents in modern thought*, **28** (1972); Pflug H. D., *Zur Fauna der Nama-Schichten in Südwest-Afrika*, in Paleontographica Abt. A, **134, 135, 139** (1970–72); Barghoorn E. S., *I fossili più antichi*, in Le Scienze (1971); Margulis L., *Origin of eukaryotic cells*, Yale (1970); Schopf J. W., *Precambrian micro-organisms and evolutionary events prior to the origin of vascular plants*, in Biol. Reviews, **45** (1970); Cloud P. E., *Pre-metazoan evolution and the origin of the Metazoa*, in: Drake E. T. (ed.), *Evolution and environment*, Yale (1968); Termier H., Termier G., *Biologie et écologie des premiers fossiles*, Parigi (1968); Harland W. B., Rudwick M. J. S., *The great infra-cambrian ice age*, in Scientific American, **211** (1964); Glaessner M. F., *Pre-cambrian animals*, in Scientific American, **204** (1961).

THE interval between the beginning of the geological history of Earth (3.8 to 4 billion years ago, as we saw in the article by Moorbath) and the start of the most ancient era considered by geology until a few decades ago, the Paleozoic (5.7 billion years ago) is important to humanity not only because it was the period in which fundamental biological events occurred such as the birth of life and the evolution of the cell, but also because man owes to the geological phenomena that took place in this period the supply of raw material on which he has built his empire: 90 % of the world's iron supplies are to be found in deposits belonging to this early period of the Earth's history, the Precambrian or Archean. Much of our gold and copper deposits are also linked to phenomena which occurred in this era. The same can be said of nickel, titanium, vanadium, chromium, platinum and a whole array of other metals and rare-earth elements. Even our future supplies of energy are linked to Precambrian events since the vast majority of uranium deposits are to be found in Precambrian rocks.

The formation of deposits of ore is thus a salient feature of this period of Earth's history. What is the reason for this special richness of Precambrian formations? It must be taken into account that this era was five times as long as all the later eras put together; it lasted about 3.2 billion years while all the others add up to about 600 million. Is the richness of its deposits due to its longer duration or to the existence of particular conditions that favored formation of great masses of mineral deposits? This theme, on which opinions are still largely divided, is developed in the article by G. Gross which follows. The fact remains, in any case, that the formation of a mineral deposit depends above all on the environmental conditions, whether or not these are of a special nature. The Precambrian deposits thus bear extremely interesting witness to the conditions that existed in the remotest past of our planet, and a study of them can be a most interesting guide to an understanding of the characteristics of the atmosphere and of the oceans at the time in which life first developed and evolved.

The history of organisms and of the environment appear to be closely linked. First of all the atmosphere: its oxygen content, now at a level of 21%, appears to be almost totally a product of the activity of organisms capable of carrying out photosynthesis by means of chlorophyll. Before these organisms became active the amount of oxygen in the air must have been negligible, even if the oxidized minerals found in extremely ancient sedimentary rocks now suggests that the increase in oxygen content was not constant in time, but went through a series of high and low levels. As regards the sea, it is very hard to find convincing evidence as to its composition in the past, or as to the variations which may have taken place. However, if we are to judge by the types of minerals that have been deposited on the sea bed, though changes must have taken place, its composition may not have been substantially different from that of the present day: at most there may have been twice as much of certain components as there is today. Here too the substantial difference appears to be due to organisms which learned to fix mineral substances present in the water to form their shells or skeletons; this took place only after the beginning of the Paleozoic. Thus certain components which in later periods were limited by organisms—for example silica, extracted by Radiolaria, Diatomaceae, sponges and other creatures with siliceous skeletons—remained in a free state in the water until they reached maturation and precipitated to form deposits. Chemically deposited flints are particularly abundant in the Precambrian era.

The fundamental factors in Precambrian metallogenesis do not, however, appear to be substantially different from present-day ones and they were, as they are today, directly linked to zones of activity in the Earth's crust, to which considerable space is dedicated in the following article.

GORDON A. GROSS

Geochemist with the Geological Survey of Canada, and an expert on the origins of metals with the Commonwealth Geological Liaison, headquartered in London. Born in Goderich, Ontario, Canada, in 1923, he received his Ph.D. in geology in 1955, and then went on to study the origins of mineral deposits. As a United Nations expert, and as a member of the Geological Survey of Canada (where he directed the division on the geology of mineral deposits), he evaluated the world's iron resources, acquiring a detailed knowledge of this subject. This led him to concentrate entirely on the Precambrian era, to which 90% of the ore deposits belong.

GORDON A. GROSS

Metallogenesis and Precambrian Environment

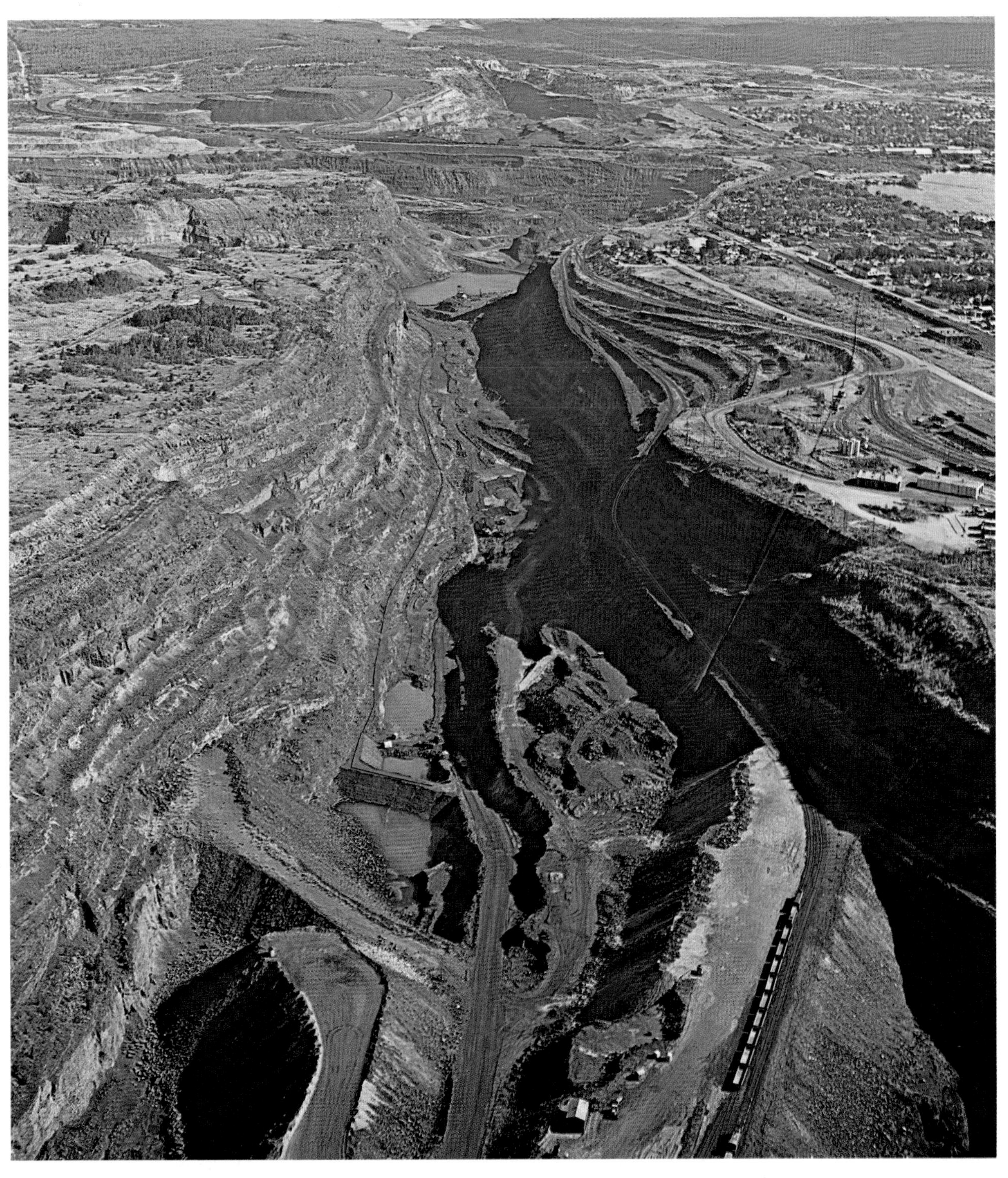

Overleaf: The deep pit of an open-cast iron mine in Minnesota. This mine, like 90% of the world's iron deposits, has opened up Precambrian strata, i.e. between 600 million and 3 billion years old. (Marka).

The age of Earth's crust is calculated to be about 4.5 billion years. Geological events and processes during the first quarter of this period of time cannot be established in a definite chronological sequence or even clearly defined, because metamorphism and alteration of the rocks has obliterated evidence of their previous history. Less than one-seventh of the Earth's history took place after the Precambrian when organic activity flourished in the seas and evolved rapidly on land. The period of time in the Precambrian considered here is about 3 billion years, a time in which the geological record is incomplete and there are long periods where no record or evidence exists of events or environmental conditions.

Mineral deposits comprising a concentration of elements in the crust in a form and location where they are usable by man are an integral part and product of the geological processes which have created and shaped the Earth. Concentrations of useful metallic minerals are not formed by exotic or sporadic diversions of natural phenomena. They occur as a predictable result and product of geological processes or the combined effects of interrelated processes which are recorded or discernible from evidence in the rocks. The processes of origin and formation of mineral deposits are referred to as metallogenesis. A very large proportion of the resources of metallic minerals in the world occur in Precambrian rocks. About 90% of the iron ore resources, and a very large proportion of the resources of other metals such as uranium, gold, copper, nickel, titanium, vanadium, zinc, platinum and the rare earth elements were formed in Precambrian time. Because of the abundance and diversity of mineral resources in these old rocks there has been intensive study of Precambrian geology and consideration given to possible unique features in the geological environment that influenced metallogenesis during this long period of the Earth's history.

Contrary to the opinions held by many geologists in the past, it is very doubtful whether there are any kinds of mineral deposits that are strictly unique Precambrian phenomena, and which have no specific comparable type in younger rocks. The banded chert iron-formation sediments are usually cited as being unique to the Precambrian, but this kind of iron-rich sediment with many similar lithological types of beds occurs in all ages of rock. A contemporary example of the formation of a banded siliceous iron and manganese mud is found in deep basins on the floor of the Red Sea. But the banded chert iron-formations which occur in continental shelf environments apparently have only been preserved in rocks of middle to late Precambrian age.

There are a large variety of mineral deposits which are most abundant in Precambrian rocks and therefore provide a basis for believing that metallogenesis and environmental factors were different or distinctive in the Precambrian. The outstanding examples of these are the chert iron-formations, the manganese-iron formations, and the stratiform base-metal sulphide deposits with gold associated, which are all related to volcanic rocks and processes. Uranium minerals with or without associated gold in ancient conglomerates and sandstones, and extensive occurrences of sandstones and black shales impregnated with copper minerals, form an important part of the world's copper, uranium and gold resources. Extensive masses of anorthosite rock containing ilmenite and titanium-bearing magnetite, differentiated masses of mafic and ultramafic rock with segregations of chromium, titanium, nickel, vanadium and platinum minerals, nickel and copper deposits in mafic to ultramafic rocks, and a number of other important kinds of mineral deposits provide major mineral resources that formed in the Precambrian.

This diverse group of mineral occurrences may be more prevalent in Precambrian rocks because conditions for their preservation were more favorable within or on the borders of the tectonically stable shield areas. They represent an accumulation of the relics of mineral belts built up over a very large part of the Earth's history.

Much of the Precambrian terrain and known shield areas represent deeply eroded parts of the Earth's crust where geological phenomena which originated at considerable depth in the crust are now exposed near the surface.

Most of the rocks which formed at or near the interface of the Precambrian land and its oceans or atmosphere are now preserved in structure infolded, or as relicts surrounded by masses of coarsely crystalline metamorphosed rock. Nearly all of the Precambrian rocks have been recrystallized or altered through metamorphic processes at least once since they were formed, and their ages as determined by the study of isotopes reflect the time of metaphorphism and not their time of origin.

In spite of these factors that are prevalent in Precambrian geology, much can be deduced about the environment and its effects on metallogeny during the early history of the Earth. Before considering the metallogeny of some conspicuous kinds of mineral deposits in the Precambrian, some consideration is given to concepts and ideas about environmental factors during this time.

Precambrian Climate

The belief is widely held that the primitive atmosphere of the Earth was deficient in oxygen and that the present atmosphere has evolved through a long complex history. Four stages in the evolution of the atmosphere have been postulated starting with primordial gases that were later lost, exhalations from the molten surface, additions of gases from volcanic activity and the addition of oxygen by plant life. It is believed that oxygen was not abundant until organisms were capable of a photosynthesis process and thereby released oxygen in quantity to the atmosphere and oceans.

Oxygen was obviously prevalent in the atmosphere in the period of the Precambrian considered here as some of the oldest sedimentary rocks which show little evidence of metamorphism have minerals in a highly oxidized state. A fluctuation in the content of oxygen in the atmosphere has been postulated considering that it may have been released from CO_2, and with the accumulation of carbon from organisms, oxygen would again become fixed with carbon and participate in another cycle of release and fixation. Vast amounts of oxygen may have been absorbed in sea water and through both inorganic and organic reactions participated in the cycle of fixation with carbon and with other elements such as sulphur, iron and mineral compounds. Speculation about the fluctuation of oxygen content in the atmosphere and oceans has played an important part in developing metallogenic concepts for the concentration of metallic elements in Precambrian sediments.

It is very difficult to demonstrate that the composition of sea water has varied greatly throughout recorded geological time. Undoubtedly, its composition has varied with respect to some elements or groups of elements, but these fluctuations are believed to have been of brief duration. Perhaps the content of any one element has not deviated by more than twice its present amount in the oceans. No doubt with a stable balance and adjustment of elemental content in the seas any areas where important additions took place either through sedimentary processes or volcanic exhalations would also be the location for precipitation or fixation of these elements in sedimentary beds. The evaporite or salt deposits provide useful evidence about the composition of the seas through geological time. Ancient evaporite deposits are not markedly different in composition to more recent ones and it is believed that salts from many early beds have been dissolved and recycled through the ocean reservoir. Because chert rocks may be more abundant in the Phanerozoic, a difference in the composition of the early and later seas with respect to silica and calcium is suggested. These differences may be attributed in part to the role of organisms and their part in controlling the composition of sea water.

Organic activity has been a dominant factor in controlling the composition of the atmosphere and of sea water. The content of oxygen and carbon dioxide is directly related to the role of organisms and biogenic processes. Through the release of oxygen in phötosynthesis, organisms have created an environment in which advanced forms of life could evolve. The oxygen and carbon dioxide content of sea water along with its inter-relationship with the atmosphere determine the oxidation potential of ocean environments, and hence the kind of minerals that will be precipitated. Organisms play a direct part in the deposition of certain elements such as silica, calcium, iron, manganese, copper and others, either through selective precipitation of them or through the secretion of elements in the formation of their shells and tests. Silica may have been more abundant in Precambrian seas if silica secreting organisms were not present at that time to extract it from the water. Without organic extraction it probably became concentrated to the point where it precipitated inorganically as amorphous silica gels. The deposition of calcium in sediments today is controlled by both the physical chemical environment and by lime-secreting organisms.

Fig. 8-1. The Lac Jeannine iron mine in northern Quebec. The hematite is extracted from a ferriferous formation which, like 90% of the world's ferriferous reserves, is Precambrian. The Lac Jeannine deposit is part of a large sedimentary ferriferous formation which extends uninterrupted for more than 1500 km along the western edge of the Labrador geosyncline, in Labrador itself and in Quebec. The ferriferous formation was formed by chemical precipitation, in the form of an iron-rich silicic ooze, on the continental shelf of 1.8 billion years ago, which consolidated and then recrystallized during the Grenville orogenesis 900 million years ago.

Fig. 8-2. Two photos of the iron mine near Schefferville in northern Quebec. This is a ferriferous formation of the Lake Superior type which contains rich deposits of hematite and goethite; these formations are associated with sedimentation in rough, shallow waters.

The fossil remains of primitive single cell organisms have been found in the oldest known Precambrian sediments and the occurrence of stomatolites in the oldest dolomite rocks indicates multicellular species and colonies of organisms. There is no doubt that organisms existed as far back in antiquity as the decipherable geological record, and that organic activity became more abundant in the Late Precambrian with a rapid evolution of complex forms in the Phanerozoic. We do not know the precise part that early primitive organisms played in the deposition of metallic elements in the Precambrian, although there has been much speculation on their probable role. To understand their part in metallogeny it is necessary to show that specific kinds of organisms had the ability to collect and deposit selectively particular elements or to influence the chemistry of their environment, and this has not been done conclusively in the case of Precambrian organisms.

Life in the Precambrian was apparently confined to aqueous environments and we picture barren land areas where rocks eroded, topography developed and sedimentation was dominated by physical factors and inorganic chemical processes. Aside from the influence of vegetation and organic activity today it is likely that environments on the Earth varied from

Fig. 8-3. A highly metamorphic and markedly deformed ferriferous formation made up of hematite and coarse quartz; this photo was taken near Mount Wright in northern Quebec.

Fig. 8-4. Lake Superior type ferriferous formation, carbonate and silicate levels with iron and flint, and fossils of very old microorganisms, part of the Gunflint Formation.

Fig. 8-5. In the Schefferville case the ferriferous formation has undergone a selective leaching of the silica from phreatic water; the iron content of the deposit has thus become steadily richer as a result of the gradual but constant withdrawal of part of the silica.

place to place in the Precambrian as much as they do today, with cold polar regions and hot equatorial belts, arid deserts and regions with moderate to intense precipitations. Glaciation took place at several different times in the Precambrian and is recorded in all of the major shield areas.

Mineral Deposits in the Precambrian

Prominent kinds of mineral deposits are considered separately to illustrate environmental factors in metallogenesis. Much discussion and controversy has related to the banded chert iron sediments which occur in various different geological environments and are most commonly of Precambrian age. These rocks called iron-formation contain 15 % or more iron, and consist of thinly bedded or laminated chert and iron minerals, magnetite, hemanite, goethite, siderite and the iron-silicate and iron sulphide minerals. They are the most abundant of all chemically precipitated sediments and are classified in four main facies groups—oxide, silicate, carbonate or sulphide depending on the kinds of minerals which predominate in the facies—and these reflect the physical-chemical environment in which sediment was deposited. The immediate

Fig. 8-6. Jasper facies with hematite in Lake Superior type ferriferous formation of Labrador syncline near Schefferville: indicating the proximity of the hematitic mineral.

Fig. 8-7. Oxidized facies of the ferriferous formation with flints and hematite in the Labrador syncline. Lower right corner, a small calamite which is attracted by the rock.

Fig. 8-8. Ferriferous formation with quartz, magnetite and siderite of the Algoma type in the Wawa zone of the Algoma district in Ontario.

Fig. 8-9. Ferriferous formation (containing magnetite and quartz), extremely metamorphic, of the Algoma type emerging near Nakina in Ontario.

environment for deposition of the different facies which make up an iron-formation unit was influenced by the following factors—pH, varying from acid to alkaline, Eh, varying from oxidizing to reducing, the concentration and number of chemical components in the water in the depositional basin, and the possible influence of organisms. The iron-formation rocks are distinctive in that they contain very little clastic or fragmental material but are almost exclusively chemically or organically precipitated silica and iron minerals, but locally they may contain facies with clastic material, or have shale, tuff or other material of volcanic origin interlayered with them.

Most of the iron-formations occur in two main kinds of depositional environment with different kinds of associated rocks. The Lake Superior type forms prominent iron ranges of middle to late Precambrian age in nearly all shielded areas of the world and is associated with quartzites, dolomite, black shale, and frequently material of volcanic origin, mostly tuff and volcanic debris. They formed in continental shelf and shallow water environments along the borders, and in embayments of the ancient continental areas. Iron formations of this type are hundreds of meters thick and may extend for more than 1000 km in length in one continuous formation. The beds have features indicative of shallow water deposition, may have prominent granular or oolitic textures, and show marked evidence of a high-energy environment with disturbance of the sediment by current and wave action. Others, such as some found in Australia, formed in very quiet water and a low-energy environment.

The second prominent group of iron formations referred to as Algoma type are widely distributed in the volcanic-sedimentary rock belts of the very old Archean parts of Precambrian shield areas. They are also found in very similar geological environments in the Phanerozoic and a modern example of this kind may be forming today in the deep basins of the Red Sea and along the rifts and tectonic ridges of the ocean floor. They are characteristically thin-bedded or laminated with interlayered bands of ferruginous gray or jasper chert with hematite and magnetite, associated with siderite and carbonate beds and iron-silicate and iron-sulphide mineral facies. Single iron-formation members of this type range from more than a hundred meters to less than a meter in thickness and usually extend from a few kilometers to tens of kilometers in length. Those iron-formations are intimately associated with various volcanic rocks, including pillowed andesites, tuffs, pyroclastic rocks or phyrolitic flows and with graywacke, gray green slates, carbonaceous shales and ferruginous cherts. All of these different lithological types of rock may be interbedded in the iron-formation forming very heterogeneous stratigraphic successions. This group of iron-formations is closely related in time and space to volcanic activity and to centers of volcanism. The sulphide and carbonate facies occur at or near the centers of volcanism, and the oxide facies are usually distributed farther away, and may be almost entirely enclosed by clastic sedimentary beds. These beds of graphitic schist, derived from black carbon-rich mudstones, occur with Algoma type iron-formations and are more common in stratigraphic successions where volcanic rocks are more abundant than graywacke sediments. The fine-grained clastic material in the black schists may be derived from tuff and volcanic ash which collected in depressions in the sedimentary basin. They usually contain pyrite and pyrhotite and some parts of them are rich in lead, zinc, copper, gold and silver.

The parts of the black schists and fine-grained clastic sedimentary beds which form the stratiform base metal sulphide deposits are closely associated with the iron-formations, particularly with the cherty carbonate facies material. They may occur in the same bed or stratigraphic member as the iron-formations in the volcanic-sedimentary rock sequence or as separate beds and lenses.

Fig. 8-10. Another view of the ferriferous formation with magnetite and quartz, highly metamorphosed and deformed, near Nakina in Ontario.

Other kinds of mineral deposits closely associated with the iron-formation and having a similar metallogenesis are the chert manganese and siliceous manganese-iron-formations. They occur in geological environments similar to the Algoma type iron-formation and in some cases form manganese rich facies in these sediments. The segregation of manganese and iron facies seems to have been more marked and distinctive in the Precambrian than in more recent deposits. Important gold deposits in the Precambrian are commonly associated with iron-formation and its carbonate and sulphide facies, and although this fact of environmental association has long been recognized, the genetic processes by which the gold has been concentrated are not well defined. Gold is found in association with iron-formations in three principal ways: disseminated in the massive iron, copper, zinc, silver sulphide deposits that occur in or near volcanic necks and vents, disseminated in the iron-formations and stratiform sulphide beds with the carbonate facies being the prevalent host rocks; and as more conspicuous impregnations or disseminated in the coarse-grained quartz veins and stockworks which formed in shear zones, faults, fracture and structurally deformed zones which disrupt the sequences of volcanic and sedimentary rocks. It is now believed that the gold was originally deposited in the iron-formation and associated sediments, and in some cases it was redistributed later in veins and fracture zones by the action of hydrothermal solutions. Occurrences of gold disseminated in stratiform deposits composed of siliceous carbonate and sulphide facies of iron-formation, as well as the vein-type deposits are now well recognized. Gold deposits of this kind and association are not to be confused with the ancient placer gold deposits such as those found in South Africa where fragments of gold were concentrated by sedimentary processes through the action of streams, and the gold-bearing sandstone rocks were later metamorphosed to quartzites.

The Algoma-type iron-formations, the manganese-iron-formations, the stratiform iron-copper-zinc-sulphide deposits, the gold deposits associated with them and massive sulphide in volcanic necks and vents, all occur in belts of volcanic and graywacke type sedimentary rocks. They form a predominant metallogenetic group of mineral deposits in the Precambrian and are all products of volcanic processes. The metallic elements are believed to have been concentrated deep in the earth in the magmas which fed or supplied the volcanic rock or from rocks in the magma source area. During crystallization and differentiation of the magmas, aqueous solutions, gases and emanations were produced which contained metallic elements and chloride, sulphide and carbonate salts of these elements. These metal-bearing solutions were discharged from their volcanic source as exhalations and fumarolic gases into the seas which covered the volcanic belts. Some of the elements were precipitated along the walls and channels of the volcanic vents through which the solutions rose, replacing the enclosing rock and forming massive metallic-sulphide bodies.

More of the exhalations carried metals from the magma and from the rocks through which they passed and discharged the metals into the overlying seas. They were later precipitated as ferruginous siliceous oozes and in the black muds that now form the iron-formations and stratified sulphide deposits.

This volcanogenic mode of formation for the Algoma type of iron-formations and their associated manganese sulphide and gold deposits is now generally accepted and the concepts related to volcanism and exhalative processes have been used very successfully in the exploration and search for ore deposits in both Precambrian and younger terrain.

A volcanogenic origin for the Lake Superior type of iron-formation is still doubted by many geologists because the continental shelf environments in which this iron-formation is

Fig. 8-11. A heterogeneous facies of the Algoma type ferriferous formation at Timigami in Ontario; this facies consists of a tight series of interstratifications of levels with magnetite and quartz with graywackes and chloritic schists derived from volcanic material.

Fig. 8-12. Quartz veins crossing an Algoma type ferriferous formation, sediments and volcanic rocks, in northern Ontario. These are the typical veins which contain important amounts of gold.

found seem to have been separated from or only remotely related to tectonic belts of active volcanism. An important question is whether the iron and silica components of the Algoma and Lake Superior type of iron formation came from the same kind of source but were transported and deposited in different environments, or whether these constituents came from different sources, volcanic exhalations or by weathering of a land mass.

Implications of the volcanogenic hypothesis are examined

Fig. 8-13. A compact network of gold-bearing veins filling the fissures of an Algoma type ferriferous formation interstratified with graywackes and chloritic schists, emerging in northern Ontario.

There are a few iron-formations, such as the one in the Rapitan formation in the Yukon Territory, Canada which occur in thick conglomerate beds which may fill depressions and grabens formed along fault or rift zones. All of the iron-formation environments indicate voluminous deposition of silica and iron-minerals with almost complete exclusion of clastic sedimentary material. The silica-iron precipitation has been superimposed over many different and varied sedimentary and volcanic-sedimentary environments. There appears to have been one direct and major source of iron and silica that was not controlled or influenced by the immediate sedimentary environment. Volcanic processes may well have provided such an independent source for large quantities of the iron and silica. It is realized, however, that modern occurrences of siliceous iron and manganese muds are not as extensive as the Lake Superior type iron-formations in Precambrian basins.

Volcanic activity in the Precambrian most likely was located along the interconnected ocean ridges on the sea floor, the rift fault systems of the continents, and the island arcs, in much the same way as it is today. The deep seated structural dislocations in the Earth's crust appear at the surface in both the sea beds and continental areas and the nature of the volcanic activity developed along these deep-seated tectonic belts will reflect the character of the continental or oceanic crust in which it developed. Where rift faults intersected, continental masses in the Precambrian and a separation of the continental plates developed, the associated volcanic activity may have supplied an abundance of silica and iron-rich emanations or exhalations for the precipitation of iron-formation. Where precipitation of these constituents occurred close to the cen-

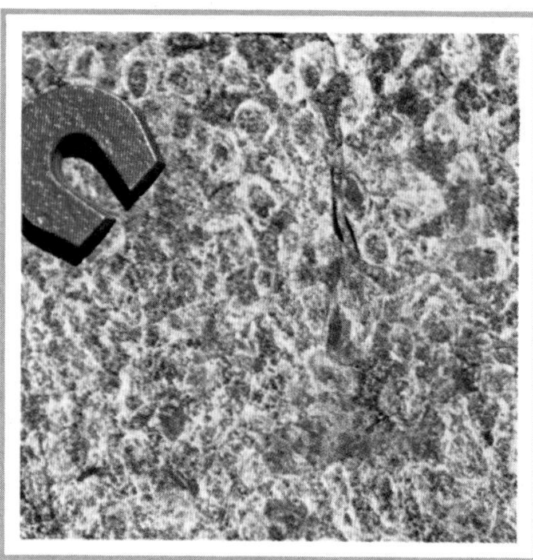

Fig. 8-14. Unidentified thimble-shaped fossil imprints, present in large numbers in a flinty Lake Superior type ferriferous formation emerging in northern Minnesota. They were probably formed by colonies of algae or similar organisms.

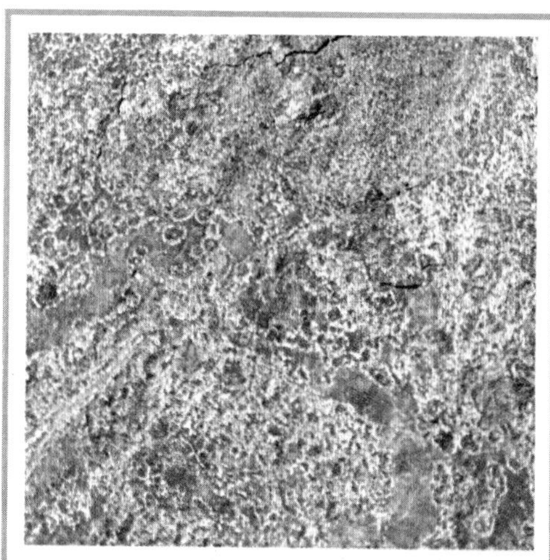

Fig. 8-15. Spheroidal forms emphasized by grains of pyrite in the flinty formation with siderite, pyrite and pyrrhotite of the Wawa Algoma type: found in rocks more than 2 billion years old, they may represent organisms which secreted sulphur or iron sulphide.

first. Since more detail about the geological environments of iron-formations has been obtained in recent years, it is realized that all of the chert iron sediments do not occur in the distinctive environments of the Algoma and Lake Superior types. Many of the rock successions composed of iron-formation, mature quartzite and dolomite sediments also have considerable volcanic rock associated with them, and there is evidence of contemporary volcanism at the time of their deposition. On the other hand, iron-formations with many of the characteristics of the Lake Superior type are found associated with rocks more typical of the Algoma type environment.

ters of volcanism we can visualize the development of Algoma type iron-formation and the associated metallogenic group of deposits. Where the sea water charged with silica and iron circulated over the continental shelf and into embayments on the shore line of the continent, precipitation of a Lake Superior type iron-formation would be expected.

The silica-iron muds in the Red Sea or the hematite-jasper iron-formation in the much older (Precambrian-Phanerozoic) Rapitan conglomerate occur in environments where volcanic activity is not well developed or defined. In these deposits the iron and silica must have been contributed from hot solutions rising and circulating in the adjacent fault zones which form part of major rift systems.

Seas filling narrow depressions or rift valleys which cut across a continent, and flooding large areas of the land which was already reduced to low relief by erosion, could easily transport and precipitate the iron and silica from solutions originating along the rift faults. The environmental conditions required for deposition of Lake Superior type iron-formation seem to be satisfied reasonably well under these conditions.

Fig. 8-16. Layers of intraformational breccia in the Yukon flint and hematite ferriferous formation. The breccias and other structures of the rock indicate a deposit in a turbulent sedimentary environment.

Silica and iron would be derived from a volcanic rather than a sedimentary source in this situation and would be deposited with normal mature quartzite and dolomite rocks in a continental shelf environment.

Many geologists have advocated a sedimentary source for the iron and silica components of the iron-formation and two schools of thought on the origin of these rocks have developed in this century. One objection offered to the volcanogenic model is that volcanic exhalations cannot provide the vast quantities of iron and silica deposited in the iron-formation and that these must have been derived from a land mass by deep chemical weathering processes. Recognizing the size of the silica-iron mud deposits forming today in deep-sea basins along rift faults, and the size and extent of many Algoma-type formations, it is not unreasonable to project this genetic concept to the larger iron-formations of the Precambrian.

Hypothesis for a sedimentary source for components of the iron-formation include the following general environmental conditions for the formation of the Lake Superior type beds: a land mass with low relief where chemical leaching of constituents predominates over mechanical erosion of the land mass, seasonal fluctuation in the proportion of the iron and silica transported in the streams, and discharge of the stream with precipitation of the constituents from solution into a shallow closed basin of the sea where clastic sediment carried by currents and tides was restricted from the basin by off-shore sand bars or barriers. Separation of the iron and silica to form thin bands in the rock of different composition was thought to be caused by the transportation of silica in alkaline ground water during the dry season of a tropical monsoon climate.

Fig. 8-17. Interstratification of a ferriferous formation with magnetite and quartz, with layers of quartz and feldspar probably derived from liparitic tuff, typical of the Algoma type ferriferous formations which are quite frequent in the Canadian Shield.

With the flushing of alkaline silica-bearing groundwater by heavy monsoon rains, the more acidic rain water entering the soil would dissolve iron and transport it either as ferrous iron carbonate or as colloidal particles of ferric iron. Such a model requires a very delicate balance and a special combination of environmental factors, and conditions would have to prevail for a very long time if the iron-formations were to form under these conditions. Iron and silica beds are not forming today where comparable environmental conditions are found.

Fig. 8-18. Detail of a sedimentary layer of copper in Zambia: bornite and calcocite are scattered between strata of Precambrian sandstone. The copper probably comes from volcanic sources; carried along by volcanic flows and submerged in the sea, it is flung into the sandy and muddy sediments which cover the edges of fairly large local sedimentary basins.

Even where dense tropical forest covers land areas of low relief, and there is an abundance of organic acids or compounds in the soil and streams to facilitate the solution and transportation of the iron, recent sedimentary beds comparable to iron-formation are not found. It is doubtful whether there was enough activity in the drainage and ground water in Precambrian time to have a marked effect on the solution and transportation of iron. If organisms played an important part in the transportation of the constituents of the iron-formation it is expected that these rocks might be more prominent in recent times than in the Precambrian.

Because iron in the ferric or more highly oxidized state has a very low solubility but is much more soluble in the ferrous or more reduced state, it has been difficult to understand how such large quantities were taken into solution and transported to form iron-formations, and many of these formations have more ferric than ferrous iron minerals. Organisms may have been effective in the transportation of the iron, but their actual role has not been demonstrated.

More serious consideration has been given to the fact that the early atmosphere of the Earth may have been low in oxygen and therefore a reducing environment would facilitate the solution, transportation and accumulation of iron in the ancient seas. With the advent of photosynthesis and the release of large amounts of oxygen into the environment by organisms, the iron already accumulated in the seas would be oxidized and precipitated. This would imply one or more major periods of widespread deposition of iron and the prevalent development of the iron-formations in a few important and restricted periods in the Precambrian. There may have been much less oxygen in the atmosphere in the early Precambrian than now but significant effects of a reducing atmosphere on the development of the iron-formations are doubted. The iron-formations have formed over a wide range of time starting in the early Precambrian, and examples of recent origin are found. Limitations in the methods for dating rocks, and the incomplete geological record, make it difficult to show that special massive precipitation of iron beds took place at restricted times in the Precambrian or that this iron deposition was directly related to oxygen content in the seas and atmosphere.

Copper-bearing sedimentary rocks are an important group of mineral deposits that formed in both Precambrian and later time. The Precambrian deposits and their depositional envi-

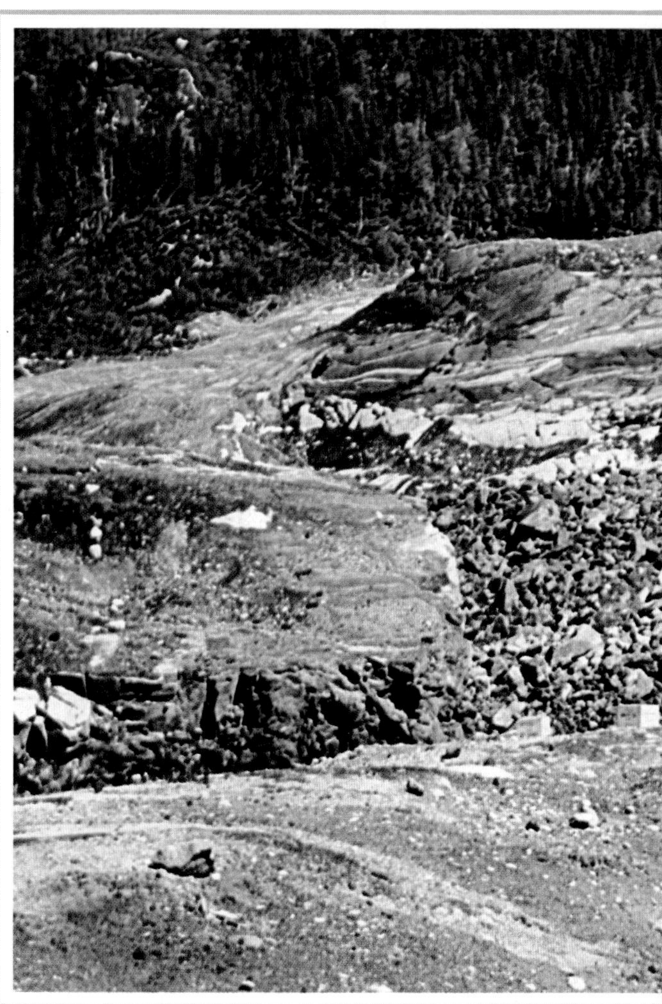

Fig. 8-19. An irregular mass of ilmenite (black) in an anorthositic rock, as seen in the Lake Tio deposit near Allard Lake in Quebec. It is thought that ilmenite is the product of a late differentiation of the same magma which has produced anorthosite. The part of the magma which is rich in iron oxide and titanium has been inserted within the anorthositic rock which was the first to crystallize in the course of the cooling of the magmatic mass. The intrusion has thrown out numerous blocks of anorthosite, separating them from the main mass; these blocks are now embedded in the actual body of the ilmenitic rock. This is one of the largest deposits of this type and for this reason is a major source of titanium and iron.

ronments are not distinctive from the younger deposits of this kind in any special way and would not suggest distinctive and different factors in Precambrian environments. Most of these deposits consist of copper sulphide minerals such as bornite, chalcocite, chalcopyrite and sometimes native copper, disseminated in a variety of different sedimentary rock facies, argillites, siltstones, arenites, sandstones and in black shales, with notable examples being deposits in the Copper Belt of Zambia. Conditions for their sedimentation are interpreted as deposition in shallow and isolated saline basins, where organic material and sulphur-producing bacteria were abundant. Conditions in these basins were highly alkaline and reducing and the influx of river waters caused changes and fluctuation in the physical-chemical environment in the basin which gave rise to development of distinctive sedimentary facies.

The stratiform shape, the dissemination of copper minerals, zoning and grade of the ore are controlled by sedimentary factors. The distribution of the copper in the beds, and later by folding and metamorphism and in some cases by secondary enrichment of the copper, but distinctive sedimentary features are usually preserved. The source of the metals in these deposits—copper, cobalt and iron—is not clearly defined, and like the iron-formations the metals may have come from erosion of the adjacent land area with local concentrations of the metals, supplied by ocean currents upwelling from ocean deeps and from volcanic exhalations.

Precambrian mineral deposits formed by igneous and metamorphic processes are basically not different from similar kinds of deposits formed in more recent time. There are some, however, such as the large anorthosite masses with associated Almenite, ilmenite-magnetite-hematite deposits containing important resources of titanium, vanadium and iron which were nearly all formed in the Precambrian. Other mineral deposits have formed by the settling and segregation of the metallic constituents from basic to ultrabasic magmas which are Precambrian in age and intruded into these old rocks. Important nickel, chromium, vanadium, titanium and iron deposits were formed in this way, but their formation is not dependent on or related to any special Precambrian environmental factors.

The study of metallogeny for the vast period of time in the Precambrian presents distinctive problems and a major challenge in the geological sciences because of the great age and complex superposition of many geological events, rather than by the combination of unique environmental factors and processes. The role of organisms in geological processes in the Precambrian has still not been defined or evaluated and therefore comparisons of all factors in the environment from the Precambrian to recent times cannot be made. Knowledge and understanding of the Earth's origin and early geological history is developing rapidly with the benefit of scientific data from the moon, other planets, and stellar bodies in the universe as well as from deep probing of the Earth itself. A better understanding of metallogenesis and the environment in the Precambrian is being gained from observation and definition of recent geological processes, and the projection of this knowledge into the past.

GORDON A. GROSS

Bibliography: Holland H. D., *The geologic history of sea water—An attempt to solve the problem,* in Geochimica et Cosmochimica Acta, XXXVI, 637 (1972); *The Precambrian environment and the origin of life,* in *Proceedings of 24th International geological congress,* Montreal, Canada (1972); *Zambia issue,* in Geologie en Mijnbauw, LI, 3, 247 (1972); Amstutz G. C., Bernard A. J. (ed.), *Ores in sediments,* Berlin (1971); Barghoorn S. E., *I fossili più antichi,* in Le scienze (1971); Cloud P. E., jr., *Precambrian,* in Science, CLXXIII, 851 (1971); *ONU Survey of world iron ore resources,* New York (1970); Tatsumi T. (ed.), *Volcanism and ore genesis,* Tokyo (1970); Degens E. T., Ross D. A. (ed.), *Hot brines and recent heavy metal deposits in the Red Sea,* New York (1969); Eugster H. P., *Inorganic bedded cherts from the Magadi area, Kenya,* in Contr. Mineral and Petrol. XXII, 1 (1969); James C. H., *Sedimentary ores, ancient & modern,* Leicester (1969); Cloud P. E. jr., *Pre-metazoan evolution and the origins of Metazoa,* in; Drake E. T. (ed.), *Evolution and environment,* New Haven (1969); Gross G. A., *Geology and iron deposits in Canada,* Geological Survey of Canada (1965–68); Siever R., *Sedimentological consequences of steady-state ocean-atmosphere,* in Sedimentology, XI, 5 (1968); James L., *Chemistry of the iron-rich sedimentary rocks,* United States Geological Survey, Prof. Paper 440 (1966); Roy S., *Syngenetic manganese formations of India,* Jadavpur University (1966); Sillen L. G., *How has sea water got its present composition,* in Svensk. Kem. Tidskr, LXXV, 161 (1963).

The idea had been in the air for a long time. Its birth can be dated back to the first surveys carried out in the 16th and 17th centuries along the coasts of the Atlantic Ocean. At that time some cartographers noticed that the coasts of Africa and South America were very similar, and that you only had to shift the two continents across the map to see that they fit one into the other. The idea was brought up several times at later dates, but it would have remained at the level of a simple geographical curiosity if the climate of research, above all in the field of paleontology, at the end of the 19th century, had not brought it back to the forefront. The paleontologists, in fact, found themselves facing a most curious situation: in certain periods of the past it appeared that the geographical barriers, consisting mainly of large masses of water, had not existed at all for certain groups of animals that were clearly and indisputably walkers and not swimmers. For example, the presence of analogous species of dinosaurs in Africa, Spain and North America, or of certain other reptiles in Africa and South America, had led paleontologists in the late 19th Century to imagine the existence of dry-land connections (continental bridges) between the various continents. The distribution of the fossils was, however, such as to require the repeated appearance and disappearance of the bridges. Meanwhile, the new science of geophysics had shown that it was by no means a mere chance that Earth's continental rocks are at a higher level than the oceanic ones. Continental rocks are in fact lighter than oceanic ones, and the position they occupy is due to this fact.

It thus became something of a problem to imagine the repeated appearance and disappearance of dry land according to the paleontologists' desires, emerging from the ocean and sinking back into it without leaving any trace.

At Christmas in 1910 Wegener, as he himself tells his future wife Else Köppen in a letter, happened to be looking through an atlas. As he turned over the pages he noticed the resemblance, which others had noted centuries before, between the opposite seaboards of South America and Africa. At this point the idea which was to determine the history of 20th century geology was born.

The idea was on the whole simple enough. If animals had passed from one continent to another in ancient times, and if the appearance and disappearance of long strips of dry land was contrary to the principles of geophysics, the only possible solution was that in the distant past the two continents had been side by side. What better proof could there be than the fact that the two continents fitted together like the two halves of a torn piece of paper? Fortunately, however, Wegener did not stop at these simple observations. Though a meteorologist by profession he dedicated the rest of his life as a scientist to the search for proof, and to the discussion of this new idea which, from being a mere intuition, gradually became a scientific hypothesis. Truth to tell, Wegener was unable to propose geophysical mechanisms capable of explaining the movement of masses as large as continents. What he proposed—the gravitational influence of Sun and Moon—in the final analysis turned out to be inadequate and was consequently rejected. The situation for his hypothesis was worsened by the additonal fact that seismology was daily providing ever more convincing proof that the ocean beds and the substratum upon which they rested were not semifluid as Wegener believed, and as appeared to be necessary to explain the movement of the continents. The behavior of seismic waves in that stratum was undisputably the same as one would have expected if they were passing through solid rock. Thus, though the idea still received occasional support, it was practically abandoned. It gained new and decisive vigor in the early sixties, 30 years after Wegener lost his life in the ice of Greenland, where he had gone to seek decisive proof of his hypothesis. Decisive proof, however, came not from Greenland but from the ocean and the magnetic characteristics of its rocks, as we shall see in this article by John Tuzo Wilson and in the following one by Takesi Nagata.

JOHN TUZO WILSON

Director of Erindale College of the University of Toronto, where he teaches geophysics. Born in 1908 in Ottawa, Canada, he became interested in geology towards the end of the 1920s, making numerous expeditions, among them the first highly mechanized one to the Canadian Arctic. He studied continental drift for the first time in 1959, when the theories were still considered unsustainable. His observations on the age of oceanic islands and the mechanics of transformer faults, particularized by him, contributed in a decisive way, first to the affirmation of the theories of ocean floor spreading propounded by H. H. Hess in 1960, and secondly, to the plate-tectonics model.

JOHN TUZO WILSON

The Mobility Of The Earth's Crust

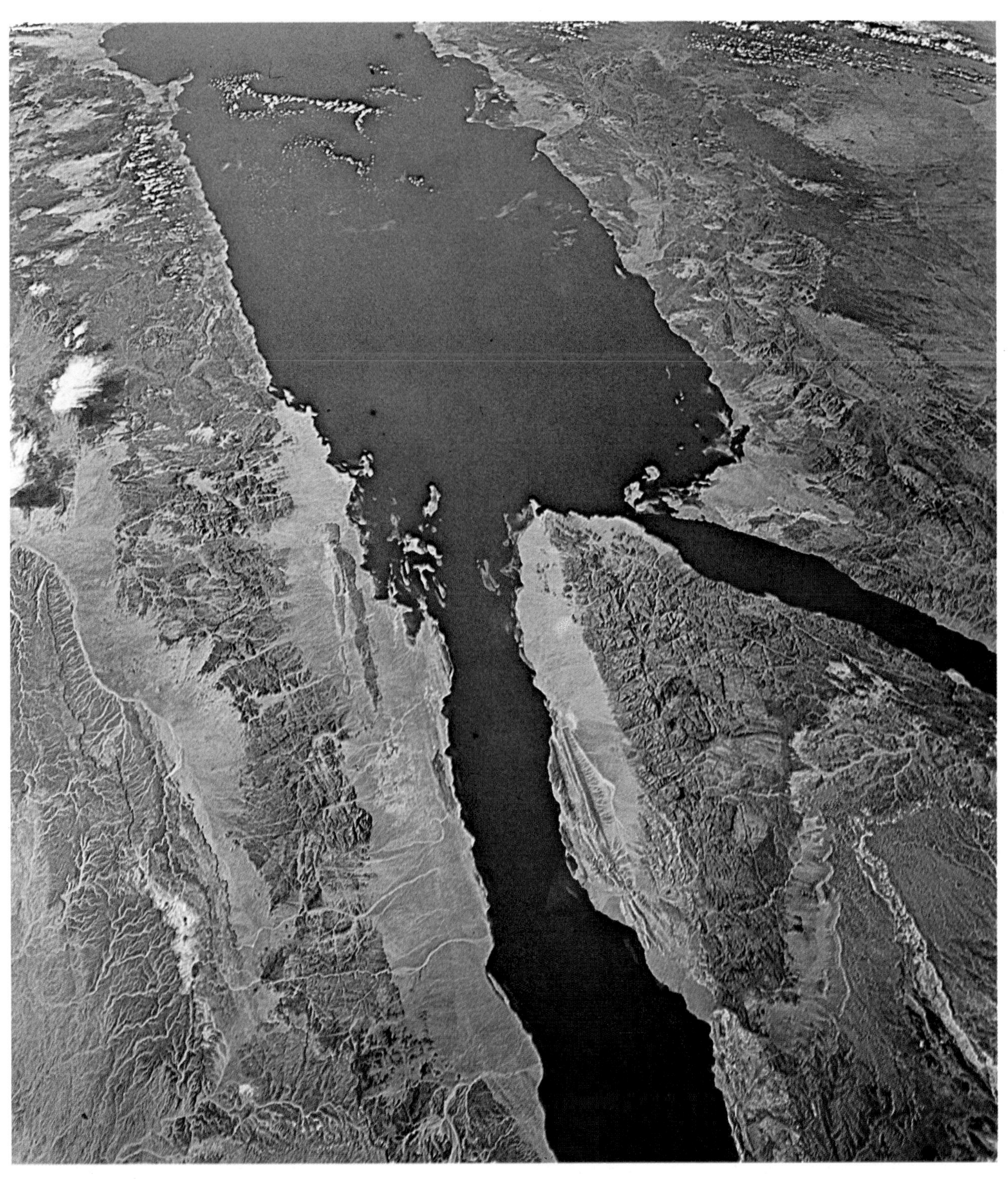

Overleaf: The Sinai Peninsula between Arabia (on the left) and Africa (on the right). The Peninsula is bounded by the Gulf of Aqaba (on the left) and by the Gulf of Suez (on the right) which merge in the Red Sea. This photograph was taken by the astronauts of the Gemini 12 mission from a height of approximately 450 km. According to recent theories, the Red Sea is a developing ocean, that is, an opening in the crust of the Earth which is destined to become wider. Approximately 150 million years ago the south central Atlantic Ocean must have had a similar configuration; Africa and South America formed its opposite shores at a distance as small as that presently between Africa and the Arabian Peninsula (NASA).

Continental drift is a theory according to which the continents move slowly on the surface of the Earth and vary their positions with respect to one another and with respect to the poles.

This article first recounts the history of this theory during the long period of time in which it was hardly credited, and then describes the recent discoveries which have led to its general acceptance (A. L. du Toit, 1937 and R. A. Phinney, 1968).

When the Atlantic Ocean was precisely mapped for the first time, scientists perceived that its two coastal lines have similar features. In 1858 A. Snider published a map which showed how well the continents conform to one another, and a book in which he proposed that they could have separated and moved apart. Other authors as well considered this idea but no one was taken seriously until, in 1912, A. Wegener began to publish more exhaustive arguments than those which had been previously advanced. He carefully studied the apparent similitude between the two Atlantic coastlines and suggested that a long time ago all continents were grouped together in one great land mass that he called Pangea. Wegener supposed that Pangea had begun to subdivide during the Carboniferous period, approximately 200 million years ago. He also demonstrated that many geological structures which are at present separated by the oceans, rejoin each other when the continents are reunited, and that a similar situation exists for many distribution areas of plants and animals present and past. As proofs of the drift theory, he also indicated that in past eras regions which are presently tropical were covered by continental glaciations, and that coal formed, or desertic conditions prevailed, in regions which are presently located close to the poles. He pointed out that in Scandinavia and in North America, where the glacial caps melted only recently, the ground is rising at a rate of up to a centimeter per year, and this vertical motion requires a horizontal influx of crustal matter which infers flow within the Earth.

Today it appears that, for the most part, the arguments proposed by Wegener are correct, and many of his contemporaries, although not convinced, were unable to contradict him completely. His publications stimulated the first serious discussions on the topic of continental drift, but unfortunately his opponents focused their attention on his errors to the exclusion of his other arguments.

He utilized outdated measurements to demonstrate a rapid increase of the distance between Greenland and Europe. More recent measurements show that the previous data were in error, although they did not preclude the possibility of slower motion. To describe continental drift Wegener proposed a mechanism which later turned out to be totally inadequate. He imagined that the continents moved on the crust underlying the oceans as ships on the sea: later scientists proved this to be impossible, although another form of the theory can be used to circumvent these difficulties. Only a few geologists accepted his views: among these were some who had studied the Alps or the southern continents.

The theory was generally rejected until after 1956 when new important facts were discovered which gave rise to a conversion process. These new discoveries stem from the study of paleomagnetism, of seismology, and from the results of the exploration of ocean floors. These discoveries have been so

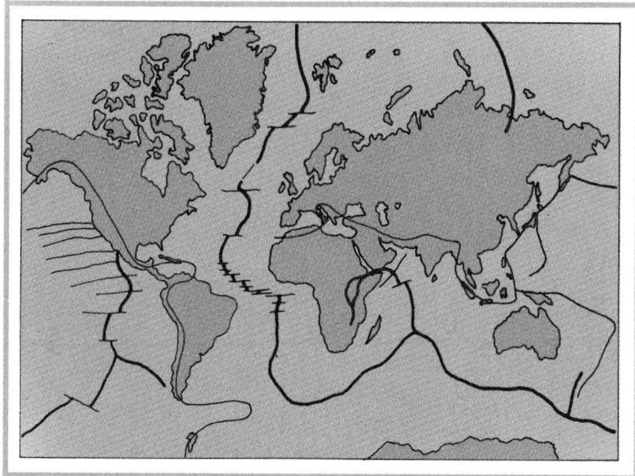

Fig. 9-1. Configuration of the Earth's midocean ridge systems (bold black line) and of continental mountain ranges and oceanic trenches (black thin line). The ridge is displaced by transverse fractures which are indicated by color lines *(from B. C. Heezen, H. W. Menard).*

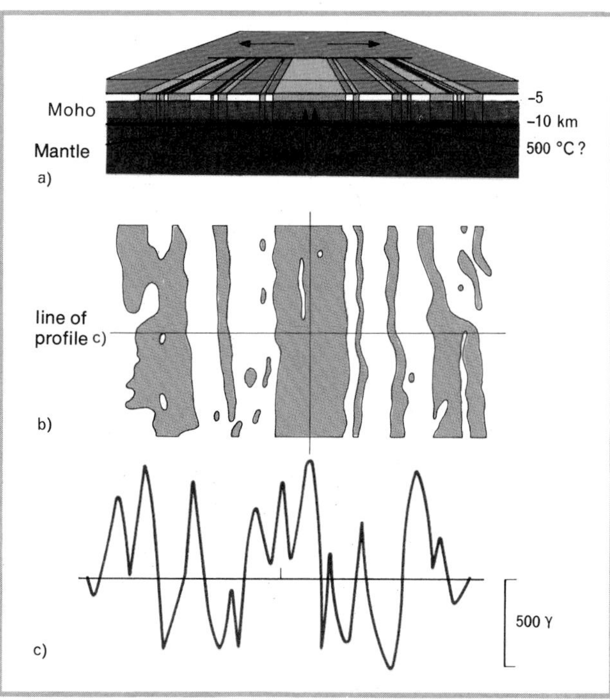

Fig. 9-2. The symmetric distribution of magnetic anomalies which results from the progressive accretion of oceanic floors from a midocean ridge (which in this case is the Juan de Fuca ridge). (a) model of the ridge; (b) map of magnetic anomalies; (c) observed magnetic profile *(from F. J. Vine).*

convincing that today most scholars accept the theory in its updated form.

Some years ago French researchers discovered that at the moment of their formation many rocks are permanently magnetized in the direction of the Earth's magnetic field. This phenomenon is called paleomagnetism, and in 1926 Mercanton suggested its use in the measurement of continental drifts. In 1956, with improved instrumentation invented by P. M. S. Blackett and applied by J. W. Graham, J. Hospers, and S. K. Runcorn, confirming results were obtained from the determinations of ancient latitudes over vast regions which made acceptable the fact that the continents have changed their positions relative to the poles and to one another. This was the first clear proof of continental drift (E. Irving, 1964).

Still more conclusive proof was discovered during the exploration of deep ocean floors. In 1956 M. Ewing and B. C. Heezen advanced the idea that a very long system of underwater mountains form a range which runs along the center of the oceans all around the globe. Other scholars, and particularly H. H. Hess and F. J. Vine, demonstrated that this range possesses characteristics which could reveal the exact history of recent expansions of oceanic floors (Phinney, 1968, D. H. Maxwell *et al.*).

A third series of arguments which favor the continental drift theory is derived from the great improvement of seismological observations obtained since 1960: the distribution and propagation direction of recent earthquakes confirm the theory (B. Isacks, J. Oliver and L. R. Sykes, 1968).

In 1906 J. Brunhes noted that some relatively recent rocks were magnetized in the direction of the Earth's magnetic field while others were magnetized in the opposite sense. Brunhes pointed out that the Sun's magnetic field inverts itself in a period of 11 years and proposed that the terrestrial magnetic field as well could undergo a similar, although slower inversion. In the late 1960s A. Cox, R. Doell, and R. McDougall established the time scale of recent magnetic inversions thus defining a very precise paleomagnetic stratigraphy.

Shortly after the notion of the midoceanic ridge was established, Hess (1962) readopted the idea that material currents rise from the interior of the Earth under these mountainous ranges and then flow away laterally. He proposed the hypothesis that these currents drag along the ocean crust and that they continuously generate new crust in correspondence to the ridge.

In 1963 Vine and D. H. Matthews thought that, when new crust is generated, basaltic lavas could pour onto the ocean floor and also receive permanent magnetization. If this were the case, zones parallel to the midocean ridge would be alternately magnetized in a normal and opposite sense; and exactly this arrangement was measured by means of magnetometers. J. R. Heirtzler (1966) made a strong contribution in demonstrating that the widths of successive bands as measured starting from the midocean ridge form ratios identical to those observed in the inversion time scales. From the knowledge of the time scales and of the width of the bands, spreading rates were computed which vary up to a maximum of 15 cm per year. The continental drift theory is also confirmed by proofs in other fields. If the ocean floor is indeed spreading, its age must be more recent next to the midocean ridge and more ancient close to the continental shelves. Although young sedimentation and new volcanoes can form anywhere on the ocean bottom, recent corings of the ocean floors prove that the most ancient sediments are adjacent to the shelves and that the sedimentation thickness increases from zero on the midocean ridge to several kilometers close to the continents.

H. W. Menard and Heezen showed that the midocean ridge system is frequently interrupted by great shifts. If these consisted of ordinary faults with horizontal displacement, the shifts would continue into the continents, frequent earthquakes would occur along the entire fault length, and the motion of the parts would be as it is in continental faults. J. T. Wilson indicated that if indeed a new lithosphere is forming, a new class of faults can exist which he called "transform," for which the continental continuation is not necessarily present, with earthquakes restricted to the central region, and for which the displacement occurs in a sense opposite to that of normal faults. Research studies by Sykes have demonstrated that this hypothesis is correct. If the Earth's volume is not increasing or decreasing appreciably, spreading of the ocean floors along the midocean ridge must be compensated for either by superimposition or reingestion of the crust at other locations. It is believed that mountain massifs of recent geological age, and deep oceanic trenches adjacent to island-arcs, are the locations where portions of the Earth's crust are subducted under others. This is possible because at depths greater than about 50 km, temperatures are sufficiently high for the mantle to become rather plastic. Sections of lithosphere can then be forced to slide on this deformable stratum and actually penetrate into it. On the basis of earthquake studies Isacks, Oliver, and Sykes (1968) determined the descending motion of overlapped regions. W. J. Morgan (1968) and X. Le Pichon (1968) showed that the distribution of earthquakes locates cer-

TABLE 9-1. STAGES OF THE LIFE CYCLE OF OCEANIC BASINS AND THEIR PROPERTIES

Stage	Example	Movements	Sediments	Magmatic Rocks
1. Embryonic	Trenches of oriental Africa	Uplift	Negligible	Platforms of tholeiitic basalts, volcanic centers of alkaline basalts
2. Juvenile	Red Sea and Gulf of Aden	Uplift and spreading	Small platforms, evaporites	Tholeiitic marine crust, islands of alkaline basalts
3. Mature	Atlantic Ocean	Spreading	Large platforms (of miogeosynclinal type)	Tholeiitic ocean crust, islands of alkaline basalts
4. Declining	Western Pacific Ocean	Compression	Island-arcs (of eugeosynclinal type)	Andesitic vulcanites, plutons of gneiss-granodiorite
5. Terminal	Mediterranean Sea	Compression and uplift	Evaporites, red mud, wedges of clastics	Andesitic vulcanites, plutons of gneiss-granodiorite
6. Residual scar	Line of the Indus River, Himalaya	Compression and uplift	Red mud	Plutonism

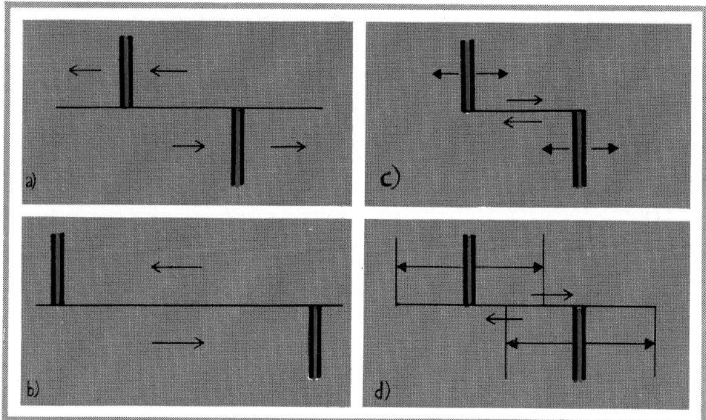

Fig. 9-3. Two classes of transcurrent faults. (a) and (b) two stages of a normal transcurrent fault, to be found in continents; (c) and (d) two stages of accretion of a transform fault in fracture zones of ocean crust. The apparent direction of displacement is the same in both cases while the direction of motion of the two sides of the fault is different *(from J. T. Wilson).*

tain fractures which separate the Earth in six major and several minor plates. For geometrical reasons the relative motion of any pair of plates on the surface of a sphere must be a rotation around a clearly identifiable axis. The magnetic marking of the oceanic crust permits the establishing of the separation velocity of any pair of plates; it also allows the calculation of the velocity with which pairs of plates approach one another under mountain massifs. It is possible to integrate these velocities and to discover at least in first approximation the location of the continents at specific moments of the past.

Arguments Against Continental Drift

Continental drift constitutes an important turning point in scientific thought concerning the Earth. Some consider this theory a true scientific revolution as important for earth sciences as the Copernican revolution was for the astronomy of the solar system.

As Galileo proved that the Earth is not the fixed center of the universe, but that it travels through space, so the Wegenerian revolution, as one could call it, demonstrates that the

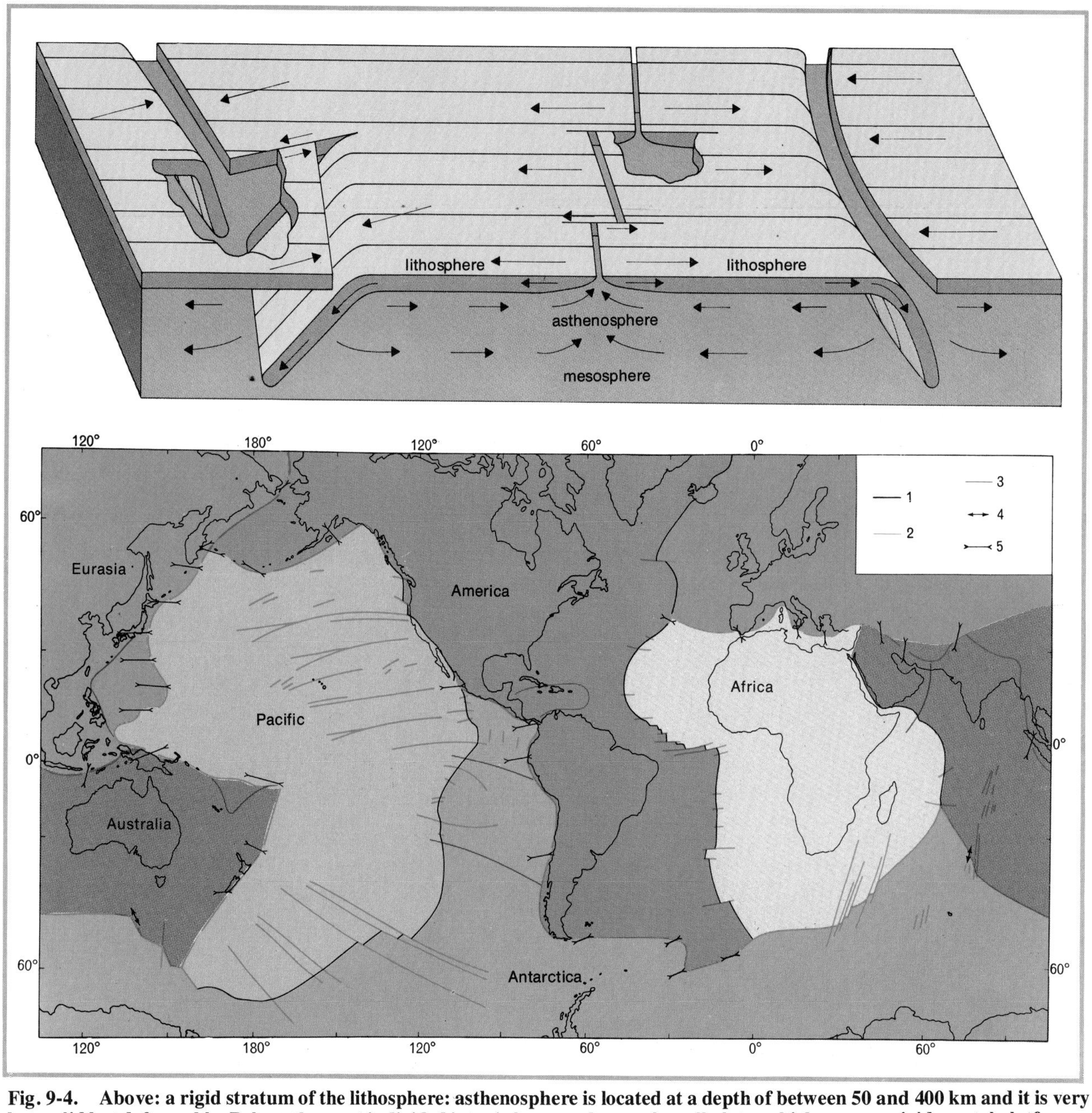

Fig. 9-4. Above: a rigid stratum of the lithosphere: asthenosphere is located at a depth of between 50 and 400 km and it is very hot, solid but deformable. Below: the crust is divided into six large and several small plates which move as rigid crustal platforms, along the borders of which many earthquakes occur. Key: 1, lines with known accretion velocity and transverse fractures; 1 and 2, boundaries of principal plates; 3, boundaries of other possible plates; 4 and 5, resulting motions of dilatation and compression *(from Isacks, Oliver and Sykes, and from W. J. Morgan, modified).*

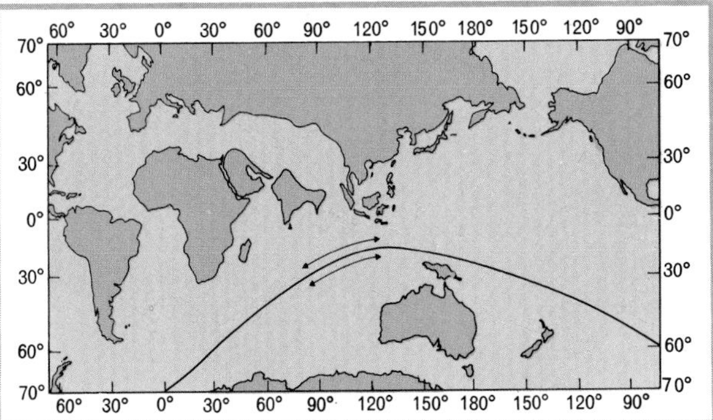

Fig. 9-5. Position of continents calculated for approximately 65 million years ago; i.e., at the beginning of the Tertiary Period; Antarctica and Africa are depicted in their present-day positions but the precise relation between the two sides of the continuous black line has not been established *(from X. Le Pichon).*

Earth's surface is not fixed. This change profoundly affects the reconstruction of the entire geological history of the Earth and many geophysical ideas. It could not be expected that this change would be accepted without several objections (V. Belousov, 1968).

Usually the objections are presented in one of the two following forms. It is either shown that a specific type of continental drift is impossible, or that a certain isolated phenomenon does not comply with the theory.

One can generally answer the first type of objection by selecting another form of the theory. For example, H. Jeffreys is correct in asserting that a force is not known that could propel the continents like ships over solid ocean crust. However, the new form of the theory overcomes this difficulty by considering that thermal currents in the Earth's mantle open the lithosphere along the midoceanic ridge and slowly drag along the continents as if they were lying on large trays or rafts.

The remaining objections seem to be founded on accidental errors. Thus A. A. Meyerhoff pointed out that one or two geological datings of rocks found in the Atlantic Ocean do not agree with the continental drift theory which he rejects: however, it is undoubtedly preferable to accept a hundred favorable observations and reject two as elements which should be carefully scrutinized and submitted to further verification.

In contrast with these objections, which many believe can be very effectively dealt with, the new theory explains a multitude of observations on which no other theory until now could shed any light.

Proofs of Prehistoric Continental Drift

The ideas of Wegener and of du Toit indicate that approximately 200 million years ago the sole supercontinent, Pangea, broke into two parts, Laurasia and Gondwana, which, in turn, later subdivided into the contemporary continents. Another point of view considers continental drift to be a very ancient process and that oceanic basins have formed and disappeared several times during the history of the Earth. This leads to the idea of the life cycle of oceanic basins which is illustrated in the table. du Toit proposed that the system of African trenches could represent the embryo of an ocean. It is connected with the Red Sea and the Gulf of Aden.

Through the Indian Ocean these basins communicate with the Atlantic, thus giving examples of three stages of oceanic spreading.

Even though the Pacific is presently the largest ocean, it must have been still larger when all continents were grouped together, therefore it is presently decreasing in magnitude. Many have suggested that Africa and India are drifting in a northerly direction against Asia; in such a case the Mediterranean Sea and the Himalayas are examples of the late stages of closing of an ancient oceanic basin. A. W. Grabau and Wilson (1968) have proposed that 500 million years ago the Atlantic Ocean opened in a position slightly different from the present one, and that it then closed and formed the Appalachian and Caledonian Mountain ranges. Supposedly the Atlantic opened again in a slightly different position, transporting along sections of North America to Europe and Africa, and vice versa. If these views are correct, the original location of major sediment-laden basins which are named geosynclines must be assumed to be along the shores of the oceans and not at their center. The closing of oceans corrugated and transformed into mountains the large piles of sediments accumulated on the continental platform.

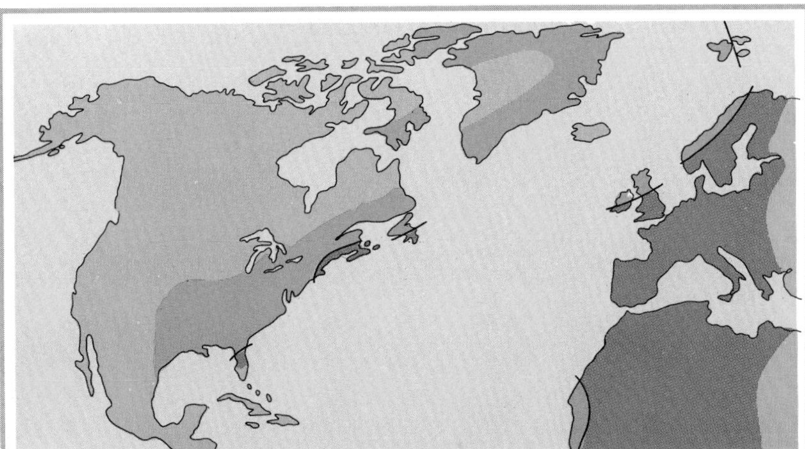

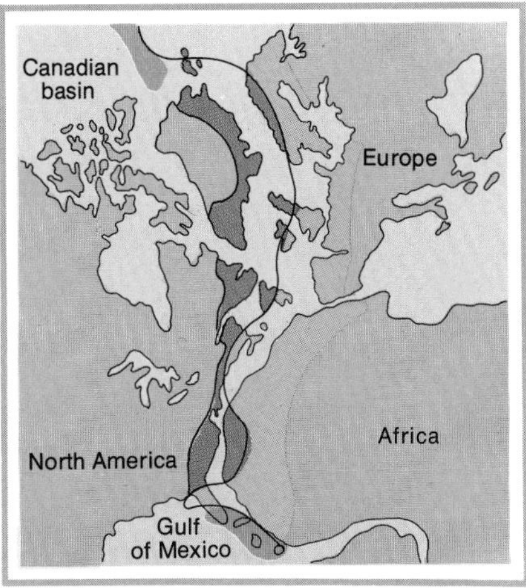

Fig. 9-6. Left: Distribution of fossils of the Cambrian Period which belong to two different faunal provinces is indicated by the two-color regions on the two sides of the black line and it shows them to be completely independent of the present geographic configuration while there is a good correlation on the opposite shores of the ocean. Right: assumed position of north Atlantic land masses in Upper Paleozoic and Lower Mesozoic times; thick line separates the faunistic provinces crossing what will become the present-day Atlantic Ocean; midocean ridge of the time is shaded in gray *(from J. T. Wilson).*

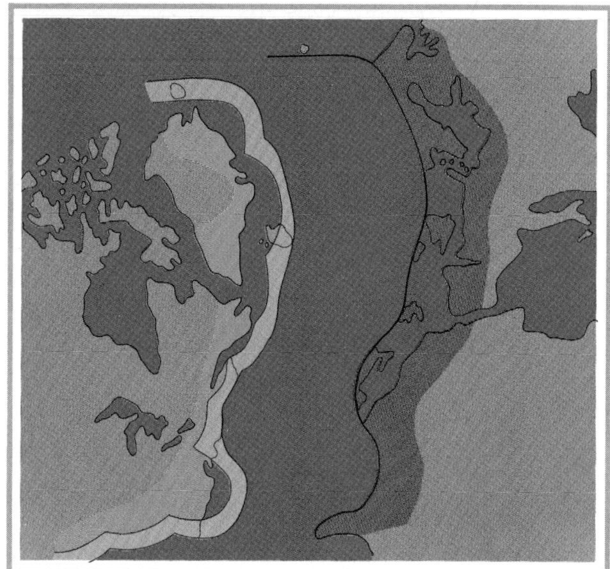

The North Atlantic in the Cambrian Period. Supposedly the Protoatlantic Ocean completely separates two faunal provinces which are considered to have evolved on separate continents. It is probable that island-arcs existed (narrow color band in the center) along the North American coast; along the trenches which were associated with them the ocean floor was gradually subducted and for this reason the continents became progressively closer to each other at the end of the Paleozoic Era *(from J. T. Wilson)*.

It is interesting to note that the public had always been attracted by the notion of continental drift which the majority of scholars had not found acceptable. It now appears that these experts were wrong and that they are in the process of changing their views in order to accept a new form of continental drift theory which will lead to the explanation of the long-known similitude between the two sides of the Atlantic Ocean.

JOHN TUZO WILSON

Bibliography: Belousov V. V., *Debate about the Earth: an open letter to J. Tuzo Wilson,* in Geotimes, XIII, 17 (1968); Heirtzler J. R. *Marine magnetic anomalies, geomagnetic field reversal, and motions of the ocean floor and continents,* in J. Geophys. Res., LXXIII, **6**, 2119 (1968); Isacks B., Oliver J., Sykes L. R., *Seismology and the new global tectonics,* in J. Geophys. Res., LXXIII, **18**, 5855 (1968); Le Pichon X., *Sea-floor spreading and continental drift,* in J. Geophys. Res., LXXIII, **12**, 3661 (1968); Morgan W. J., *Rises, trenches, great faults and crustal blocks,* in J. Geophys. Res., LXXIII, **6**, 1959 (1968); Phinney R. A. (ed.), *The history of the Earth's crust,* Princeton (1968); Wilson J. T., *Static or mobile Earth: the current scientific revolution,* in Proc. Amer. Phil. Soc., CXII, **5**, 309 (1968); Wegener A., *The origin of continents and oceans,* New York (1966); Irving E., *Paleomagnetism and its application to geological and geophysical problems,* London (1964); Vine, F. J., Matthews D. H., *Magnetic anomalies over ocean ridges,* in Nature, CIC, **4897**, 947 (1963); Hess H. H., *History of ocean basins,* in A. E. J. Engel (ed.), *Petrologic studies,* New York (1962); du Toit A. L., *Our wandering continents,* Edinburgh (1937).

DEEP down in its interior the Earth behaves like an enormous magnetic bar pointing in much the same direction as its axis of rotation. How it varies and what influence this terrestrial magnetic field has upon the surface is the subject of the article by Nagata which follows. As we shall see, its existence and past behavior have in recent years made it possible to resolve one of geology's most important problems, and its analysis has been directly responsible for the revival of the continental drift hypothesis which, with its support, has become an accepted model for the evolution of our planet.

The origin of Earth's magnetic field is one of the unsolved problems of modern geophysics. Various models have been proposed, but the one which today is considered to be most probable is that of the existence of currents within the body of the Earth itself. These currents are thought to be linked to the slow movement of convection currents which continually displace material inside the mantle or the outer nucleus of the planet. In whatever way it is generated, the Earth's magnetic field has highly important effects upon the surface conditions of our planet. First of all it must be remembered that the whole of the solar system is continually being struck by a flow of charged particles emanating in all directions from the Sun. This so-called solar wind could certainly have very serious effects on living organisms if it reached the Earth's surface, as it does that of the moon. The Earth's magnetic field is of fundamental importance for this phenomenon: the charged particles arriving from the Sun are deflected by a magnetic shield and forced to flow away in a long tail, which is known as the magnetosphere.

Could it be possible that a rock exposed to a given magnetic field might, through some internal mechanism of its own, become magnetized in a direction opposite to that field? Or was it not more likely that the polarity of the Earth's field had been inverted several times? As soon as scientists began to analyze a sufficiently large quantity of rocks, it was discovered that this inversion, if found in a given rock, was also found in other rocks of the same age but of different types. The conclusion was immediate: clearly the magnetic field had undergone an inversion during that period. Meanwhile oceanographic research, which had begun during World War II, was advancing. This research included a survey of the magnetic anomalies of the ocean beds. It had been observed that certain areas of the ocean bed present strange anomalies set out in a linear fashion: there are narrow bands of positive magnetic anomaly alternating with other bands of negative anomaly. The same oceanographic research had also shown that there are extremely long mountain ranges on the ocean bottoms. The curious fact was that long stretches of these ranges run midway between the continental seaboards on the opposite sides of the ocean. Furthermore, new depth-sounding techniques had shown the existence along their crests of a long deep valley running all the way along the range. At this point Hess suggested that the submarine ranges might represent the point at which great internal convection currents of hot material reached the surface of the Earth's crust, and then split up, traveling on in opposite directions. According to Hess the underwater ranges must have been the points at which these hot substances came to the surface, only to cool down, while day by day new material was added to the oceanic crust. The crust, he suggested, moved laterally as a block with respect to the mountain range, and this automatically led to the drift of the continents away from one another.

It must be remembered, in fact, that eruptive and volcanic rocks receive their magnetization while they are cooling down. While they are in a molten condition, by a law of magnetism, they are unaffected by the magnetic field, which comes into play only below a certain temperature, called the Curie point, which differs for different substances. Hess' hypothesis offers an explanation of the linear magnetic anomalies observed on the ocean floors; for if the magnetic field had in the past inverted its polarity several times whilst hot material was reaching the surface from the Earth's interior, the oceanic crust formed by this material might well be characterized by bands of normally magnetized material alternating with bands of material magnetized in the opposite direction. The magnetization of these rocks, either added to the normal field or subtracted from it in the case of opposite magnetization, might well give rise to the anomalies that had been observed.

TAKESI NAGATA

Director of the Institute for Polar Research in Tokyo. Born in Tokyo in 1913, his interest in terrestrial magnetism goes back to the end of the 1930s when he began studying electrical and magnetic phenomena connected with earthquakes and volcanic eruptions. In 1952, he assumed the directorship of the Geophysics Laboratory of the University of Tokyo, where his studies, including those of an experimental nature, have been concentrated on the various types of magnetism in rock residue. At the same time, he has continued to conduct research in the field of current terrestrial magnetism. Of particular importance is his contribution to the confirmation of paleomagnetism as an investigative technique.

TAKESI NAGATA

The Earth's Magnetic Field and Its Variation

Overleaf: The amazing display of a polar aurora, one of the most interesting natural phenomena which reveals some of the characteristics and variations of the Earth's magnetic field. Auroras, in fact, are caused by the interaction of the solar wind, the flow of particles which the Sun emits into space, with the upper part of the Earth's atmosphere. As a rule, the Earth's magnetic field protects our planet from this flow, but the protection is less effective at high altitudes, at which, as it happens, auroral phenomena are commonly observed. (Max Planck Institut).

The Earth as a Great Magnet

Everyone who has studied elementary science today knows that our Earth acts like a great magnet which interacts with a small magnetic compass, aligning it approximately toward the north at most places over the Earth's surface. This observation was clearly described by William Gilbert in his famous book "De Magnete" which was published in 1600.

Although Gilbert asserted that "the globe of the earth is a great magnet," the problem "why the earth can be a great magnet" had long been unsolved. In the past, a large number of hypotheses have been proposed in connection with this problem.

K. F. Gauss was the first to describe the distribution of the Earth's magnetic field in precise mathematical form. He published "Intensitas vis magneticae terrestris ad memsuram absolutam revocata" in 1932 and "Allgemeine Theorie des Erdmagnetismus" in 1938. Gauss established exact physical concepts of "force" and "a field of force," in general on the basis of an elegant mathematical scheme which is now called "potential theory." As a result of his studies he came to the important conclusion that more than 97% of the magnetic force observed on the Earth's surface originates within the Earth's interior, only a small portion of the field being attributable to some extraterrestrial source. Thus, any theory which allocates the main origin of Earth's magnetic field to the Earth's exterior space must be rejected. As is well known, a magnetic field can be produced either by a permanent magnet or by electric currents. There is no doubt, therefore, that either a certain kind of natural permanent magnet or an electrical current exists within the Earth.

The measurement of the distribution of the Earth's magnetic fields is made not only on the surface but also on artificial satellites orbiting in the wide space surrounding the Earth. Fig. 10-1 shows the recent summary of the observed distribution of the magnetic field in the vicinity of the Earth, where heavy lines represent selected lines of magnetic force. As seen in this figure, the lines of the magnetic force converge into the Earth, suggesting that the magnetic field is primarily produced within the Earth. If no particles exist in outer space, these lines of magnetic force must be extended over the entire space surrounding our Earth. In reality, however, it seems that the lines of force are enveloped by some matter flowing from the direction of the Sun, and are displaced by it in an antisolar direction. Various particle counters on artificial satellites have shown that the flowing matter is a kind of gas the major components of which are electrons and protons; namely, an almost perfectly ionized hydrogen gas, now called a plasma. The plasma gas-flow encloses the Earth's magnetic field in a cavity in the vicinity of the Earth, and further drags the field away with it, because the plasma has a very high electric conductivity which does not allow any invasion of magnetic lines of force into it, and because the kinetic energy of the plasma flow is sufficiently high to compress the field on the front side and drag out the field on the rear side.

Evidence indicates that the plasma emanates from the Sun, formerly an outer extension of the solar corona, and that the speed of the plasma flow is 300–400 km/sec when conditions

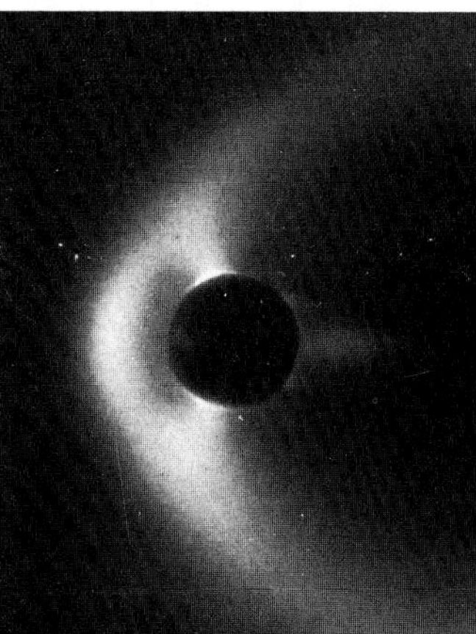

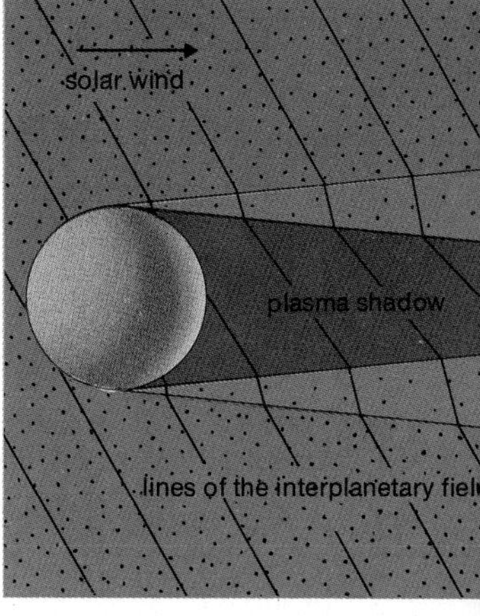

Fig. 10-1. Relations between the solar wind and the magnetic field. Top right, these relations are studied in the laboratory using a flow of plasma made up of electrons and helium ions and a sphere containing an electromagnet: the jet of plasma coming from the left of the photo is deflected and leaves an obvious cavity beside the surface of the sphere, except in the regions of the magnetic poles, precisely as happens on Earth. Opposite: a diagram, reconstructed on the basis of data gathered by man-made satellites, which illustrates the real relationship between the solar wind and the Earth's magnetic field; the numbers indicate the distances, in earth radii, from the center of the Earth; the red dots represent the flow of particles coming from the Sun; the lines of force of the Earth's magnetic field are flattened on the side turned towards the Sun and deflected to form a long tail on the other side; the overall phenomenon is what geophysicists call the magnetosphere (yellow area). On this page, in contrast, we see the moon, affected by the same solar wind which reaches our planet, but without a magnetic field of the same strength; the stream of particles reaches the surface of the moon directly from the Sun without any interference; on the far side it becomes elongated, in the simple form of a conical shadow.

are quiet, and 800 km/sec when conditions become more stormy. The flow is known as the solar wind. The interaction between the solar wind and the Earth's magnetic field has been studied extensively from both the observational and theoretical points of view. An example of a visual study of the solar wind-magnetic field interaction is demonstrated in Fig. 10-1. In this photograph, plasma consisting of helium ions and electrons is streaming at a speed of 40 km/sec from left to right in a vacuum chamber; this represents the solar wind. The Earth is represented by a dark sphere which contains an electromagnet. When the model Earth has no magnetic field, the stream of helium plasma hits its surface directly. When the model Earth becomes a strong magnet, the helium plasma stream is deflected by the magnetic field, and a cavity is formed around the model Earth, as shown in the figure. Thus, the plasma gas streams only outside the cavity surface. The scale of the cavity becomes larger for a stronger magnetic field, and it becomes smaller for a greater speed or a larger density of the plasma stream.

The strength of the magnetic field, measured at a point within the cavity, when it is formed by the plasma stream, can be compared with that at the same position when no plasma is streaming. The result of such an experimental test shows that the magnetic field is strengthened when the cavity is formed, the field becoming greater when the scale of cavity is smaller. These experimental results indicate that the magnetic field of the model Earth is compressed by the plasma stream.

Just inside the boundary surface of the cavity, the outward magnetic pressure normal to the surface is given by $B^2/8\pi$ (where B denotes the intensity of magnetic field), whereas just outside the surface, the inward kinetic pressure of plasma stream (velocity normal to the surface = $v_\perp$, number density of positive ions and electrons = N, mass of positive ions = m_+, mass of electrons = m_e) normal to the surface is expressed as $N(m_+ + m_e)v_\perp^2$. In the steady state, the outward pressure $B^2/8\pi$ and the inward pressure $N(m_+ + m_e)v_\perp^2$ should balance at the boundary surface: $N(m_+ + m_e)v_\perp^2 = B^2/8\pi$. This simple equation gives a reasonable explanation of all observed results from the above-mentioned model experiment.

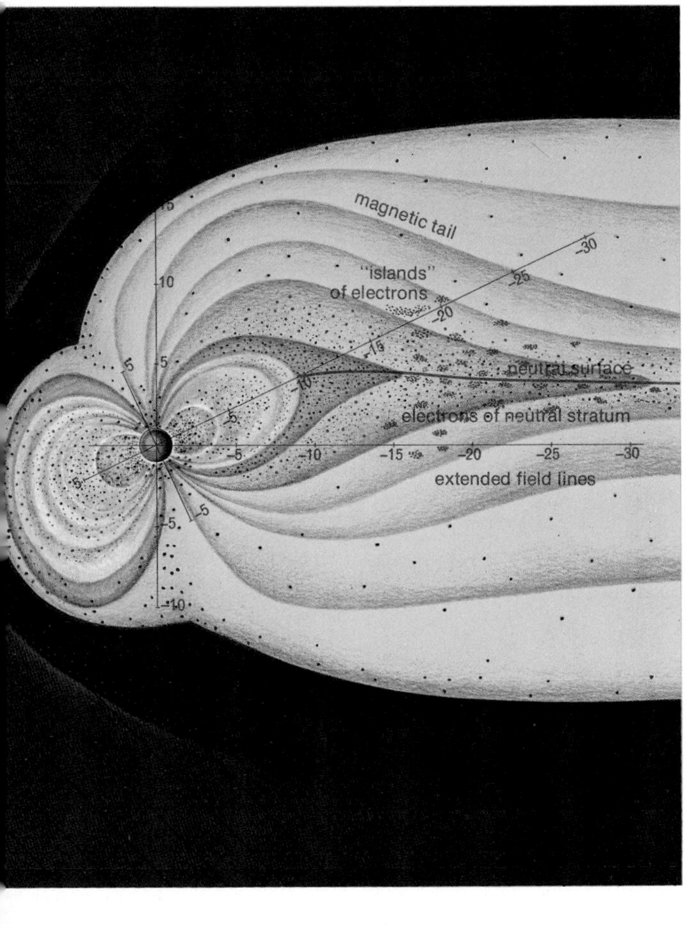

A direct comparison of Fig. 10-2 with Fig. 10-1 may clearly show how the cavity of the Earth can be formed as a result of the interaction between the solar wind and the Earth's magnetic field. The huge natural cavity around the Earth is now known as the magnetosphere, and the long tail of the magnetosphere elongated towards the antisolar direction by the solar wind, is now called the geomagnetic tail. It is obvious, as discussed above, that the Earth's magnetic field protects our Earth from the direct attack by the solar wind.

If there were no magnetic field of the Earth, the solar wind particles would reach the Earth's surface. The number of protons and electrons reaching the Earth in such a case is estimated to be about 4×10^8 particles/cm²/sec. under quiet solar condition. Such a direct attack by the solar wind actually occurs on the moon's surface. The solar wind directly hits the surface of the moon, forming the shock wave fronts and the solar wind shadow region behind it. On the other hand, the observed magnetic field very close to the moon's surface is smaller than 1/30,000 of that on the Earth's surface. This means that the moon has practically no magnetic field. Fig. 10-4 indicates, therefore, that the moon is openly exposed to the solar wind because it has no magnetic field.

Then why does the Earth have an appreciable magnetic field, whereas the moon does not? The moon is very different from the Earth in many ways. Thus, the moon's mean radius is only 1738 km whereas the Earth's is 6378 km: the mean density of the moon is about 3.34 which is much smaller than the Earth's, 5.52. These data indicate that the interior structure of the moon is fundamentally different from that of the Earth. From a large amount of data from the propagation of seismic waves and other geophysical sources, we now know a fair amount about the internal structure of our Earth. According to these data, the Earth consists approximately of three layers: (a) the Earth's crust, about 5 km in thickness under the oceans and 20–30 km thick in the continental areas; (b) the Earth's mantle extending from the bottom of the Earth's crust to a depth of about 2900 km; and (c) the Earth's core below the mantle. Various kinds of scientific evidence indicate that the density of the core is greater than 9, which is identical to the density of compressed metallic materials such as iron or nickel, and that the core material is in a fluid state. These aspects of the Earth's core are in conspicuous contrast to those of its mantle which have been ascertained from data regarding its density and elasticity to be similar to the solid state of metallic oxides. On the other hand, the moon's mean density is a little larger than 3.0, which is the mean density of the uppermost part of the Earth's mantle subject to atmosphere pressure. When the mantle rock is compressed by a hydrostatic pressure, equal to that in the moon's interior, its density increases with pressure up to 3.4 or more. Thus, it seems most likely that the whole interior of the moon is composed of solid metallic oxides: namely, compressed rocks, and that the moon has no central core of fluid metal. We may therefore conclude that the Earth's magnetic field is most likely generated within its iron-nickel core.

An Electrodynamic System Within the Earth's Core

Although the major component of the Earth's core seems to be iron and/or nickel, which is ferromagnetic at ordinary temperatures, the estimated temperature of the core is definitely far greater than the critical temperature (Curie point) at which these metals are ferromagnetic. Hence, one cannot assume that the core behaves like a permanent magnet. On the other hand, the metallic core material should certainly have a high electric conductivity so that an electric current can easily flow within the core if some electromotive force is present to main-

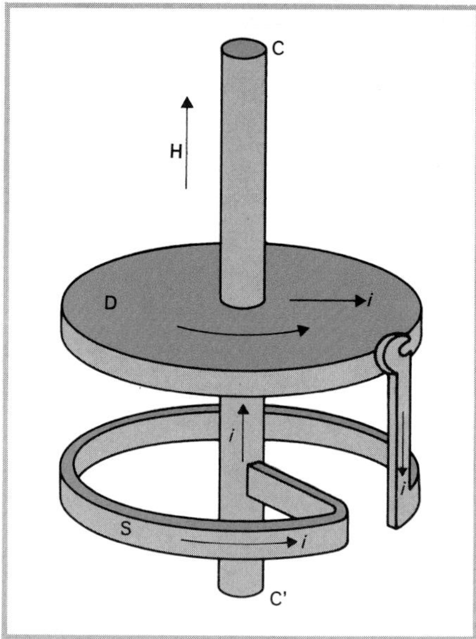

Fig. 10-2. Diagram of the simple model used to illustrate the principle of Bullard's self-excited disk dynamo ***(from T. Nagata and M. Ozima).***

tain the current flow. W. M. Elsasser (1946) and E. C. Bullard (1948) actually proposed hypotheses to attribute the origin of the Earth's magnetic field to electric currents generated in the core. Elsasser assumed thermo-electricity as the electromotive force, whereas Bullard considered a dynamic electric generator which is energized by a rotation and convection of the fluid material in the core. A number of later studies on the mechanism that generates the Earth's magnetic field have since been made mostly in England, Japan and Canada; based on various observed facts, these studies seem to support strongly Bullard's idea. Fig. 10-2 shows a simple model to illustrate the basic idea of Bullard's self-exciting dynamo. A circular disc (D) is revolving counter-clockwise around an axis (CC′) in the presence of a small magnetic field H which is parallel to CC′. In accordance with Faraday's law of electromagnetic induction in such a case, an induced electromotive force (E) is generated radially in the circular disc. Here the electromotive force at a point of the disc is given as $E = v \times H$, where v denotes the lateral velocity of the point. As seen in the figure, the circular disc is attached by an electric brush at its edge, the brush leading to a circular solenoid (S) around CC′, and the solenoid's other end is attached to the rotating axis CC′ through the other brush. Assume now that all these devices are made of electrically high conductive metals. Then, because of the induced electromotive force E, electric currents flow through D, S, CC′ and D, making a closed electric circuit. The electric current flowing in the solenoid (S) produces a magnetic field which is parallel to the initial H, the magnetic field thus being enhanced. The enhanced field then induces a larger electromotive force. If the driving torque that rotates disc D together with axis CC′ is sufficiently large, the magnetic field along CC′ becomes larger and larger as the rotation continues. Thus, a small magnetic field, which may initially arise by chance can be enhanced by the self-exciting dynamo system. Because of loss of current by the electric resistance of the device material, however, the magnetic field cannot be infinitely large, but reaches a final steady state, at which the gain and the loss balance each other. If this kind of self-exciting dynamo were formed within the Earth's core, a magnetic field would be maintained even if there were inevitable loss of electric current. Of course, the existence of such a simple dynamo itself cannot be imagined in the real Earth's core. Instead, Bullard proposed a chainlike combination of four different types of dynamo driven by two different patterns of dynamic motions as the power supply, which can all occur reasonably within the core. The two patterns of the assumed dynamic motions in the core are illustrated in Fig. 10-4. One pattern of motion denoted by v_1 is a relative revolution of the core fluid around the axis of rotation of the Earth. This motion represents a possible relative motion between the fluid core and the solid mantle, which may easily occur. The other pattern (v_2) represents the simplest possible configuration of self-consistent convection systems, which may take place because of a difference in temperature between the lower and the upper parts of the core. Fig. 10-4 shows the patterns of the magnetic fields of the proposed four different types of dynamo. In the

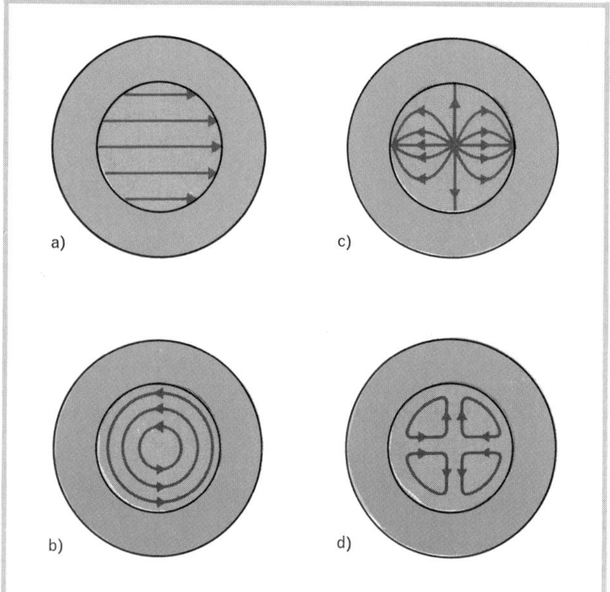

Fig. 10-3. Flow lines (red) inside Earth's core. (a) and (b) the v_1 type motion; (c) and (d), the v_2 type motion; (a) and (c), normal cross-sections at equator; in (b) and (d), the flow lines in the cross-section parallel to equator ***(T. Nagata).***

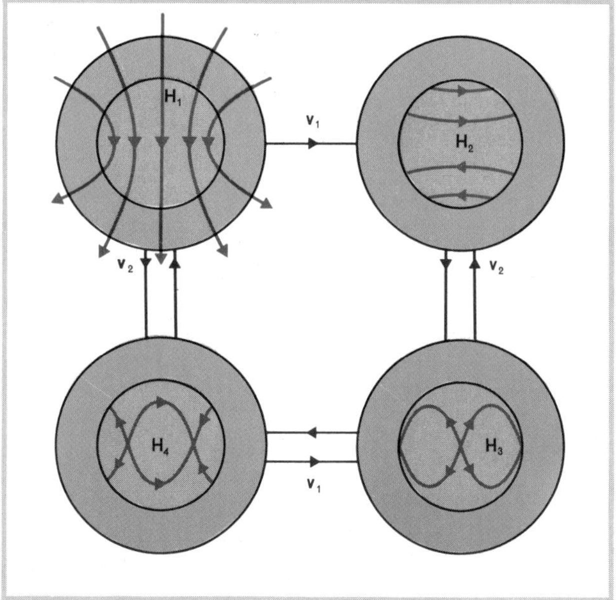

Fig. 10-4. Four magnetic fields in the Bullard-Gellman self-excited homogeneous dynamo system inside the Earth's core and coupling of H_1, H_2, H_3, and H_4 with the dynamic motions v_1 and v_2. Red lines show lines of magnetic force ***(T. Nagata).***

figure, H_1 represents a magnetic field which can be produced by uniform westward currents in the core and which resembles very closely the main part of the Earth's magnetic field. A coupling of the H_1 field with the v_1 motion produces an electromotive force $E_2 = v_1 \times H_1$, and the electric current produced by E_2 generates a magnetic field H_2 shown in the figure. The other coupling of the H_1 field with the v_2 motion produces a magnetic field pattern H_4. A coupling of the H_2 field with the v_2 motion produces an H_3 field, whereas the other coupling of the H_2 field with the v_1 motion results in only a rotation of the H_2 field. The above-mentioned processes may be expressed in a simple way as $v_1 \times H_1 \rightarrow H_2$, $v_2 \times H_1 \rightarrow H_4$, $v_1 \times H_2 \rightarrow H_2$ and $v_2 \times H_3 \rightarrow H_2$, $v_1 \times H_4 \rightarrow H_3$ and $v_2 \times H_4 \rightarrow H_1$. A summary of these processes is shown schematically in Fig. 10-5. We may conclude therefore that a cyclic chain process expressed by $H_1 \rightarrow H_2 \rightarrow H_3 \rightarrow H_4 \rightarrow H_1$ can exist within the Earth's core provided the types of fluid motion, v_1 and v_2, are present there. This is nothing but a self-exciting dynamo system composed of four types of generators and two types of engines. It must be remarked here that the H_1-field only can emerge from the core, the other three fields being confined to its interior. Thus, the H_1-field alone can be observed on the Earth's surface.

Actual mathematical calculations involved in examining these processes are so complicated, that they can be carried out only with an electronic computer of high capacity. Bullard and Gellman actually carried out the numerical calculations for this problem, and have proved that the four-generator dynamo system operates continuously within the Earth's core. The Bullard-Gellman model of a steady dynamo has been widely supported by experts of geomagnetism because it can account for various observed phenomena stemming from variations of the Earth's magnetic field with time.

It must be remembered, however, that approximate expressions are used in the chain couplings of the fields and the motions in this theory. The argument that the couplings of fields $H_1, \ldots, H_4$ with motion v_1 and v_2 produce only the same field patterns as $H_1, \ldots, H_4$ is valid only as the first approximation. As seen in Fig. 10-5, all of the four field patterns have relatively simple forms; H_1 and H_2 are axially symmetric around the H-S axis and H_3 and H_4 can be mathematically represented as proportional to $\sin^2 \theta \,{}^{\cos}_{\sin} 2\lambda$ where θ and λ denote co-latitudeand longitude respectively. When H_1 is coupled with v_1, for instance, H_2 is certainly produced as the major result, but other fields of more complicated pattern are also produced as additional minor effects. Hence, the result of the H_1-v_1 coupling must be expressed as $v_1 \times H_1 \rightarrow H_2 + h^*$, where h^* denotes the sum of additional minor fields. The situation is the same for all other couplings. Thus, a correct symbolic expression of the result of mutual couplings among the four types of field and the two types of motion may be given as $(v_1, v_2) \times (H_1, H_2, H_3, H_4) \rightarrow (H_1, H_2, H_3, H_4) + \Sigma h^*$. It was once thought that the effect of Σh^* is not very significant. With the rapid increase in the capacity of electronic computors, however, more exact examinations of the stability of the Bullard-Gellman dynamo, taking Σh^* into account, has become possible.

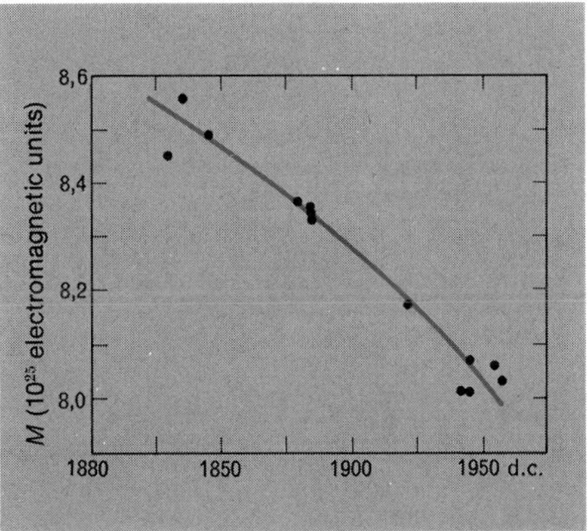

Fig. 10-6. Reconstruction of the centennial variation of the magnetic moment (M) of the Earth in the last 130 years *(T. Nagata)*.

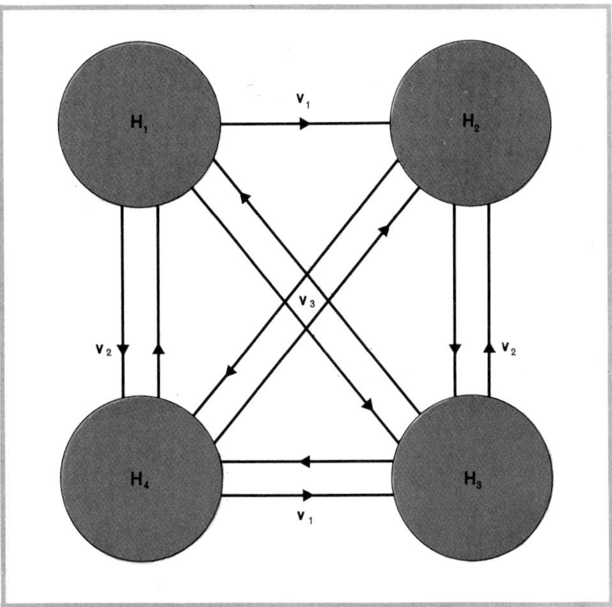

Fig. 10-5. The couplings of the magnetic fields H_1, H_2, H_3 and H_4 with dynamic motions v_1, v_2 and v_3 which occur in the modified dynamo which Bullard, Gellman and Lilley developed to simulate the Earth's magnetic field *(T. Nagata)*.

The result of such re-examinations was unfortunately negative; the Bullard-Gellman dynamo cannot achieve a steady state because the input energy coming from v_1 and v_2 is dissipated mostly in eddy currents through Σh^*, so that no balance among H_1, H_2, H_3, H_4 is established. However, further development of the electronic computor has recently resulted in a complete recovery of the Bullard-Gellman theory. In 1968, F. E. M. Lilley, a former student of Bullard's, slightly modified the Bullard-Gellman model and found the steady state of the (H_1, H_2, H_3, H_4) system even taking the Σh^* effect into consideration. In his modified model a third type of fluid motion (v_3) is introduced, though the four types of field remain as they are in the original model. The third motion v_3 has a pattern similar to that of v_2, but its phase with respect to λ is ahead of v_2 by $\pi/4$ and its amplitude differs from that of v_2.

In other words, Lilley has assumed, for the fluid motion in the core, a combination of the v_1 pattern and a distorted pattern of the convection mode. Using NASA's computer, the largest in the world, he has proved that the modified dynamo system can maintain the magnetic fields (H_1, H_2, H_3, H_4) in their steady state. In the new modified dynamo, the main reactions among (H_1, H_2, H_3, H_4) and (v_1, v_2) are such as illustrated schematically in Fig. 10-5.

History of Variation in the Earth's Magnetic Field

It is over 130 years since the direction and intensity of the Earth's magnetic field was first systematically measured in many different places. Techniques to measure the magnetic field have been revolutionized during this period. Today, the distribution of the Earth's magnetic field is measured not only on land, but also by special survey planes and artificial satellites flying over every part of the Earth's surface. It can be said therefore that the description of the Earth's magnetic field is almost perfect at present. Up to 10 years ago, the magnetic field had been measured mostly on land, and magnetic surveys of the oceans were carried out on only a few occasions when specially constructed nonmagnetic craft were used. Nevertheless, the general distribution of the magnetic field over the Earth during the past 130 years is fairly well documented. Results of analyses of the distribution of the Earth's magnetic field at each period have shown that the observed field can always be approximately represented by a magnetic field of a hypothetical bar magnet, which is located at a certain distance from the Earth's center, and whose axis deviates slightly from the direction of the axis of rotation of the Earth. Here the representation by a bar magnet is used simply for the sake of convenience of expression. Physically speaking, a bar magnet should represent a circular electric current system flowing around the axis of magnet, for example the dynamo current necessary to produce the H_1 field.

The strength, location and direction of the hypothetica magnet have changed appreciably during the past 130 years. Fig. 10-9 shows the secular change in the moment of the hypothetical magnet analyzed for various different epochs during the period. In short, the strength of the Earth's magnetic field is decreasing, as a whole, at a rate of about 5% per century. If it continues to decrease at this rate, the Earth's magnetic field will disappear within 2000 years. On the other hand, we know that the Earth is about 4.5×10^9 yeras old. On this timescale, one century is a passing moment. We must thus consider the fact that the Earth's magnetic field is extremely variable. We thus face a serious question—How has the Earth's magnetic field changed throughout its long life? Unfortunately, we had no method of obtaining even a rough value of the intensity of the Earth's magnetic field until 150 years ago, although the direction of the Earth's magnetic field had been measured at various places since the end of the 16th century.

Fig. 10-7 illustrates an example of such observations made in London since A.D. 1600. From this figure, we notice that the direction of the Earth's field is changing appreciably; the range of change of declination with respect to the N-S meridian being from 16°E to 24°W, whereas that of the inclination with respect to the horizontal plane as measured in London, has changed over 400 years from 66° to 74°. However, no record of the intensity of the field is available before 1830, and it must be remembered that even the direction of field was not known before the 16th century.

There was a promising hint, however, which suggested a possible method of estimating the intensity of the Earth's magnetic field even for earlier times. In 1853, Melloni, working in Italy, found that volcanic rocks usually have a fairly strong permanent magnetization. He believed that the permanent magnetism of some volcanic rocks is imparted to them by the inductive action of the Earth's magnetic field at the time of their cooling. Following Melloni's idea, Forgheraiter examined the remanent magnetization of still more volcanic rocks stemming from the Earth's magnetic field, and he concluded in his reports (1894–1897) that the direction of the permanent magnetization of these rocks might be an indication of the Earth's magnetic field when they are formed.

One must remember that this early work was done before the first steps of the physics of magnetism were established in the early years of the 20th century, founded upon the discovery of Curie's law (1895), Langevin's theory of paramagnetism (1910), Weiss' theory of ferromagnetism (1911) and others.

In later years after the physics of magnetism of solid materials had been fairly well developed, Thellier (1938) came back to this problem and studied the remanent magnetization of bricks and tiles, which are made from baked earth. At about the same time, Nagata (1939) re-examined, on the basis of recent developments in the physics of magnetism, the early work on natural rocks done by Forgheraiter and his successors. Thellier found that the permanent magnetization of baked earth is acquired by cooling the samples in the Earth's magnetic field. Nagata also found that the strong remanent magnetization of recent volcanic rocks is exactly reproducible by cooling them from a certain high temperature in the Earth's

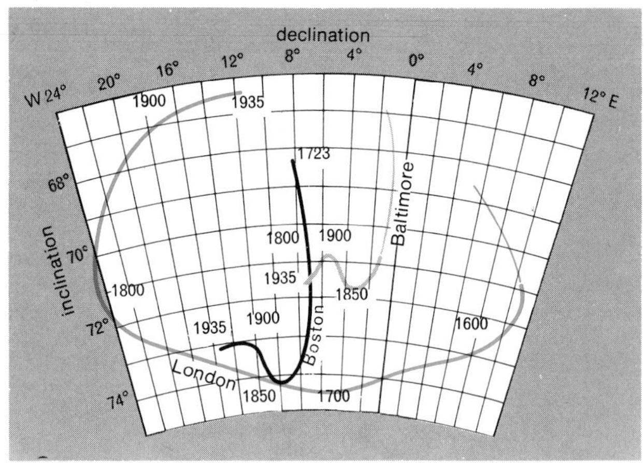

Fig. 10-7. Centennial shift of magnetic declination and inclination measured at London, Boston and Baltimore ***(from D. A. Flemming).***

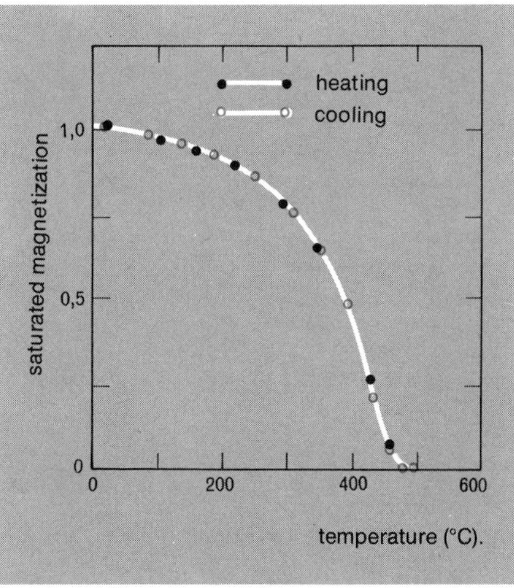

Fig. 10-8. Variation of the saturated magnetization of a typical magmatic rock on the basis of temperature. The intensity of saturated magnetization at 0°C equals 1 ***(T. Nagata).***

present magnetic field, and that the intensity of acquired remanent magnetization is proportional to the intensity of the applied magnetic field. Since the remanent magnetization of these volcanic rocks acquired by cooling in a magnetic field, is remarkably larger than the remanent magnetization of the same samples acquired in the same field without temperature change, the former magnetization has been called specifically "Thermo-remanent magnetization." A very important characteristic of thermo-remanent magnetization is that it is extremely stable against mechanical, thermal, magnetic and other external perturbations. Why the thermo remanent-magnetization is so strong and so stable has been extensively studied by Nagata, Néel *et al*. Their answer may be briefly summarized as follows: Rocks contain a large number of very fine particles of magnetic minerals. In the case of igneous rocks, the magnetic minerals are mostly magnetite with some portion of Titanium oxides. The magnetization of such minerals and rocks containing them decreases with increasing temperature and disappears at a certain critical temperature called the Curie point, as illustrated in Fig. 10-12. Obviously, these rocks and minerals can be magnetic only at temperatures below their Curie point. When particles of magnetic minerals are very fine, say a few microns in diameter or less, they have special magnetic properties: if they are magnetized by a strong magnetic field, they become stable permanent magnets. In technical terms, such a magnetic property is described by saying that "fine magnetic particles are magnetically hard," or that "their magnetic coercive force is large." With an increase in temperature, the magnetic hardness or coercive force also decreases, becoming zero at the Curie point. Hence, fine magnetic minerals in rocks can be easily magnetized even by a comparatively weak field at temperatures just below Curie point. With a decrease in temperature, both the intensity and hardness of this acquired magnetization increases remarkably, remains as strong and stable remanent magnetization at room temperature. Thus, when a rock is cooled down from a temperature higher than its Curie point in the presence of the Earth's magnetic field, it becomes a relatively strong and stable permanent magnet at room temperature whose intensity is proportional to that of the Earth field and whose direction is parallel to the field. An experimental example given in Fig. 10-16 shows how the thermo-remanent magnetization is strong and stable. In the figure the ordinate gives the intensity or remanent magnetization and the abscissa indicates the strength of the alternating magnetic field which can demagnetize the remanent magnetization.

Curve A shows how the thermo-remanent magnetization of a rock fragment acquired in the Earth's field (about 0.5 Oersteds) decreases with an increase of alternating demagnetization field. Even for demagnetization by the alternating field of 200 Oersteds, 85% still survives. Curve B represents the ordinary remanent magnetization acquired by applying a magnetic field and is almost fully demagnetized at 100 Oersteds. It has been concluded thus that the remanent magnetization of igneous rocks which contain a fair amount of magnetic minerals can be considered as "fossils" of the Earth's magnetic field when those rocks were cooled down.

The mechanics of the acquisition of and the characteristics of thermoremanent magnetization of bricks and tiles, such as studied by Thellier, should be the same as for igneous rocks, because bricks and tiles also contain a large number of fragments of magnetic minerals. Using the remanent magnetization of igneous rocks and baked earths as the fossils, one can estimate the direction and the intensity of the Earth's magnetic field in earlier times, provided the ages of these samples can be measured with some degree of accuracy. In recent years the C^{14} method and K-A or Sr-Rb method have been widely used, with a reasonable degree of accuracy to determine the ages of samples. Therefore, we can now estimate variations in the intensity and direction of the Earth's magnetic field in the past.

A summary of such measurements of the intensity of the Earth's magnetic field in the past 9000 years is shown in Fig. 10-10 where the ratio of the intensity in the past to the present intensity is plotted against time.

Some points in the figure were obtained from bricks or pots whose ages are recorded, or from volcanic lavas whose ages of flow are known.

Because of limited availability of such samples, the data shown in the figure come from Europe (France, Italy, Czechoslovakia and the Soviet Union) from Asia (Japan) and from Central and South America, so far. As seen in Fig. 10-10, the

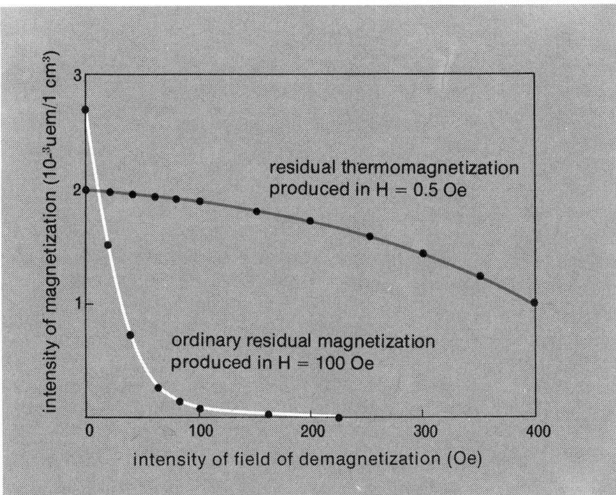

Fig. 10-9. Stability of the residual thermomagnetization and ordinary residual magnetization of a typical magmatic rock when subjected (alternate H) to demagnetization *(T. Nagata)*.

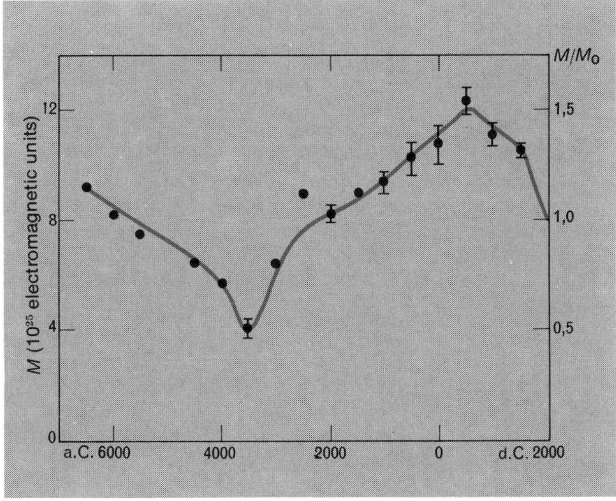

Fig. 10-10. Movement of the Earth's magnetic moment (M), expressed in terms of its daily value (M_o), during the last 9000 years *(based on a synopsis by A. Cox, 1968)*.

intensity of the Earth's magnetic field has changed in a wavy pattern during the past 9000 years. We may expect therefore that the Earth's magnetic field will probably not vanish in 2000 years, but will continue to follow a varied course as it has done in the past.

Another question which is directly concerned with this study is when was the Earth's magnetic field first produced. Even specialists are surprised to see that very old rocks, $10^8 \approx 2 \times 10^9$ years old, indicate that the order of magnitude of the Earth's magnetic field in those times was approximately the same as now, though there were fluctuations similar to the curve given in Fig. 10-18.

Reversals of the Earth's Magnetic Field

The determination of intensity and direction of the Earth's magnetic field in the past by studying natural rocks of various ages is now being carried out all over the world to clarify the distribution of the field over the whole Earth and to determine its change during a long period of the Earth's life. This branch of scientific research is called paleomagnetism. In the course of the paleomagnetic study, it became very clear that the Earth's magnetic field was frequently reversed in the past. It is geologically obvious that an upper layer is always younger than a lower layer. Actually, the K-A ages of selected lava layers, given in units of a million years (My) in Fig. 10-12, show that the age of lava became older with an increase of depth measured from the land surface. The direction of remanent magnetization of the two top layers (3.62 and 3.68 million years in age) and the bottom one (4.5 million years in age) is almost parallel to that of the Earth's present magnetic field. However the remanent magnetization of all the other lava layers in between is oriented almost antiparallel to the present Earth's field. Since these lavas are reversely magnetized they are noted by R in the figure. Table 10-1 illustrates more precise behaviors of these "normal" and "reverse" remanent magnetizations. If we assume that the distribution pattern of the Earth's magnetic field at the geologic epochs was not appreciably different from that at present, we can estimate the position of the pole of the Earth's magnetic field at those epochs from the location (latitude and longitude) of the samples and the direction of their remanent magnetization. The pole position thus estimated has been named "the virtual magnetic pole position." Fig. 10-13 gives the virtual magnetic pole positions corresponding to each lava flow. Here full and hollow circles represent respectively the north and south poles. Com-

THE POLAR LIGHTS AS A MEANS OF STUDYING THE EARTH'S MAGNETIC FIELD

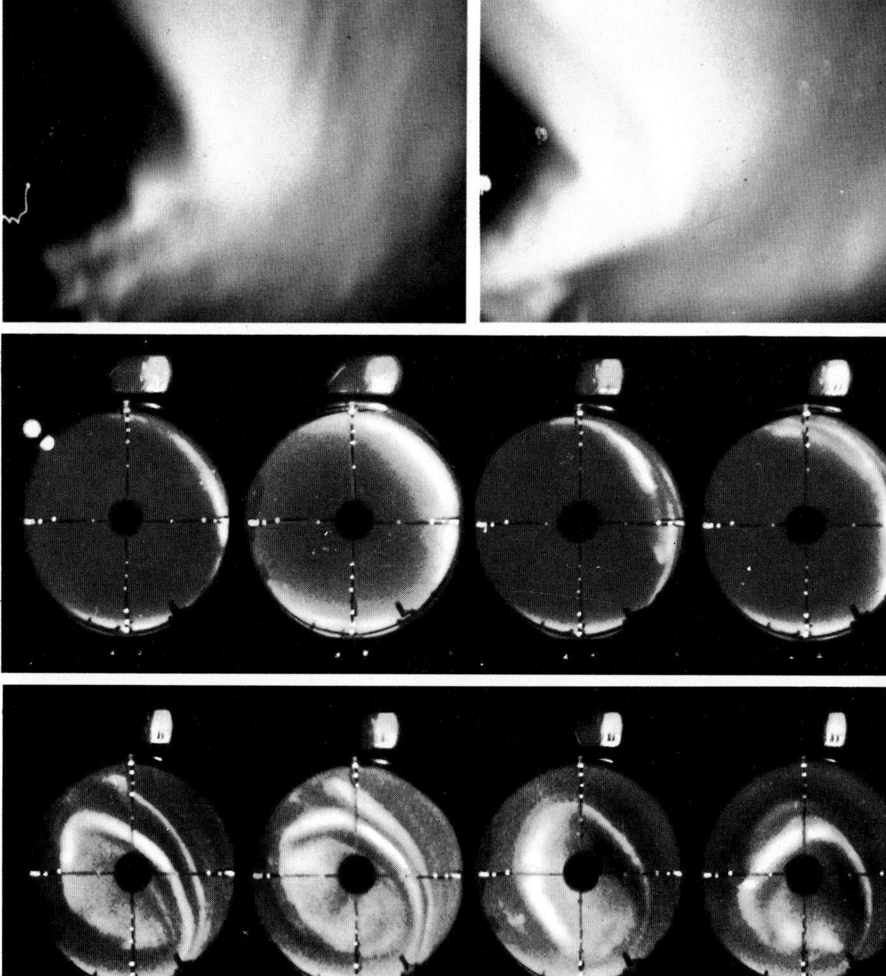

Fig. 10-11. As well as being measured or perceived with the help of specific instruments, the Earth's magnetic field can also be studied in some of its particularly important manifestations directly with the naked eye and a good camera. The phenomenon which permits this approach to the study of the Earth's magnetic field is the polar aurora or the polar lights—extraordinary "light-shows" which can be seen in the hours of darkness at high latitudes (between 65° and 70° north and south). How are these lights related to the Earth's magnetic field? As well as emitting light, the Sun is constantly emitting in all directions a stream of particles (protons and electrons), the solar wind which affects all the planets in the solar system. But the Earth's magnetic field prevents these particles from reaching the planet's surface and usually forces them to veer off and spiral along the lines of force. But near the magnetic poles these lines of force are directed towards the Earth and the particles which reach this area, via complex spiraling trajectories, penetrate the atmosphere and excite its molecular and atomic components, thus giving rise to the typical lights of the aurora. The frequencies and variations of the polar lights are therefore indicative of the actual state of the Earth's magnetic field, as well as the state of the solar wind. The two sequences of pictures to the left and below are photos taken respectively of an aurora in the

pared with the north pole position at present, indicated by a cross in the figure, it may be concluded that the virtual pole positions deduced from lavas' magnetizations, are either close to or almost completely opposite to the present position. Fig. 10-14 shows the intensity of the Earth's magnetic field corresponding to each lava layer.

Provided that the magnetic properties of those rocks of normal and reverse magnetizations are all the same, the results may indicate that the Earth's magnetic field was suddenly reversed from N to R during the period from 4.50 to 4.38 million years ago and again from R to N during 4×10^4 years from 3.72 to 3.68 million years ago. Here the absolute magnitudes of the field intensity before and after the reversals are roughly of the same order of magnitude. When such a sharp reversal of remanent magnetization was observed, a theoretical question was raised by Néel (1951) and an experimental one by Nagata and Ueda (1951) regarding the possibility of self-reversal of the thermo-remanent magnetization of some rocks. According to the experimental results, the thermo-remanent magnetization of some igneous rocks is oriented exactly antiparallel to the direction of an applied magnetic field. Since then, a number of igneous rocks having similar magnetic characteristics have been found. However, detailed analytical studies of these peculiar rocks have shown that a very close interaction between two different magnetic phases in minerals plays an essential role in the self-reversal phenomenon, and that this particular characteristic can be clearly identified by experimental tests in laboratories. Thus, those abnormal rocks which have the character of self-reversal are eliminated by these laboratory tests from the paleomagnetic studies on the ancient geomagnetic field. It is proved that none of lavas shown in Fig. 10-12 and Fig. 10-13 has such an abnormal character. Hence, those results given in the figures may certainly suggest the reversals of the Earth's magnetic field itself.

In Fig. 10-18, a summary of the latest information regarding the reversals of the Earth's field during the past 4.5 million years is illustrated (Cox 1969). As seen in the figure, the reversal of the field took place 24 times during 4.5 million years, that is, with a frequency of five times per million years. The time required to complete a reversal also has been studied. In the case of Fig. 10-14, it can be estimated to be less than 4×10^4 years for the reversal whose age is about 3.7 million years. This kind of estimate depends essentially on the accuracy of determining the ages of rocks. A tentative conclusion so far obtained is that a complete reversal of the Earth's field takes about $10^3 \sim 10^4$ years.

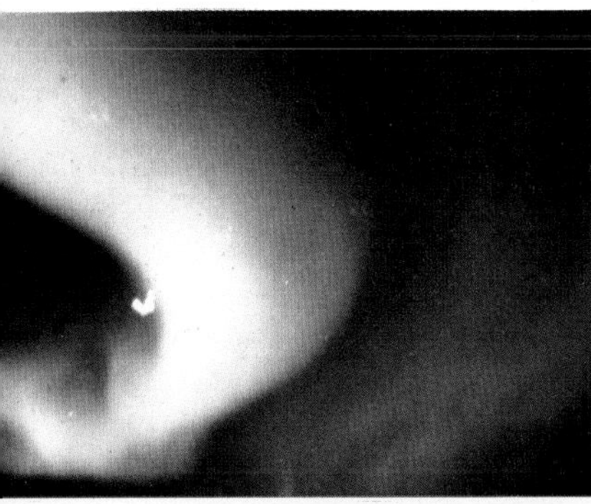

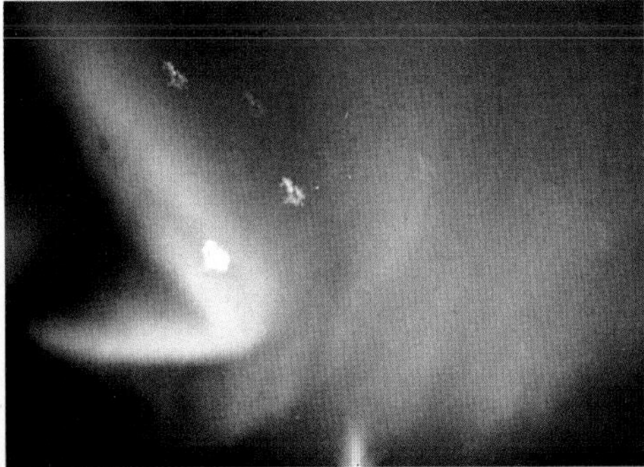

northern hemisphere and an aurora in the southern hemisphere. The first was taken over the Arctic by NASA researchers. All the photos were taken within 10 minutes: this gives a measurement of the speed at which this type of phenomenon develops in time *(photo NASA)*. The sequence below, on the other hand, was taken from Earth; the photos were taken using a special parabolic mirror capable of reflecting the entire sky and thus producing an overall picture of the phenomenon. The whole sequence was taken within 24 minutes *(photo H. Morozumi, Stanford University)*.

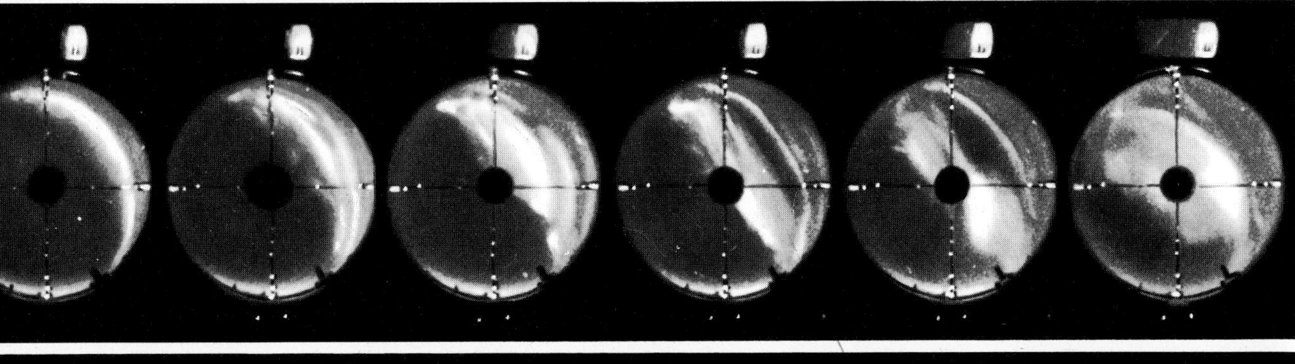

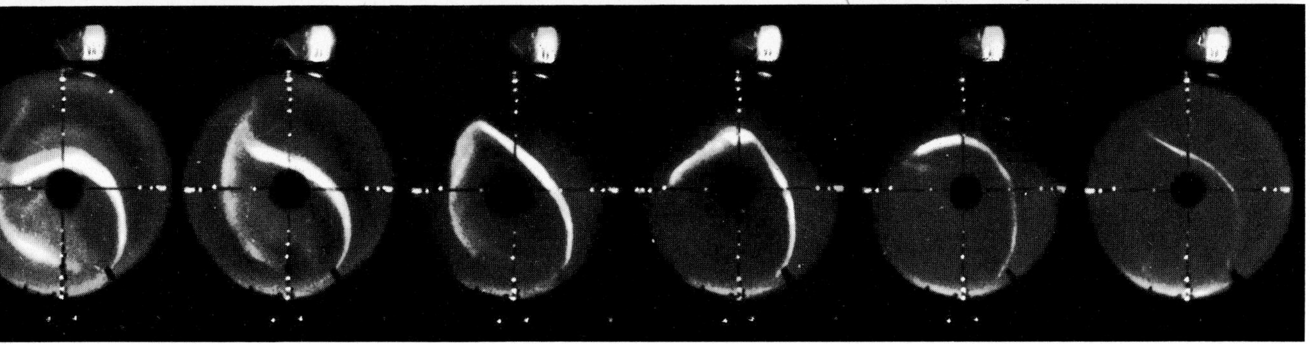

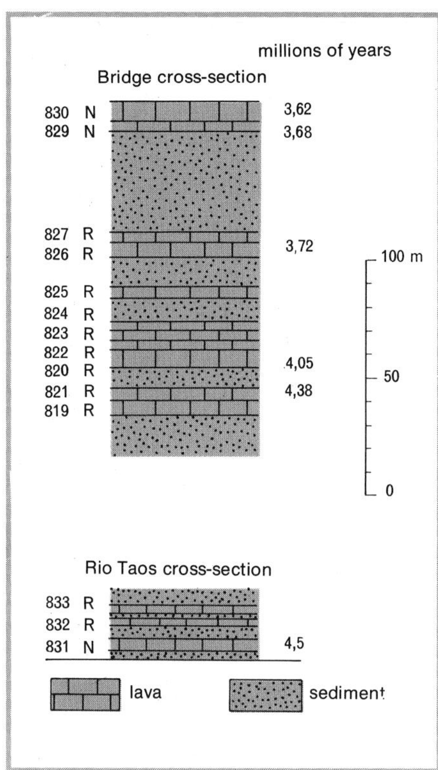

Fig. 10-12. **Magnetic polarity and ages of a series of lava flows. The numbers on the left identify the sample; *N* indicates normal magnetic polarity and *R* polarity reversal. The numbers on the right are the ages of the lava (with the K-Ar method)** *(T. Nagata)*.

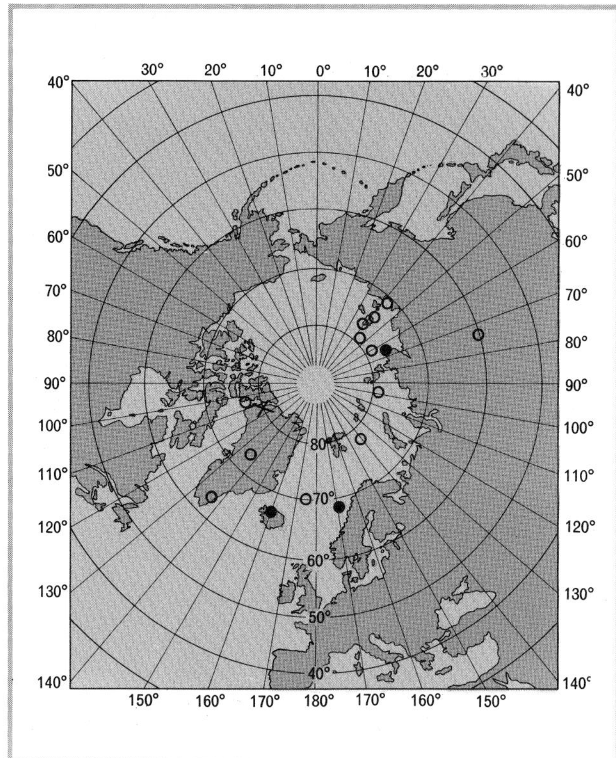

Fig. 10-13. Distribution of the north (●) and south (○) potential magnetic poles around the actual geomagnetic pole (+), derived from measurement data shown in Fig. 10-12 *(T. Nagata)*.

We must now ask why the Earth's magnetic field has been reversed so frequently? This question has not yet been fully answered, although it has been attacked by a number of scientists. Two simplified models which have the characteristic of the field reversal have been proposed. One is Rikitake's model (1958), shown in Fig. 10-5, and another is Herzenberg's model (1958) shown in Fig. 10-16. In both, two dynamos are electromagnetically coupled. Rikitake's model is an extension of Bullard's self-exciting dynamo shown in Fig. 10-2, whereas the Herzenberg model is a further extension of Rikitake's model effected by replacing lead wires, brushes and solenoids by more realistic continuous blocks. Numerical computations based on both of these models predict intermittent reversals of the magnetic field after repeated oscillations as shown, for example, in Fig. 10-19. Actually Lews and his colleagues (1967) have experimentally demonstrated the field reversals using an apparatus with a design based on the Herzenberg model. The studies may lead to a general conclusion that an electromagnetic coupling between two self-exciting dynamo systems can cause repeated reversals of the magnetic field under appropriate conditions. It is difficult, however, to accept such a simple electromagnetic system as given by Fig. 10-18 for the Earth's core. In this sense, the problem of reversals of the Earth's field is a very difficult one in physics of the Earth at present. As a way to solve this problem, this author calls attention to the instability of the original Bullard-Gellman dynamo and to the stability of Bullard-Gellman-Lilley dynamo. If we can assume that the convective motions within the Earth's core are gradually changing their patterns, the H_1-field can be stable and can increase to a certain extent during the period that the pattern has an asymmetric configuration. When the configuration becomes symmetric as assumed by Bullard and Gellman, the four fields become unstable and they may collapse, but a certain portion of the fields may remain as the residual field.

When the asymmetry of the convection is recovered, the whole dynamo system must become stable again and the four fields will grow. Here, one must keep in mind that the behaviors of the four magnetic fields are essentially the same in the basic equations as those of a system of fields in which each of the four magnetic fields is reversed, provided that the pattern of the fluid motions is conserved. If a component of the H_1-field in the small residual field has a normal direction, then the normal field will grow when the whole dynamo system is recovering its stability. If, on the contrary, the residual field has a reversed H_1-component, then the reverse field will grow. We can assume thus that the recovered field can either be normal or reversed equally by accident. Actually, in Table 10-1, the sums of the normal and reverse periods during the past 4.5 million years are 2.17 and 2.33 million years respectively, or approximately two halves, though the duration of individual magnetic polarities varies over a wide range of time.

Remanent Magnetization of Ocean Sediments and Long Lineation Pattern of Geomagnetic Anomalies over Oceans

It has long been known that continental sedimentary rocks also have remanent magnetization, and that the uppermost parts of ocean sediments have the remanent magnetization whose direction coincides approximately with that of the Earth's magnetic field now. We can demonstrate experimentally such a magnetic polarity of sediments in a laboratory by dropping a large number of fine particles of sedimentary rocks through water in a vessel. The magnetic polarization of the deposit thus obtained is in exact agreement with the direction of the Earth's magnetic field at the location of the experiment. When the water depth is sufficiently large, say more than several tens of centimeters, the intensity of remanent magnetization approaches a constant saturated value. The remanent magnetization of sediments thus acquired has been named deposital remanent magnetization. The detailed mechanics of acquisition and the characteristics of the deposition of remanent magnetization have also been closely studied. According to those studies, small particles of magnetic minerals in sediments are oriented along the Earth's magnetic field while they are slowly deposited through the water. Thus deposits coagulating together with nonmagnetic particles at the ocean bottom have the characteristic remanent magnetization.

Recently, on the other hand, extensive core sampling of various ocean sediments have been carried out. Fig. 10-22 shows an example of the results of measurements of the depositional remanent magnetization of a deep core of ocean sediment. In this figure, only the polarities of remanent magnetization, namely, normal or the reverse, of the sediments are indicated in black or white in the columns. Comparing the sequence of the magnetic polarity of ocean sediments with that of igneous rocks shown in Fig. 10-18, one notices that their patterns are extremely similar to earh other. One must take into consideration, in the comparison, that tracing the resolving power with respect to time for the magnetic polarity of sediments is much harder that for igneous rocks, because the rate of deposition of ocean sediments is extremely small, say of the order of magnitude of 1 mm per 1000 years and the linear dimensions of a test sample necessary for the magnetic measurements are several centimeters. Since all the magnetic polarity patterns of ocean sediment cores collected from the bottoms of various oceans, North and South Pacific, North and South Atlantic and Indian Oceans, always show similar configurations of N and R magnetic polarities, we may conclude that the pattern represents the change in the direction of the Earth's magnetic field at the time when the particles of the magnetized sediments were depositing. Then, the time scale of deposition of those sediments can be estimated by a comparison of their magnetic polarization patterns with the standard scale such as that given in Fig. 10-18. The speed of deposition thus estimated ranges from 0.5 mm to 3 mm per 1000 years. In this way, the measurement of magnetic polarity of ocean sediments has become an extremely valuable tool for the study of marine geology and geophysics.

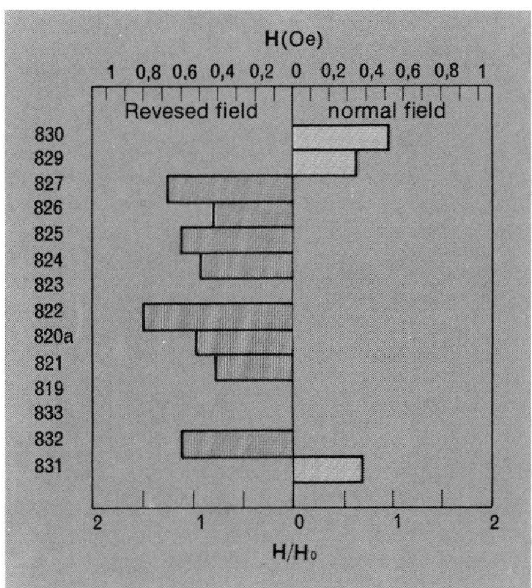

Fig. 10-14. Intensity and course of the Earth's magnetic field as deduced from the residual magnetism of the lava flows taken into account in Fig. 10-12. On the right, course of the field in the last 4.5 million years *(T. Nagata).*

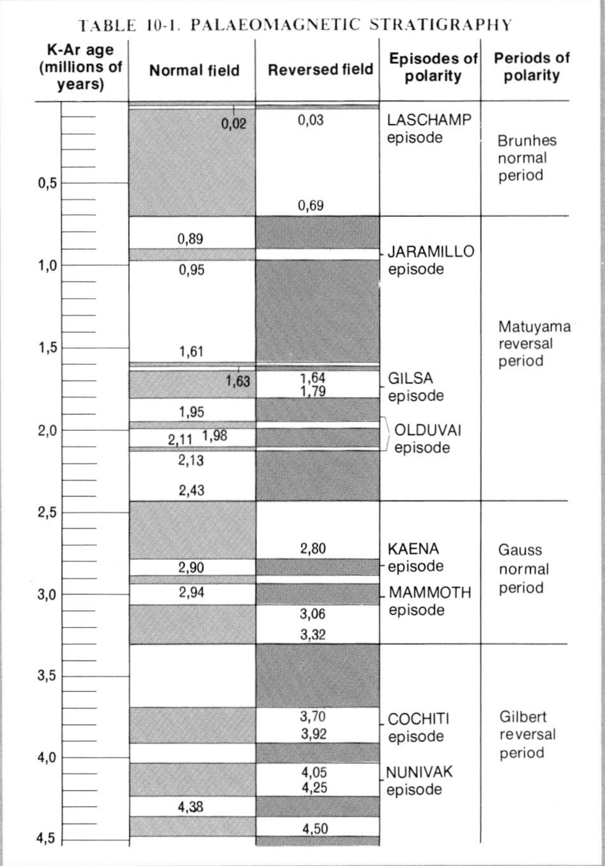

The effect of the regular pattern of remanent magnetizations of ocean sediments cannot be detected on the sea surface, because their intensity is too weak. Nevertheless, regular local anomalies of the distribution of the Earth's magnetic field are observable on the sea surface. Fig. 10-19 illustrates, for example, a group of profiles of the intensity of Earth's magnetic field along eight lines crossing the Mid-Atlantic Ridge perpendicularly. The area of this survey indicated in Fig. 10-19 is just on the line of the Mid-Atlantic Ridge shown in Fig. 10-20, and the dotted trends of all other positive peaks (indicated by numerical figures) on both sides of the ridge are parallel to the ridge line. Moreover, the distribution of positive and negative anomaly trends on both sides is approximately symmetric with

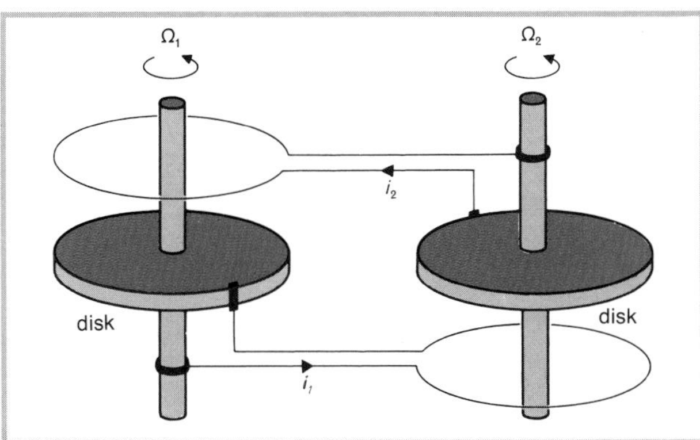

Fig. 10-15. System with two coupled disk dynamos used to simulate Earth's magnetic field *(from T. Rikitake, 1958).*

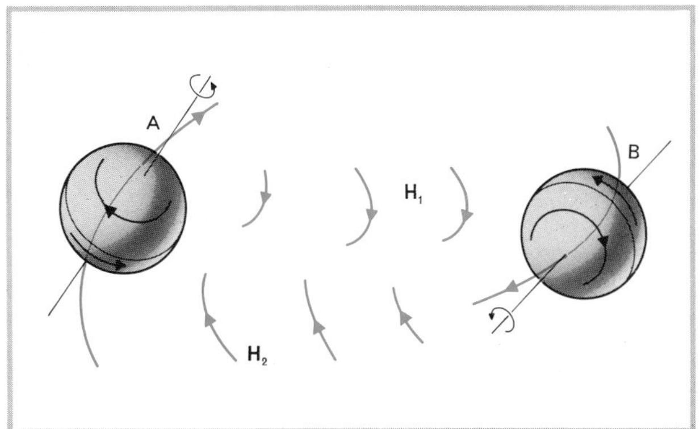

Fig. 10-16. Herzenberg's dynamo. Rotors A and B are coupled electromagnetically by a very conductive medium which fills the space between them *(from Herzenberg, 1958).*

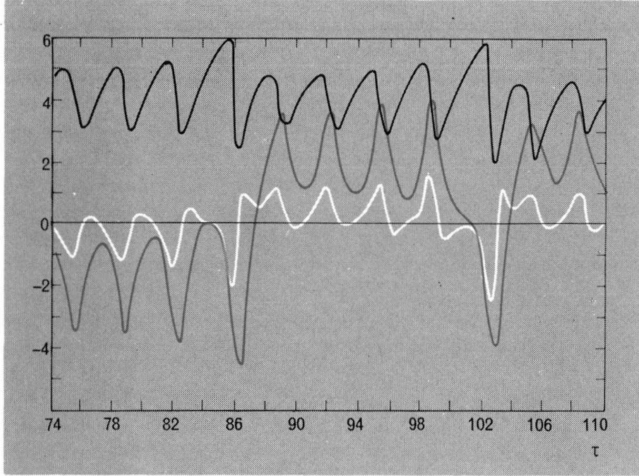

Fig. 10-17. An example of the solution of the possible movements in the pair of Rikitake disk dynamos: the black curve is a parameter proportional to the angular speed of the rotors; the white curve is the electrical current i_2 in Fig. 10-15; the blue curve is the current i_1 in Fig. 15. *(from Allan).*

respect to the ridge line. Similar features of the magnetic anomaly pattern have been found in the neighborhood of the Mid-Atlantic Ridge in other latitudes. As seen in Fig. 10-19, fluctuations of the magnetic anomalies amount to several hundreds of gammas in intensity. (For reference, the total intensity of the Earth's magnetic field in this area is about 50,000 gammas.) Then, our next problem may be how to interpret such a regular long lineation pattern of magnetic anomaly which is parallel to the ridge line. With our present knowledge of magnetic properties of natural rocks the observed regular pattern of magnetic anomalies can be explained if we assume that basaltic dikes having thermo-remanent magnetization intrude towards the ocean bottom within an interval of 40 km on both sides of the ridge line. However, results of seismic exploration of this area do not indicate such a subterranean structure, but instead they show that, below the uppermost layer of sediments of 0.4 ~ 1.0 km in thickness, a horizontally continuous surface of igneous rocks is exposed. Dredged rock samples from this area suggest that these igneous rocks are mostly basaltic, as deduced from the magnetic data. The possible structure based on this data would be that the horizontally extended basaltic layer is magnetized alternately into normal and reverse directions as shown in Fig. 10-21. In this figure, the normal and reverse magnetizations occur alternately every 20 km in distance from the ridge line center towards both sides, and the central normal magnetization is assumed to be stronger than the others. Insofar as the observed pattern of magnetic anomaly is concerned, the model structure shown in Fig. 10-21 can well explain all observed aspects of the phenomenon.

On the other hand, the Mid-Atlantic Ridge has various remarkable features. Topographically, the ridge is very mountainous especially near the crest; a deep valley splits the crest of the ridge, and a similar topography is extended along the crest line from the north end to the south end of Atlantic Ocean. Epicenters of shallow earthquakes are almost exactly confined to the long, narrow belt of the valley in the ridge, suggesting that only this belt is seismically active in the Atlantic Ocean. Another important suggestion regarding the structure and activity of the ridge comes from volcanic activity and the geologic structure of Iceland, which is located just on the ridge line as shown in Fig. 10-20. Iceland is composed of a large number of lava flows and injections of dikes, and geological evidence suggests that old lavas there have been carried apart more than 400 km by crustal drift since the time when they were initially formed.

Summarizing all these observed data, we may state with a certain high degree of confidence that the Mid-Atlantic Ridge represents a long crack in the Earth's surface through which magma is continually intruding from below. A reasonable deduction from the magnetization pattern shown in Fig. 10-21 is then the following. The crust under the crest of the ridge is being extended by the intrusion of numerous dikes, and consequently the older crustal rocks are shouldered aside. The continuous intrusion of dikes may cause earthquakes under the crest, and the youngest crustal block consisting of recent dikes and lava flows there, may be magnetized in the direction of the present Earth's magnetic field when they are cooled from their initial high temperature. The older crustal blocks adjacent to the youngest block on both sides consist of reversely magnetized rocks which were emplaced at the crest

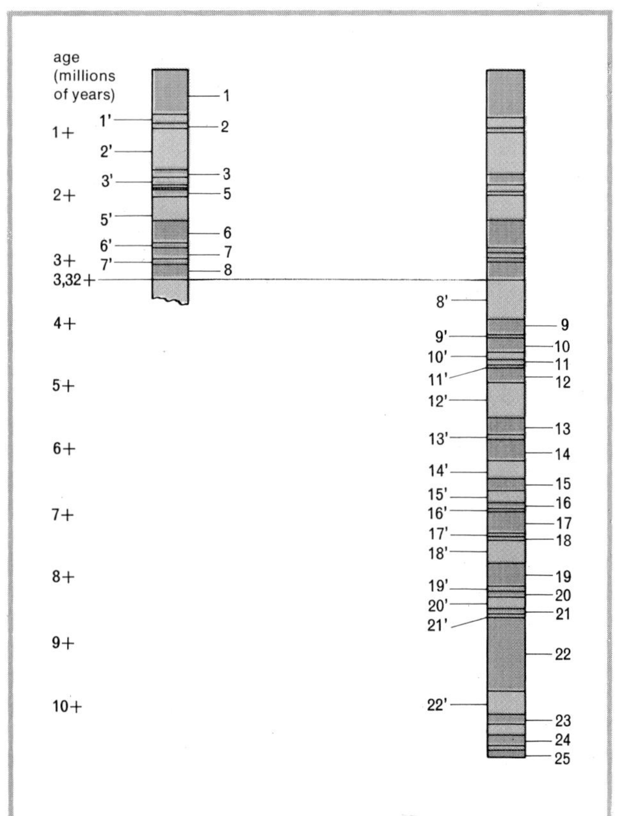

Fig. 10-18. Distribution of magnetic polarity over a long vertical field taken from inside oceanic sediments. The areas of normal polarity are those in color *(from A. Cox, 1968).*

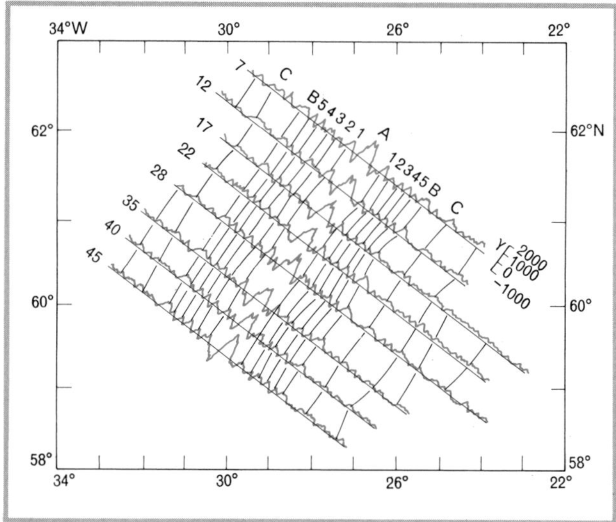

Fig. 10-19. Behavior of anomalies in the total intensity of the Earth's magnetic field taken along eight routes traced at right angles to the mid-Atlantic Ridge *(from Hirtzler et. al., 1966).*

Fig. 10-20. Morphology and tectonics of the ocean-bed of the Atlantic ocean, as shown by the oceanographic research carried out in the 1960s and 1970s. From the point of view of magnetization, the sea-bed on either side of the mid-Atlantic Ridge which runs down the middle typically shows symmetrical anomalies of the same age *(Rand McNally and Co., R.L. 75-GP 1-2).*

of the ridge when the Earth's magnetic field was reversed. Because the adjacent blocks may well be invaded by some new dikes and covered by new lavas, their total magnetization (of the reverse direction) may be weaker than that of the youngest crustal block in the center. With this process occurring repeatedly in the presence of the Earth's magnetic field, which has varied with time as shown in Fig. 10-18, a pattern of alternately normal and reverse magnetized blocks, such as is illustrated in Fig. 10-21, could have been built up in the oceanic crust. This is a hypothesis of ocean-floor spreading which suggests that the Atlantic Ocean is being spread continuously towards both east and west by the pressure along the Mid-Atlantic Ridge. A hypothesis that the west coasts of Europe and Africa were once connected with the east coasts of North and South America has been frequently discussed since the first proposal made by Wegener. If we could rotate North and South America eastward around the north pole, their eastern coasts, including their continental shelfs, would

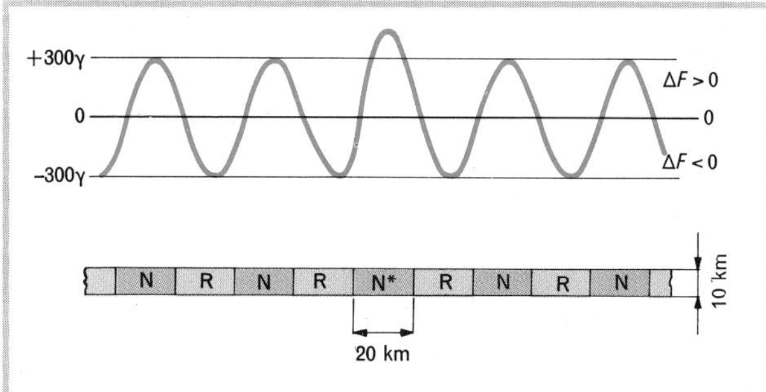

Fig. 10-21. Alternate distribution of the course of magnetism in the crust on the ocean-bed (below) and consequent anomalies ΔF of the Earth's magnetic field (above) on the surface above the mid-Atlantic Ridge *(based on the idea originally put forward by Vine and Matthews, 1963).*

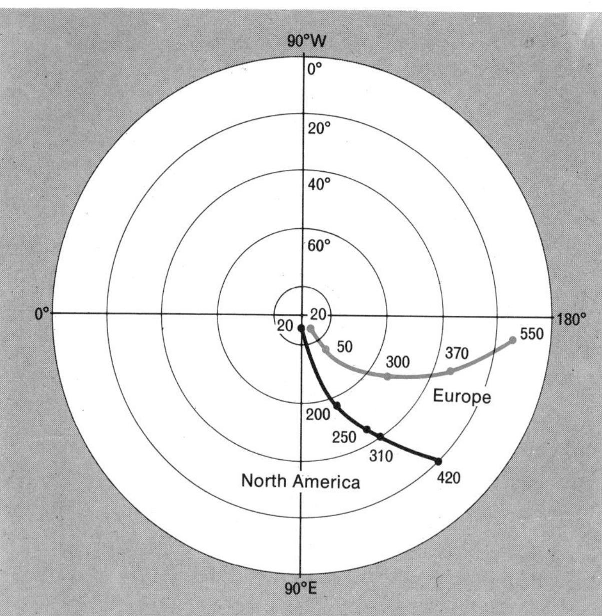

Fig. 10-22. Reconstruction of the migration of the potential magnetic pole based on the analysis of a series of ancient rocks in Europe (in color) and North America (in black). The numbers give the age in millions of years *(based on original data obtained by Runcorn, 1956).*

fit the boundary lines of the continental shelfs of the east coast of the Atlantic Ocean. Moreover, the old geologies on both sides of Atlantic Ocean would become continuous by this rotation. The age of ocean bottom rocks dredged from various locations in the Atlantic are indicated in Fig. 10-22, where the rocks near the Mid-Atlantic Ridge are 1 ~ 10 million years old, whereas those near both sea coasts are 100 million years old or older. That is to say, the ocean bottom is youngest near the median ridge and oldest near the coasts. This evidence also seems to support the hypothesis of ocean floor spreading. There is other evidence to support this hypothesis from paleomagnetic studies. For geologically young rocks of an age less than about 50 million years, the north or south virtual magnetic poles, estimated from their magnetic polarization, are located near the present north pole, as in the case of Fig. 10-13. For older rocks, however, the similar paleomagnetic studies have indicated that the virtual magnetic poles are deviated from the present pole and are inclined toward the present equator. Fig. 10-22 shows the positions of virtual magnetic poles thus estimated from a sequence of European rocks and from a sequence of American rocks. In both sequences, the older pole position is more inclined towards the equator. However, if these virtual pole positions represent the real pole positions of the Earth's magnetic field at various geologic epochs, they must lie on a single locus. Runcorn (1956) suggested that the two loci, one from European data and the other from American, can be made to coincide if the American locus is rotated counterclockwise around the north pole by about 30°. If we assume that the European and American continents were connected on the line of the Mid-Atlantic Ridge more than 100 million years ago and that since then they have been gradually separated by continental drift, the paleomagnetic result of virtual pole positions in Fig. 10-22 can be well understood. Thus the hypothesis of ocean floor spreading and continental drift is favorable from many different viewpoints.

Epilogue

We have seen in Table 10-1, Figs. 10-5, 10-18, 10-21 and 10-22 that repeated reversals of the Earth's magnetic field in the past geologic times have been imprinted on magnetic fossils, on piles of volcanic lavas on land, on the sediments of the ocean bottom and on the laterally spreading ocean floor basaltic crusts. On the basis of the observed fact that the Earth's magnetic field has been repeatedly reversed, the evolution of the Earth has been analyzed with fruitful results, such as the hypothesis of ocean floor spreading and continental drift.

One must remember that the significant information about the behavior of the Earth's magnetic field during the past long life of the Earth has been obtained from studies based on the strength and direction of remanent magnetization of very tiny particles of magnetic minerals in rocks. When the remanent magnetization of rock was discovered, nobody could predict such an extensive development of its application to earth science.

Magnetic methods are widely used for studying the ocean floors and the mid-ocean ridges not only in the Atlantic but also in the other oceans. However, we have not yet fully discovered the mechanics of the maintenance and reversals of the Earth's magnetic field, which have long operated within the Earth's core. We do not yet have details of a reversal of the Earth's magnetic field, because the period of reversal is too short to leave reliable evidence on rocks which can be determined by existing methods of research. Several research groups throughout the world are attacking this difficult problem using various new techniques, and we shall undoubtedly have an answer to this question in the future. If the Earth's magnetic field vanishes completely for a short time during field reversal, the solar wind must hit the Earth directly and affect all lives on the Earth. Fortunately, the present Earth is surrounded by a thick atmospheric layer which protects these lives from direct attack by the solar wind though there may be secondary effects.

TAKESI NAGATA

Bibliography: Cox. A., *Geomagnetic reversals,* in Science, CLXIII, **3864**, 237 (1969); Cox. A., *Length of geomagnetic polarity intervals,* in J. Geophys. Res., LXXIII, **10**, 3247 (1968); Bullard E. C., *Reversals of the Earth's magnetic field,* in Phil. Trans. (GB), A, CCLXIII, 481 (1968); Nagata T., Ozima M., *Paleomagnetism,* in Physics of Geomagnetic Phenomena I, 103 (1967); Rikitake T., *Electromagnetism and the Earth's interior,* Amsterdam (1966); Blackett P. M. S., Bullard E. C., Runcorn S. K. (ed.), *A symposium on continental drift,* in Phil. Trans. (GB), A, CCLXVIII, 1 (1965); Nagata T., *Main characteristics of recent geomagnetic secular variation,* in J. Geomag. Geoele., XVII, **3-4**, 263 (1965); Nagata T., *Rock magnetism,* Tokyo (1961); Bullard E. C., Gellman H., *Homogeneous dynamos and terrestrial magnetism,* in Phil. Trans. (GB), A, CCXLVII, 213 (1954); Elsasser W. M., *The Earth's Interior and geomagnetism,* in Rev. Mod. Phys., XXII, **1**, 1 (1950); Nagata T., *The mode of causation of thermoremanent magnetism in igneous rocks,* in Bull. Earthquake Res. Inst., Tokyo Univ., XIX, **1**, 49 (1941); Thellier E., *Sur l'aimantation des terres cuites et ses applications géophysiques,* in Ann. Inst. Phys. Globe, Univ. Paris, XVI, 157 (1938).

THE Atlantic is becoming wider by about 5 cm every year, and the Americas are traveling away from Europe and Africa by the same amount. These conclusions were rapidly reached at the beginning of the sixties thanks to studies of the magnetic anomalies of the ocean floors described in the article by Nagata. However, whole continents cannot drift around, and oceans cannot get wider or narrower without other important effects taking place. In this mechanism some scientists immediately saw a way to provide a unified global explanation of all of the planet's activity: its seismic as well as its volcanic activity, which is at the root of the mountain ranges. A new model of the Earth's dynamics based on plate-tectonics or global tectonics had been constructed on this foundation. The development of this model depends on a special method of analysis of earthquakes. An earthquake occurs at a particular point in the Earth's crust, when a certain critical amount of elastic energy has accumulated. If this energy increases beyond the limit of the resistance of the rocks, these break up and an earthquake occurs. It is obvious that the direction of the fragmentation of the rocks depends on the direction of the stresses acting on the rock. A group of seismologists carried out a systematic analysis of the earthquakes taking place, for example along the oceanic ranges, and observed that in this case the stresses were all perfectly coherent, with a tendency toward a lateral displacement of the oceanic crust with respect to the range. Thus, after having proved that such displacements had taken place in the past as evidenced by the magnetic anomalies, here was evidence that the same kind of movement was still taking place today.

Having established that fragments of oceanic crust were moving with respect to one another, one was confronted by the problem of discovering their size and shape. An examination of maps indicating the distribution of earthquakes on a world-wide scale, revealed that these events were by no means casual: there were regions in which almost all seismic activity was concentrated and others in which it was quite sporadic.

These limited high-activity belts coincided partly with the oceanic ranges, and also followed other regions which, in one way or another, were remarkable for other geophysical characteristics. For example they followed belts of high volcanic activity, insular arcs and ocean deeps. Using the distribution of the seismic characteristics of these other structures, one observed that the Earth's crust could be subdivided into a series of "plates" which moved in relation to one another.

This layout of the Earth's crust appeared to be highly convincing. If it was to become a model in the modern sense of the word, all that was lacking was a mathematical system to describe it. This task was undertaken by the young French geophysicist Xavier le Pichon. Taking into consideration the fact that these rigid plates were moving upon a spherical surface, he simply had to apply the principles of Euler's spherical geometry. The model he thus obtained gives an adequate description of past movements and forecasts what can be expected in the future. Tests so far made with the plate-tectonics model have all given positive results, and the model is now accepted by a majority of geologists and geophysicists.

JOHN F. DEWEY

Teaches geology at the State University of New York at Albany, and is Senior Research Associate at Lamont-Doherty Geological Observatory. Born in London, England in 1937, he received his Ph.D. at the age of 23 from Imperial College, London. He is one of those responsible for the confirmation of the plate-tectonics model, on which he has worked since 1967, when he moved to the United States. He developed the model, applying it to the reconstruction of the history of the Alpine and Appalachian mountain chains. He is one of the leaders of the conceptual revolution in the earth sciences.

JOHN F. DEWEY

Plate Tectonics—A Global View of the Dynamics of the Earth

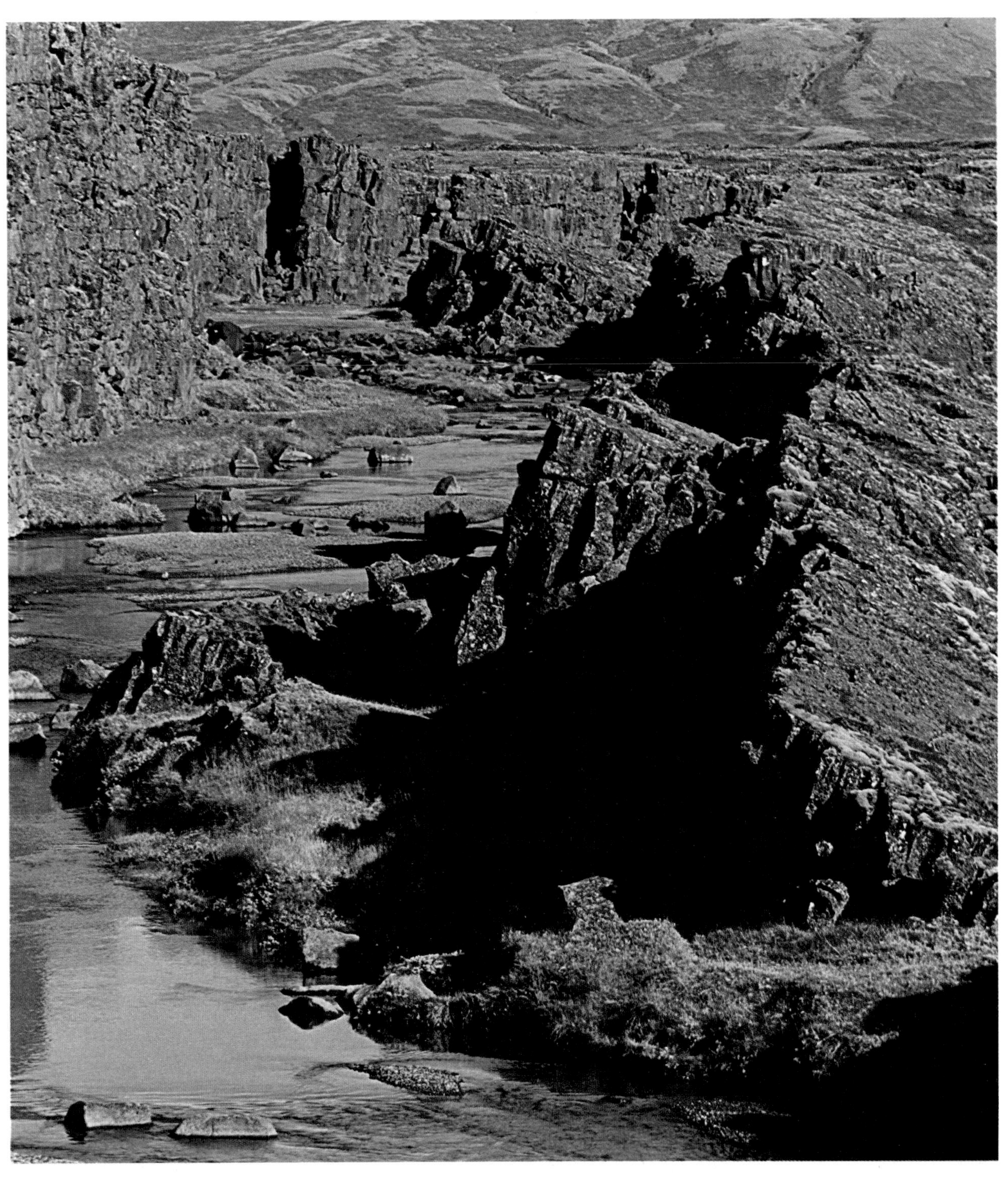

Overleaf: In the Thingvellir region of Iceland. This is a long crack which runs through the island in a roughly north-south direction, along a line which is a direct extension of the mid-Atlantic Ridge. In fact many people consider it to be an uncovered section of the deep rift valley which runs along the ocean-bed in the middle of the ridge. On the basis that it is precisely along this underwater valley that the hypothesis of the expansion of the ocean-beds entails the drawing-apart of the opposed plates which would form the slopes, Ronald Mason, of Imperial College, London, has had the idea of verifying this same process on dry land as it relates to Thingvellir. On both slopes he has set up measuring stations which are diasymetrically extremely accurate, and from these he gathers data once a year. The measurements are encouraging: it seems that the two slopes are actually drawing apart from each other, even if only at the rate of half a centimeter a year, i.e. less than has been supposed (I.C.P.).

In a paper published in Nature in 1967, D. P. McKenzie and R. L. Parker created a paradigm change in the earth sciences. McKenzie and Parker argued that well-defined earthquake zones of the Earth mark the boundaries of a series of essentially rigid "paving blocks" forming spherical caps of lithosphere (hard outer shell) which interlock, without gaps, as a global mosaic of plates (Fig. 11-1). They proposed that the plates are in constant relative motion and are bounded by one of three kinds of margin: oceanic ridges where new lithosphere is created, oceanic trenches where lithosphere is consumed and returns into the mantle; and transform faults where plates slide past one another and surface area is conserved. Like most new conceptual models in science, the theory of plate tectonics was developed by looking at data in a fresh way. It incorporated and explained the idea of continental drift and sea-floor spreading in a simple elegant theory. Continental drift is an old idea first put into a logical scheme by Taylor and Wegener in the early 1900s, although precursive suggestions had been made by Francis Bacon in the 17th century and Antonio Snider in the last century. A. Holmes used the concept of drifting continents to explain mountain belts using, as an analogy, D. Griggs' experiments on simulated convection. In a series of incisive papers in the 1950s Warren Carey linked the tectonic evolution of the Mediterranean and Caribbean regions, among other areas, to large horizontal continental displacements, and developed his magashear theory, the embryo of the transform fault concept. These ideas, however, failed to make any real impact on earth scientists because of their vagueness. Until 10 years ago continental drift, versus a static earth, seemed more an article of faith than an impersonal assessment of the data. The other problem was that geologists had worked for well over a century on the continents with the basic principles worked out by William Smith and Charles Lyell. The principles were good but it turned out that the continents were the wrong place to find the evidence for a general global dynamic theory. It is now clear that the complex geology and evolution of the continents is due, at least in large part, to the operation of much simpler processes in, and around the edges of the oceans. Attempting to unravel continental geology to unfold a mechanism for global tectonics is like trying to precisely reconstruct a building to its original design after a major earthquake.

During the 1950s Maurice Ewing and his coworkers at Lamont Geological Observatory were exploring the oceans with the new geophysical techniques they had developed. They discovered a world-encircling system of oceanic ridges characterized by earthquakes. Then they showed, from seismic refraction experiments, that the crust of the oceans is considerably thinner than that of the continents and is made up, in its upper part, of basalt capped by sediment.

A further difference between continents and oceans emerged in 1959 when R. C. Mason and A. D. Raff described a series of linear magnetic anomalies in the northeast Pacific; a pattern wholly unknown on land. These new data led H. H. Hess in 1960 to put forward the sea-floor spreading hypothesis. Hess proposed that the ocean floor is being pulled apart to allow the injection of new igneous material at the axis of the ridges. He argued that the surface area of the Earth is conserved by the destruction of sea floor in the deep oceanic trenches at a rate equal to its generation at the ridges.

A partial qualification of sea-floor spreading came in 1963 with J. Tuzo Wilson's analysis of the increase in age of volcanic islands away from the ridges. However, continental drift and sea-floor spreading were still essentially qualitative ideas and lacked a general reference framework. The idea which was to eventually lead to this framework was Vine and Mattews' suggestion in 1963 that linear oceanic magnetic anomalies are the result of an alternately normal and reversed geomagnetic polarity frozen as a remanent magnetic "tape recorder" into the spreading oceanic crust. Although P. David and B. Brunhes had shown in the early 1900s that the magnetization direction of a Lower Pleistocene lava flow is opposed to the direction of the present magnetic field, it was not until 1968 that A. Cox, C. B. Dalrymple and R. R. Doell had unequivocally established a time-scale for geomagnetic polarity epochs during the last 4 million years. By 1968 magnetic anomaly patterns had been mapped over large tracts of the oceans. Thus, sea-floor spreading was at last given a firmly quantitative framework for the past 4 million years (Heirtzler et al., 1968; Vine, 1968). J. Heirtzler and his coworkers (1968) extrapolated the geomagnetic time-scale back to about 80 million years using a magnetic profile from the South Atlantic and assuming a constant spreading rate. This extrapolated time-scale has received striking confirmation from the JOIDES deep-ocean drilling program.

The central argument of the theory of plate tectonics is that, if plates are rigid, the instantaneous relative motion of two plates must be described, in Eulerian terms, by a rotation about an axis passing through the center of the Earth. Thus, the velocity vector across a plate margin depends upon the orientation of the margin with respect to circles of rotation about the rotation axis. Furthermore, the rate of sea-floor spreading at a ridge, or the penetration of a plate in a trench, varies systematically according to distance from the pole of rotation. Confirmation of this came with J. Heirtzler *et al.'s* analysis of spreading rates. They showed that spreading rates systematically decrease north and south from the equator along the Mid-Atlantic Ridge, increase systematically south-eastwards along the Indian Ocean Ridge and systematically decrease along the Pacific/Antarctic Ridge from north to south. X. Le Pichon used this data to determine poles of rotation to describe the relative motion of the six major global plates and the slip vectors across their margins (Fig. 11-1). X. Le Pichon's slip vectors were matched from seismic studies by B. Isacks, J. Oliver and L. Sykes. They showed how the motion of rigid plates explains in remarkable detail the stress fields deduced from seismic first motions along the plate margins.

Plate tectonics is not simply another name for continental drift and sea-floor spreading. Continental drift and sea-floor spreading are consequences of plate evolution. Plate tectonics is a theory with rigorous geometric and kinematic rules. It is emphasized that the rules and the various corollaries of plate tectonics are not dependent for their verification on knowing the mechanism that drives plates around. Plate tectonics stands as a theory which, alone, convincingly explains much of the major tectonics of the Earth and accounts for earthquakes, volcanoes and mountain belts in a rational way.

Geometry and Kinematics of Plate Tectonics

It is assumed that plate evolution takes place on an Earth of roughly constant size. This assumption is justified not only from paleomagnetic evidence, but by the fact that continental margins can be fitted together very precisely. The basic tenets and corollaries of plate tectonics as presently formulated and understood are:

1. Plates are essentially rigid, spherical, caps of lithosphere bounded by the Earth's major earthquake belts (Fig. 11-1). At present, there are six major plates and numerous minor ones. Although large earthquakes occur in plate interiors, particularly in the continents, they are scattered about, are infrequent, and cannot be arranged to form coherent plate margins.
2. Plate margins may lie along a continent/ocean boundary, within an ocean, or within a continent (Fig. 11-1). Thus, plate margins are commonly not continental margins and a single plate may contain continental and oceanic portions.
3. Plate margins are of three types (Fig. 11-2). (a) Accreting—oceanic ridges where plates are separating and growing by the addition of new crust and mantle along their trailing edges; (b) Consuming (subduction zones)—where one plate is "eaten" by sinking into the mantle beneath the leading edge of another plate; (c) Transform—where two plates slide past one another and the surface area is conserved. Seismic first motion studies indicate tension across ridges, strike-slip motion along transforms and compression across consuming plate margins.
4. The lithosphere, of which the plates are made, is probably a thermal boundary conduction layer overlying a weaker layer, the asthenosphere some 600 km thick (Fig. 11-2). The lithosphere probably varies considerably in thickness, being thinnest at the high heatflow regions of the ridges and thickest under the continents. The lithosphere is thus composed of oceanic crust and mantle under the oceans and continental crust and mantle beneath the continents. It can be seen from Fig. 11-2 that the continents are passengers on moving plates and, therefore, that continental drift is simply a consequence of plate motion just as sea-floor spreading is

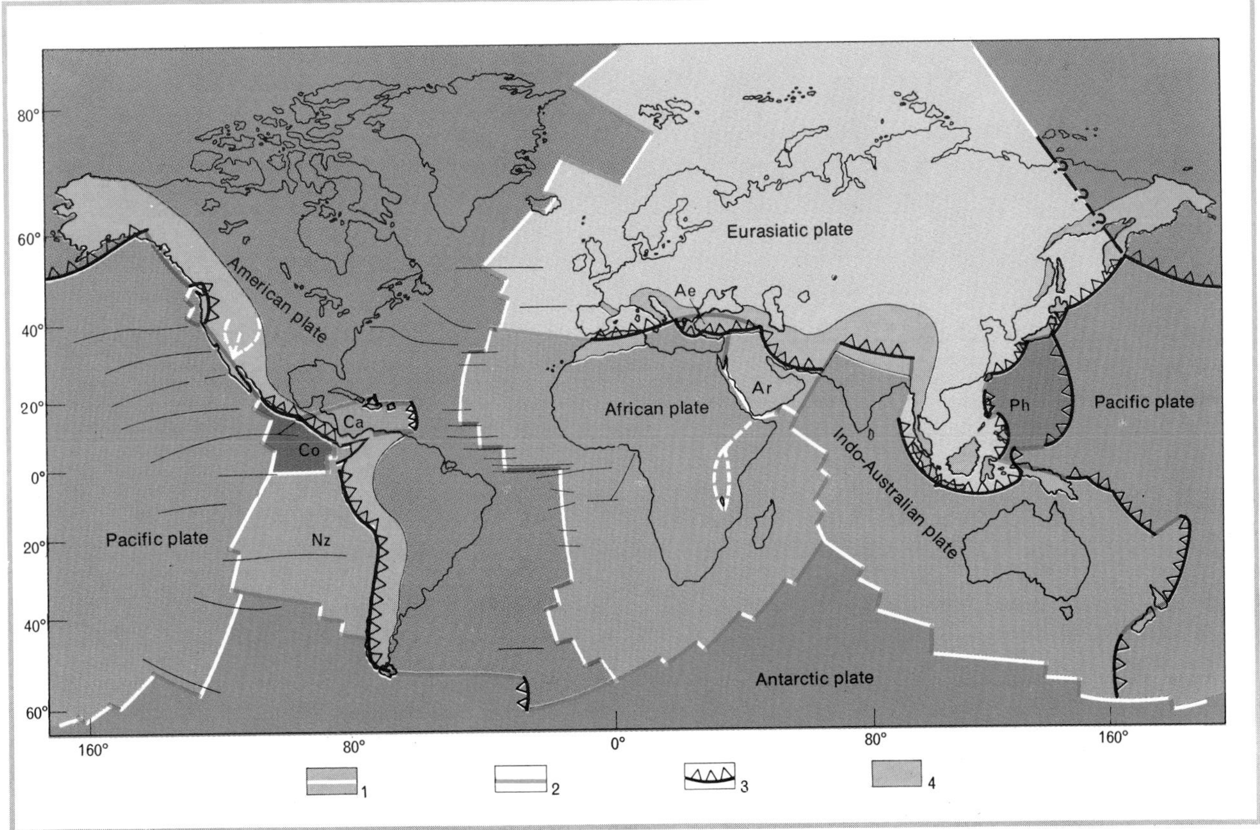

Fig. 11-1. A map of the world showing the present-day plates of the lithosphere. The six main plates are indicated by name, and the six minor ones with the following symbols: *Nz*, Nozca, *Co*, Cocos; *Ca*, Caribbean; *Ae*, Aegean; *Ar*, Arabian; and *Ph*, Philippines. The joins between the plates are schematic; various other smaller plates have been omitted. The joins between divergent plates are represented by the lines indicated by 1 in the key; these delineate the global system of oceanic ridges. The transform limits are represented by the lines indicated with 2 in the key. The joins between convergent plates are indicated by the toothed lines (3 in the key), with the toothing on the edge of the plate beneath which the subduction of the adjacent plate occurs. The zones indicated by 4 in the key represent the essential contours of the actual relief of the Earth, formed substantially by mountain systems concentrated mainly along the western edge of the American plate in the Andean-Cordillera system, and along the entire southern edge of the Eurasiatic plate in the Alpine-Himalayan system.

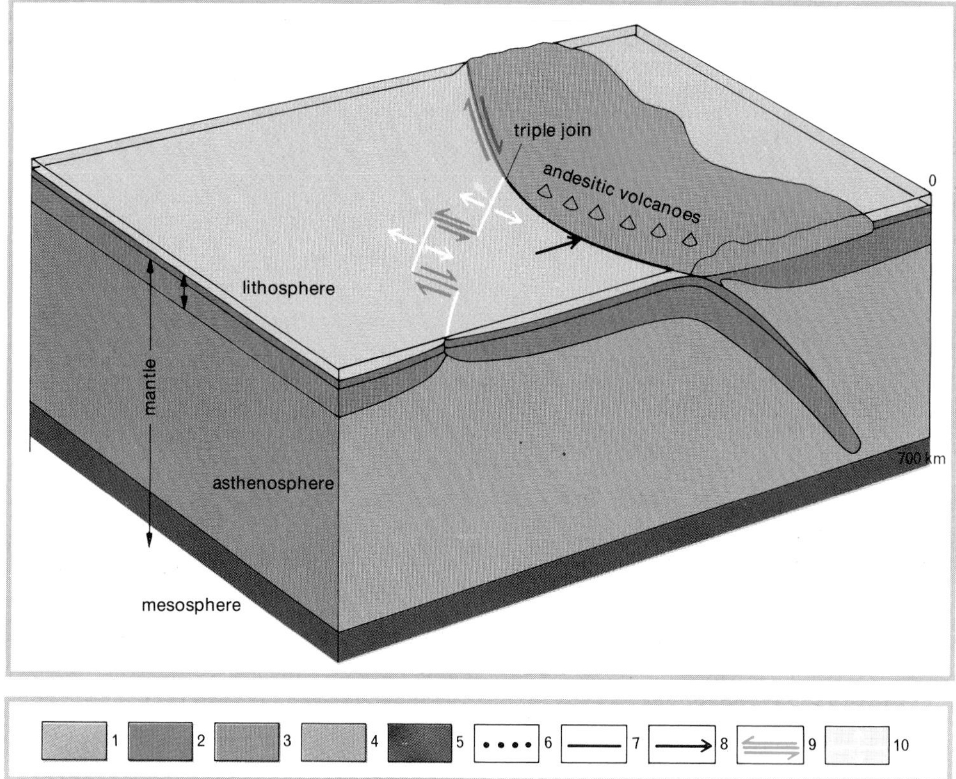

Fig. 11-2. Triple join between three different types of plate edges. In the diagram: (1) continental crust; (2) oceanic crust; (3) section of the lithosphere (rigid stratum) lying beneath the oceanic and continental crust; (4) asthenosphere (less rigid stratum); (5) mesosphere (rigid stratum); (6) area in which deep earthquakes originate; (7) oceanic trench; (8) direction of slip; (9) edge marked by transform faults and course of reciprocal slip of the two plates; (10) ridge and course of the ocean-bed expansion. The edge of the continental plate adjacent to the trench is the one in front, based on the direction of the shift: parallel to it a chain of typically andesitic magmatype volcanoes develops.

a consequence of plate separation. Continents may separate across an expanding ocean (e.g., Africa and South America) or move towards one another (e.g., Australia and southeast Asia) and eventually collide (e.g., peninsula India and Tibet).

5. Accreting and transform plate margins are characterized by shallow earthquakes. The deep inclined earthquake zones sloping beneath island-arcs, from oceanic trenches mark the position and shape of the descending slab of lithosphere (Fig. 11-3). The study of the primary seismic motion indicates tension as the plate bends to sink beneath a trench, and either compression or tension along the slab depending on how far the plate has descended.

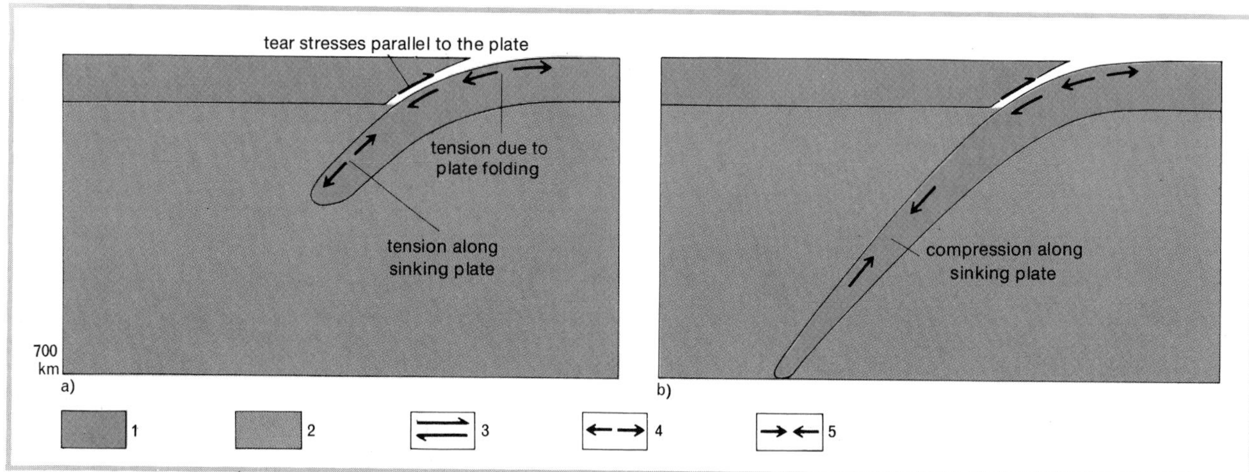

Fig. 11-3. Forces acting on a sinking plate, determined by the time of arrival of the first seismic impulses. In (a) the plate folds and starts its downward movement; down to a depth of a few hundred kilometers, its outer part is subjected to tension; in any event the zone where the plate folds is subjected to tension; at low depths the friction of the sinking plate against the adjacent one seems to generate considerable tear stresses. In (b) the edge of the sinking plate has now dropped to 700 km; while the situation remains unchanged in the secondary fold zone and the friction zone, thus producing tension and tear stresses respectively, the part of the plate which sinks is subject to compression, because of the increase in the resistance by the asthenosphere with the increase in depth. In the diagram: 1, lithosphere; 2, asthenosphere; 3, tear stresses; 4, tension; 5, compression.

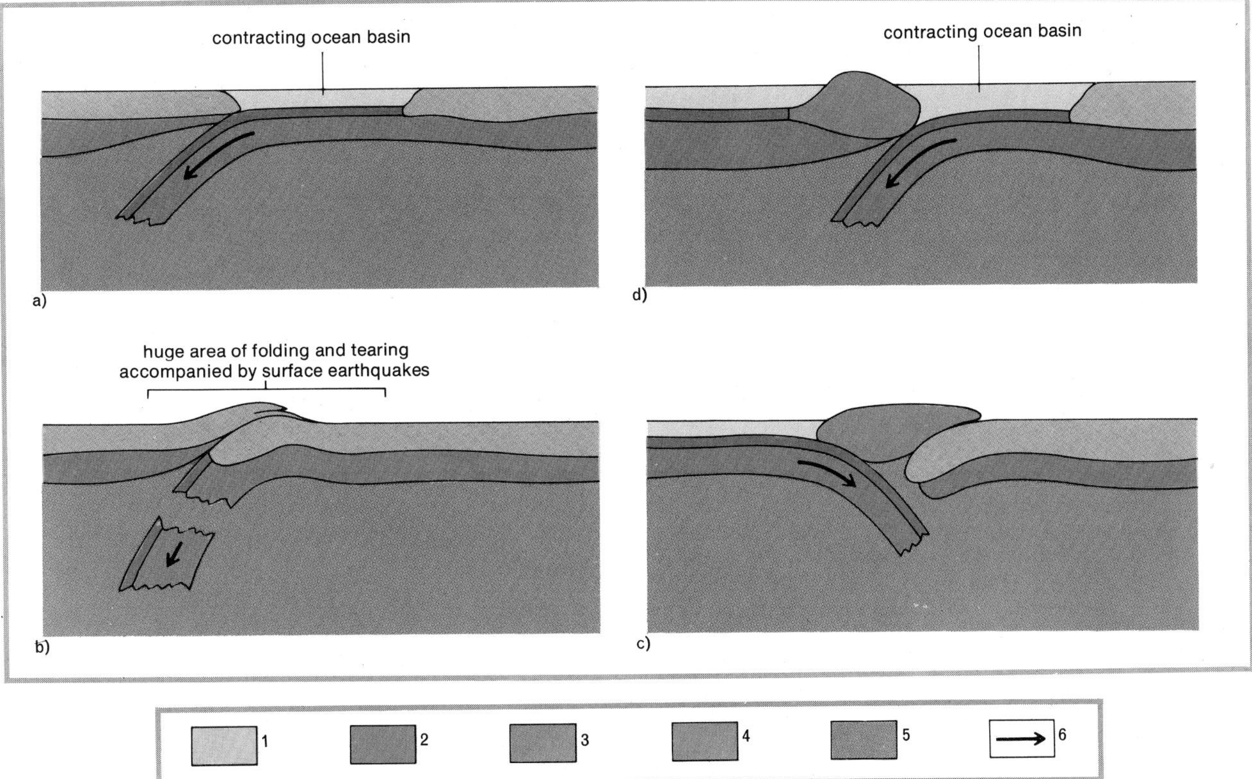

Fig. 11-4. Two cases of crust contraction beneath an ocean basin. Left: two plates, the one on the right moving right to left (top diagram), collide giving rise to a large tear and folding area and the development of large nappe folds (bottom diagram). Right: the collision between a continental plate, moving right to left, and an island arc (top) can give rise to a reversal in the direction of the subslip (below). In the diagrams: 1, continental crust; 2, ocean crust; 3, lower stratum of the lithosphere; 4, island-arc crust; 5, asthenosphere; 6, course of the slip.

6. Where consuming plate margins involve the underthrusting of an oceanic portion of a plate beneath an island-arc or continental margin, the earthquake pattern appears relatively simple and well-defined. Where, however, a consuming plate margin cuts into a continental region for example, along the Himalayas and in the Middle East (Fig. 11-1) the situation is more complex. Most of the earthquakes are shallow and commonly spread out over a relatively wide area suggesting compressional deformation over a wider zone than that of island-arc regions. This may be due to events following the arrival of a continent at a trench. Oceanic lithosphere is heavier and therefore better able to descend into the asthenosphere than the continental

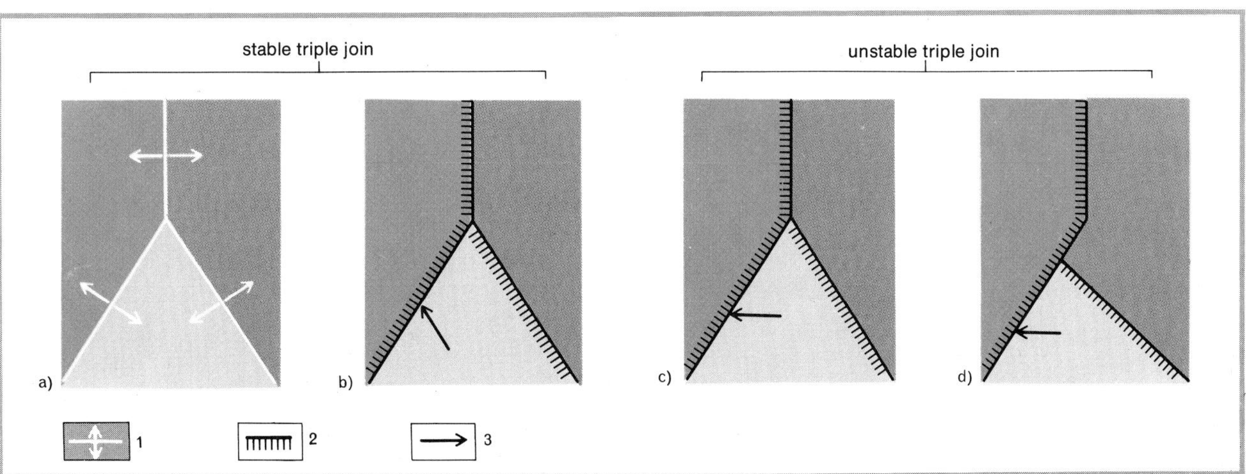

Fig. 11-5. Some of the situations which can occur in the join zone between different crust plates. In the broad play of embedding processes of the Earth's crust, as we have seen in the model of global tectonics, the extremities of each plate edge, i.e. of each line along which two plates come into contact, end up in the triple join between three different crust plates. In (a) all the plates are delimited by ridges; in (b), (c), and (d) they are delimited by oceanic trenches. In the last three cases the orientation of the slip direction is decisive for the stability of the join. Cases (a) and (b) represent situations in which the triple join is stable; (c) and (d), cases in which it is unstable. In the diagram: 1, ridge with indication of the course of expansion; 2, oceanic trench delimiting two plates, where the one towards which the hatching is directed tends to shift towards the other; 3, slip direction.

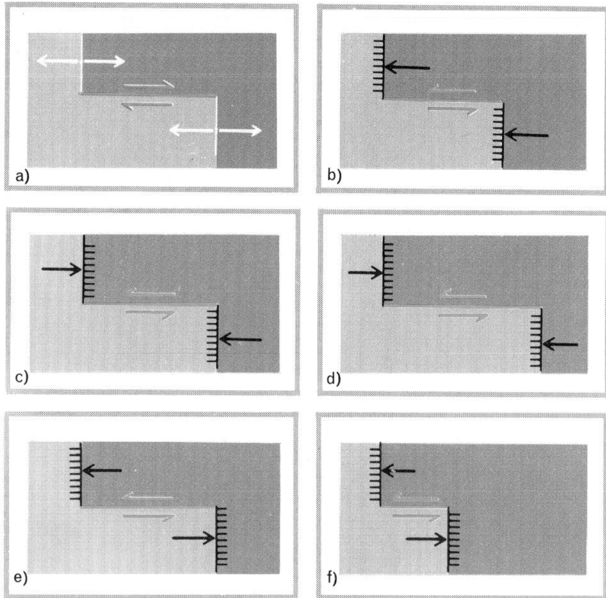

Fig. 11-6. Greater or lesser stability of a transform fault (blue line and arrows) which dislocates two parts of the same ridge (in white) or of the same oceanic rock (black). In cases (a) and (b) the length of the transform fault is not modified in time because expansion (white arrows) or slip (black arrows) occur in the same direction; in (c) and (d) the length of the fault increases because the edges of the zones draw apart; in (e) and (f) it decreases in time because the edges of the plates come together.

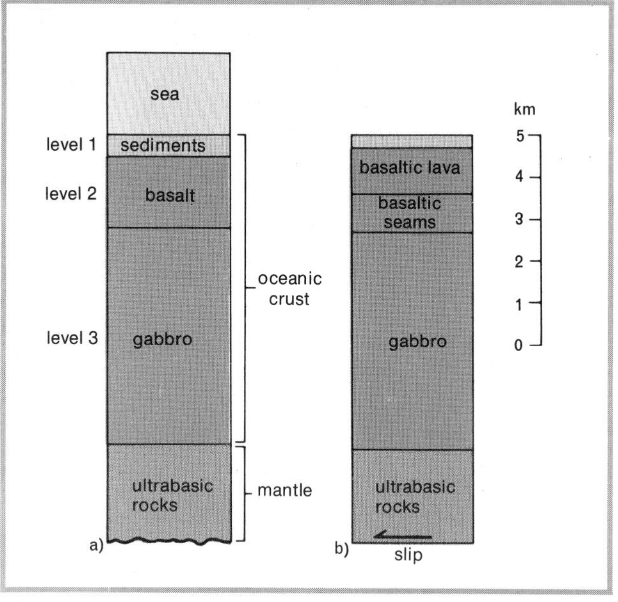

Fig. 11-8. In (a) a typical section through oceanic crust and the Earth's mantle; the various strata have been singled out on the basis of data gathered from seismics by refraction; their composition has been defined on the basis of the analysis of samples taken from the ocean-bed by dredging or boring and on the basis of the experimental calculation of the speed of the compression waves in the different rocks. In (b) a section through the Semail ophiolitic complex in Oman which shows a close similarity to that in the oceanic crust.

lithosphere (Fig. 11-4a). Upon collision the buoyancy of the continent prevents extensive subduction beneath the leading edge of another plate (Fig. 11-4b). The heavy dangling slab of lithosphere may break off and sink and a simple zone of deformation associated with oceanic underthrusting is replaced by a wide zone of intercontinental crushing. If a continent arrives at an island arc with oceanic lithosphere on the other side (Fig. 11-4c), plate consumption may continue if a new trench forms on the other side of the arc, with a consequent flipping of direction of plate underthrusting (Fig. 11-4d).

Fig. 11-7. Relative movement of two crust plates in a certain time span. The relative movement of two elements on the surface of a sphere is always a rotation around an axis passing through the center O, the intersection of which with the surface of the sphere is called the pole of rotation; for example, the relative movement between the two points A and B is less than that between C and D. The rate of expansion of the ocean-beds increases from a minimum in E to a maximum in F, as the speed at which the plates are worn away along the trench GH increases from a minimum in G to a maximum in H. In the diagram: 1, transform faults; 2, ridges, 3, trenches.

7. No plate margin may end except by joining two others in a triple junction. D. McKenzie and J. Morgan showed how the evolution of plates, and whether they may be classified as stable or unstable depend upon whether they can retain their geometry as the plates move. Examples of stable and unstable triple junctions are shown in Figure 11-5.
8. Transform faults are plate margins parallel to the slip vector between adjacent plates but are of several types and may be stable or unstable depending upon the way they join other kinds of plate boundary. A transform fault joining two ridges (Fig. 11-6a), or two trenches which eat lithosphere on the same side (Fig. 11-6b), are stable because they remain the same length as the plates move. Transform faults which join trenches which consume lithosphere in opposite directions will elongate or shorten depending upon whether the plates' leading edges face away from or towards (Fig. 11-6e) one another.
9. The relative motion of one plate with respect to another is described by a rotation about an axis through the center of the Earth (Fig. 11-7). The intersection of this axis

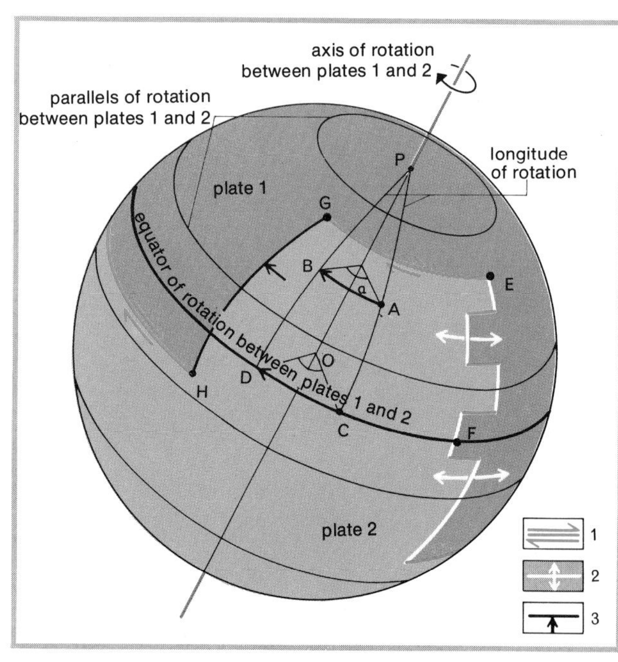

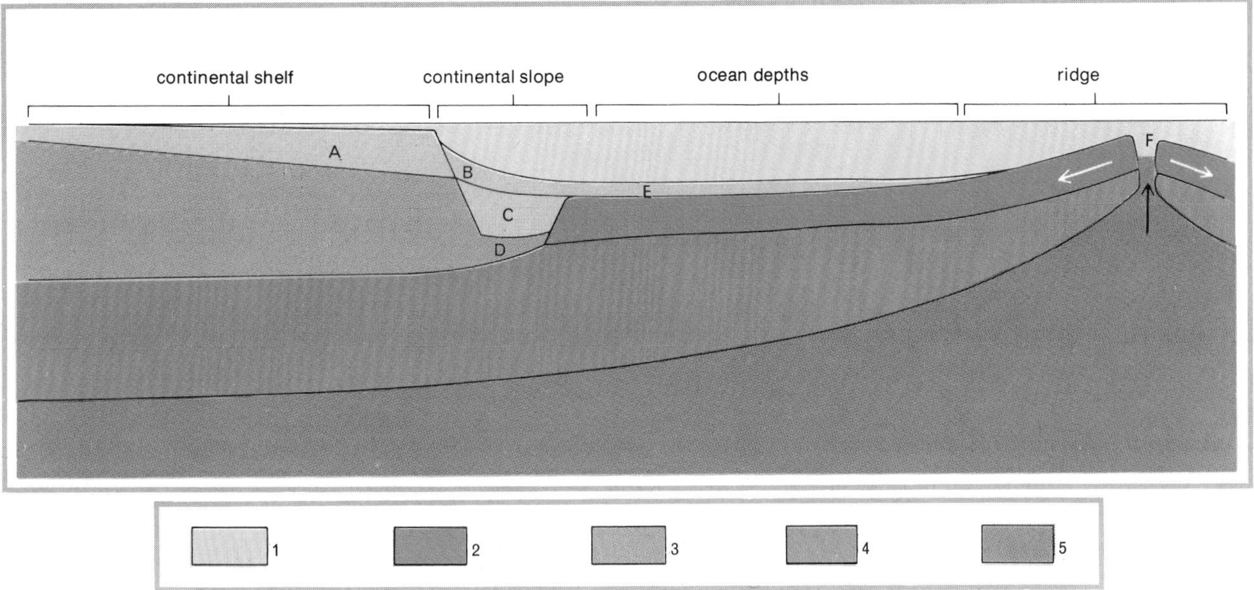

Fig. 11-9. A geological section through an Atlantic-type continental edge formed by the initial rift between two continental masses and their subsequent separation by expansion between them of new ocean depths; the direction and course of the expansion of the depths is indicated by the white arrows. In every zone characteristic lithological associations are formed which then crop up in the mountain chains and help reconstruct their genesis and history; for this reason the development of marine geology is causing rethinking about continental geology too. In the diagram: 1, sedimentary layer; 2, oceanic crust; 3, continental crust; 4, deepest stratum of the lithosphere; 5, asthenosphere. In addition, in A, neritic limestones; in B, ooze and deposits due to submarine landslides; in C, coarse sediments formed during the initial fracturing period; in D, transitional type crust; in E, fine grain oceanic sediments; in F, axis of the ridge along which mantle matter re-emerges (indicated by the black arrow) and formation of new oceanic crust.

with the Earth's surface is called the pole of the rotation. While the angular velocity of movement between two points on any circle of rotation is constant, the linear velocity of approach or separation increases from zero at the pole to a maximum at the great circle, or equator, of rotation. E. C. Bullard, J. E. Everett and A. C. Smith used rotation axes to achieve a fit of the margins of the Atlantic continents. They were, therefore, the first to regard the continents as moving rigid spherical caps, although they did not explicitly describe the continents as parts of rigid plates bounded by spreading ridges. The slip vector across the margin between two plates thus clearly depends upon its orientation with respect to the circles of rotation. In Fig. 11-7 two plates are shown with a margin along which the slip vector changes from pure extension (accretion) to horizontal slip and to compression (subduction). The slip velocity on transform is constant along their length while the rate of separation and approach across ridges and trenches respectively, increases towards the equator of rotation. Transforms define circles of rotation and, therefore, if they are sufficiently long and well-defined, poles of rotation may be deduced from them. Rotation poles may also be deduced from the variation in the spreading rate along ridges. There is as yet no independent method for accurately defining the rate and direction of lithosphere consumption in trenches.

10. The rules of plate geometry and kinematics do not dictate that plates or plate margins are fixed with respect to any set of co-ordinates except a set arbitrarily chosen for a particular plate or plate margin. J. Tuzo Wilson pointed out that, since Antarctica is completely surrounded by a spreading ridge system, the ridge must be an expanding ring, moving outwards from Antarctica. There is no simple one-to-one correspondence between ridges and trenches. Recently, however, J. Morgan has suggested that loci of strong volcanism such as Hawaii and Iceland are "hot spots," above rising mantle plumes, that provide fixed reference points for describing plate motions.

11. The history of plate motion may be worked out by fitting together matching magnetic anomalies symmetrically disposed across ridges using the same least squares technique that E. C. Bullard, J. E. Everett and A. C. Smith used for fitting continental margins together. Clearly, if a particular pair of matching anomalies were generated as a single axial anomaly at the ridge axis, and the plates have moved apart as rigid caps, it is possible to fit them together and thus find the successive relative positions of continents carried by those plates. W. C. Pitman III, who devised this motion, has successfully applied it to the evolution of the Atlantic Ocean. He has progressively fitted magnetic anomalies back together between North America and Europe and between North America and Africa and thus deduced the relative movement of parts of Africa with respect to Europe for the last 180 million years since early Jurassic times. This analysis is fundamental to understanding the closing history of the ocean (Tethys) that originally lay between Africa and Europe and the structural evolution of the mountain belts of the Alpine/Mediterranean/Middle East region. Tanya Atwater has similarly applied the simple geometric rules of tectonics in relating the geological evolution of the western United States to the interactions between the Pacific, American and several intervening plates. Analyses of this kind must of necessity form the basis of serious studies of the tectonic evolution of mountain belts younger than about 200 million years. This does not mean that older mountain belts are not the result of similar and equally complex plate interactions, but we have as yet no way of working out their plate motions in a quantitative way. Virtually all the oceanic lithosphere we see today has been generated during the last 200 million years with the possible exception of a major area of the Indian Ocean between the 90E Ridge, the Java Trench, Australian and Broken Ridge (Fig. 11-1). The only solution to working out plate motions before 200 million years ago seem to be a careful and systematic deduction of paleomagnetic pole positions for sed-

iments laid down on opposite sides of the older mountain belts, such as the Appalachians and the Urals, during the time of their evolution. Even this will probably only give the relative motions of the larger plates. D. P. McKenzie has shown that the Mediterranean region at the present day is a mosaic of small plates with complex relative motions, sandwiched between the African and Eurasian plates. The knowledge that such a complicated plate system exists in a mountain belt system, that is still evolving and already has a complex structure and history, is a daunting prospect to geologists attempting to unravel the older mountain belts in terms of plate tectonics.

Implications for the Evolution of Mountain Belts

The mountain belts of the continents are complex linear/arcuate zones in which rock sequences are strongly deformed and metamorphosed. They are characterized by a large amount of volcanic rock and very often have, locally, very

---

The Geodynamics Project

The Geodynamics Project is an international program of research on the dynamics and dynamic history of the Earth with emphasis on the foundations of geological phenomena. This includes investigations of movements and deformations, past and present, of the lithosphere, and of all relevant properties of the Earth's interior and especially any evidence for motions at depth. The program is an interdisciplinary one established by the International Council of Scientific Unions, at the request of the International Unions of Geodesy and Geophysics and of Geological Sciences. It is an outgrowth of the Upper Mantle Project, which had been a highly successful venture in coordinating and promoting international efforts to increase understanding of the nature of the outer 700 km of the Earth. The research carried out during the period of the Upper Mantle Project has revealed many details of the properties and the structure of the Earth's outer shell that had not previously been recognized. It has further resulted in exciting and provocative ideas about the history and development of its surface features. The Geodynamics program is thus based on the important developments realized during the Upper Mantle program and is oriented towards major problems of the Earth's interior.

Major Developments During the Upper Mantle Program. *One important discovery during the Upper Mantle project was that the Earth is not simply a radially symmetrical sphere but that lateral inhomogeneities of considerable extent exist in the upper 700 km at least. In particular, these inhomogeneities are often related to surface areas which have been in the recent geological past, or still are, tectonically active. During the same time, major progress was made in the studies of the deformations occurring in the upper layer of the Earth.*

It had long been known that this layer is affected by short periodic motions, associated in particular with earthquakes and volcanoes which are most probably the manifestation of longer-term motions which may take tens of millions of years, and are responsible for the present configuration of the Earth. Vertical uplift due to the melting of the glaciers over tens of thousands of years is a proof that geologically rapid flow does occur in the upper mantle below the upper rigid layer. And mountain belts, whether fossil or still active, show that all parts of the Earth's surface, even those which are at present quite stable, have been affected by violent deformation related to motion at depth. Thus, it is quite clear that the pattern of surface motions and deformations is produced by mechanical energy resulting from motions within the upper mantle.

But, perhaps, the most important result of the Upper Mantle program was the progressive discovery that horizontal relative motions of up to several thousands of kilometers have occurred on the Earth during recent geological time. This type of motion is known from seismological and geodetic studies to occur now along large systems of strike-slip faults, like the San Andreas system in California and the Anatolian system in Turkey. There, movements averaging several centimeters per year are well established and their relation to seismic activity well known. However, it is from the study of the oceans that the most startling discovery came: that the floors of the oceans, which cover two-thirds of the Earth, are geologically very young, no part of them being more than 200 million years (i.e. one-twentieth of the age of the oldest continental rocks).

This realization led to the elaboration of a unifying hypothesis known as plate tectonics, or global tectonics. While still a hypothesis, and not a proven theory, it has brought together all the different disciplines that constitute the earth sciences, and has given considerable impetus to research in many new directions. The bases for this hypothesis had been progressively set over more than 60 years and include the concepts of crustal dispersion and continental drift of Taylor (1910) and Wegener (1929) as well as the sea floor spreading hypothesis of Hess (1962). The hypothesis is based on the simple fact that most of the Earth's mechanical energy dissipated at its surface is spent within a few narrow orogenic belts that surround large areas of relative quiescence. Most of the world's earthquake and volcanic activity, and important portions of its mineral resources are concentrated in these active zones. The blocks or plates, which are surrounded by these active belts may contain continental as well as oceanic surfaces and can be considered as rigid to a first approximation. Thus orogenic belts are considered as zones of differential movements between rigid plates. Where plates separate, the boundary is a rift zone where new oceanic crust is created. Where they collide, it is a zone of underthrusting, or of major compression and uplift. Where they glide against each other, it is a zone of strike slip.

Consequently, this hypothesis supposes that the uppermost surface of the Earth consists of a few rigid plates, which form what has been called the lithosphere. The lithosphere, which is about 70 to 100 km thick is characterized by the fact that it has significant strength on a geological time scale whereas the underlying asthenosphere has effectively no strength over the same time scale and flows under an applied shearing stress.

It is not possible here to develop all the major advances in the different branches of the earth sciences which have been brought about by this discovery. Most important, perhaps, have been the interpretation of the seismological activity on a global scale and its integration within the framework of a general tectonic pattern. Petrology and geochemistry have been renewed by the fact that any vertical column of upper mantle and crust can no longer be considered a closed system. Paleomagnetics, paleontology and paleoclimatology have been revived by the new interpretations within this framework. New ideas have swept geology, and models still schematic, are proposed to account for the formation of most mountain chains. The problems of the origin of the continental crust and of the lithosphere are put on new bases. Some advance has been made in the direction of the comprehension of the most difficult problem of the driving mechanism. In particular, we know much more about the pressure, temperature and time dependence of the different physical and chemical parameters within the mantle. Yet, we are still dealing with a hypothesis, not with a complete theory, and it still leaves unexplained many phenomena occurring over the Earth. Thus, for example, the prob-

thick sequences of sedimentary rock. Distinctive rock assemblages (facies belts) occur in a number of distinct zones broadly parallel to the length of the mountain belt and have complicated, often structural, interzone relationships.

The most recently developed mountains such as the Himalayas and the Andes have clear spatial relationships with present-day consuming plate boundaries (Fig. 11-1) and it now seems certain that the evolution of mountains for at least the last 2 billion years has resulted from plate evolution. Mountain belts, however, show considerable variations in the nature and distribution of different rock sequences (geosynclinal sequences) and the way in which they are deformed and metamorphosed. These differences can be related to the way in which continental margins, island arcs, and oceans, interact.

Possibly the most significant rock suite in mountain belts is the ophiolite suite (Fig. 11-8b). This consists of a sequence usually comprising dense dark ultrabasic igneous rocks passing into basic igneous rocks (gabbros). The gabbros pass into so-called sheeted complexes consisting of fine-grained basic intrusive igneous dikes (dolerite) and the dikes pass into basic

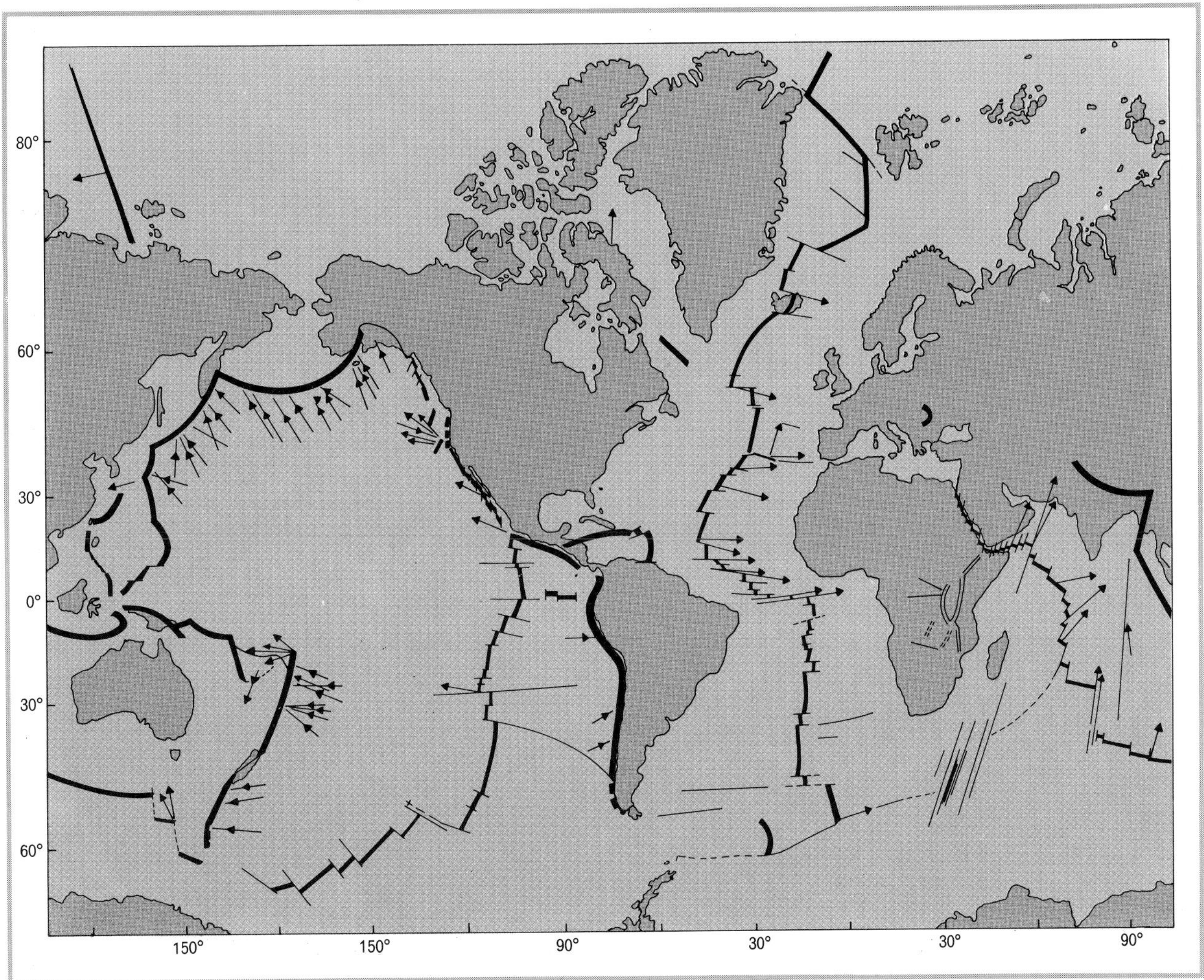

A map of the distribution of slip tendencies (arrowed) extracted from the study of earthquake mechanisms. It shows that seismicity can be interpreted as the result of horizontal movements relating to just a few large crust plates *(after B. Isacks et al.)*

lem of epeirogenical movements occurring over millions of years on very large surfaces is totally unexplained. And the very success of the hypothesis makes us realize that it is still very schematic and that it actually brings to light many new problems which were not even considered before.

Focus of the Geodynamics Program. *In the light of these results, there are two obvious directions which seem especially relevant to our understanding of the dynamic evolution of the Earth.*

The first is a better determination of the pattern of surface motions and deformations, which can now be measured by several different methods.

The second is the search for the mechanism by which the mantle provides the mechanical energy for surface motions. This supposes a much better knowledge of the physical state of the Earth's interior. The relation between these two lines of studies is the central problem of geodynamics.

Contemporary Geodynamics: A Comparison with Past Geodynamics. *To obtain a good integrated picture of the pattern of deformation of the Earth, all geological, geophysical and geochemical methods should be used simultaneously. The emphasis, of course, should be put on contemporary deformations, because seismicity and geodetic studies provide a physical framework for other relevant observations. This contemporary pattern of deformation should then be extended as far as possible toward the past to see whether it is the continuation of a process which may have been active during most of the Cenozoic time. Once the kinematic pattern is established in an active seismic belt, it is possible to relate it to the geological history of the region to deter-*

lavas erupted under water (pillow lavas). The pillow lavas are then capped by sediment. It is virtually certain from what we know of the structure and composition of the crust and upper mantle of the oceans (Fig. 11-8a) that ophiolites are slices of oceanic crust and mantle now within the continents. This means that oceans must have existed along, for example, the axis of the Ural Mountains and along the axis of the Appalachian Mountain belt. These older oceans were driven out as the western USSR approached and collided with Europe some 350 million years ago. It is not yet clear, however, to what extent ophiolites represent chunks of oceanic crust and mantle generated at the main spreading ridges of the oceans as opposed to in small marginal oceanic basins such as the Sea of Japan (Fig. 11-1). D. Karing has cogently argued that such marginal oceanic basins are generated by a slow "pulling-away" of island arcs from continental margins. The characteristics of ophiolite suites in some mountain belts suggest that they were generated in marginal basins.

A simplified view of the course of evolution of mountain belts is as follows. If a continent ruptures and a new accreting

---

mine variations of the kinematics in time and space. In this way, past history of the present-day movement pattern can be related to geology, and a closer understanding of the origins of a variety of major structures would be obtained. Such an approach requires that the timing, amount and sense of movements and deformations be carefully established by use of seimological, geodetic and geological techniques. Of course, the aim is to arrive at a single geodynamic pattern over the Earth, in which all active belts interrelate with each other. This is the main rationale for keeping within the same program structures as different as the oceanic rifts and the island arcs.

In studying these active seismic belts, we are studying the near-surface manifestations of fundamental processes occurring in the deep crust and upper mantle. We consequently should determine carefully the distribution of physical parameters within the deep crust and upper mantle underlying these areas, which requires systematic intensive structural studies.

The techniques to be used are very diverse and wide ranging. They include principally the study of seismicity, which is a measure of present-day tectonic activity. Microseismicity should not be neglected as it tells something about the activity within a short period of time. Studies of the focal mechanism of earthquakes, measurements of the amount of energy expended, are specially important. In situ measurements of stress patterns provide essential information. Geodetic measurements extend these instantaneous methods toward the past by giving rates of slippage along faults, and overall surface deformations through repeated triangulation, leveling and geodimeter surveys. These measurements should be integrated within the pattern of recent deformations obtained from geomorphological, geochronological and geological studies. One should also study the reaction of the crust to man-made perturbations: like fluid injection at depth or a water loading dam. It might eventually be possible to arrive at gradual release of crustal stresses to prevent destructive earthquakes. A major effort should be made to relate volcanic to tectonic activity.

In many areas the reconnaissance stage is coming to a close and should be replaced by an era of systematic integrated surveys, oriented towards those areas where special problems need to be solved. Complete geophysical coverage should in general include seismic refraction and reflexion, magnetism and gravity. In magnetic investigations, long wavelength magnetic anomalies provide important information, as well as the study of the geomagnetic variations which can be related to the distribution of conductivity at depth and should be coupled to heat flow interpretations.

In geology, much information already exists which should be synthesized and reinterpreted. In particular, attention should be paid to the relation of various plutonic, metamorphic and magnetic processes to tectonic activity.

As soon as we are dealing with past geodynamics, we are losing the guide provided by the physical measurement of contemporary deformation. We should try in this study to develop keys to an understanding of the evolution of orogenic zones in earlier times. It is evident that, if there have been large displacements of plates throughout the past a first requirement is to obtain a knowledge of these displacements through paleomagnetic, paleoclimatic, paleontological and chronological studies. The second requirement is to localize precisely the boundaries where this differential movement has been absorbed. Clearly, the interpretation of mountain belt structures cannot be made in the same way if they are the result of deformation occurring at plate margins, where thousands of kilometers of differential motion has occurred, and if they are the result of motion such that all the preorogenic surface can still be accounted for in the mountain belt.

Whatever the working hypothesis, we should relate the occurrence, composition and distribution of volcanic rocks to mountain belts. The same study should be made for plutonic bodies and attempted for metamorphic mine assemblages. These rock sequences are found in both youthful and old mountain belts and thus provide a common key to the structures of both. Particular attention should be paid to the distribution and significance of ophiolites, this association of ultra mafic and mafic rocks with radiolarites. To many research workers, those are remnants of oceanic crust within the mountain belts. Similarly, propositions have been made which relate low-grade metamorphism to rapid burial of sediments in the active trenches, and grade metamorphism to high temperature in the volcanic zone behind the trench. It is important to test such propositions as they provide a direct relationship between active orogenic zones and past mountain belts.

Finally, it is obvious that we should use the great quantity of new information recently acquired, through coring and drilling in the ocean, on the nature of deep sea sediments, to examine whether and which sediments in mountain belts may be the analogues of deep sea facies: pelagites and turbidites of abyssal plain, sediments of continental rise etc.

The emphasis on horizontal motions should not lead to neglect of vertical displacement. Throughout the history of the Earth, we know of many basins like the North Sea which have been continuously subsiding during hundreds of millions of years, next to plateaus which were continuously uplifted like the Scandinavian platform. These important vertical displacements occur within otherwise quiescent blocks and seem to ignore the variation in the configuration of orogenic belts. Other important vertical movements are well known in the continental margins which have been regions of continuous subsidence since their formation. Yet we know very little about these movements and no convincing physical model has been proposed which relates them in any way to a global geodynamic pattern. We should aim at obtaining a detailed picture of the duration, wavelength, amplitude, and velocity of movement in the course of time for each of these areas. Clearly, the problem is of major economic significance since most of the oil deposits have accumulated in these regions. The solution of this problem will involve the understanding of mass transfer at depth.

Rheology and Physics of the Deep Interior. *It is difficult, but essential, to relate this superficial geodynamics to motion occurring in depth. A major effort should be devoted to relating surface tectonic activity to upper mantle inhomogeneities. The only direct evidence for motion in depth is deep and intermediate earthquake activity. Studies of such earthquakes, in particular of their exact location, of their first motions and of travel time residuals,* plate margin develops, the continents move apart on separate plates as a new ocean grows in between. As the ocean grows, a series of distinct geological environments evolve (Fig. 11-9). These are (1) the continental shelf on which shallow water sediments accumulate on older continental crust; (2) the continental rise where deeper water sediments accumulate on, perhaps, partly oceanic crust or distended and thinned continental crust. The rise sediments often overlie a thick accumulation of coarse shallow-water sediments developed in fault bounded basins in the early phases of rifting before the generation of oceanic crust; (3) the abyssal plains where fine pelagic sediments accumulated on oceanic crust; and (4) the ridge where new oceanic crust is generated and little sedimentation occurs. The essence of this is that a continental margin is formed as an aseismic junction between continental and oceanic crust with different sedimentary rock assemblages. The key to mountain building appears to be the way in which such a continental margin is converted into, or arrives at, a consuming plate margin. Consuming plate margins at the present day are the essential loci of mountain building and lie

The diagram shown here illustrates the principal mechanisms which help to explain the global tectonic activity of the Earth as seen by the model of plate tectonics. In the center we have a submarine ridge: corresponding to this structure the hot matter coming from the Earth's interior rises to the surface and cools to form the new oceanic crust. On the right, two oceanic trenches where the old oceanic crust is swallowed up inwards. This phenomenon can give rise to an island arc or a mountain chain *(Crown copyright, MHSO, Inst. Geol. Sci. London.)*

are essential. As a gravitating body in a static state can have vertical variations but cannot support horizontal variations, one should search for evidence of lateral inhomogeneities in distribution of any measurable properties in the mantle, in particular seismic velocity, density and electrical conductivity. Rheological properties can be obtained from a study of the uplift of previously glaciated areas.

Another important line of studies is the examination of instabilities in the vertical stratification in the mantle. In particular, it has been proposed that there is a density minimum near the base of the lithosphere, which is associated with a shear velocity minimum, and probably results from a zone of partial melting. Such a zone corresponds to a gravity minimum and too much reduced viscosity, and must play a critical role in any displacement of the lithosphere, and its study should be a first priority within the geodynamics project.

It is now recognized that there are other deeper zones in the mantle where there are jumps in density caused by phase transformations. All these instabilities, density minimum, partial fusion and solid-solid phase transformations must profoundly affect motion in the mantle, and should be taken into account in theoretical studies which examine the role of the mantle as a heat engine. To proceed with these theoretic studies, it is also required to have a good knowledge of the variation of certain parameters with temperature and pressure, in particular, the coefficient of viscosity, the thermal expansivity, the shear elastic constants and the thermal diffusivity. Thus, high-pressure laboratory experiments should be pursued and extended. Rheological experiments should lead to a better understanding of creep of geological materials. Finally, it is quite clear that the distribution of temperature with depth is the most critical parameter concerning motion in the mantle. It is unfortunately one of the most difficult to study.

Significance of the Program

The Upper Mantle Project led to major advances in earth science and prepared the way for the Geodynamics Project. Yet, it was dominated by research on a disciplinary basis and often carried within a national framework. This research benefited tremendously from international exchange of data and ideas but was limited by a difficulty in correlating different researches in different disciplines on a global level. Today, owing to the advances made during the UMP, the Geodynamics Project provides the possibility of truly international interdisciplinary research on a global level. Understanding the geodynamic pattern of deformation requires the cooperative efforts of many nations. Thus the Alpine-Himalayan seismic zone is a whole and cannot be understood completely except as a whole. Yet its area comprises tens of different countries and internal seas, and research there can only proceed through cooperation. This research is necessarily interdisciplinary and requires that frontiers between geology, geochemistry and geophysics be abolished. It is probably reasonable to suppose that at the end of the Geodynamics Project, we will be in possession of a unifying theory, incomplete yet, but sufficiently tested to serve as a framework to all researchers in earth science. It will undoubtedly lead to much closer cooperation between the different disciplines of earth sciences and perhaps make the hopes of Prof. Tuzo Wilson come true, that there will not be any more several earth sciences, but one earth science comprising many disciplines.

Social and economic aspects of this program are important, as major developments appear in the direction of better protection against earthquakes but also prediction, and eventually control of earthquakes. Much better understanding of the relation between mineral deposits and geological structure has already been obtained, both for mineralizations concentrated in orogenic zones and for oil deposits in subsiding sedimentary basins. One may hope that the search for these deposits will progressively change from its present empirical stage and become truly predictive within a not too distant future.

XAVIER LE PICHON

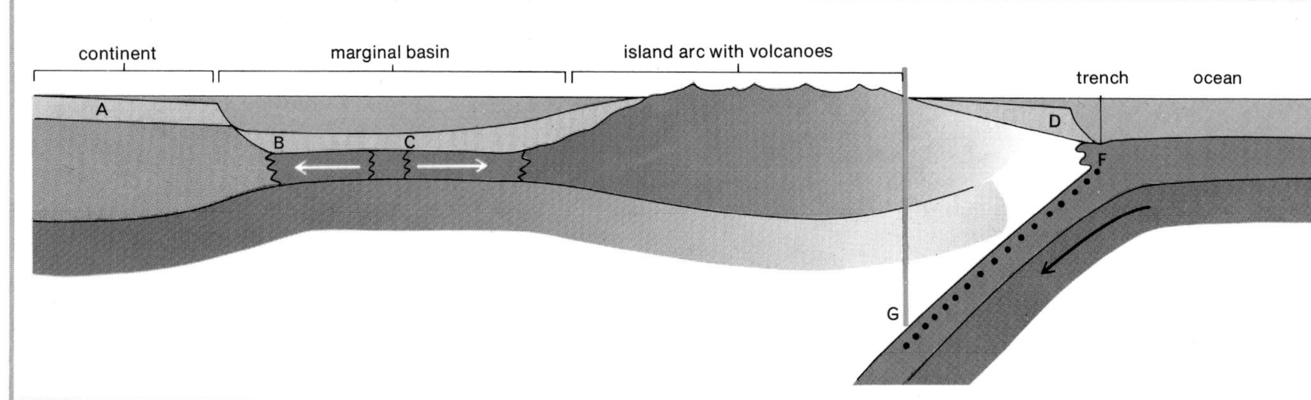

Fig. 11-10. Mountain ranges at the edges of a plate bordered by an island arc. In the diagram above it delimits a marginal basin caused by crust expansion phenomena. In the diagrams on both these pages: 1, sedimentary layer; 2, continental crust; 3, oceanic crust; 4, crust of the island arc formed by volcanic and metamorphic rocks; 5, lower level of the lithosphere; 6, zone of origin of deep earthquakes; 7, zone of possible expansion of ocean depths; 8, subslip; 9, marginal over-slip; 10, volcanic front.

either at continental margins (e.g., Tonga-Kermadec trench, Fig. 11-1). In island arc systems a series of distinct zones are developed on the plate leading edge (Fig. 11-10). The trench marks the junction of two plates and is a deep furrow characterized by low heatflow and negative gravity anomaly. Many workers have argued that oceanic trenches are the site of high temperature/low temperature metamorphism in a mélange of deformed sediments and ophiolites scraped off, and ripped from, the descending plate (Fig. 11-10a). This kind of assemblage is developed in the Franciscan Formation along the coast of California, where it is thought by W. Hamilton to represent the position of a Mesozoic trench. Behind the trench, a wide zone of little deformation and no volcanic activity is developed. This is called the arc/trench gap and its inner margin is sharply defined as the volcanic front where volcanics appear abruptly. The arc itself is a zone of andesitic and basaltic vulcanism and high temperature/low pressure deformation and metamorphism. The rate of extrusion of volcanic rocks in Japan suggests that the piling up of the volcanics in these regions may be an effective way of growing new continental crust. Island arcs such as Japan, the Marianas and the Tonga-Mekadec chain (Fig. 11-1) have marginal oceanic basins on their inner sides (Fig. 11-10a) which may be due to sea-floor spreading behind the arcs. Islands arcs such as the Java-Sumatra chain and volcanic chains such as the Andes (Fig. 11-1) lack marginal ocean basins and instead appear to form the continental margin. They possess many of the characteristics of intra-oceanic island arcs in having a tripartite volcanic chain (arc/trench gap, trench) arrangement. They differ, however, in showing evidence of strong thrusting towards, and over, older stabilized continental crust or foreland regions (Fig. 11-10b).

There seem to be five basic ways in which a continental shelf/continental rise assemblage becomes involved in mountain building. The first involves the formation of a consuming plate margin near the continental margin with the continental

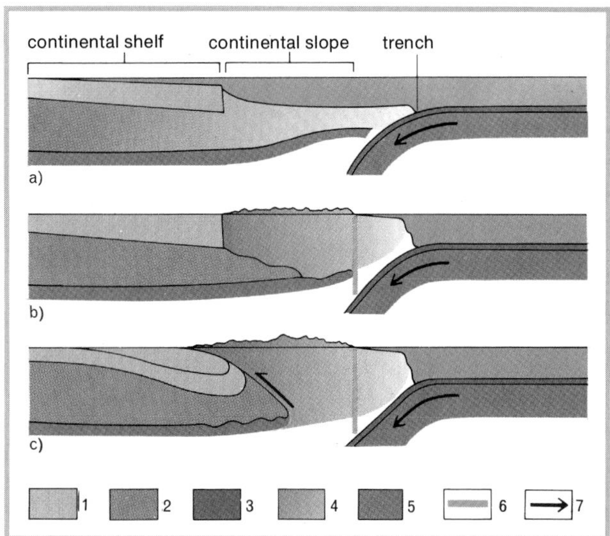

Fig. 11-11. In (a) along the zone of contact between two plates an oceanic trench develops; in (b) the interaction between the two plates causes the first fold and the development of the type of crust typical of archipelagos; in (c) we have the development of huge overslip phenomena along the continental edge. In the diagrams: 1, sedimentary layer; 2, continental crust; 3, oceanic crust; 4, island arc crust; 5, lower level of the lithosphere; 6, volcanic front; 7, subslip.

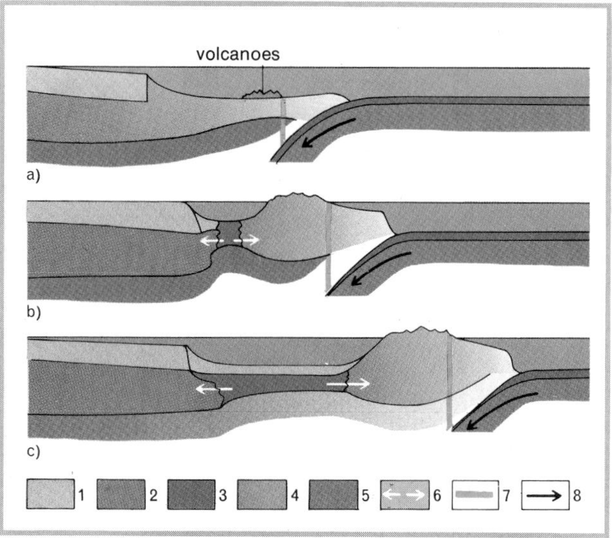

Fig. 11-12. In (a) on the edges of a plate a trench develops, together with active submarine volcanism; in (b) an island arc has developed which delimits a marginal basin partly formed by oceanic crust subject to expansion; in (c) the marginal basin extends through the progressive expansion of the ocean depths. In the diagrams: 1, sedimentary layer; 2, continental crust; 3, oceanic crust; 4, island arc crust; 5, lower level of the lithosphere; 6, course of expansion; 7, volcanic front; 8, subslip.

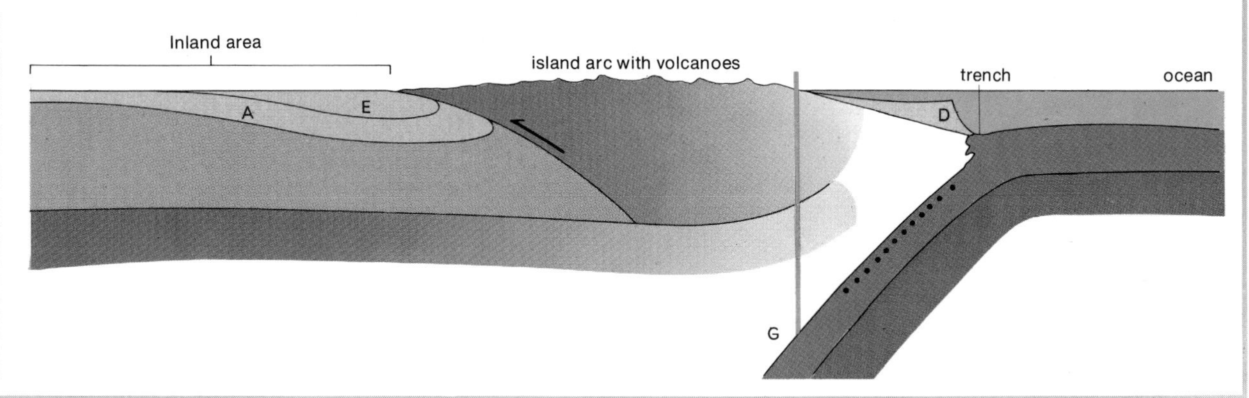

The island arc here is in direct contact with the continental landmass and is characterized by extensive phenomena of folding and overslip. Opposite and above: A, neritic limestones; B, ooze and deposits due to submarine landslides; C, volcanic sediments derived from the archipelago; D, nonvolcanic sediments of the zone between the archipelago and the trench; E, coarse sediments; F, deformation zone with low-temperature, high-pressure metamorphism; G, partial melting of the basalt and re-emergence of andesitic magma.

margin on the plate leading edge (Fig. 11-11). As the plate descends, volcanism begins on the leading edge, and sediments of the continental rise, and eventually the edge of the continental shelf, are deformed. Eventually a wave of deformation extends into the continental shelf and deformed rocks are thrust onto the foreland.

The second mechanism involves a similar sequence but a marginal ocean basin opens, so that the volcanic belt, perhaps, with a core of older continental rocks moves away from the continental margins (Fig. 11-12). Both of these methods of mountain building are a consequence of the continental margin rocks being transformed into a deformed and metamorphosed complex by dominantly thermal processes on a plate leading edge. The other three involve collision between the continental margin and a plate leading edge. If a trench develops near the continental margin so that the plate leading edge faces the continent, the continental margin is driven towards and eventually meets the trench (Fig. 11-13). Since continental rocks are lighter and, therefore, more buoyant than oceanic rocks, very little of the continent can be consumed and this prevents further underthrusting. Thus, a great thrust wedge of oceanic crust and mantle (ophiolite suite) is pushed out across the continental margin. Wedges of this kind are known in Papua and New Caledonia.

Once the ophiolite wedge has been pushed so far out that the continent can no longer be consumed, a new trench may form oceanward of the wedge and andesitic and basaltic volcanoes may develop in the new continental margin (Fig. 11-13). If a continental shelf/continental rise assemblage, on an aseismic continental margin, travels towards a fully established island arc, a similar collision occurs except that island-arc volcanic and metamorphic rocks, rather than a simple slab of oceanic crust and mantle, are thrust across the continental margin. However, sheets of oceanic crust and mantle may be caught up in the collisional suture or join-line between arc and continent. The last and most effective mountain-building

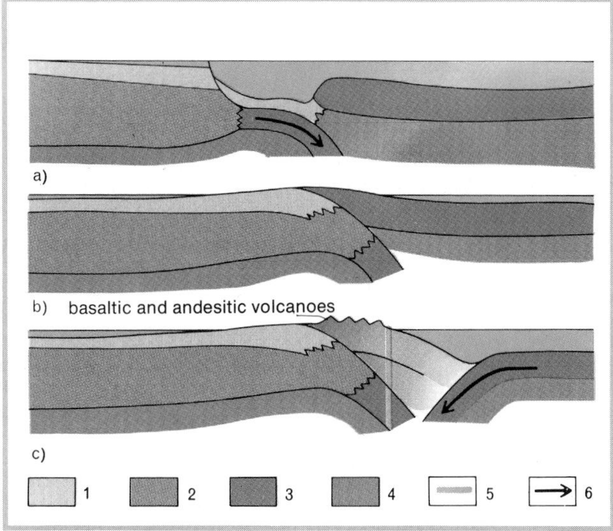

Fig. 11-13. In (a), with the contact between two plates, an oceanic trench develops; in (b) the oceanic crust slips over the continental crust forming a marginal mountain chain; in (c) the course of the subslip is reversed, while, on the continental edge, a characteristic andesitic and basaltic type of volcanism develops. In the diagrams: 1, sedimentary layer; 2, continental crust; 3, oceanic crust; 4, mantle; 5, volcanic front; 6, subslip.

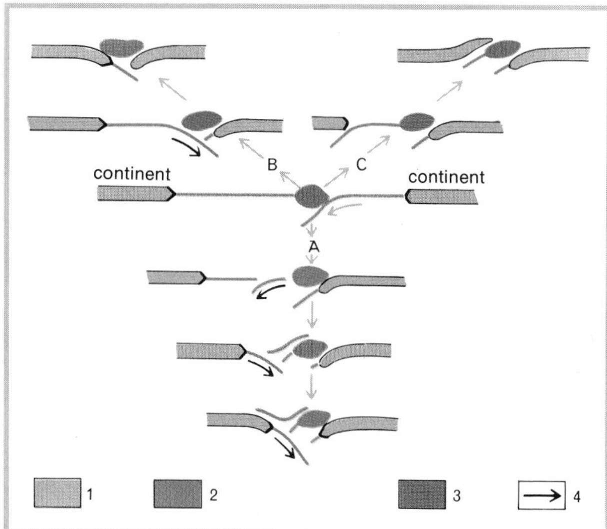

Fig. 11-14. Evolutionary line A: the oceanic crust slips beneath an island arc; plate, left, advances until it encloses basin and causes largescale overslips. Line B: the clash of the island arc with the continental plate on the right causes the reversal of the course of the subslip and the juxtaposition of the two plates via island arc crust. Line C: because of trench, left, the two plates come into contact via island arc. 1, continental crust; 2, oceanic crust; 3, island arc crust; 4, subslip.

mechanism occurs when two large continental masses collide and cause the piling up of giant thrust sheets, or nappes (Fig. 11-4b lower left). The Himalayas were generated in this way by the collision of peninsula India with Tibet.

Mountain belts, in reality, stem probably from complex combinations of some or all of these five basic methods. The evolution of mountain belts such as the Urals involved the driving out of a major ocean basin. Judging by what we see in the Pacific at the present day, it is hard to believe that the Ural ocean did not have a complex form involving marginal basins, island arcs, and various kinds of continental margin. In Fig. 11-14 three alternate geometric ways of driving out an ocean

Plate Tectonics and Orogenesis: the Alp-Appennine Limits

The Alps and Appennines form two large mountain systems which are apparently without any continuity, and are in fact substantially different from a geological viewpoint. An indepth analysis of the border area between the two enables us to reconstruct the paleographic evolution of the region and to incorporate the genesis of the two systems in the model of global tectonics, as a result of the interaction between the African and European plates.

It is standard practice among geologists to put the border zone at the Cadibona pass. But this does not coincide with the zone which can be deduced from the geological natures of the two chains; in fact, geologically speaking, from the turn of the century, we have seen the establishment of a basic divergence in the stratigraphic and structural features of the Alps and Appennines, starting from an alignment developing between the townships of Sestri-Ponente and Voltaggio in the Lemme valley. This alignment has been studied by well-known geologists of various schools (in particular E. Argand, R. Staub, L. Kober, P. Crettaz and A. Tollmann; and the Italians A. Boni and L. Ogniben); there are a variety of interpretations, but they all agree in interpreting it as a large scar formed by numerous subparallel fractures reaching the deepest strata of the Earth's crust and affecting a whole belt between 1–4 km wide.

On either side of this belt, known as the Sestri-Voltaggio Zone there are thus two opposing and geologically different regions: the Alpine region to the west with the Piedmont part of the Pennidic complex, and the Appennine region to the east with the Ligurian complexes.

The Pennidic complex *occupies the innermost part of the western Alpine range and is formed by sedimentary, volcanic, plutonic and metamorphic rocks (the latter partly ophiolitic, as will be made clear below), which, because of the high level of plasticity attained in the course of the various orogenic deformations, have usually given rise to large horizontal folds, astride one another, and pushed towards the outside of the range.*

The Ligurian complex *occupies all the western sector of the northern Appennines and is largely formed by sedimentary rocks of the Jurassic-Eocene period, and to a lesser extent by igneous and metamorphic rocks (mainly ophiolites); it lies eastwards across the adjacent Tuscan region and to the north it shifts towards the Adriatic-Paduan depression in the form of a layer of extremely confused heterogeneous materials.*

Similarly, in the Sestri-Voltaggio Zone, the sedimentary sequence is accompanied by igneous and metamorphic rocks. The sedimentary sequence is formed by: Upper Triassic dolomites and limestones, Lower and Middle Jurassic clayey-phyllitic schists and limestones, Middle and Upper Jurassic jaspers and limestones, lower Cretaceous limestones and argillites; igneous and metamorphic rocks are also incorporated on the ophiolitic group. These rock formations have been sharply dislocated by orogenic stresses; they usually form tectonic flakes, i.e. elongated and flattened masses limited by fracture surfaces, with a subvertical position or sunk towards the eastern quadrants and describing a wide arc oriented NNW/SSE to north-south from the northern tip of the Sestri-Voltaggio Zone to the southern end.

Geological Evolution of the Sestri-Voltaggio Zone

The stratigraphic and petrographic features of the rock formations described offer some extremely interesting indications. As far as the Triassic and Upper and Middle Jurassic formations are concerned, they enable us to affirm an Alpine affinity, in other words a fairly conspicuous analogy with contemporary formations in the Alpine region; from the Middle-Upper Jurassic period, on the other hand, they provide a definite comparison with typically Appennine formations. All this may show that down the geological ages the Sestri-Voltaggio Zone has had an extremely important paleographic function; at the edges of the two regions it would have undergone first the influence of the Alpine system, and then that of the Appennine system. The text below gives a synthesis of how this could have come about from the Upper Triassic period onwards.

Upper Triassic. *On both the Alpine and Appennine areas, across the Sestri-Voltaggio Zone, there is a more or less uniform and shallow sea.*

Upper Triassic and Lower Jurassic. *Between the end of the Triassic and beginning of the Jurassic periods the continental crust, formed by the previously deposited sedimentary layer and a thin horizon of schistose-crystalline and granitic rocks, was torn by stretching movement affecting the paleo-Mediterranean area; its ophiolitic substratum, forming a transition to the mantle beneath, was laid bare in the adjacent Alpine-Piedmontese and Appennine-Ligurian regions which become gradually "oceanized."*

It is as well, at this point, to draw the reader's attention to the ophiolites, given the increasing importance of these lavas in the evolution of this area. Also known as greenstones, they form associations of basic and ultrabasic rocks (gabbros, serpentinites, peridotites, diabases etc.) which recur in the Alpine-type chains and in the midoceanic ridges with features that strikingly resemble one another even in places far removed from one another; this confirms their deep-down provenance, while some volcanic-type manifestations (e.g. diabasic) connected with them can be attributed to the surface reactivation of the same material.

Middle Jurassic. *The oceanic area previously defined has two distinct parts: a trench in the Piedmontese Alpine zone with clayey-calcareous sedimentation, and a relatively higher area (ridge-like) in the Appennine Ligurian zone. The passage from one to the other probably occurred across the present-day Sestri-Voltaggio Zone, which was then already structurally unstable. In these conditions the ophiolites form a dome-shaped mass in the Ligurian region and the substratum of the Piedmontese basin; in this latter, what is more, the ophiolites themselves collapse in the form of rocky fragments (olistholiths) which slipped from the highest eastern edge and re-emerge as diabasic volcanic "newcomers."*

Middle-Upper Jurassic and Lower Cretaceous. *The high Ligurian region grows deeper while the Piedmontese trench is gradually raised.*

The deepening of the Ligurian ridge is accompanied by the expansion of diabasic flows and gives rise to the creation of open-sea conditions in the area.

The raising of the Piedmontese basin is linked with a very wide arching of its ophiolitic substratum with, in particular, the extrusion of powerful serpentinelike plates, upheaving the upper schistose-crystalline and sedimentary layer.

In the time-span in question, there would have been a definite and important reversal of subsidence caused by vertical movements of the ophiolitic substratum. Regarding this phenomenon, it is thought that the Sestri-Voltaggio Zone was once again in the key position; its structural instability would have favored the lowering of the Ligurian-Appennine area on the east side, and the raising of the Piedmontese Alpine area on the west.

basin are illustrated. These are but two of the many combinations possible. The elucidation of which combination was responsible for a particular mountain belt will depend on a tremendous amount of careful synthesis of existing geological data and detailed geological field work in the critical areas.

Role of Plate Tectonics in Continental Evolution

It seems a logical assumption that, wherever we find a mountain belt, however old or deeply eroded, forming a linear or arcuate zone with some or all of the characteristics described above, a plate tectonics mechanism was involved in its evo-

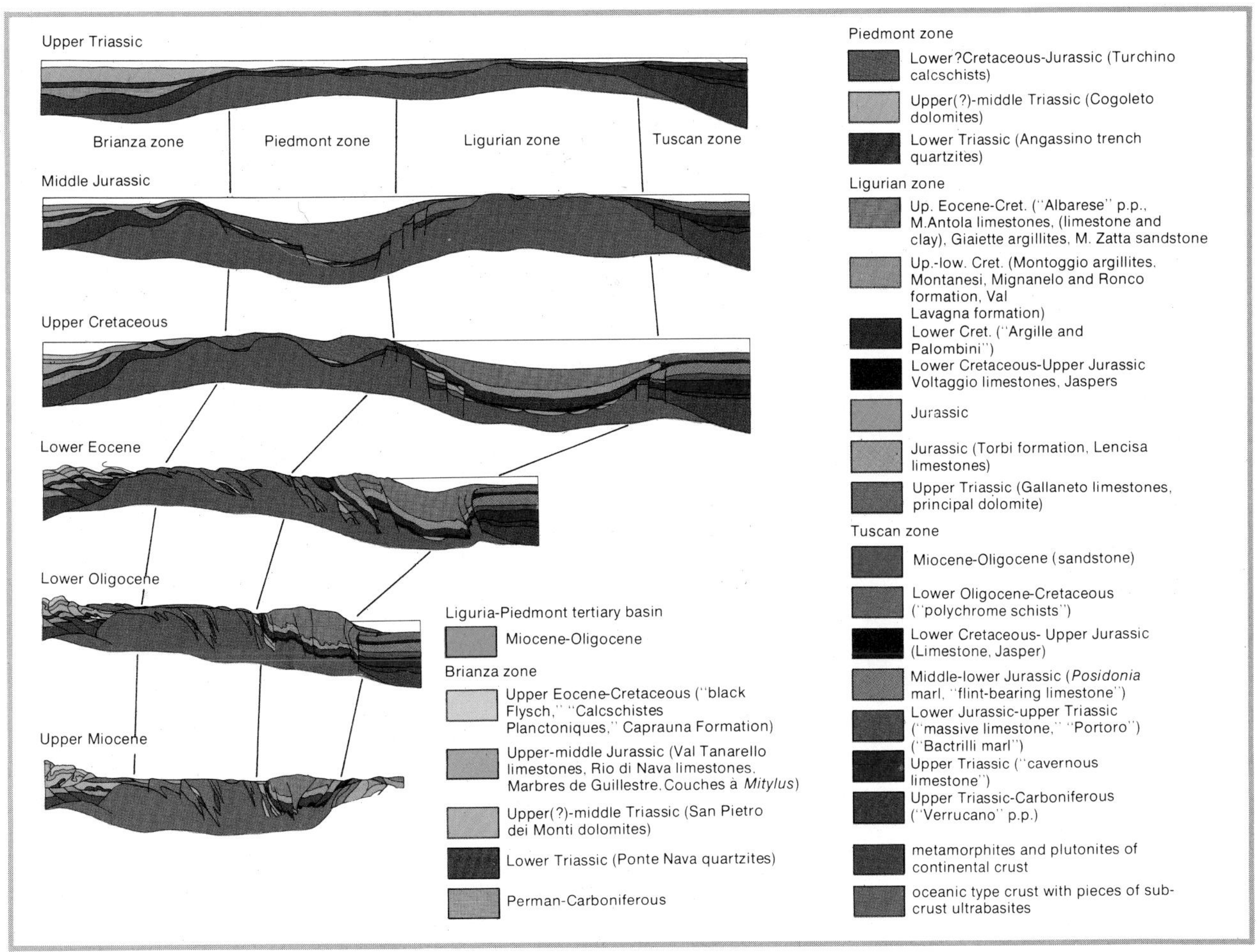

Paleographic and structural evolution of the land at the limits of the Alpine and Appennine chain in Liguria (Sestri-Voltaggio Zone).

Middle-Upper Cretaceous and Paleocene

The phenomena described become more accentuated from the Lower Cretaceous period onwards. The Ligurian area definitely takes on the nature of a quickly subsiding trench with the very extensive (depth-wise) accumulation of detrital sediments (2–3000 m), which were active in this way until the Paleocene; the Piedmontese area finally emerged once and for all, exposing vast areas of the present-day Voltri Group with ophiolites from the original oceanic crust.

The raising of the Voltri Group heralds the replacement of the role of the ophiolites as a driving force in the crust evolution with a passive participation in subsequent tectonic events.

Eocene and Lower Oligocene

The whole region became the site of intense compression oriented south-east to north-west. This compression is shown first and foremost with splintering in the numerous ophiolitic masses of the Voltri Group and the consequent ejection of the same outwards from the Alpine chain together with the upper sedimentary layer of the Triassic and Jurassic which then formed the Calcschist Fold. Likewise, from the Sestri-Voltaggio Zone, the Appennine area at this point follows the fate of the Alpine structure, joining the movement towards the old European continent or landmass. Significant evidence of this is represented by the rocky plates in Appennine facies which are found in various parts of the Alps and which probably form the remains of larger layers or folds which formed the great Pennidic structure: this applies to the Montenotte Fold in the Ligurian Alps, the Helminthoid Flysch in the Maritime Alps, and the Simme in the Swiss Pre-Alps.

Crust Plates and Alpine Orogenesis

The stretching movements to which the crust plates in the Mediterranean area were subject in the Jurassic led to the formation of the ridges and basins mentioned above. The subsequent Eocene-Oligocene contraction caused the drawing-together of the plates themselves, often with "engulfment" (subduction) of belts of land at their edges. This drawing-together is part of the general advance of the African continent towards the European. In the same context, the Appennine area and its extension in the Southern and Austrian Alps would form a plate between the two continents and astride the European margin. The collision of this plate against the European margin and against the Swiss Alps and Pennidic system integrated with it, occurred along a large scar partly buried beneath the Po Valley, and most evident in the Sestri-Voltaggio Zone.

ROMANO GELATI
GIORGIO PASQUARE

lution. Since volcanism on leading plate edges is such an effective method for increasing the volume of the continental crust, much of the continental crust of the world may have been generated in this way. This does not mean that the area of the continents has increased to the same extent. Continental accretion in the past has often been accepted as a concentric growth of continents by the successive addition of new mountain belts with North America as the type example. Mountain belts eventually form on continental margins produced initially by continental rifting; thus some of the younger mountain belts lie on continental margins. Mountain belts formed by continental collision, however, lie within continents with only a shred of ophiolites marking the former existence of an ocean. The continents, may, therefore, be regarded as mosaics resulting from the perpetual splitting and recombination of continental segments riding on an evolving plate mosaic.

A full understanding of mountain belt evolution as a function of plate tectonics is, at present, only potentially possible for the last 200 million years. This is because plate kinematics can only be accurately worked out using oceanic magnetic anomalies. If we can learn to recognize the current distribution and evolution of plate margins from geological criteria, it should be possible to make full integrated analyses of continental geology for the past 200 million years.

Interpretations of Paleozoic mountain belts is more difficult. Although we can recognize all the sedimentary and volcanic rock assemblages, which indicate a plate-tectonics origin for these older mountain belts, we cannot yet work out the plate motions quantitatively. The Precambrian presents an even greater problem. Linear belts of deformation which are fairly certainly the result of plate tectonics date back to about 2.0 billion years. The older Precambrian regions of the continents, however, such as the Superior and Great Slave Provinces of the Canadian Shield are very different. They are characterized by complex irregular zones of ultrabasic, basic and sedimentary rocks "swimming" in a sea of acidic intrusive rocks. The geological evidence suggests that plate tectonics began about 2.0 billion years ago and that the old continental nuclei evolved by some other mechanism.

Present Trends and Future Work

Plate tectonics has provided the earth sciences with a challenging new model for Earth behavior. It has opened up a new era in geology and invigorated a science that was becoming progressively more introspective and concerned with minutae. We need to reassess what has gone before and what we now ought to be doing. There is a particular need to re-examine and synthesize existing geological data. Data storage and retrieval systems, using high-speed computers, make this a realistic task in spite of the necessary drudgery of selection and storage. New and meaningful patterns will certainly emerge from such analyses and will also point to the major gaps in our knowledge.

A complete magnetic anomaly map of the oceans will provide the necessary data for making systematic maps of plate evolution and the consequent continental drift for the last 200 million years. Intensive studies are in progress to investigate the sedimentary, volcanic and structural environments associated with present-day continental margins, plate margins, and oceans. This will provide us with the necessary criteria for investigating the extinct plate margins of older mountain belts and to make viable plate models for these belts. Plate motions older than 200 million years, however, will not be properly worked out until a great many careful paleomagnetic studies are made on the older rocks. A particularly important line of investigation is neotectonics: careful and systematic mapping of present day displacements and deformations on the Earth's crust in active earthquake zones. These studies will show how mountain belts develop structurally in relation to the evolution of consuming and transform plate boundaries, at least at shallow levels in the Earth's crust.

Plate tectonics should give new impetus to the more traditional geological studies of continental rocks. For example, the correlation between the ophiolite suite of mountain belts with oceanic crust and mantle means that research on ophiolite complexes is the best way to study the structure and composition of oceanic crust and mantle. Detailed structural and petrological studies of ophiolites should provide many of the answers to the evolution of spreading ridges and marginal basins. Studies of the structural style of deformed rocks in mountain belts, with careful analyses of the strain suffered by these rocks, will help us to understand to what extent the continental crust and the sedimentary rocks of continental margins shorten during mountain building. One of the main aims of geology should now be an attempt to make global reconstructions, supported by carefully collated local detail, working systematically back from the present day. This is a difficult task and will take decades of painstaking geological mapping and a great deal of integration of geophysical and geological data.

Much of the basic data which led to the emergence and refinement of the theory of plate tectonics was collected during the International Upper Mantle Project. The International Geodynamics Project (see article by X. Le Pichon) was set up as its successor to encourage the kind of studies outlined above. It is to be hoped that it will be given the national support it needs and deserves.

JOHN F. DEWEY

Bibliography: Dickinson W. R., *Plate tectonic models of geosynclines,* in Earth and Planetary Science Letters, X, 165 (1971); Hamilton W., *The Uralides and the motion of the Russian and Siberian platforms,* in Bulletin of the Geological Society of America, LXXI, 2553 (1970); Dewey J. F., Bird J. M., *Mountain belts and the new global tectonics,* in Journal of Geophysical Research, LXXV, 2625 (1970); Beloussov V. V., *Against the hypothesis of ocean-floor spreading,* in Tectonophysics, IX, 6 (1970); McKenzie D. P., *Speculations on the consequences and causes of plate motion,* in Geophysical Journal of the Royal Astronomical Society, XVIII, 1 (1969); Isacks B., Oliver J., Sykes L. R., *Seismology and the new global tectonics,* in Journal of Geophysical Research, LXXIII, 5855 (1968); McKenzie D. P., Parker R. L., *The north Pacific; an example of tectonics on a sphere,* in Nature, CCXVI, 1276 (1967).

According to the plate-tectonics model described in the article by Dewey, it took only 150 million years for a mere fracture in a large continent to turn into an ocean such as the Atlantic, 6–7000 km wide. The opening up of an ocean implies the continuous formation of new oceanic crust along the central ridge. If new crust is constantly being formed along the ridges, it follows either that the Earth's radius is constantly growing, or that while the new crust is forming along the ridge, an equivalent quantity of old crust is being consumed somewhere else. This realization has drawn geophysicists' attention away from the ocean ridges where new crust is forming, to other structures, also part of the ocean floor, where the crust is being consumed. These structures are the ocean deeps.

Seismological analysis has shown that seismic activity in the ocean troughs is substantially different from that in the ridges. The most obvious difference is that while earthquakes under the ridges are mainly superficial, seismic activity under the trenches reaches down to a depth of as much as 700 km. Furthermore, their distribution is by no means random: they are most superficial when they are closest to the trough, and deepest when they are farthest away. In practical terms, when they are plotted in a diagram, they appear to be distributed along an inclined plane (Benioff plane). Those who were looking for a sinking-point in the Earth's crust saw in the trenches the most evident proof of its existence.

The fact is that to admit a progressive disintegration of the oceanic crust leads to important consequences. Imagine for a moment two continents separated by an ocean in which there is a trough. Slowly, the trough swallows up the oceanic crust and the two continents, carried along by the crust which is being consumed, move progressively closer together. The first conclusion is that, while the ridges were a point of progressive separation of continents, the ocean deeps have become points in which continents move together. The second conclusion is that a collision between the two continental masses is inevitable. There has been speculation as to what takes place when two continents collide, even though this takes place extremely slowly. To answer this question, it was necessary to find some examples that could be observed. According to paleogeographic reconstructions, about 180 million years ago India was wedged between Antarctica and Africa in the midst of Wegener's immense supercontinent of Pangea, at a distance of 6–7000 km from the continental mass of Asia. Afterwards, India began a rapid drift which led it towards an oceanic trough situated along the edge of the Asian continent, and about 50 million years ago India collided with Asia resulting in the growth of the Himalayas. These mountains, like many other mountain ranges, contain fragments of typical basaltic rocks, called ophiolites, closely resembling those that today form the ocean floors. The general conclusion is that mountain ranges containing ophiolite fragments are the result of the swallowing-up of an ancient ocean and, in many cases, proof of the collision of two continents. This conclusion has a far-reaching significance for the plate-tectonics model: not only is this model an instrument with which to study the present dynamics of the planet—it is also a way to study its entire history.

WILLIAM R. DICKINSON

Teaches geology at Stanford University, California. Born in 1931 in Nashville, Tennessee, his interest in geology began when, at the age of 15, he moved with his family from green and wooded Tennessee to arid and rocky California. He has experienced, step by step, the development of the plate-tectonics model, foreseeing its application to mountain chains, through ophiolites, and thus the possibility of obtaining the vital key to the reconstruction of the geological history of the Earth.

WILLIAM R. DICKINSON

Plate Tectonics and Ancient Mountain Belts

Overleaf: The new world-map of the 1970s can no longer overlook data concerning the 70% of the Earth's surface which lies beneath the oceans. These data are now becoming as familiar as is the geography of the continents. Knowledge of this area of the Earth was acquired during the 1950s and 1960s and has caused a major change in the way we conceive the whole evolution and history of our planet. In fact, the structures which appear in this illustration have made it possible to gain a clear understanding of the mechanisms which are changing our planet today. Applied to the past, these same mechanisms enable us to understand the reason for many structures known to us, but not explained satisfactorily before; first are the mountain chains, which are today considered to be the remains of huge collisions between continents in perpetual and reciprocal motion (Geol.Mus., London, Photo L. Pizzi).

The geometric arrangements of most rock masses in the continental crust are apparently a reflection of processes related to the origin and history of mountain systems, or orogenic belts. All mountains are but temporary irregularities on the surface of the solid earth. With time, any ground which stands high above sea level is reduced to lowland by processes of erosion. No mountain chain can survive long beyond the time that forces responsible for its elevation cease. The rocks that occupy the roots of mountain systems are gradually exposed by the erosion that destroys the highlands. These roots of mountains have structural and petrologic characteristics that can be preserved for all time as a plain record of the mountains that once towered above them. The great volume of sediments washed off mountain chains as they are destroyed in the topographic sense may also be preserved as additional evidence for the former existence of mountain systems.

The geologic record of truly ancient mountain systems is preserved only within the continental crustal blocks or within insular blocks that have the thick and diversified crustal structure typical of continents. The whole sea floor, meaning the areas above thin oceanic crusts, is comparatively young. Only insignificantly small patches of oceanic crust may be as old as 250 million years, whereas some rocks of the continental crust may be nearly 4 billion years old and there is a rich record going back nearly 3 billion years on several continents. Plate tectonic theory explains this startling circumstance by postulating that oceanic lithosphere, in which crust is thin and dense, is systematically destroyed in time as slabs of cool lithosphere are consumed by the deeper and hotter region of the mantle called the asthenosphere. On the other hand, lithosphere in which thick blocks of light continental crust are imbedded is too buoyant to descend and, therefore, is conserved in an accessible surficial position. In a general view, the rocks of the continents are a concentrate of flotsam gathered together through time. The main processes that cause the construction of continental masses involve mountain-building, or orogenesis.

Most rocks within the continents have the nature and the structure of those formed in mountain roots. Plate-tectonic theory provides a view of the reasons for the existing pattern of mountainous belts on the globe, and a means for the analysis of orogenic episodes in the past using data from the rock record.

Plate Tectonics and Modern Mountain Belts

The central idea of plate tectonics is the concept that the Earth's rigid outer rind, the lithosphere, is segmented into a number of spherical caps, or plates. There are six giant plates, which together account for most of the Earth's surface, and an additional six or more of regional size. These plates are all in relative motion with respect to one another and with respect to an underlying hotter and softer layer, the asthenosphere. The base of the plates is probably at or near the low velocity zone, where seismic waves travel at anomalously slow speeds in the upper mantle. This feature is everywhere deeper than the base of the Earth's crust at the M discontinuity for velocities of seismic waves. Blocks of both oceanic and continental crust are passive passengers riding imbedded in the tops of the plates of lithosphere.

The junctures between the plates are delineated by the world's active seismic belts, which serve to mark where the motions between plates are concentrated. The plates behave individually as intact spherical caps, and any internal deformation of the plates occurs at strain rates that are orders of magnitude below those represented by the relative movements at plate junctures. The rigidity of the plates is less perfect near some of the junctures, and broad bands of marginal deformation affect the edges of some plates. Similar behavior may lead to the development of a number of small plates distributed along a major plate juncture with movements and internal behaviors that are more complex than the ideal plate model implies. In the limiting case, even individual fault blocks may behave structurally as discrete units, but only blocks of combined crust and mantle whose sides break entirely through the lithosphere can be described legitimately as small plates.

Most mountainous belts on the continents, and major linear trends of unusually shallow or unusually deep sea-floor under the oceans, are conspicuously associated with junctures between lithosphere plates. In short, most anomalous elevations of the surface of the lithosphere are aligned along the zones where lithosphere is undergoing deformation at or near plate junctures. These linear junction regions contrast geometrically with the broad expanses of continental lowlands and ocean basins in plate interiors. This line of thought is in harmony with the knowledge that the rocks of mountain systems show evidence of strong deformation in the form of folds and faults. The mapping of the linear trends of such structures in the rocks of old mountain roots reveals the former existence of grand mountain chains that are now subdued or nonexistent as topographic features.

Plate junctures are of three main kinds where the relative motions of the two plates involved are (1) divergent away from the juncture; (2) lateral by shear as plates slide past one another parallel to the juncture, which is then called a transform; or (3) convergent toward the juncture. Each kind of juncture has characteristic geologic and topographic expressions. Processes at fully developed divergent junctures along the intraoceanic rise crests, which mark axes of sea-floor spreading, are responsible for the formation of new oceanic crust and lithosphere. This fresh lithosphere fills the gap created by the spreading motions, and is accreted to the receding margins of both diverging plates. Convergent plate junctures are marked by the linear deeps of oceanic trenches and the parallel volcanic chains of magmatic arcs, which may rise as island arcs above oceanic crust or as marginal arcs above continental crust. Convergence of the plates is accommodated by the descent of one plate into the mantle. Surficial materials are carried downward first into a crustal subduction zone associated with the trench, and then along an inclined seismic zone that dips into the mantle beneath the arc and stimulates the production of magmas to feed the crustal magmatism of the arc itself. Complex processes associated with convergent plate

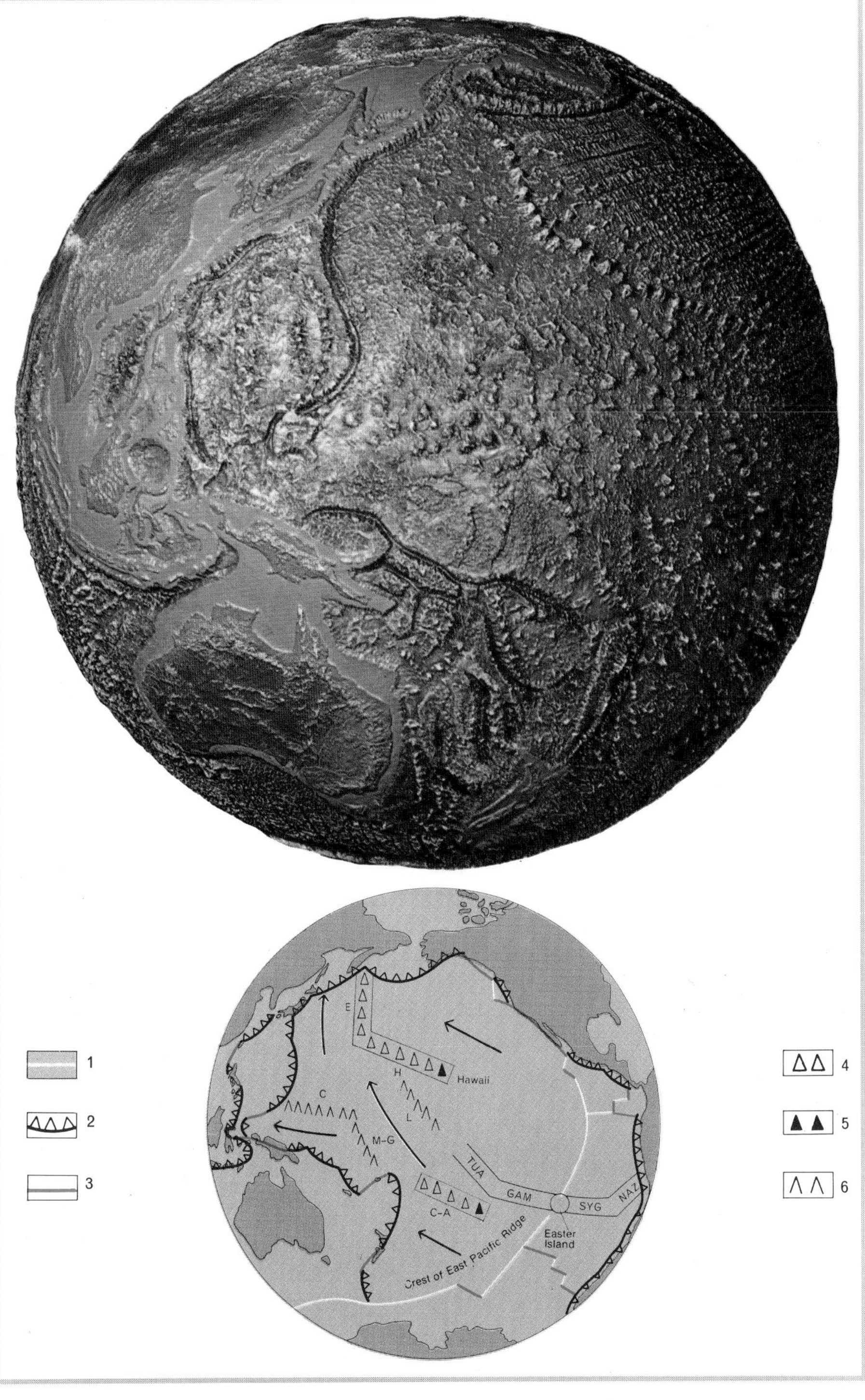

Fig. 12-1. A physical model (top) and sketch (below) of the Pacific hemisphere. The crest of the East Pacific Ridge lies along the join between the divergent plates (1 in the key) corresponding to the edge (which is undergoing concretion or moving) of the Pacific plate, where the North American, Eurasiatic and Australasiatic edges are hallmarked by a series of arc-trench systems positioned along parts of joins between converging plates (2 in the key) linked by transform faults (3 in the key). The curved chain of Emperor(E)-Hawaii(H) submarine peaks is the supposed trace of the hot-spot represented by the Hawaiian Islands on the Pacific plate. The chain formed by Cook-Austral (C-A) submarine peaks seems to be analogous with the Hawaiian chain and has an active submarine volcano (5 in the key) at the correct end as far as the hypothesis formulated above is concerned. The Line Islands (L) and the Marshall-Gilbert system (M-G) may be analogous with the Emperor chain and thus formed by presumed volcanic peaks (6 in the key). The black arrows indicate the apparent approximate movement of the Pacific plate in respect to the hotspots in the Hawaiian Islands and in the Austral system to form the Hawaiian and Cook-Austral chains. The white arrows indicate the movement necessary to give rise to the Emperor chain and analogous chains between 25–75 million years ago. The solid arrow indicates a possible older movement relating to the Caroline Islands (C). The aseismic chains, connected with the Easter Island hotspot on the East Pacific Ridge, include the linked "wings" of the Sala y Gomez (SYG) and Nazca (NAZ) chains on the Nazca plate and the Gambier and Tuamotu chains (GAM and TUA) on the Pacific plate. The TUA-GAM wing may be correlated with the E-H wing and reflects the same variation in the movement of the Pacific plate in respect to the semistable hotspots, between 25–50 million years ago. In this case, the symmetrical wing in the aseismic chains on the Nazca plate reveals analogous variations in the movements of the plates on either side of the East Pacific Ridge *(Geol.Mus.London., photo L. Pizzi)*.

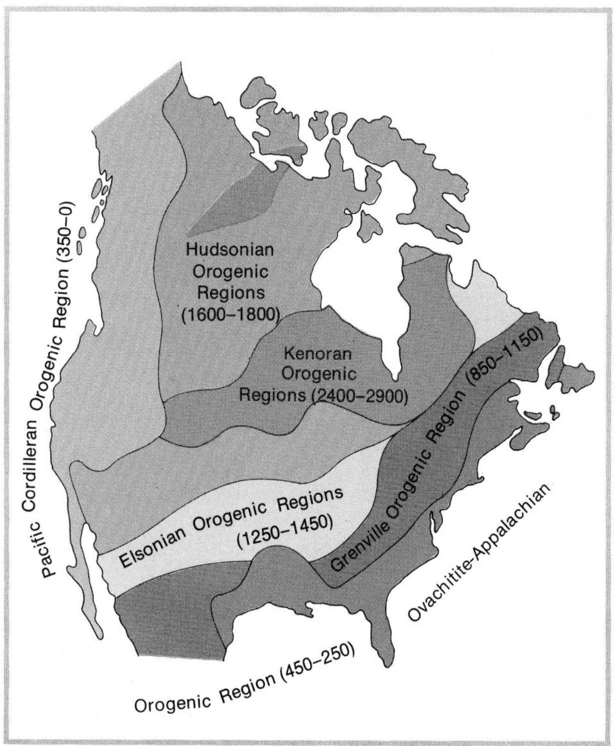

Fig. 12-2. The presumed ages, indicated in millions of years by the numbers in parentheses, of the recrystallization of the eruptive and metamorphic rocks of the orogenic roots of North America reflect the distribution of orogenic processes in the past. The example shown here suggests a concretion of continental blocks produced by the envelopment of successive orogenic regions around a pre-existing core. The irregular interruption of certain regions of differing ages possibly reflects periods in which there was a break-up or separation of the continents by fracturing into a number of continental fragments.

junctures are those mainly responsible for the associations of rocks and structures characteristic of major mountain systems and recognized in the geologic record as indicative of ancient mountain belts.

Volcanic Mountain Chains in Plate Interiors

In the interior of plates, mountains are rare with one prominent exception. In the ocean basins where erosion is largely ineffective below the wave base, the large submarine volcanoes called seamounts may persist for long periods after the eruptions that built them cease. These are peculiar sorts of mountains, for they rise no higher above the general level of the sea floor than does the general level of the continental lowlands. Still, they are features of impressive bulk and some rise well above sea level, most notably in the case of Hawaii where the big island reaches elevations near 5000 m above sea level and 10,000 m above the sea-floor. Seamounts are made almost entirely of basaltic lavas similar, but not identical to the lavas that form the upper tiers of the broad expanse of oceanic crust. Some stand as isolated and randomly spaced eminences, just as there are solitary volcanoes scattered here and there across the faces of the continents, although erosion tends to cut these

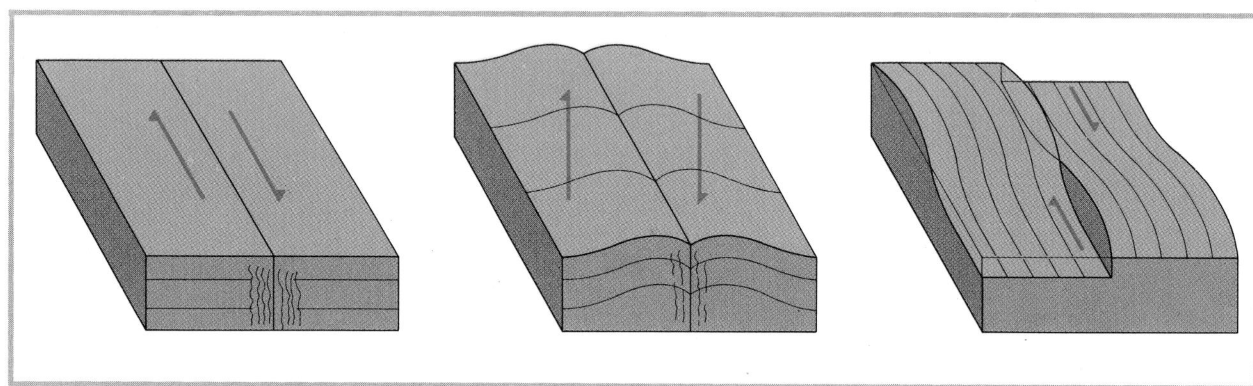

Fig. 12-3. Left an ideal transform fault with relative movement (shown by the arrows) parallel to a simple slip fault following the direction with a vertical fault plane. Middle, an oblique transform fault in respect of the element of convergence between the plates in relative movement; faults and folds due to deformation by contraction give rise to mountains as found in the Coastal Ranges of California along the San Andreas Fault. Right, the activity of a transform fault causes mountain chains and valleys with a transversal orientation; the result is the juxtaposition of upland plateaus gradually eroding to deep basins, separated by steep escarpments, like the Transverse Ranges of California, hewn from the San Andreas Fault.

down to the prevailing level of surrounding continental regions soon after their formation.

Many seamounts, however, are parts of linear chains that march across ocean basins, especially the Pacific, in orderly rows. Years ago, A. Wegener, the great champion of continental drift, suggested that these rows of seamounts, which he knew only as lines of volcanic islands, might somehow be streamlines of crustal movement past deeper parts of the Earth. J. T. Wilson gave substance to this thought by suggesting that each line of volcanic islands and submerged seamounts built across the surface of the lithosphere was fed from a common magma source below. Relative motion between the deep source of magma and the surficial lithosphere would then account for the development of a linear volcanic chain. This idea is in apparent harmony with the fact, discussed by L. J. Chubb, that each of the volcanic chains in the Pacific basin, with the sole exception of Samoa, has its most active or most recently dormant volcano at the more southeasterly end of the chain.

This pattern is to be expected if the Pacific plate of lithosphere is in motion past each magma source on a course laid out generally from the East Pacific Rise toward the Asian continent with its fringing arcs and trenches. W. J. Morgan has dubbed the subterranean magma sources hotspots, which he assumes to be located above the semifixed points of emergence of conventional plumes of material rising from deep in the mantle. Most hotspots he recognizes lie on or near rise crests where they are responsible for topographic culminations like Iceland and other mountainous islands along the Mid-Atlantic Ridge. Such hotspots located on axes of sea-floor spreading are also thought to be responsible for aseismic ridges that extend as linear submarine features into ocean basins within plate interiors. These ridges, like the Sala y Gomez and Gambier ridges which meet near the Easter Island hotspots on the East Pacific Rise, are formed as linked pairs oriented at high angles to the rise crests. Sea-floor spreading continually splits the renewable rise culminations formed above the hotspots. The paired segments are rafted progressively away on either side as they are accreted to the plate margins receding from the rise crest, and in this way semicontinuous ridges are constructed with a common origin on the rise crest.

The seamount chains and aseismic submarine ridges maintain their form and character only until the oceanic lithosphere upon which they stand is consumed at a convergent plate juncture. No one is certain what becomes of them when consumption occurs. Their whole mass of basaltic rock may be inverted to the dense rock called eclogite, as higher confining pressures are imposed during descent into a subduction zone. If so, their whole bulk could return into the mantle. On the other hand, parts of the structures might be stripped off the oceanic plate to be lodged at crustal level in the subduction zone. In this case, they might be incorporated into continental crust together with other materials brought together at convergent plate junctures.

Rises and Rifts Along Divergent Plate Junctures

Broad thermal upwarps, extensional block-faulting, and massive basaltic volcanism are characteristic of regions along divergent plate junctures. The effects of thermal arcing are especially well shown by the intra-oceanic rises, whose crests lie along the axes of sea-floor spreading and which stand more than 2500 m above the general level of the sea-floor. D. P. McKenzie and others have shown that the elevation of the rise crests can be attributed almost solely to thermal expansion of underlying mantle owing to the high heat flow through the rise. The high heat flow results, of course, primarily from the upwelling of hot asthenosphere where the lithosphere is disrupted by plate divergence, and secondarily from the ascent of basaltic magmas produced by partial melting of upwelling

Fig. 12-4. Strata revealed by erosion on the Diablo Canyon Reef Ridge along the Californian coast. The upward projection of the inclined strata may show a fairly wide arc-type structure crossing the picture; but erosion has interrupted the strata leaving just the slightly sloping lower part: in this way only the roots of the fold affecting the zone remain.

Fig. 12-5. Contorted and faulted strata laid bare by erosion in the Death Valley desert. The structure of these strata has been decisive for the deformation undergone in the course of an orogenic phase occurring about 100 million years ago, probably because of plate interactions with reciprocal movement; but all that remains of this mountain system is the odd trace of it in the topography of the region.

mantle rock. The melting is induced by release of confining pressure on upwelling material moving from deeper to shallower depths beneath the rise. J. G. Sclater and his associates have shown that the depth of the sea floor is mainly a simple function of the age of the oceanic crust. As the crust moves away from a spreading rise crest, it cools and subsides at a regular rate. This means that the past or future elevation of any part of an ocean basin can be estimated for all times since the crust in that part of the basin formed, so long as the lithosphere of the basin is not consumed. Similarly, every part of the open ocean basins was once part of an elevated rise crest, yet every part of the present rise crests will eventually subside. The very existence of a piece of sea floor marked by the linear magnetic anomalies created in axisymmetric pairs by sea-floor spreading is thus at once the rock record of a portion of a rise crest, and yet at the same time it is merely a segment of a featureless ocean basin. The real importance of oceanic rocks in a discussion of mountain-building lies, however, in the information they give about the former positions of ocean basins when the rocks themselves are caught in subduction zones and incorporated within continental blocks. The rocks of the oceanic crust form a diagnostic layered assemblage which can be correlated with the ophiolite sequences of orogenic belts.

The continual extension of attenuated lithosphere along the trend of divergent plate junctures can cause normal faulting which delineates blocks of crust separated by steep escarpments. On rapidly spreading rises, where the heat flux and thermal softening of lithosphere close to the divergent juncture is marked, this effect is largely suppressed, presumably in favor of stretching by plastic flowage, rather than by brittle breakage. On more slowly spreading rises, the crestal region is marked by a complexly faulted rift valley, sunk like a keystone along the trend of the thermal arch. The escarpments caused by the faulting are smoothed by adjustments during subsidence of the sea floor and also by the gradual accumulation of marine sediments as the faulted sea floor moves away from the rise crest. The details of such structures and bottom relief would be destroyed, in any case, when the oceanic plate is consumed.

Where incipient divergent junctures break across continental blocks, thermal, structural, and volcanic effects are all prominent. The rift valleys of East Africa and the Great Basin of the western U.S. are both examples. Thermal arching is reflected in Africa by a series of topographic domes of subregional size strung along the trend of the rift belt, and in the Great Basin by a broad structural upwarp of the whole region. Steep escarpments from normal faulting are spectacular in both regions, and rugged topography of jumbled fault blocks is prominent, especially in the Great Basin, where local relief of 2500 m between uplifted ranges and down-dropped valleys is common. Basaltic volcanism is widespread in both provinces, but most notably in Africa. The character of the basaltic rocks contrasts markedly with those erupted to form the sea floor along intra-oceanic rises. However, D. H. Green has shown that the differences between the two types of magmas can be reconciled to the overall concept of plate divergences as the cause of volcanism in the two cases. On the continents where the lithosphere is ruptured, but not yet separated, the magmas separate from the parent mantle at comparatively great depths below a large thickness of cool and brittle lithosphere, and so they have a character which bears the stamp of partial melting at high pressures. On the other hand, under the rises, the parent mantle rises in coherent diapirs to a much shallower level beneath the lithosphere, and separation of magma takes place under much lower pressures.

If fragmentation of a continental block by plate divergence is arrested during an incipient stage, the fault structures and possibly the volcanic strata remain as a permanent record. More interesting is the case where fragmentation proceeds to completion and an expanse of ophiolitic oceanic crust is produced between two continental fragments. The two newly formed edges of the new continents will face one another, at least initially, across a new ocean basin. Each edge will be scarred by complex fault escarpments produced by extension of the once intact continental lithosphere, and by eventual foundering of the continental edges as the continental crust itself was attenuated. Each raw edge can then be masked by sediment derived from the interior of each continental fragment and deposited as a thick wedge draped over the raw continental edge where it rests on much older rocks exposed by erosion of the upwarp that proceded final continental separation.

The resulting continental terrace of sedimentary strata

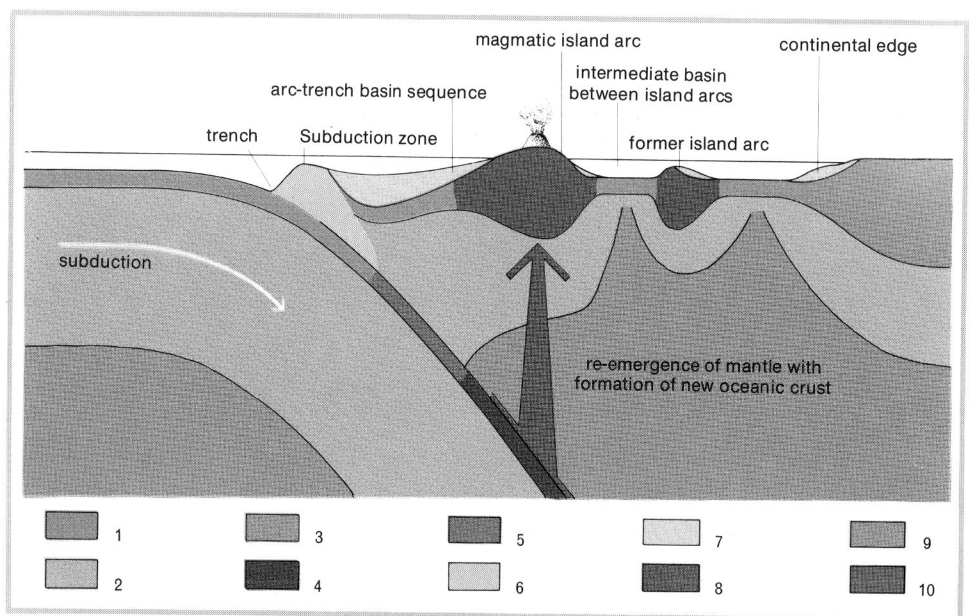

Fig. 12-6. Diagrammatic cross-section through the typical arc-trench system of island arcs with stretch-tectonics in the area behind the arc; in the subduction zone near the trench we find the mélange and metamorphism of green and blue schists; in the roots of a magmatic arc with a volcanic covering we find batholiths and metamorphism of green schists and amphibolites; wedge-shaped prisms of clastic sediments are found beside active arches, relics of arcs and fractured continental edges; the volcanic-plutonic orogeny of the arch is an isolated crust element, as is the case in Japan. In the diagram: 1, basaltic oceanic crust; 2, bottom of the lithosphere; 3, asthenosphere; 4, eclogite plate; 5, transition from basalt to eclogite; 6, deformed crust; 7, marine sediments; 8, crust of the arch; 9, continental crust; 10, magma.

flanking and overlying older rocks along the whole length of the rifted continental edge is the feature termed a miogeocline by R. S. Dietz. Inshore portions of continental terraces are characterized by basal redbeds and lavas deposited in terrestrial basins formed by the initial rifting. These may be succeeded by evaporites deposited in an initial restricted seaway. After continental separation is achieved, a thick and rapidly deposited wedge of coarse clastic strata buries the sloping edge of the continental block until a sedimentary platform is constructed upon which finer clastic and carbonate strata are deposited on a slowly subsiding shelf. Offshore, finer clastic and turbidites accumulate as a continental rise at the toe of the slope along the edge of the ocean basin.

Like the ophiolitic sequences formed by an oceanic crust, the miogeoclinal sequences that mark rifted continental margins are important components of orogenic belts formed along convergent plate junctures. We may turn now to the nature of these most important junctures, to the indigenous assemblages of rocks and structures that are formed within the arcs and trenches associated with them, and to the various means by which such foreign elements as ophiolitic and miogeoclinal assemblages come to be joined with the indigenous assemblages by orogenic processes.

Petrotectonic Assemblages Along Convergent Plate Junctures

The roots of eroded mountain systems are especially characterized by structures indicating contractional deformation of crustal rocks. Folds and faults of various kinds show that the lateral dimensions of crustal rock masses within the mountain belts have been reduced by shortening in direction oriented transverse to the trends of the belts. The same structures show also that the crustal elements within the belts are thickened by the orogenic processes responsible for the crustal shortening. It is this increase in the total thickness of light crust, buoyant with respect to the underlying mantle, that causes the highland masses of major mountain systems to stand so uniformly high above adjacent plains where the crust is thinner. The rocks of the deformed mountain roots typically have been subjected to metamorphism involving both recrystallization to mineral assemblages that can form only at elevated temperatures and pressures, and also the production of crystalline fabrics indicative of strain during recrystallization. The apparent evidence for lateral crunching, vertical stacking, widespread heating, and deep burial, imprinted on the rocks of typical orogenic belts suggests that processes at convergent plate junctures are mainly responsible for mountain-building. Combined lateral and vertical movements, as well as major thermal anomalies, are characteristic aspects of this plate tectonic setting.

Major steps toward relating metamorphism and orogeny to the arc-trench systems along convergent plate junctures were taken in Japan where A. Miyashire showed that each well-known circum-Pacific orogenic province 100 million years old or older can be divided into two parallel paired metamorphic belts. In one member of each pair, the ratio of pressure to temperature was lower than normal for the crust during metamorphic recrystallization, while in the other member of each pair, the P/T ratio was higher than normal. T. Matsuda and others later argued successfully that the metamorphic belts representing inferred high P/T ratios in the crust can be interpreted as old subduction zones where cool crustal rocks descended to great depth beneath or near trenches. Similarly, the belts representing low P/T ratios recrystallized within or beneath the arcs themselves where the crust is unusually hot. These analogies can be expanded to include a whole spectrum of petrotectonic assemblages, or characteristic associations of particular rock types and structural styles, in relation to the main geologic features associated with convergent plate junctures.

Trenches, Mélanges, and Ophiolites

The wave trains of earthquakes radiating from seismic shocks that originate along the shallower parts of the inclined seismic zones in the regions between arcs and trenches, are compatible with the favored interpretation that the movement of the Earth blocks that cause the earthquakes is related to the descent of oceanic lithosphere beneath the flanks of the arcs. This underthrusting of an oceanic plate beneath the arc on the lip of the overriding plate is the consumption of lithosphere required to balance the creation of lithosphere along rise crests on an Earth of constant surface area. Until recent years, the presence of flat-lying and undisturbed sediment layers on the floors of many trenches appeared to negate the idea that any

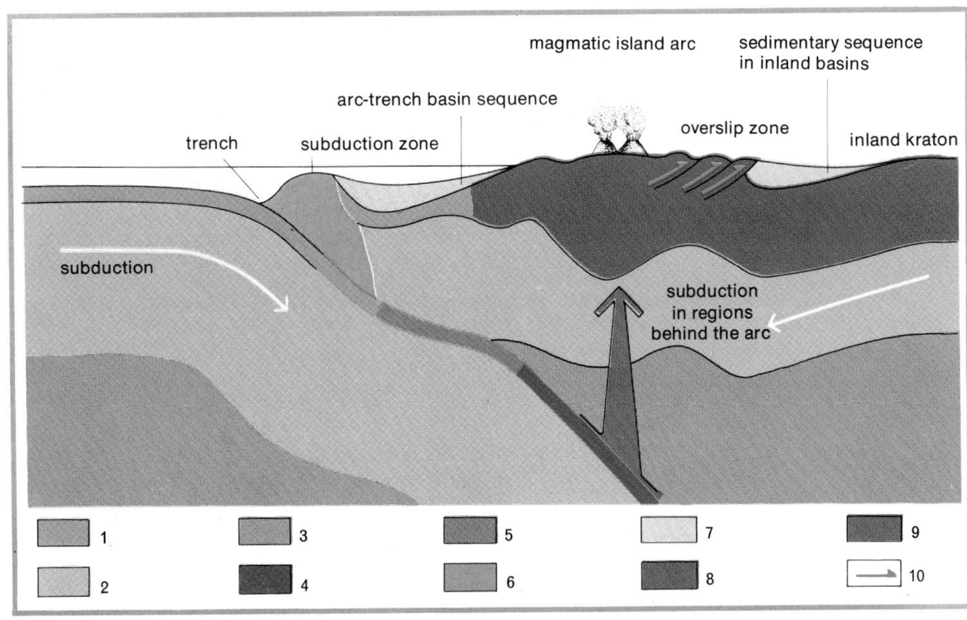

Fig. 12-7. Diagrammatic cross-section through the typical arc-trench system of edges with contraction tectonics in the area behind the arc; mélange and metamorphism of green and blue schists are present in the subduction zone near the trench; batholiths and metamorphism of the green schists and amphibolites are found in the roots of the magmatic arc with volcanic covering; the phenomena of overslip and metamorphism behind the arc reflect the partial subduction of the continental lithosphere which forms an inland basin behind the arc; the orogeny has merged with the continent, as in the case of the Andes. In the diagram: 1, oceanic crust; 2, bottom of the lithosphere; 3, asthenosphere; 4, eclogite; 5, transition from basalt to eclogite; 6, deformed crust; 7, marine sediments; 8, continental crust; 9, magma; 10, overslips.

striking, as isolated, intact and resistant blocks of varied rock types are distributed about the landscape in apparently random fashion imbedded in a scaly matrix of pervasively sheared sediments. The blocks themselves are most commonly graywacke, chert, greenstone, serpentinite, and blueschist. The blueschists are indicators of metamorphism under high P/T ratios. The other rocks were probably incorporated within the mélanges by the fragmentation of oceanic crust. As A. V. Peyve has noted, the presence of typical mélanges in selected zones within the interior mountain systems of Eurasia is evidence for the former presence of ocean basins that have closed by contraction during orogenesis.

In some subduction zones, and particularly adjacent to them and forming the basal part of the overriding crustal block, intact slabs of oceanic crust have been preserved locally. These are the ophiolite sequences, and typical mélanges are made largely of dislocated ophiolitic scraps in various mixtures. Intact ophiolites have a pseudostratigraphic layering which can be correlated with the layering of oceanic crust as detected by seismic refraction studies. At the base is peridotite, which corresponds with the mantle beneath the oceanic crust. In many

Fig. 12-8. Typical mélange in the Franciscan Formation of the California Coast Range. The outcrops are formed by graywacke fragments, plus fragments of greenstones, flints, serpentinites and other rocks belonging to an ophiolitic Mesozoic sequence. The blocks are sunk in a schistose matrix of markedly piezo-metamorphosed argillites and graywackes, and cataclasites. This easily changeable matrix produces gentle slopes. The slip surfaces, in the mélange, have a slight slope to the right; some blocks have their longer slopes oriented in the same direction. The mélange formed in a subduction zone active 75–125 million years ago and was exposed to erosion by the lift process along the San Andreas Fault which has occurred in the last few million years.

such dynamic process was associated with the formation of trenches. The technique of continuous subbottom profiling by seismic reflection has now shown that such a cautious judgment was faulty. The subduction zone, or region where the shallow layers of the oceanic crust are drawn down and crumpled as they are led under the flank of the arc by the descent of underlying lithosphere, is located not right under the axis of the trench, but beneath and beyond the inner wall of the trench on the arc side of the axis. This position coincides roughly with the axis of a negative gravity anomaly long known as a characteristic feature of arc-trench systems. The anomaly probably reflects the presence of a light subducted wad of deformed sediments scraped off the descending oceanic plate and thus attached to the flank of the arc.

No active subduction zone of this kind is exposed fully to view, although one core has been obtained from crumpled sediments above the inner wall of a trench during the deepsea drilling project. The rocks of old subduction zones related to trenches appear in the geologic record within orogenic belts as partly metamorphosed zones of extremely complicated structure typified by the terrains called mélanges, an old term revived by K. J. Hsu. A mélange is a heterogeneous rock mass formed by pervasive shearing and slicing. Rocks of diverse types are juxtaposed in nearly chaotic arrangements across fault surfaces that form a subparallel field of shear planes, but in detail can be quite disorderly. The outcrop expression is

Fig. 12-9. Upper glacial valley in the Trinity Mountains in northern California. The granitic rocks of the irregular peaks were originally injected as melted magma into the roots of a magmatic arc. The arc formed a volcanic-tectonic mountain chain connected with a subduction zone, which formed a marginal trench along a join between convergent plates parallel with the west coast of North America. The intrusions solidified, recrystallizing, in the period between 125–150 million y.a. They were then exposed by the erosion of a vast upland plateau which formed several million years ago as the result of a lift-process connected with the more recent interactions between the plates in contact on the separation surface between continent and ocean basin.

ophiolites, the peridotite was largely hydrated to serpentinite during tectonic emplacement of the ophiolite sequence within or above a subduction zone. Above the mantle peridotite, and corresponding to "Layer 3" of the oceanic crust, is a complex of intrusive gabbros, dike swarms of dolerite, and altered basalt flows. These rocks are the product of the injection of basaltic magmas into the lower part of the growing oceanic crust along rise crests. In this setting, the rocks low in the crust floor metamorphism are of low P/T type. The effects of this kind of alteration must be distinguished carefully from episodes of metamorphism that occur later as the rocks encounter a convergent plate juncture, and the potential for confusion is great. Next in sequence, and representing "Layer 2" of the oceanic crust, are unaltered or less altered basaltic pillow lavas like those dredged from sediment-free parts of the ocean floor today. "Layer 1," the sediment layer of the oceans, is represented in ophiolite by a variety of rocks ranging from the sequences of chert and argillite with some layers of tuff or limestone, to great piles of turbidites, sandstones or graywackes deposited by turbidity currents that carried continental detritus well out into subduction zones. The sedimentary components of ophiolite sequences are disproportionally over-represented, for the basaltic igneous components can more readily invert to the dense phases of eclogite and descend into the mantle with the rest of the sinking lithosphere.

The recognition of ophiolites is a key to the interpretation of ancient mountain belts, for each zone of ophiolite slabs or ophiolitic mélange marks a former ocean basin. Their common occurrence within orogenic belts implies that the closure of ocean basins is an important facet of mountain-building processes. The timing of such events is tricky to establish, for the time that the rocks of an ophiolitic suite were formed in an ocean basin is not the same as the time they are later emplaced against other rocks as part of a tectonic crustal stack of continental thickness by activity along a subduction zone.

In arc-trench systems, the crest of the inner wall of the trench is marked commonly by a bathymetric ridge or a sharp, liplike formation in bathymetric slope. This feature apparently marks the top of a pile of crustal materials crammed into the subduction zone to form a buoyant crustal stack. In some places, notably the Mentawai Islands off Sumatra, young mélanges of this newly constructed crustal pile actually peek above water. In most places, the drag from continental descent of lithosphere beneath the subduction zone holds the pile down and maintains the negative gravity anomaly. Clearly, the materials of the subduction zone would rise rapidly, under isostatic influences, as soon as plate consumption is arrested for any reason. In this sense, the subduction zones of trenches harbor potential mountain ranges which must form as soon as the materials are allowed to rise.

Arc-Trench Gaps and Volcano-Plutonic Orogens

The bathymetric ridges or breaks in slope above the inner walls of trenches also act as barriers to the transport of sediments into the trenches from the highlands associated with the magmatic arc. Sediment thus ponds in sediment traps within the spaces between arcs and trenches. Sedimentary basins of these arc-trench gaps in modern examples include terrestrial lowlands, subsiding shelves or shallow basins, transverse slopes or deep terraces, and deep marine troughs which flank the highlands or islands of the arc trend. Sequences deposited in old arc-trench gaps appear among the rocks of orogenic belts as zones of relatively undeformed and unmetamorphosed sedimentary rocks sandwiched between the metamorphic and igneous rocks of the subduction zones and the magmatic arcs. The "arc-trench gap" sequences are commonly in contact with the mélanges along thrust faults marking the upper tectonic limits of the deformed subduction zones. On the arc side, such a sedimentary sequence of the arc-trench would be at odds with the plutons forming the roots of partly dissected arcs or grade laterally into volcanic or volcaniclastic strata covering the arc region.

The axis of the arc itself is marked at surficial levels by the eruption of vast quantities of volcanic debris, much of it pyroclastic, belonging to the orogenic basalt-andesite-dacite-rhyolite association. Some accumulates as volcanic cones above eruptive centers but much more is dispersed as volcaniclastic sediment over a much broader belt along the arc trend. The line of the arc is thus a positive topographic feature all during the magmatic activity. Even so, subsidence at the surface as the erupted materials founder into the roots of an arc leads to a net accumulation of an elongate prism of volcanogenic materials.

Fig. 12-10. Erosion slopes in the Sweetwater Mountains in western Nevada. The topmost volcanic rocks erupted 25–50 million years ago as part of the linear volcanic-tectonic mountain region of a magmatic arc, linked with a subduction zone which formed a marginal trench along a join between convergent plates, parallel to the west coast of North America. These were subject to rapid erosion in the last few million years, because of the raising in a fault block during the crust stretching of the Great Basin, a fracture region in the embryonic stage. With time, the surface layer of volcanic rocks will be carried away altogether; only the intrusive and metamorphic roots of the old magmatic arch will remain.

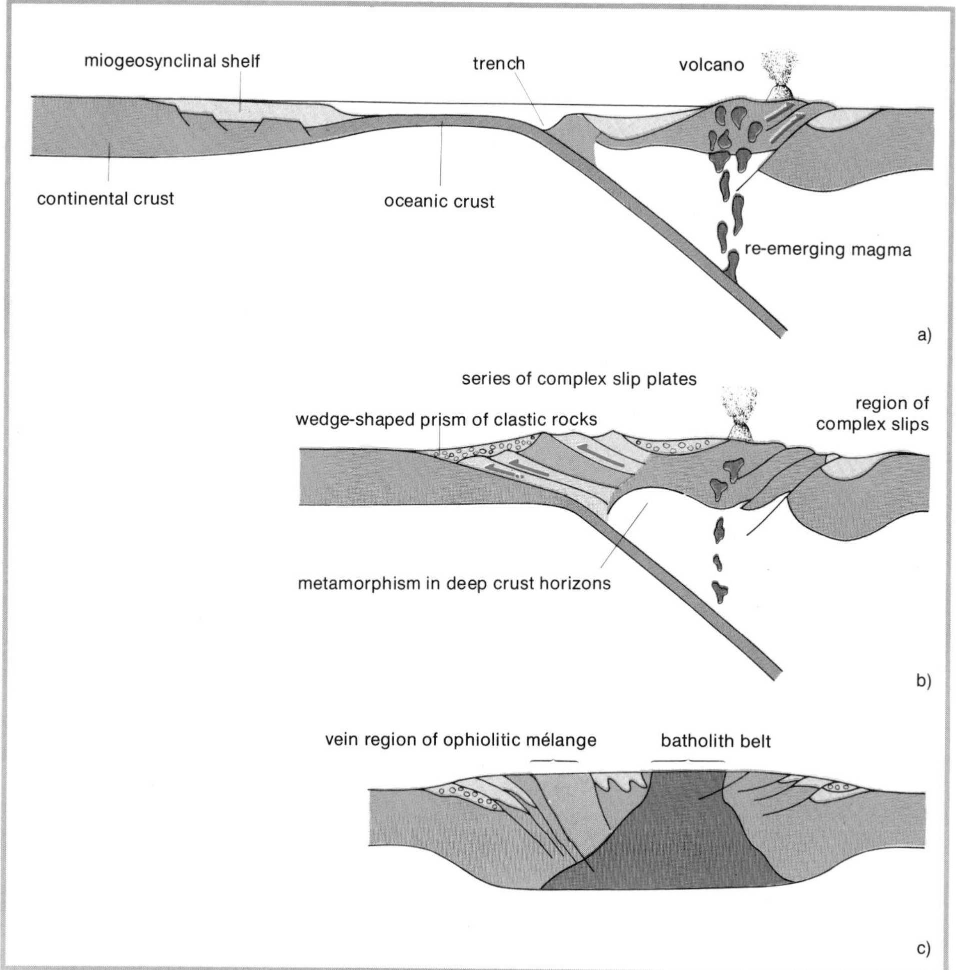

Fig. 12-11. Diagrammatic drawings representing the development of a collision orogeny. In (a) an inactive sunken continental margin (left) approaches the arc-trench system and the volcanic-plutonic orogeny of an active continental margin (right). In (b), during the collision, overslipping occurs and the sedimentary prisms coming from a re-emerging orogeny take the place of the trench. In (c), after the collision and erosion a bilateral orogeny remains, with overslip regions and filling with clastic sediments of the deformed troughs on each side of the axis of the volcanic-plutonic orogeny, while the vein region remains only on the one side.

These are fed to crustal levels by magmatic transfer from sites of melting along the inclined zone in the mantle below. At deep crustal levels, some magmas are intruded into the roots of the arc as plutons of the orogenic tonalite-granodiorite-adamellite-granite association. The combined plutonism and volcanism forms a characteristic crustal element that can be termed a volcano-plutonic orogen. Such a feature is characterized by thick sequence ratios in the hot arc roots and cut by belts of granitic batholiths.

Volcanic-plutonic orogens are themselves the record of a kind of mountain chain. Modern examples of magmatic arcs range from nearly submerged island chains like Tonga or the Marianas to volcanic chains like that of the Andes along the margin of South America. Intermediate types include the complex island arc of Japan with thick continental crust but with basins of oceanic character to either side, and the island arc of Sumatra and Java which really stands on a continental margin south of the barely submerged Sunda Shelf.

The igneous activity along the axis of an arc is not, however, the only mountain-building process associated with magmatic arcs. Related tectonic events are also significant. In fact, many arcs both ancient and modern do not display intense contractional deformation within the arc itself. Block-faulting and local extensional deformation related to volcano-tectonic subsidence may be the dominant structural features along the axis of an arc. It is the region behind an arc, on the side away from the trench, that more commonly displays folding and thrust-faulting of continental rocks on a grand scale. The sub-Andean region of Cenozoic folding and thrusting behind the Andes is a modern example in South America and the Mesozoic Cordilleran belt of folds and thrusts that extended for 5000 km from Alaska to Central America behind the Mesozoic volcano-plutonic orogen that extended along the whole of the western margin of North America. In both cases, crumpled and broken masses of strata behind the arcs were overthrust away from the arcs across continental crust along the margin of the stable continental interior, or craton. Mountainous tracts of the rumpled and overthrust strata shed voluminous detritus into linear "foreland" basins aligned like moats just beyond and parallel to the flanks of the uplands. At the same time, thrust sheets of the folded strata crept across the very margins of the basins themselves.

Back-arc thrusting is not easy to explain, for the region immediately behind some arcs is marked by wholesale extension involving the development of truly oceanic basins by the creation of new lithosphere along back-arc spreading centers. This process can cause a marginal arc to detach from the continent of which it was initially an integral part, and thus, become an island arc as Japan has done. D. Karig has discussed the development of marginal seas in this fashion, and also how repeated episodes of back-arc spreading can cause the successive calving of crustal slivers from the rear side of evolving arcs to leave a succession of what he calls remnant arcs separated by a coordinate succession of interarc basins.

The two contrasting tectonic environments in back-arc areas can probably be explained best as expressions of two

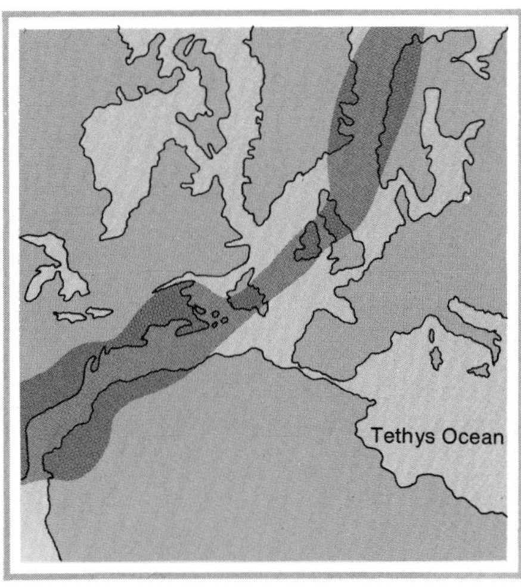

Fig. 12-12. This figure shows the Bullard reconstruction of the positions of the continents with the continental shields joined together before the opening-up of the North Atlantic ocean which occurred within the last 200 million years. The dark band indicates the Appalachian-Caledonian orogenetic region formed by a complex series of orogenic events between 450–250 million years ago. Various and numerous orogenetic episodes culminated in continental collisions during the late Paleozoic period; these gave rise to a vein region within the orogenic complex by closing-off the pre-existing ocean basins which dated back to the Lower Paleozoic. The dark band encompasses ophiolitic zones and subduction mélange, volcanic-plutonic orogenies, metamorphic areas and post-arc overslip complexes, all linked to the old joins between convergent plates. The movements which occurred between these latter led to the collision which gave rise to the long vein region.

different patterns of bulk motion between the lithosphere of the over-riding plate, upon whose edge the arc is built, and the underlying asthenosphere. Where the relative motion of the lithosphere with respect to the asthenosphere is away from the convergent plate juncture, back-arc spreading occurs. The plate of the lithosphere is driving, as P. J. Coney puts it, past underlying asthenosphere, and is thus actively pressing against the flexure in the descending plate of lithosphere beyond the trench, then the tectonic setting behind the arc is contractional, and back-arc thrusting occurs. In this interpretation, the thrusting is not driven entirely by gravitational gliding, but is due in large part to partial subduction of the continental craton beneath the rear of the arc. The separation of lithosphere in the one case and the partial subduction in the

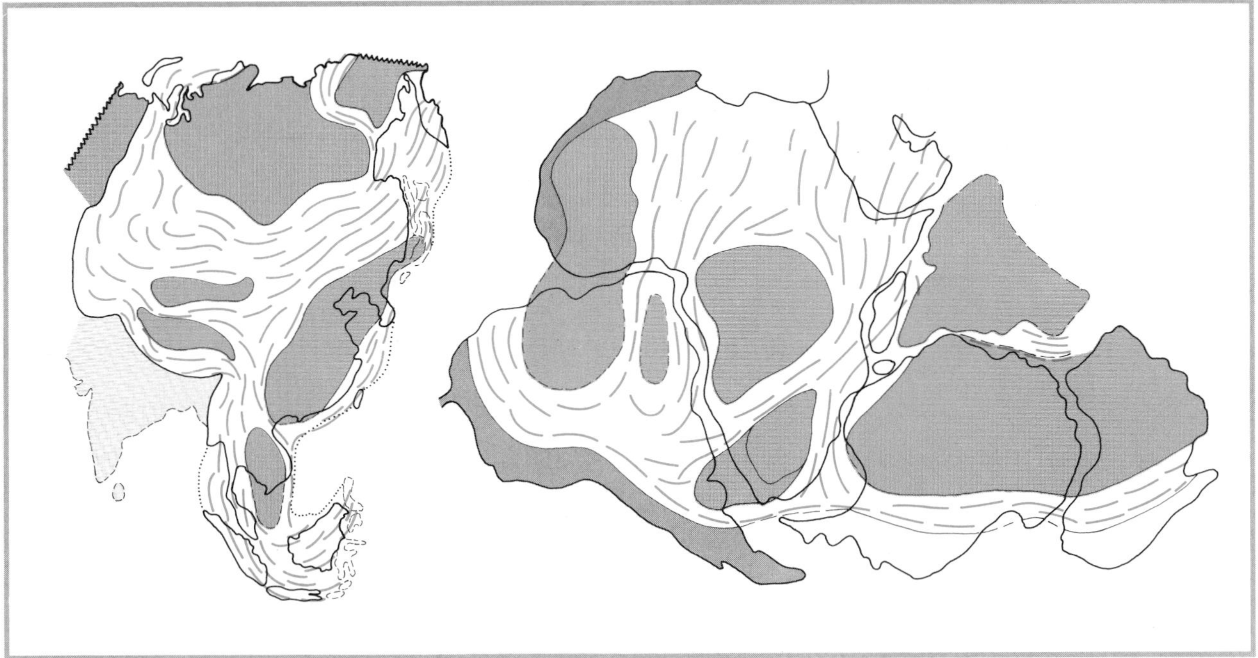

Fig. 12-13. Diagrams illustrating the concept of composite continental grouping due to the "welding" together of the different continental fragments along orogenic regions with median veins. Left, Asia about 100 million years ago; blocks of mainly Precambrian rocks (colored areas) are sunk in a vortex complex of Mesozoic and Paleozoic orogenic regions (colored lines); the Cainozoic addition of India is indicated (in light gray) while the marginal island arcs (hatched lines) are roughly in the positions they occupied prior to the opening-up of the marginal seas in the area behind the arc during the Cainozoic; the dotted line indicates the edge of the shelf and represents the approximate limit of the shield off the east coast. Right, the Gondwana in Smith and Hallam's reconstruction; the gray areas are Phanerozoic marginal orogenic regions; the colored areas indicate cratonic blocks more than a billion years old; the colored lines indicate Panafrican orogenic regions (650–450 million years), which are considered to be of the same date as the "reunion" of the Gondwana continent, which then split up in the last 200 million years.

other presumably are localized at the rear flank of the arc because the high heat flow associated with arc magmatism creates a thermal curtain which heats and softens the lithosphere. In effect, there is a thermal "saw" at work.

Crustal Collision and Suture Belts

The continued operation of an arc-trench system is possible only so long as oceanic lithosphere continues to be consumed. If thick crustal elements arrive at a trench plate, consumption is terminated by crustal collision as crustal elements ram into the subduction zone. In the extreme case where a continent crashes into the subduction zone of a marginal arc bordering another continent, the continental collision welds the two together as a composite continent. Only an ophiolitic mélange zone along the suture belt and the roots of a volcano-plutonic orogen remain as evidence of the previously active continental margin. Intense deformation and thrusting of the miogeoclinal sedimentary rocks along the previously passive continental margin afford evidence of the attempted subduction. The whole assemblage of volcano-plutonic orogen, mélange belt, and thrust belt are seen where India collided with a subduction zone lying south of Tibet during the Cenozoic. The present Himalayan mountain belt is thus a collision orogen. Continental collisions may be preceded or accompanied by downbowing of the passive continental margin, and deposition of strata shed from the growing orogen within a linear basin beside it may closely mimic the deposition in "foreland" basins behind marginal magmatic arcs.

As discussed by J. F. Dewey and others, many types of crustal collisions are possible; and the degree of reorganization of plate motions required by different types is variable. If small crustal elements, such as subsea turbidite fans or seamounts or even small remnant arcs, arrive at subduction zones, they may simply be accreted to the flanks of arcs as part of complex mélange belts. Arc activity may even be shifted by such accretions, so that the rocks of subduction zones with the special petrotectonic imprint of that setting then may become part of the roots of the arc and receive a distinctive petrotectonic overprint. Where an active intraoceanic island arc collides with a quiet continental margin it may lodge against the margin after rucking up the miogeoclinal wedge in the subduction zone. The polarity of plate consumption may then flip, but continue in the same region with the accreted arc now a marginal arc consuming oceanic crust formerly behind it but now offshore in front of it. Major continental collisions probably force global reorganizations of plate motions, and may even trigger the activation of marginal arc-trench systems along previously inactive continental margins. If so, miogeoclinal sequences will be disrupted or modified in place by subduction and arc activity.

The relative motions of plates involved in the development of collision orogens may not be strictly normal to the suture belt. Oblique components of the relative motion may cause transform or shear motions parallel to the suture belt, in which case the structural features caused by collision may be exceedingly complex. It is also apparently true that oblique plate convergence along simple arc-trench systems prior to any collision may cause lateral or transform motions generally parallel to the trend of volcano-plutonic orogens.

The logic of plate tectonic theory implies that major mountain systems, if they are allowed to evolve fully, will eventually become parts of complex collision orogens. Varied crustal elements will be caught up as scraps within broad suture belts. Similarly, rifted continental margins with their miogeoclinal cover will eventually be incorporated into orogenic belts. The fragmentation and assembly of continents by separations and collisions probably has pursued a kaleidoscopic course of ever-changing patterns throughout most of Earth's history. There is thus no unique sequence of events in the evolution of mountain belts. The variable associations of diverse assemblages of rocks within each belt must be used to infer the particular sequence of successive plate tectonic events that took place. Any large and complex orogenic belt harbors a complex array of orogenic elements that record a long progression of individual orogenic events.

WILLIAM R. DICKINSON

Bibliography: Burchfiel B. C., Davis G. A., *Structural framework and evolution of the southern part of the Cordilleran orogen, western United States,* in American Journal of Science, CCLXXIII, 97 (1972); Nicolas A., *Was the Hercynian belt of Europe of Andean type?* in Nature, CCXXXVI, 221 (1972); Boccaletti P., Elter P., Guazzone G., *Plate tectonic models for the development of the western Alps and northern Apennines,* in Nature Physical Science, CCXXXIV, 108 (1971); Coney P. J., *Cordilleran tectonic transitions and motion of the North American plate,* in Nature, CCXXXIII, 462 (1971); Davies H. L., Smith I. E., *Geology of eastern Papua,* in Geological Society of American Bulletin, LXXXII, 3299; Dickinson W. R., *Plate tectonics in geologic history,* in Science, CLXXIV, 107 (1971); Atwater T., *Implications of plate tectonics for the Cenozoic tectonic evolution of western North America,* in Geological Society of American Bulletin, LXXXI, 3513 (1970); Dercourt J., *L'expansion océanique actuelle et fossile; ses implications géotectoniques,* in Bulletin de la Société de France 7e série, XII, 261 (1970); Dewey J. F., Bird J. M., *Mountain belts and the new global tectonics,* in Journal of Geophysical Research, LXXV, 2625 (1970); Dewey J. F., *Evolution of the Appalachian-Caledonian orogen,* in Nature, CCXXII, 124 (1969); Hamilton W., *Mesozoic California and the underflow of Pacific mantle,* in Geological Society of America Bulletin, LXXX, 2409 (1969); Gansser A., *The Indian ocean and the Himalayas,* in Eclogae Geologicae Helveticae, LIX, 831 (1966).

IN believing that such gigantic units as entire continents might move with respect to one another, Wegener, with his continental drift theory, attempted the impossible. But as we have seen, reality always goes beyond human imagination; the plates in motion with respect to one another, the moving plates of the plate-tectonics theory are much larger than the single continents whose motion was imagined by Wegener. In the plate-tectonics model the continents are like passengers, passive and of no special importance, on a raft of inconceivable dimensions.

The plate-tectonics model has altered another deep-rooted belief of geophysicists in the first half of this century, since 1900 when the Yugoslav seismologist Andrija Mohorovičić, analyzing an earthquake that had taken place in the Balkans, had identified a sharp discontinuity in the propagation of seismic waves at a depth of about 40 km beneath the Earth's surface. This discontinuity (later called the Mohorovičić discontinuity, or Moho for short) has become symbolic of a separation between the continental crust and everything below it. Wegener too (and this was one of the reasons for his initial success) had felt that this could be the surface of lower-lying materials over which the continents were slipping. Progress in seismological studies led to the realization that the Moho, though very deep down under the continents, rose a depth of less than 5–6 km under the oceans. Furthermore seismologists have proved that at that depth the material which forms the Earth's interior behaves as if it were solid. Thus it was no longer possible to assert that the Moho was the surface over which all the crust movements occurred.

Laboratory experiments show that rocks in this particular state, if subjected to slow and long-lasting stresses, behave as if they were fluid, while in the face of sudden and violent stresses, such as the unleashing of a seismic wave, they behave as if they were rigid. In the plate-tectonics model, this stratum, called the asthenosphere, or "low-velocity-layer" (LVL) is of fundamental importance; it is the surface over which the plates slip with respect to the Earth's interior.

But what motor keeps them in endless movement? Research workers have not yet reached final agreement on this point. Theories under discussion at the present time, however, attribute much of the cause to the Earth's inner heat. The classic theory is that of convection currents, on which Hess' hypothesis of the expansions of the ocean bed is based. The Earth is supposed to have hotter regions in its interior, and around these, because of the extra heat, warmer and less dense materials with respect to those of surrounding regions tend to rise to the surface. This hypothesis postulates that when these materials reach a position close to the surface, they lose part of their heat by conduction and thus begin to move parallel to the outer surface of the planet, tending to descend until they are cooler than the surrounding material and sink back into it.

Thus great convection cells similar to those formed in a fluid when it is heated might be formed. These currents would, on the one hand, bring the hot material to the surface in regions corresponding to the oceanic ridges and at the same time would carry along the tectonic plates which, in their turn, might already be subject to movement in the same direction by the force of gravity. An alternative hypothesis, also based on the presence of thermal anomalies in the mantle, is that of "plumes" of heat in practically stationary positions with respect to the Earth's interior. Very hot material is supposed, according to this hypothesis, to rise rapidly to the surface, producing a hot spot, usually marked by considerable volcanic activity, and there to expand rapidly. This movement of lateral expansion should be sufficient to shift the plates. One of these hot spots might be the Hawaiian volcanic region. The most fascinating part of this hypothesis is that the positions of the hot spots would remain practically fixed while the plates move over them; as a result they would act just like punches, boring holes through the crust and creating volcanic regions that would enter into activity and then become extinct as soon as the plate moved away from that position. Indeed, various volcanic centers in the Pacific have this characteristic: they are found at the tips of long chains of extinct volcanic islands, in which the oldest islands are farthest away from the end which is active today, and which run in the same direction in which the Pacific plate has moved. But for the moment it is still hard to reach a full explanation and geophysicists are still studying the question.

STANLEY K. RUNCORN

Director of the School of Physics of the University of Newcastle-upon-Tyne, where he teaches physics. Born in 1922 in Southport, England, he has been interested in the problems of geophysics during his entire scientific career. Through the study of geomagnetism, he was convinced of the reality of continental drift even before the models of ocean floor spreading and plate tectonics appeared and were confirmed. As a geophysicist, he has always paid particular attention to the mechanics through which continental drift can manifest itself. He is also concerned with the evolution and internal composition of the planets and the moon.

STANLEY K. RUNCORN

Crust-Mantle Relation in Plate Tectonics

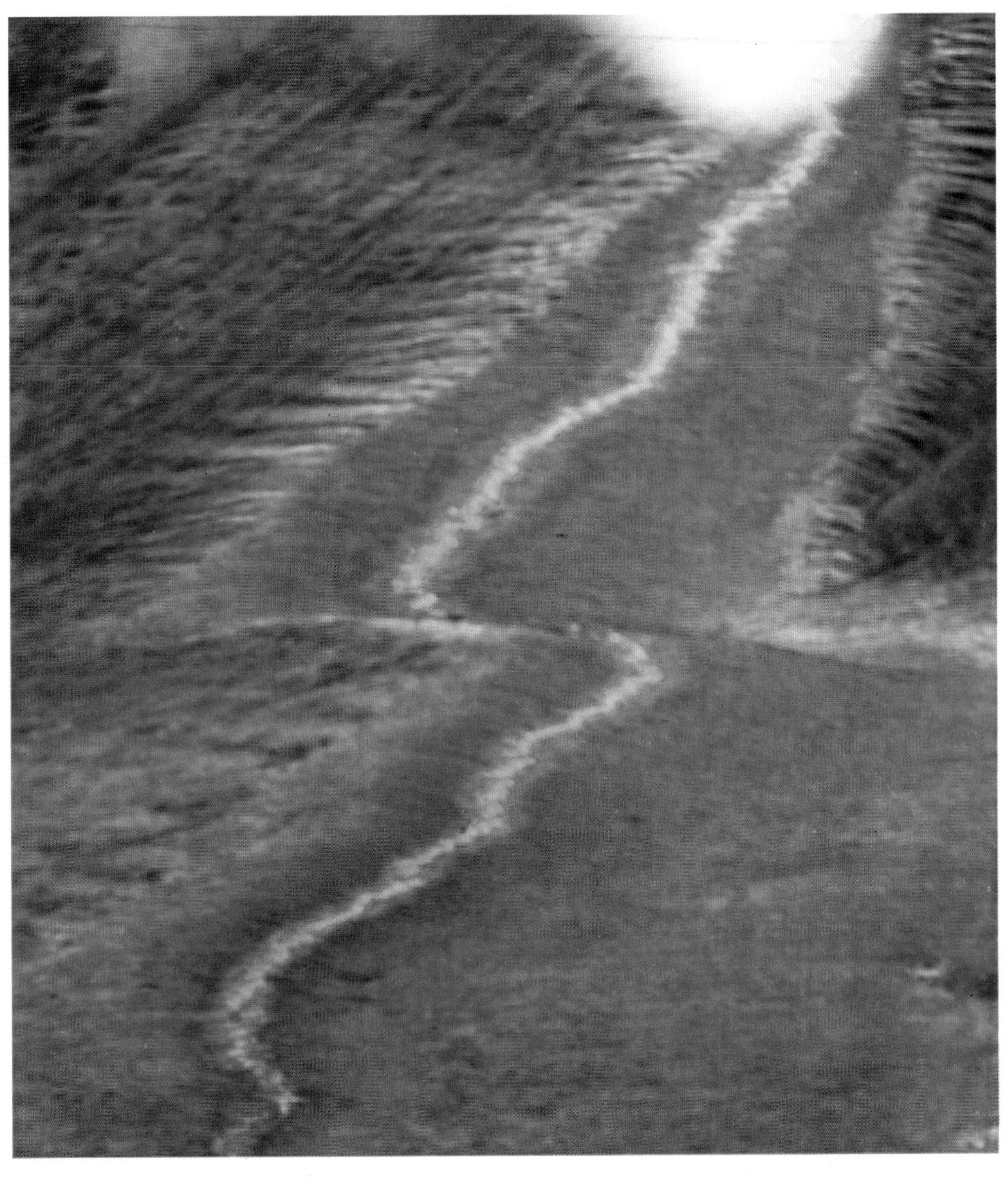

Overleaf: A long fracture in a lava crust observed in the Kilauea volcano in the Hawaiian Islands; out of it flows the molten lava which, once it reaches the surface, cools and is deposited at the edges of the fracture, causing a lateral expansion of the lava crust as a whole. This mechanism simulates the phenomenon of ocean-bed expansion along the submarine ridges. (Geological Survey, photo W. A. Duffield).

In 1909 Mohorovičić discovered a discontinuity below the continents where the velocity of longitudinal earthquake waves (P waves) jumps from 7 to 8 km/sec. The Moho is now known to lie at an average depth of 35 km below the surface of the continents and the velocity contrast is consistent with the identification of the continental blocks with granite and the underlying material with olivine, the most closely packed ferromagnesian silicate commonly found as phenocrysts in volcanic rocks and thus derived from depth. The density of the outermost part of the mantle inferred from seismological considerations, for the velocity of earthquake waves varies with density as well as with the elastic constants, is about 3.3, similar to olivine. Development of seismic explosion techniques at sea has determined that the Moho in the ocean basins lies at a depth of about 5 km only, and oceanographic surveys suggest that most of the material above is basaltic lava of density about 3 with perhaps only 0.5 km of ocean sedimentary deposits.

Thus the global picture of gravity and these seismic determinations of the thickness and nature of the outer layers of the Earth were in accord: the total mass of a unit column of continent equals 2.75×35 in hundreds of kilograms per sq cm while a corresponding column through the ocean, the average depth of which is 5 km, equals $1 \times 5 + 3 \times 5 + 3.3 \times 25$ in the same units. Similar arithmetic shows the two nearly equal, in accordance with the gravity data.

In the middle of last century, Pratt and Airy formulated the theory of isostasy, to explain why the gravitational attraction of elevated topography (mountains) was not as expected: the plumb line was deflected away rather than towards the Himalayas. They inferred the presence beneath mountains of lighter "roots," which we know now in many cases are the result of a downwarping of the Moho, thickening the continental blocks locally to perhaps 60 km. The main idea of the theory of isostasy was that the less dense roots of the mountains protruding into the more dense mantle provided the upthrust, according to Archimedes' principle, necessary to support the elevations. Thus at least over times of the order of those involved in mountain building, tens and hundreds of millions of years, the Earth's mantle was supposed to have the property of flow. Thus at a certain depth known as the depth of isostatic compensation, the condition of hydrostatic equilibrium would obtain. The gravity data over the oceans showed that isostasy was true over the broadest scale of thousands of kilometers, as well as over hundreds of kilometers, but the depth of compensation remained poorly determined 50–100 km, and it was perhaps rather too easily assumed to be the Moho.

The development of seismology half a century ago, however, gave no support at first to the assumption of a region in the upper mantle where over geological time fluid properties could be assumed. The shear (S) waves from the earthquakes propagated through the mantle but not through the truly liquid core, appeared to be evidence for a rigid mantle, and departures from isostasy on the large scale were sought and obtained by H. Jeffreys. Of course on the scale of tens of kilometers departures from isostasy arising from inhomogeneities in the outermost layers, were revealed by gravity anomalies and were supported by the evident strength of these rocks, which under ordinary temperatures and pressures were expected to possess finite strength.

The hypothesis of continental drift was first put forward by A. Wegener in 1912 as a result of an assessment of geological evidence in the different continents which strongly indicated that they had moved relative to each other.

Anomalous paleoclimatological evidence in the Mesozoic and Paleozoic, e.g. the Permo-Carboniferous glaciation in the southern continents and India, suggested considerable changes in latitude. Similarity of paleontological records before the late Mesozoic suggested that continents now separated by large oceans had once been in contact, e.g. Gondwanaland was the original continent, the break up of which had produced the southern continents and India and Laurasia had separated into N. America and Europe. Tectonic trends and lithologies in the various continents fitted better, it appeared, if the continents were reconstructed in this way. The fit of the coastlines of South America and Africa and the demonstration of a better fit of the 500 fathom isobath in the continental shelves were used by S. W. Carey and later by E. C. Bullard and colleagues. The excellent fit of North America, Greenland and Europe and in the southern hemisphere Australia and Antarctica has now been demonstrated.

The mechanism of these great horizontal displacements of the crust remained obscure, yet the key importance of isostasy was realized by Wegener as showing that flow in the mantle beneath the continents was possible. However, it was tacitly supposed that the granitic continental blocks were moving through the ocean floor and Jeffreys pointed out that the latter were too rigid and could not be assumed to be "fluid" as they possessed considerable topography.

But while the Moho represented a transitional surface chemically different below from that above, it has become clear that this is not the transition of special significance to the theory of continental drift. From the study of the amplitudes rather than the travel times of seismic waves, Gutenberg found that the amplitude of compressional waves traveling from an earthquake reached the surface much decreased in amplitude if the distance from the epicenter was between 100 and 1000 km, in the latter case the amplitude being 1% of the former. Beyond 1000 km the amplitudes increased again.

Gutenberg therefore postulated the existence of a subsurface layer between 100 and 200 km deep in which the velocity is 6% less than it is just below the Moho, i.e., the rock is less rigid. The velocity only reaches its value below the Moho again at a depth of about 300 km below the surface.

In addition to the P and S body waves which travel in refracted paths through the deep interior of the Earth, earthquakes produce surface, Love and Rayleigh waves, which pass around the surface but especially the latter penetrate the depths comparable with their wavelength. Great earthquakes, such as the Chilean earthquake of 1960, set the whole Earth into oscillation making it "ring" like a bell. The surface waves so produced travel around it several times and the presence of this low velocity layer has been conclusively proved.

The concept of an asthenosphere is supported by the storage of earthquakes over this range of depths: they necessarily involve the storage of strain energy which is suddenly released by some trigger. It is also in agreement with our better knowledge obtained from the quantum theory of the solid state of matter. This shows that a solid under constant stress will flow if given sufficient time especially if its temperature is elevated say above half its melting temperature. This occurs through diffusion and movement of grain boundaries and the effectiveness of such processes depends on the exponential of temperature. Thus as the temperature of the Earth increases downwards at the considerable gradient of 25°C/km, a surface is reached below which the material over long periods of time becomes plastic. Temperature therefore, rather than chemistry, controls the depth at which the mechanical properties of the outer layers of the Earth change from those of a classical solid to those essentially of a fluid of very high viscosity.

The depth is isostatic compensation, and the depth at which the horizontal movements of the "crust" in continental drift occur, lie well below the Moho, say at 70–100 km depth. This proves to be a most vital idea in understanding continental drift, for it is now clear that the relative motion between the continents and the rest of the Earth's interior occurs at this boundary and not at the base of the continental block, as was once thought. Although Wegener has supposed that the westward drift of the Americas had resulted in the crumpling of the western part of these continents to form the Rockies and the Andes, Jeffreys had pointed out that this idea is hard to reconcile with either the hypothesis that the continents were stronger or weaker than the ocean floor.

In 1961 Hess postulated the idea of sea floor spreading and this has proved to be complementary to the idea of continental drift, explaining exactly how the ocean floor is created as the continents separate. Similarly Dietz clarified the consumption of ocean crust implied if, as is certain, the surface of the Earth does not expand in area. The study of the topography of the ocean floor had shown that the mid-Atlantic ridge was part of a worldwide ridge system which in many places was equidistant from the continental slopes on either side of them.

Fig. 13-1. Lava crust, crossed by an active fracture, in the Kilauea volcano: molten lava flows out of the central fracture and cools along the edges on either side of the already consolidated places, thus causing outward expansion; this overall process is a natural model of various phenomena occurring near oceanic ridges; note, for example, that the direction of the expansion, indicated by the striations on the already consolidated crust, is often not perpendicular to the fracture, and produces slips and transform faults *(U.S. Geological Survey, photo W. A. Duffield)*.

It was then found that along much of the crest of these rises there is a central valley, which was interpreted as a rift. The world ridge system showed that the oceanic crust was in tension, and was parting at the rifts. Thus the theory of mountain building, long favored, that the Earth's crust was contracting due to the cooling of the interior, was disproved: equal evidence for global "expansion" had been found in the ocean floor. The theory of sea-floor spreading supposed that the drifting apart of the Americas from Europe and Africa, India from Africa and Australia and Antarctica, was accompanied by an equal motion of the ocean floor adjacent to these various continents, the new ocean floor being entirely generated by the rise of basaltic lava at the crest of these ridges. The fact that there is no mountain building contemporaneous with the continental drift, described in Wegener's theory, and no present indication of geotectonic activity, e.g. earthquake, associated with the east coasts of the Americas or the west coasts of Europe and Africa shows that this concept is valid. The Earth's lithosphere is thus broken up into six lithospheric plates, the American, African, European and Indian plates containing these continents and in addition the adjacent ocean floor and the Pacific and Nazca plates having only ocean floor, the latter lying between the South American trench and the East Pacific Rise. Within each of these plates there is little relative movement and the main geotectonic activity occurs at their boundaries.

Where plates are being pushed together ocean floor descends into the mantle, thus compensating for the new ocean floor being generated at the ocean rises. Where the converging plates are both oceanic crust, the lithosphere of one plunges down at an angle of about 45° into the mantle, and an island-arc system is generated with an associated deep oceanic trench. The Tonga trench and the Japan trench are thus the results of the northwestern motion of the Pacific plate. Where the plate edge is continental as in South America the ocean plate disappears below the continental block, and the latter "floats" above it due to its lesser density.

These ideas explain the following phenomena concerning the distribution over the world and the depth distribution of earthquakes, in a word, seismicity. It was found that most strong earthquakes are concentrated into narrow bands: arc groups all shallow, follow the ocean ridge system and when accurate epicenters were obtainable they were found to coincide with the central rifts on the ridges. The other group of earthquakes were found to be concentrated at various places in the great world-girdling system of Tertiary mountain building, which is still proceeding, the great east-west belt of the Alps and Himalayas and the north-south belt of the Rockies and Andes, and in marked concentration near the oceanic trenches. In addition to shallow arcs, many deep-focus earthquakes occur in this group, the deepest down to 700 km. It was always paradoxical that earthquakes could occur to this depth in a mantle which otherwise was thought to be plastic. The paradox is resolved if the lithospheric plate edges, 100 km thick, descend into the mantle down to this depth thus providing the rigid material in which stress differences can build up, as a result of the plate motion, until the breaking stress is reached and rupture and an earthquake occurs.

Fig. 13-2. Triple join between three active fractures in a lava crust observed in the Kilauea volcano: molten lava flows from each of the three fractures; as it cools it causes the lateral expansion of the various already solidified plates; a situation comparable with the one shown here in miniature is the one found on the bed of the Indian ocean where three ridges converge *(U.S. Geological Survey, photo W. A. Duffield).*

The proof of the theory of sea-floor spreading was furnished by paleomagnetism. Magnetic surveys had long been done over continents as a technique of applied geophysics to ascertain the structure of the rocks of the crust. As a result of the difference in their magnetic susceptibility and in some cases a strong remanent magnetization acquired at their origin, anomalies in the geomagnetic field above rock structures are found. In general they are most complex. The development of the technique of measuring the magnetic field at sea by towing a magnetometer behind a ship and measuring the total field intensity, revealed that the ocean magnetic anomalies are considerably simpler: the lines of equal anomaly being elongated "ridges" and "valleys." Vine and Matthews suggested that these alternations of positive and negative anomalies were caused by the successive reversals of the geomagnetic field discovered by the study of the directions of remanent magnetization in continental igneous bodies, especially of the Cainozoic. The axes of the anomalies paralleled the ridge crest cooling to form a strip of new ocean floor, the lava rises at the ridge crest cooling to form a strip of new ocean floor, the lava becomes magnetized in the direction of the geomagnetic field at that time.

When subsequently the field reverses, the strip of new ocean floor which then begins to form becomes magnetized along the opposite direction to that of the former strip now divided on either side of the ridge crest. The magnetic anomalies over the ocean thus are alternately positive and negative reflecting whether the geomagnetic field was normal or reversed when the floor beneath them was formed. Further, the breadth of each anomaly equals the rate of ocean floor spreading multiplied by the time interval between successive reversals of the field. As the latter is known from the radioactive age dates on the continental lavas from which paleomagnetic data was obtained, the rate of spreading can be found. In general it is 1–10 cm/y and similar to that deduced for the motion of the continents from paleomagnetic work on the rocks of Phanerozoic time. Thus the theory of sea-floor spreading was verified by quantitative methods.

Fig. 13-3. An active fracture ending at two extremities in a triple join, as seen in a lava crust observed in the Kilauea volcano; this perfectly simulates the situation of any given expanding ridge where this comes into contact with another plate delimited by another ridge; the relative speed of expansion determines the size and shape of the plates ***(U.S. Geological Survey, photo W. A. Duffield).***

The symmetry of the magnetic anomalies over the ocean ridges is remarkably good, each magnetized strip of ocean floor dividing equally on either side of the rift and moving uniformly away from it. This curious mechanical phenomena can only be explained by supposing that the lithospheric crust over the ridge is only a thin veneer of rigid material over a plastic substratum, while away from the ridge the lithosphere is thick and strong. This can be readily understood if it is recalled that the difference between the rigid lithosphere and the plastic mantle is a temperature effect. As the lithospheric plates part, the "fluid" mantle which rises to fill the gap decreases a little in temperature due to the release of pressure and partial melting which occurs to produce lava. The temperature immediately below the ridge crest is thus about 1000° higher than at the corresponding level beneath the rest of the ocean floor and continents. This column of hot mantle beneath the ridge essentially accounts for the elevation of the ridge, although doubtless chemical differences play a part. As the sea floor moves away from the ridge the hot material beneath it loses its heat by conduction through the ocean floor and the isotherm corresponding to the division between rigid lithosphere and plastic mantle slowly descends. Cooling reaches 100 km deep in about 100 million years.

At the trenches the ocean lithospheric plate is being pushed into the mantle at a rate of 1–10 cm/y. By the converse process, this lithospheric plate is heated by the surrounding hot mantle: heat being conducted into it at the same rate of 100 km in 100 million years. This explains why this plate edge retains its strength until it is at a depth of 700 km. Again temperature rather than chemistry seems to be the dominant process at the trenches. However, the volcanism occurring in the island arcs which are associated with the trenches is different chemically from the basalts of the ocean ridges and appears to be partly composed of the ocean sediments carried down into the mantle by the descending plates.

The comparatively thinner sedimentary layer on the ocean floor of just a few hundred meters, has long been a puzzle: it is several orders of magnitude less than that deposited on the continents, even though these have only been covered by water for a small fraction of geological time. However, the above picture shows that most of the ocean floor is young, less than about 100 million years—only about 2 % of the Earth's life. As we have seen some of the ocean sediment is continually being disposed of in the ocean trenches.

It is now believed that some too is transported by the plates to the edges of the continents and added to them. The identification of such continental sedimentary deposits is an important task for the geologist.

STANLEY K. RUNCORN

Bibliography: Runcorn S. K., *Dynamical processes in the deeper mantle,* in Tectonophysics, XIII, 1-4 (1972); Elsasser W. M., *Convection and stress propagation in the upper mantle,* in: Runcorn S. K. (ed), *The application of modern physics to the Earth and planetary interiors,* New York (1969); Knopoff L., *On convection in the upper mantle,* in Geophys. J., XIV, 341 (1967).

WHATEVER its future destiny may be, it is today clear that the plate-tectonics model which the preceding articles have examined from every point of view, has brought about a full-scale cultural revolution, above all insofar as it has introduced into geological science a quantitative approach which formerly appeared to be impossible. On this basis the geologist no longer just says "the continents move," he can also specify the number of centimeters they move each year.

Simultaneously the development of geochronolgy has taken place. This science makes it possible to determine the age of many types of rocks. Together these two new possibilities permit us to face the task of reconstructing the Earth's history in a fundamentally new way. Within this general framework, the article by Dickinson showed us how mountain ranges and many structures characteristic of them have taken on new meanings. Thus it is necessary to review all the data so far accumulated by traditional geology and to reinterpret it in the light of the new model. Geochronology then permits us to establish the precise schedule according to which the events determined by this review took place.

The path to be followed goes far beyond Wegener's Pangea. In the last 60 years geologists have debated whether the continents were once really united to form this super-continent. When at last they agreed that the continents are really drifting, and that it is highly probable that in the past they formed a single vast continent, they realized that Pangea was a relatively recent episode: the formation of Pangea dates from about 200 million years ago, while the ecological history of the planet is at least 17 times longer.

The data available to provide pointers to this history is extensive. A. G. Smith, the author of the article which follows, has gathered it together and fed it into a computer specially programmed to produce an output of maps of the geography of the past, indicating not only whether two continents were joined together at a given moment, but also in what position they were located on the Earth's surface.

This is a fundamental approach to obtaining an understanding of ancient geography, and to completing the mere geographical outline with an adequate description of environmental and climatic conditions characterizing particular continents in a given period. Certain observations which traditional geology had been able merely to record without comment can in this way alone be explained in a satisfactory manner—for example the traces of the passage of gigantic glaciers in the Sahara, the coal seams formed from the remains of tropical plants in the Antarctic, are no longer data of only local interest but can be coordinated instead into a logical explication in a global context.

What is being reconstructed today, described in an up-to-date version by Smith, is the external appearance of our planet in the past. We can compare this to a film; we are beginning to have a pretty clear understanding of the last few scenes—say the last 200 m over a total length of 46 km of film. However, we have also already found a few frames dating from earlier times. In this article Smith moves back to describe the last 570 m. The more than 4 km that are missing will be dealt with in the article by A. E. Engel.

ALAN GILBERT SMITH

Teaches geology at St. John's College, Cambridge University. Born in Watford, England, in 1937, he studied physics at Cambridge and geology at Princeton University. He developed an original and vital technique to reconstruct the latitudinal and longitudinal positions of the continents in the past. The method is based on the insertion of geological data into a specially programmed computer of data in this reconstruction. This research has shed new light on the position the continents had before they were joined in Wegner's Pangea.

ALAN GILBERT SMITH

Continental Distributions Before and After Permian Time

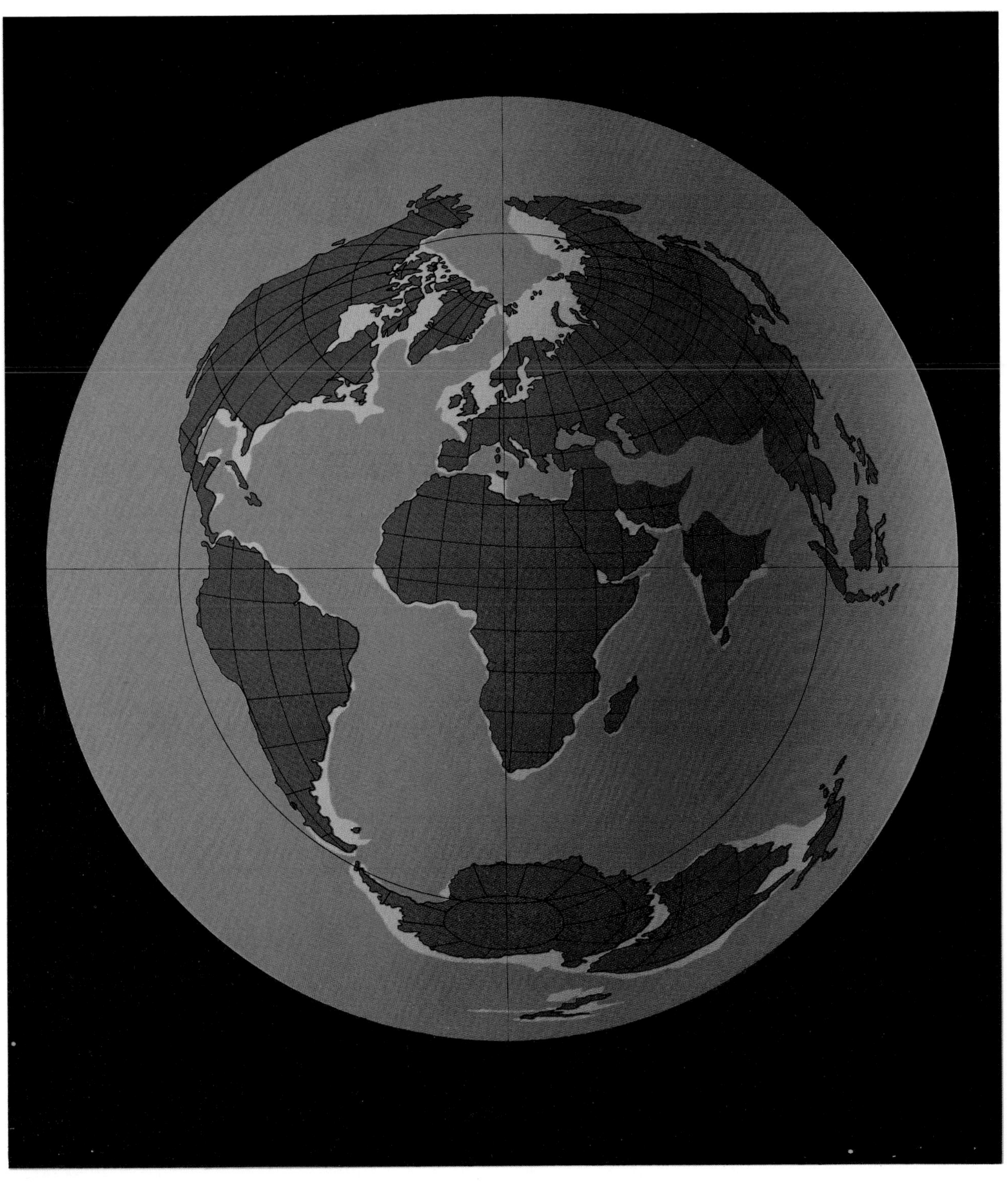

Overleaf: A computer reconstruction of the arrangement of the continents 50 million years ago. Reconstructions of this type are nowadays possible thanks to large banks of data available and to the existence of models, such as the plate tectonics model and the model of the expansion of the ocean-beds, which make it possible to extrapolate movements and tendencies for the past which can also be observed today, and thus to have a basic reconstruction in which to place the observations at hand. But these latter will determine either the confirmation or rejection of the reconstruction.

In 1965, Bullard, Everett and the writer published the first computer-based maps showing the fit of the continents around the Atlantic Ocean. They were made at a time when continental drift was not generally accepted as a reality, even by the three authors concerned. Five years later the first computer-drawn reassembly of the southern continents was published by Hallam and the writer. Today the drawing of maps showing the distributions of the major continents at any time during the past 250 million years—Permian and later time—is mostly a routine matter. For periods prior to about 250 million years only "composites" of the world can be drawn at present: the making of maps showing the distributions more than 250 million years ago is not yet possible. This article does not relate how these dramatic developments have come about.

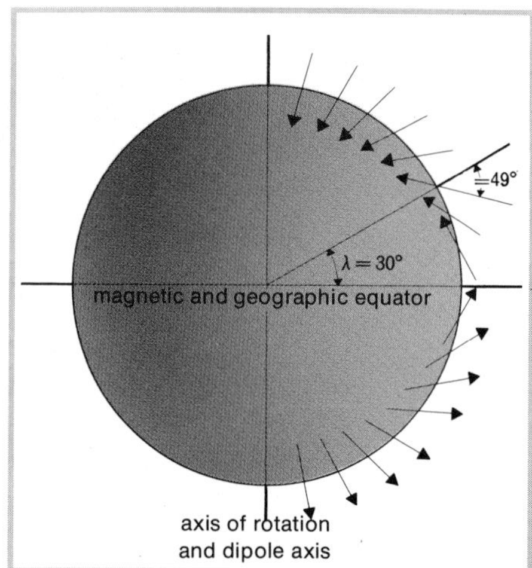

Fig. 14-2. Variations in the magnetic inclination in a centered dipole magnetic field parallel to the axis of the Earth's rotation.

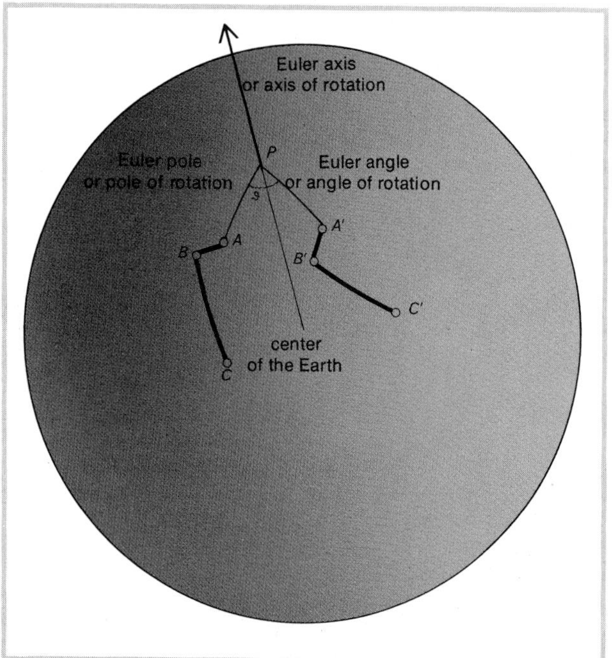

Fig. 14-1. A line *ABC* drawn on the Earth's surface, when rotated at an angle on it, moves in *A'B'C'*. The axis which links the center of the Earth with point *P* at the top of the angle is called the Euler axis or axis of rotation. Point *P* at which this intersects the Earth's surface is called the pole of tectonic rotation or Euler pole. These are the main geometric elements which govern the plate tectonics model.

Instead, it discusses what the present state of knowledge is of this rapidly evolving field. It outlines how such maps and composites are made, gives the assumptions on which they are based, and discusses their implications.

The essential requirements for accurate, routine work are a map-making computer program and data. The data fed into the program consists of the geographic coordinates of points on the present-day globe. The instructions to the program tell it how to process the data, the kind of map to be drawn, its size, and so on. The program in most general use is a version of Parker's Supermap program. Output from the computer is a map drawn by the computer's plotter according to the instructions provided.

Typically, the data for any of the maps shown in this article might consist of 10,000 points. It would take less than 10 seconds to process and a few minutes of plotter time to draw the map from the processed data.

How are the boundaries of past continents determined? What information allows us to suggest that a former continent lay in polar, rather than in tropical regions? How can we be sure that the separation between say, Europe and North America, has changed in time, and by what amount? To answer these questions it is necessary to summarize the main conclusions of plate tectonics.

Plate Tectonics

The present-day distribution of earthquakes may be accounted for by the view that the Earth's outer layers consist of rigid spherical slabs, or plates, that move relative to one

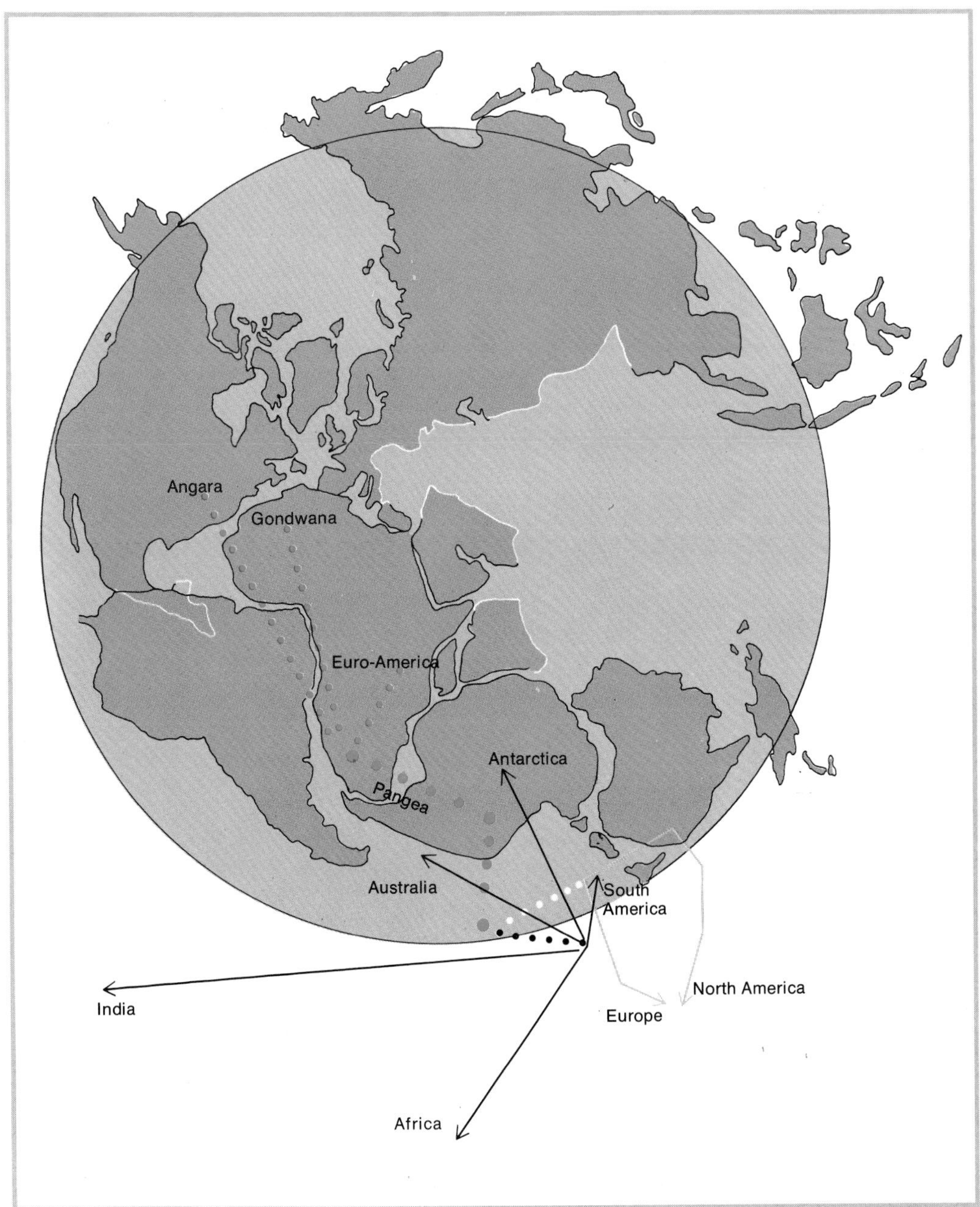

Fig. 14-3. Reconstruction of Pangea existing in the Permian and Triassic; a slightly modified stereographic projection, used by Briden *et al.* The reconstruction is totally independent of the paleomagnetic data and shows the surface of one hemisphere. The regions outside this hemisphere are delineated outside the circle, instead of being projected inside it, as is the usual procedure. This reconstruction shows the schematically delineated curves of apparent polar migration. The small blue circles indicate the curves relating to the Lower and Middle Paleozoic of the Euro-American continents, Angara (Eurasia east of the Urals) and Gondwana. There are no data on China and southeast Asia. The three curves converge into one (large circles) with the arrival of the Upper Paleozoic and Lower Mesozoic. Initially Pangea broke into two parts: Laurasia (white circles) and Gondwana (black circles); the southern continents (black arrows) and northern continents (gray arrows) broke off from these.

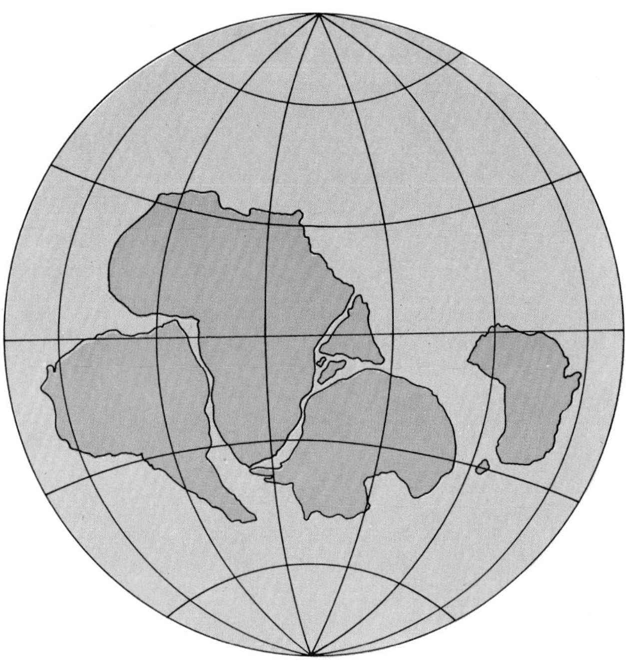

Fig. 14-4. **Reconstruction of the continent of Gondwana, based entirely on paleomagnetic data and made by postulating that between the Cambrian and Permian there were no relative movements between the continents. This is an equivalent projection. The discrepancies with the figures that follow can be attributed to uncertainties about the positions of the individual fragments determined on the basis of paleomagnetic data.**

another. The very slow motions among the plates—commonly a few centimeters a year—gradually alter the distribution of the continents. In time, even the shapes and sizes of the continents are changed by these slow motions. If the motions are known, and the plate boundaries are known, then past and future distributions of the continents can be estimated. In general, we do not know all these parameters. But fortunately, we do not need all of them to make maps that show past continental distributions.

The essential feature of plate theory is that significant motion and deformation of the Earth's surface can occur only at plate boundaries. No significant movement occurs within a plate.

Three ideal kinds of plate boundary exist: extensional, translational and compressional. At extensional plate boundaries new ocean floor wells up into the space created by the separating plates. Along translational boundaries two plates move past one another without separating or moving closer together. At compressional plate boundaries one plate, generally oceanic, sinks into the mantle underneath the edge of a second plate, which is commonly continental. At the present time these boundaries give rise to distinctive surface features and seismic activity. Extensional plate boundaries are associated with submarine mountain chains known as midoceanic ridges. Earthquakes are small in size and limited to the up-

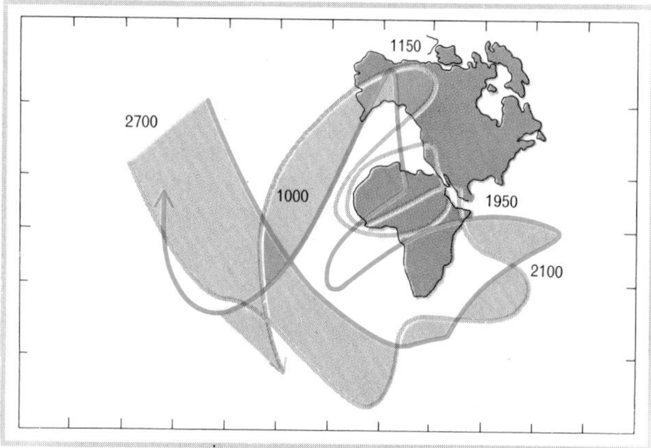

Fig. 14-5. Apparent polar migration between 2.7 billion and 900 million years ago, determined for Africa and North America. The ages are given in millions of years. The lighter color indicates the areas of uncertainty for the periods 2.7–1.95 billion years ago and 1.15 billion and 900 million years ago. The colored line indicates the apparent migration for North America, the gray line for Africa. The marginal lines indicate 30° intervals of latitude and longitude.

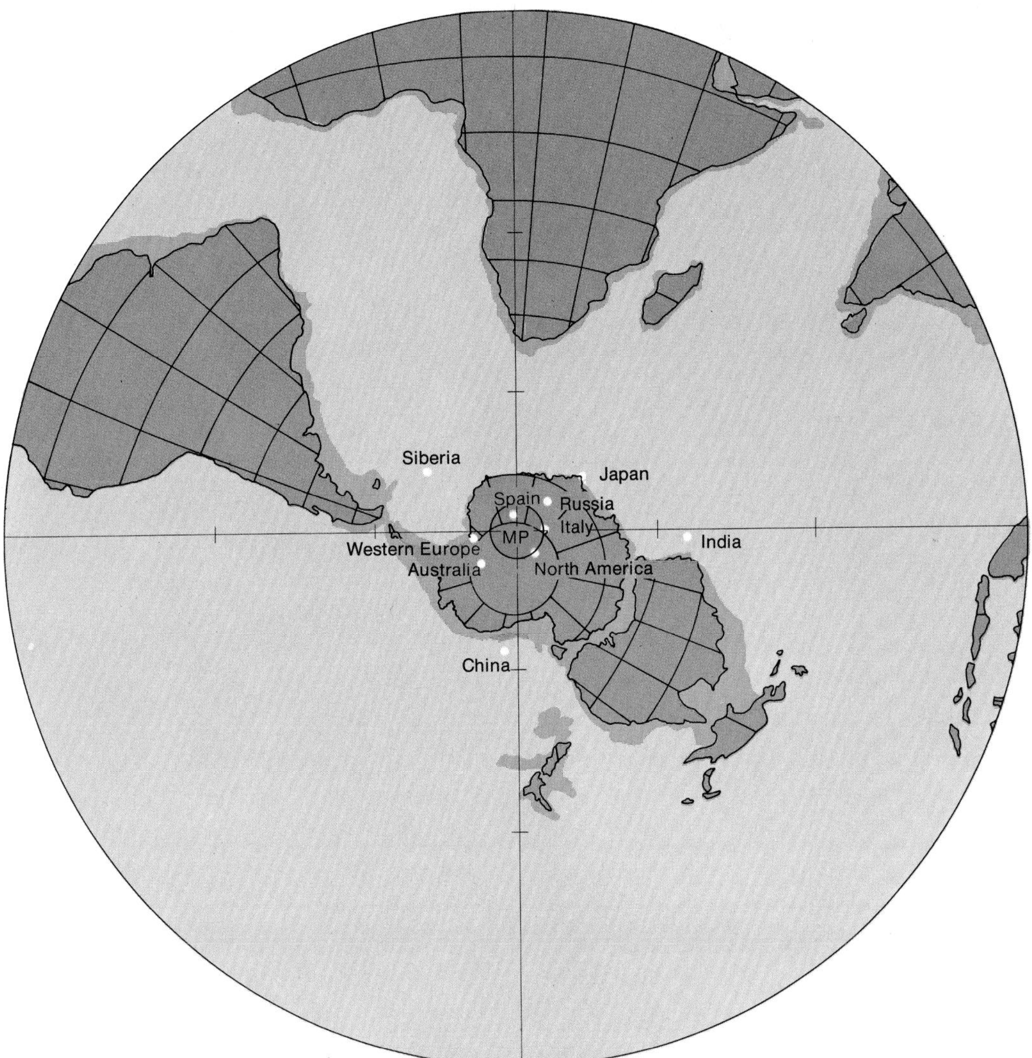

Fig. 14-6. The position of the South Pole in the Eocene. This reconstruction is based solely on data gathered from the ocean-bed expansion process. The white dots indicate the position of the pole on the basis of individual calculations in different countries. Note how these become frequent around the blue circle which defines the area within which there is a 95% certainty of calculation. *MP* is the average position of the pole.

permost layers. Translational plate boundaries are marked at the surface by zones of crushed rock. Earthquakes are generally much stronger but are still confined to the uppermost layers. Compressional plate boundaries are associated with deep-sea trenches, behind which lie chains of andesitic volcanoes formed above the descending slab. Earthquakes are strong and lie on inclined zones believed to mark the approximate position of the descending slab.

If two continents belong to the same plate, then there is no relative motion between them. Only when a new plate boundary comes into existence and separates two continents that previously belonged to one plate can any relative movement between the continents occur. The converse is also true. If a plate boundary separated two continents, then relative motion must occur between them.

For example, at the present time there is no plate boundary between India and Australia. Australia is part of the Indian plate. No midocean ridge; no actively crushed rocks; no andesitic chains or deep-sea trenches and no earthquake zones separate the two continents. Whatever the relative motion is between India, and say, North America, is also the relative motion between Australia and North America. Similarly, at times in the past, Europe, Greenland and North America all belonged to the same plate—they were in fact joined together. During this period, any motion of North America relative to any other plate was identical to that of Europe and Greenland. At the present time this is not so. A plate boundary—the mid-Atlantic ridge—separates North America and Greenland from Europe. Greenland today belongs to the North American plate. The two continental areas west of the mid-Atlantic ridge must move relative to Europe at the rate at which new ocean floor is created at the midocean ridge.

Thus it is incorrect to equate continents and tectonic plates. Parts of one continent may belong to two or more plates, and more than one continent may belong to the same plate.

Width of Plate Boundaries

Fundamental to any attempt at finding the distributions of continents in past time is a knowledge of how wide plate boundaries are. According to plate theory, plate boundaries are the only zones in the Earth's surface along which significant deformation occurs. Unless some method is available for unraveling the deformation, it will be impossible to know the shape and size of the area involved in the deformation before the deformation has occurred. No reliable map could be drawn of the area prior to its deformation, or while it was in progress. If plate boundaries are very wide, and if unraveling of the deformation is impossible, no maps of past periods could be drawn.

On a global scale, the width of extensional and translational plate boundaries is remarkably small. For example, the sharpness of the change from normally magnetized rocks to reversely magnetized rocks near the East Pacific Rise suggests that, when averaged over a period of several thousand years, new ocean floor is formed here in a zone that may be less than 1 km wide. Similarly narrow zones probably exist at all other major oceanic ridges. Surveys of the translational boundaries in the oceans—transform faults—commonly show a zone of disturbance that may be less than 10 km wide, even though these faults may be traced across the ocean bottom for thousands of kilometers. On land, direct observation of transform faults, like the San Andreas Fault in California, shows that motions averaged over a period of million years or so have caused crushing in zones that may be about a kilometer wide.

The width of compressional plate boundaries is uncertain. There is no doubt that most of the movement between the descending oceanic slab and the overriding slab takes place in a relatively narrow zone, at most a few kilometers wide. But as the slab descends, part of it, or part of the adjacent slab, or both, may melt. The molten rocks rise through the overriding plate. They erupt at its surface as volcanoes, or crystallize at depth in the crust. The rocks at deep levels in the crust are deformed and recrystallized in the solid state. Eventually their cover is eroded and they form large tracts of regionally metamorphosed rocks, typically made of slates, schists and gneisses. These metamorphic terrains parallel the trench systems but lie some distance from them. These deformed, regionally metamorphosed and intruded zones are called orogenic belts, or mountain belts, by geologists. In some cases they may be thousands of kilometers long and up to 1000 km wide.

It is not yet known how much of the deformation in orogenic belts owes its origin to compressional stresses transmitted from the compressional plate margin near the orogenic belt, how much is the byproduct of uplift, and how much is due to the formation of the metamorphic and igneous rocks. Perhaps the deformation is created by a combination of these and other, still unknown processes at the plate margin.

There is also no agreed way in which the structures and deformation in orogenic belts can be unwrapped to arrive at the original shape of the belt. Thus no maps of past periods can show the shape of an area that has been involved in orogenesis or mountain building, until after its completion. The simplest way to represent orogenic areas in past time is to draw them with their present-day boundaries. We know the original shape was different, but at least the area affected is recognizable.

It may well be that orogenesis affects only the upper part of the crust in a significant manner. This would be so if the deformation is caused by the sliding downslope of warm, viscous rocks under the influence of gravity. In such a case, then the original extent of continental crust in an orogenic belt may not be very different from that which the crust has after the orogenesis is completed. However, at the present time we cannot be definite about such matters.

In summary: extensional and translational plate boundaries are very narrow. Compressional plate boundaries may be much wider, depending on whether or not orogenic belts are directly or indirectly related to them. But even the widest, about 1000 km, are small on a global scale, and take up less than 3% of the Earth's circumference. All the other parts of the Earth's surface remain free from significant deformation. They belong to rigid spherical caps.

Evidence for Rigidity

A rigid body is one that does not change its shape. Evidence for rigidity within plates is well known by the structure of the oceans and of the continents. Pairs of magnetic anomalies that originally formed at a midocean ridge and have since moved away from it, still have the same shape, even though their present separation may be several thousand kilometers. In some cases the present-day ridge has the same shape as the anomalies formed at it millions of years ago. This close similarity of shape shows that once a piece of ocean floor has moved away from the plate boundary—the midocean ridge—it does not change its shape. The ocean floor behaves like part of a rigid body. The edges of continents that were once joined together also have very similar shapes. For example, the misfit between South America and Africa averages less than 90 km over a stretch that is more than over 300 km long. Within continents we commonly find very large areas covered with sediments that show no evidence of faulting, flowage or other deformation. They demonstrate that the part of the continent on which they rest has been rigid since the sediments were deposited, in some cases over 1000 million years ago. The only areas of possibly significant deformation on continents are orogenic belts.

The only ways in which a continent can change its shape are by being broken into smaller pieces, or by having orogenic belts developed along its edges, or perhaps by colliding with a second continent. In the first case the individual pieces are themselves rigid bodies. They can be reassembled into the original continent provided that the joins have not had younger orogenic belts developed on them. In the second case, the only parts affected by shape changes lie near a downgoing slab. The slab itself lies near the continental margin. In the third case, a continent that has grown by collision will have an orogenic zone crossing it. This zone will separate two stable undeformed areas that represent the continental parts of two plates prior to their union.

Geometry of Plate Motions

Ocean and continents are rigid except at plate boundaries. Plate boundaries are very narrow, except perhaps compressional plate boundaries. These two properties allow us to treat plate motions very simply. All rigid body motions on a sphere may be described by rotations. This result is embodied in a theorem due to Euler, sometimes referred to as the fixed point theorem. In essence it states that the net motion between two rigid bodies on a sphere, whatever its complexity, may be described by turning one or other of the bodies about a suitable axis passing through the center of the sphere (Fig. 14-1).

In plate tectonics this axis is known as the Euler axis. It cuts the Earth's surface at two diametrically opposite points known as the Euler poles. For example, Africa and South America are parts of rigid bodies. They have moved apart from one another in a complex manner as the plates to which they belong have grown at the mid-Atlantic ridge, moving the continents away from it. Euler's theorem says that they may be brought together again by turning one or other of them about a fixed axis that cuts Earth's surface at two points. One such point lies near the Azores; the other at a diametrically opposite point. The angle that brings them together to their best-fitting position is about 57°. The easiest way to see this is to use a piece of tracing paper on a globe. If the outline of the edge of South America is traced onto the paper it can be turned about the Azores and brought into a position that fits it against the edge of Africa. Note that the important line is not the coastline, but the edge of the continent. This lies somewhere on the continental slope. The best computer fit for these two continents is at the 500 fathom (about 1000 m) submarine contour, though the fit at the 1000 fathom (2000 m) contour is almost as good. To command a computer to do the same thing, all we need to do is to feed the outline of South America into the map-making program, read in the latitude and longitude of the Euler pole, and define the Euler rotation angle. The program rotates South America accordingly and makes a map.

For some periods in the past Antarctica was also joined to Africa and South America. It too can be brought into its original position against Africa by a single rotation. For part of this period India, Australia and Madagascar were joined to Africa, South America and Antarctica. Together with some other much smaller continent fragments they formed a supercontinent known as Gondwanaland, or the southern continents. Gondwanaland may be reassembled simply by applying a single rotation to each of its component continental fragments in turn. Similarly, any arrangement of the southern continents that is intermediate in time between the time that Gondwanaland existed as a single supercontinent and the present-day dispersed arrangement of its former components, can be made by applying single rotations to those present-day continents that originally made up the supercontinent. The only knowledge needed to reassemble former arrangements of any continents is the shape of each continent at the time and the rotations that must be applied to each of them in turn to place them in their original relative positions.

Boundaries of Former Continents

Fortunately, some of the boundaries of former continents approximate to those of present-day continents. For example, most of the continental edges around the Atlantic and Indian Oceans have been created by the break-up of the northern and southern continents. Changes in shape have been small, even where large rivers like the Niger River have added a large volume of sediment to the continental margin.

Significant changes in shape may have occurred in those areas affected by orogenesis that has taken place during or after the break-up. The only orogeny to have affected South America after its separation from Africa is that which created the Andes. Similarly, Africa has been involved in orogeny since its separation from South America only in the northernmost part. The area affected by orogenesis in the northern continents since break-up is also small. Thus it is a relatively simple matter to find the shapes of most of the continents throughout the life-span of the Atlantic and Indian Oceans.

For a short period of time, North America, Greenland and most of Eurasia formed a second supercontinent collectively known as Laurasia. It is believed that Laurasia and Gondwanaland were themselves joined together in Permo-Triassic time—about 270 to 180 million years ago—to form a supercontinent embracing nearly all the present-day continental areas. This single supercontinent was first postulated by Wegener over a century ago (Fig. 14-3) and named Pangea (literally, the whole Earth). Did Pangea always exist prior to Permian time? Is continental drift a relatively recent phenomenon as some geologists have suggested? If Pangea did not exist before Permian time, what were the boundaries of the pre-Permian continents?

Some clues are provided by the structure of Laurasia. Though it was a continuous continental mass in Devonian time—about 440 to 400 million years ago—it is crossed by an older orogenic belt known as the Caledonian mountain chain. Parts of this chain lie today on different continents as a result of the opening of the Atlantic Ocean. For instance, the chain can be seen in eastern Greenland, western Spitsbergen, western Scandinavia and Britain. Its continuation is known in North America as the Appalachians, where it forms a chain stretching from Newfoundland to the southern United States. By analogy with present-day processes, this chain is thought to have been created by the consumption of old ocean floor that originally separated the continents which lay on either side of the Caledonides. Most of this ocean-floor has vanished into the mantle, but its former existence is suggested by the regional metamorphism, igneous activity and intense deformation in the Caledonian-Appalachian mountain chain. The chain is currently interpreted as roughly marking the edges of two former continents that were originally separate from one another and have subsequently collided.

In a few cases the scar or "suture" along which the continents have been joined together may have been identified. Generally, it is concealed or is a matter of conjecture or controversy. Thus in most cases the precise boundaries of continents that may have been joined together by continental collision is not known. As noted above, some decision has to be made about where the boundaries of areas involved in orogenesis originally were. The boundaries shown on the maps are mostly arbitrary. The main guide used to draw them is that such boundaries have to lie within orogenic belts.

Obviously, as we go back in time, more and more of the continental crust becomes involved in orogenesis. The area of stable crust decreases and there is a corresponding increase in the area of those parts whose positions are uncertain. The practical effects of these changes is that at present it seems virtually impossible to draw the boundaries of the continents in most of Precambrian time—earlier than about 600 million years. This conclusion may change if some intriguing recent paleomagnetic results from Africa and elsewhere are substantiated by further work (see below).

Rotation Needed for a Reassembly

During the past 250 million years the continents were either joined together to form Pangea, or they were in various stages of break-up and separation from one another. The Euler rotations needed to make a reassembly are readily found, at least in principle. All that we need are the rotations that bring the continents into the correct relative positions for the time concerned. These are given either by the rotations that match appropriate pairs of magnetic anomalies in the Atlantic or Indian Oceans, or by the rotations that produce the best fit between the corresponding continental edges. The only difficulty in applying this method is that some parts of the Atlantic and Indian Oceans have yet to be surveyed in sufficient detail to find the trace of the magnetic anomalies on the ocean bottom. The magnetic time-scale prior to the Upper Cretaceous time—about 100 million years and earlier—is also not yet generally agreed upon. Nevertheless, plausible reassemblies of the major continents may be made for the past 250 million years. They will require some modifications as the pre-Upper Cretaceous to Jurassic spreading history of the oceans is worked out.

The rotations needed to reassemble the pre-Permian continents are not yet known. Virtually all the pre-Permian ocean floor has been recycled into the mantle. The small fragments that are preserved are trapped within orogenic belts and give little information that might enable the rotations to be discovered. There may well have been still older Atlantic-type oceans bordered by continental edges whose shapes did not change during the life-span of such oceans. But these edges, if they ever existed have subsequently been deformed and the continental fragments have been incorporated into the material that made up Pangea.

The only present guides to the rotations needed to make continental reassemblies older than about 270 million years are those provided by the distributions of pre-Permian sediments and by pre-Permian fossils, pre-Permian orogenic belts, by pre-Permian fossils, pre-Permian sediments and by pre-Permian paleomagnetic measurements. Of these, only pre-Permian paleomagnetic measurements usefully constrain the positions of the continents. In the future the pre-Permian fossil and tectonic data may provide useful additional constraints, but do not yet do so. Paleomagnetism is also important in that it allows the Permian and younger continental reassemblies, which do not depend on paleomagnetism for their reconstruction, to be turned into world maps.

Paleomagnetism

To a good approximation the Earth's magnetic field behaves as if it were caused by a bar magnet buried at the Earth's center. Over a long period of time this imaginary magnet is aligned parallel to the Earth's spin axis; that is, it parallels the Earth's geographic poles. There is a simple and unique relation between the inclination of the magnetic field of a centered dipole and the geographic pole (Fig. 14-2). It is given by tan (I) = 2.tan(L), where I = the inclination of the field (the angle between the lines of force and the horizontal) and L = the geographic latitude.

Most rocks acquire magnetism when they first form and preserve a record of the former magnetic field. Thus they may be used to estimate the former geographic latitude, or paleolatitude, of the rock when it acquired its magnetism. By taking sufficient samples paleolatitudes may be estimated to within 10° or better. Rather than estimating latitude it is customary to estimate the position of the geographic pole of the time, which can be either the north or the south pole, as convenient.

At first sight it would appear that ambiguities would arise when the magnetic field reverses. After reversal a north pole would appear to become a south pole, and vice versa. However, after a reversal the position of the pole relative to the continent does not change, only its sign alters. Provided that the time interval between sampling is small compared with the time required for a continent to drift from one hemisphere to another, then no ambiguity arises. The time interval between pole determinations could be as much as 100 million years without giving rise to any problems of interpretation. If it is much greater than this, ambiguities may well arise. Quite commonly intervals greater than 100 million years exist between successive positions of the poles in Precambrian time—earlier than about 600 million years—and the true polarity of a pole is not known. But data is accumulating so rapidly that this problem is likely to be transient.

A line that connects the present positions of successive magnetic poles for a particular continent is known as a polar wandering curve. The movement of the pole is only apparent: it is the continent that moves relative to the geographic pole. Polar wandering curves provide useful checks on the plausibility of continental reassemblies. For example, if two continents are joined together for a period, then their polar wandering paths will be identical. Geological evidence suggests that all the continents were joined together in Permo-Triassic time to form Pangea. As noted above, the orogenic belts crossing Pangea suggest that Pangea was formed by the collision of continents whose boundaries differed from those of the continents formed by the break-up of Pangea.

This interpretation is supported by the polar wandering paths for the continents in Phanerozoic time, that is, during the time period from the Cambrian to the present day, or the last 600 million years. Imagine the polar wandering paths plotted relative to Pangea (Fig. 14-5). When the continents were joined together in Permo-Triassic time the paths should all be identical within the limits of experimental error. In post-Triassic time—that is, in the past 180 million years—the paths should gradually diverge and separate into distinct trends that represent poles determined from the present-day continents. In pre-Permian time the polar wandering paths should also gradually diverge as we go back in time, but the grouping of the poles should be different. It should correspond to the grouping of the pre-Permian continents. This is precisely what is observed (Fig. 14-5). Thus paleomagnetism provides completely independent support for the deductions made from plate theory that were outlined above.

Obviously, if two continents remain fixed relative to one another for a period, that is, if they belong to the same plate, then their polar wandering paths will be identical. If these paths are sufficiently distinctive, it may be possible to suggest that two continents belonged to the same plate solely from the paleomagnetic data. We simply examine the polar wandering paths for the same distinctive shape in the same time interval. If such shapes exist, then we may also uniquely reposition the continents during that interval. The relative positions of the two continents at the time is merely that position that causes their polar wandering paths to coincide. This potentially powerful method has been used to reassemble Gondwanaland entirely from paleomagnetic data (Fig. 14-4). It is the only method that presently allows the relative longitude separation of continents to be determined in pre-Permian time. But it does depend on two important and not very common conditions:

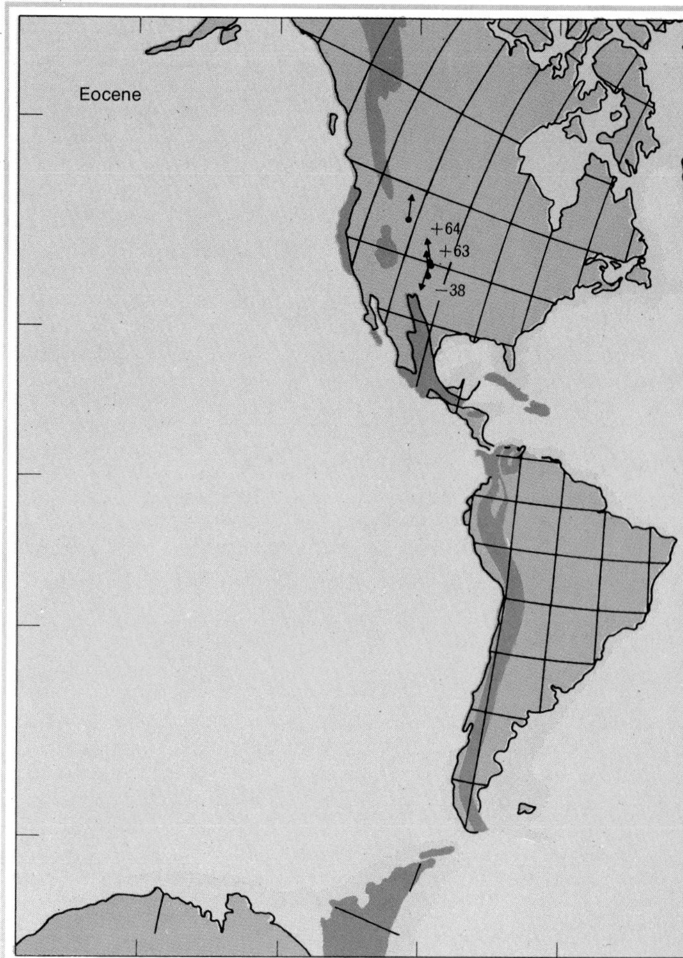

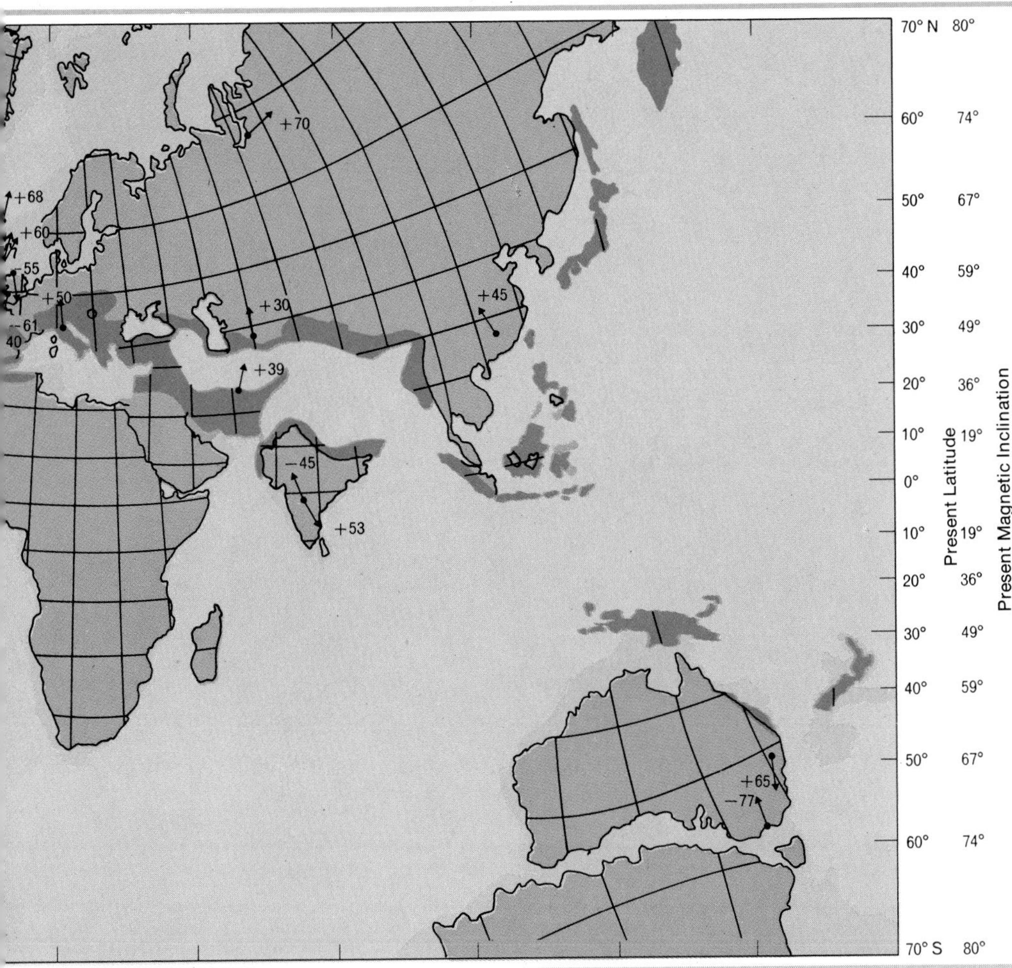

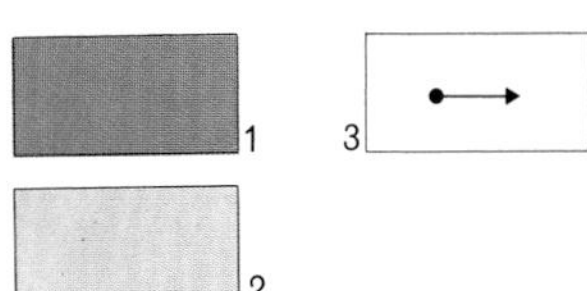

Fig. 14-7. The world in the Eocene. The series of figures starting with this one creates a journey back in time, in search of the palogeographic features which the Earth's surface assumed little by little. These geographical maps of the past have been compiled by using the Mercator projection. Inside the continental areas the present-day geographical grid has been retained; this permits a more immediate comparison of the shifts in respect of the present situation. Key: 1. zones affected by orogenic movements in the Cainozoic; 2. extension of the continental shelf; 3. direction and course of residual magnetism, measured in rocks in various regions; the numbers indicate the unit of magnetic inclination. In all the figures (Figs. 14-7–14-14) the marginal markings top and bottom indicate 30° intervals of longitude, but not absolute values which, unlike the latitude values, cannot yet be reconstructed.

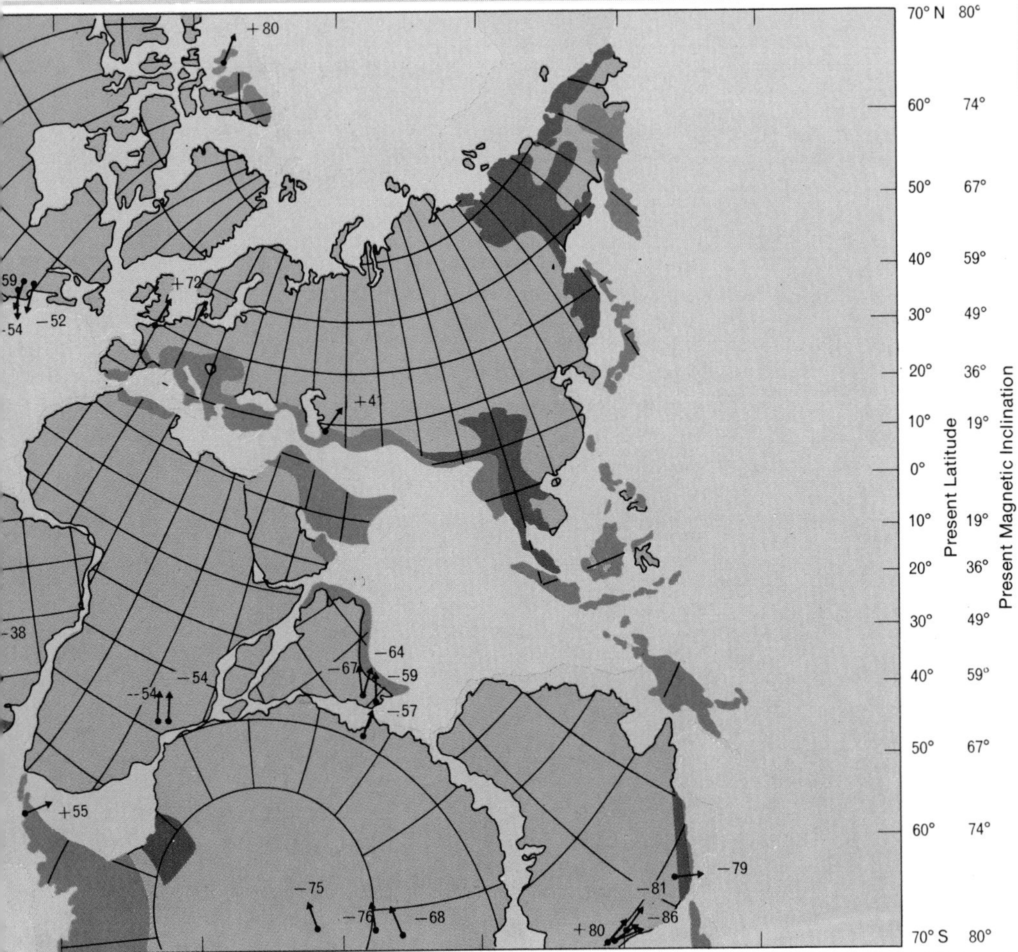

Fig. 14-8. The world in the Cretaceous, i.e. about 100 ± 10 million years ago. The Mercator projection used for this reconstruction has the considerable advantage of being a fairly common projection in present-day atlases, and of reducing to a minimum the various types of distortion which are inevitable when a spherical surface is reduced to a flat one: but this applies most of all in the belt which extends around the equator, while the distortion increases the nearer one gets to the poles. In fact, with a cylindrical, though modified, projection, the poles in the Mercator projection are placed at an infinite distance. The Mercator projection does not extend beyond Lat.70° north and south, but it is quite suitable for giving a picture of the paleographic evolution of our planet. Key, plus the three symbols in Fig. 14-7: 4: regions affected by orogenic movements in the Mesozoic.

a distinctive shape to the polar wandering curve and no relative motion between the continents.

The polar wandering paths of North America and Africa have been determined as far back as 2-7 billion years ago, or earlier Precambrian time (Fig. 14-4). In general they differ from each other. This difference shows that relative motion must have occurred between them in that period. That is, continental drift is not merely a post-Triassic (later than 180 million years) or a Phanerozoic (later than 600 million years) effect. Furthermore, the rates of change of apparent pole position on the Earth's surface are at about the same rates as those of Phanerozoic time. This suggests that continental drift has been at about the same rate as the present rate from the past 2-5 billion years or more. There is as yet no evidence to suggest that it has been significantly faster or slower. There have also been periods when the African and North American polar wandering curves have been roughly the same shape (Fig. 14-6), which may indicate periods when the two continents did not move much relative to each other.

Some geologists have questioned the usefulness of paleomagnetic results as paleolatitude indicators. For example, it has been suggested that the Earth's field may at times be a nondipole field or that the magnetic axis of the field, whether dipole or not, may differ significantly from the Earth's spin axis. It would not then be coaxial. Since the assumption of a geocentric coaxial dipole field is basic to the drawing of the world maps shown in this article, the evidence for such a field will be briefly discussed. A markedly nondipolar field would lead to a scatter of pole positions on a reassembly, because the correct estimates of the pole would not be given by the equation $\tan(I) = 2.\tan(L)$. If, in addition, the field were noncoaxial, then tropical or polar features could appear at any magnetic latitude.

Evidence for a Predominantly Dipole Field

Obviously, if a strong nondipole component existed in the Earth's magnetic field it would cause estimates of the pole positions to differ from place to place. To test this possibility we need reliable reassemblies that are independent of paleomagnetic information. During the time for which the reassemblies apply, we need good paleomagnetic data scattered over a large range of magnetic latitudes. The two best reassemblies are those of Tertiary (Eocene) time—about 50 million years ago—which is based entirely on ocean-floor magnetic anomaly data; and the Permo-Triassic reassembly which is based entirely on the best fit among the continental edges.

The scatter of Eocene paleomagnetic poles on the reassembly is remarkably small (Fig. 14-7). The scatter of Permo-Triassic poles is greater: in some cases the scatter exceeds the confidence limits assigned to the poles. Thus the Eocene data in no way support the existence of significant nondipole components in the Eocene magnetic field, whereas the Permo-Triassic data could mean that such components existed in the Permo-Triassic field. Alternative explanations, which are all equally likely in the present state of knowledge, are that the data are in error, that their magnetic ages are incorrect, or that the fit of the Permo-Triassic continents is wrong.

The largest departures from an apparent dipole field would cause at most errors of about 30° in the estimated paleolatitudes. On a global scale these are relatively small. For global map-making purposes there is little evidence to suggest that the Earth's magnetic field differs from a dipole field by an amount that would cause errors of more than 30° in paleolatitude, and these errors appear to be exceptionally large. Such errors could also have several other explanations.

Evidence for a Coaxial Field

The evidence against any significant and long-sustained differences between geographic and magnetic poles is provided by sediment distributions. Climatically sensitive sediments include tillites and evaporites. Tillites are fossilized boulder clays laid down by ice sheets or glaciers. Evaporites are salts precipitated from evaporating saline water, usually the sea. Provided that the climatic belts have been more or less in the same position relative to the Earth's spin axis as they are today, we would expect tillites to be concentrated in former

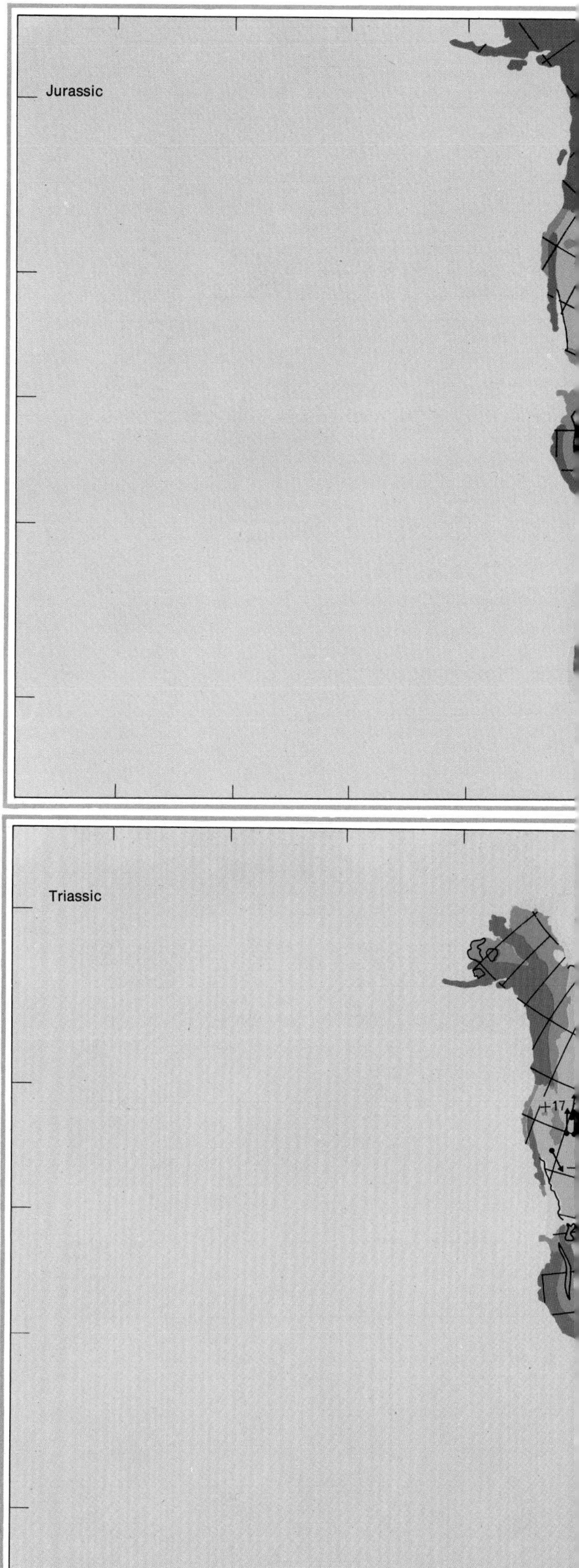

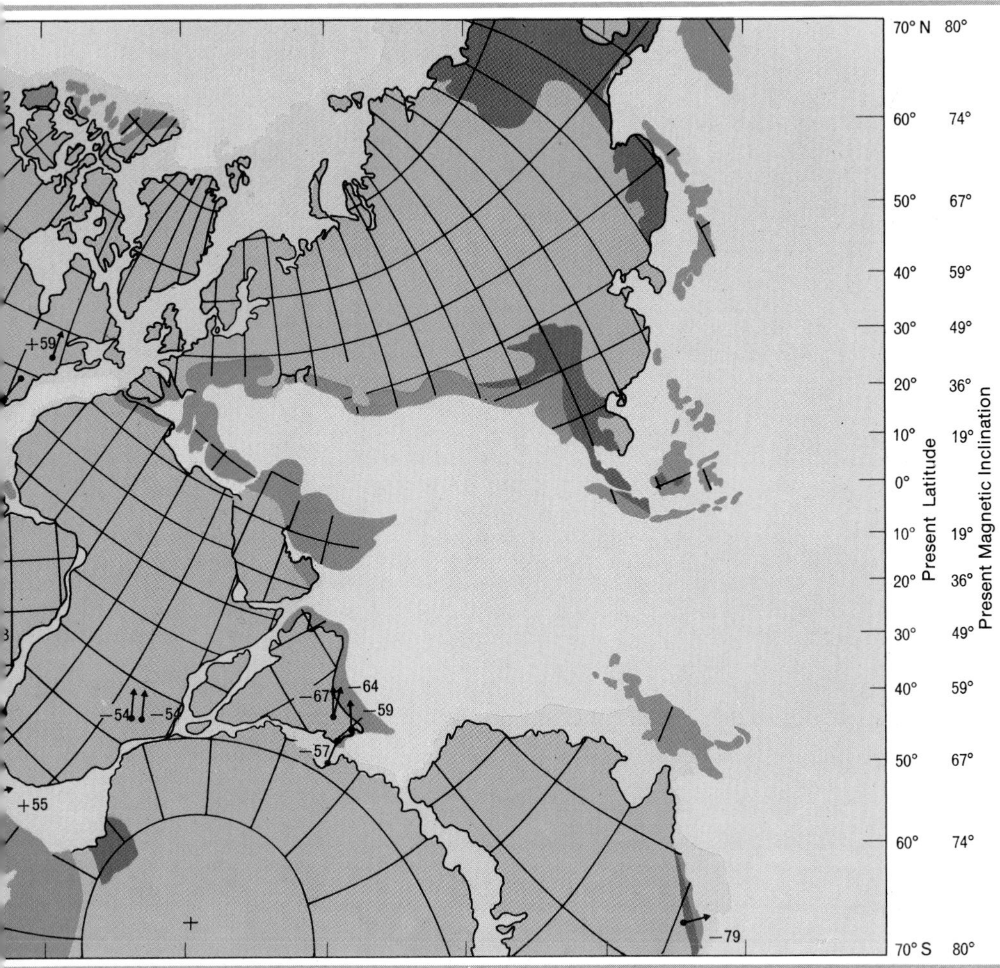

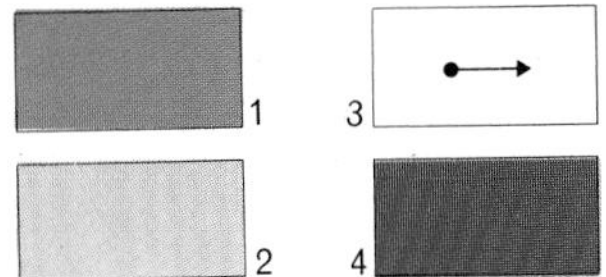

Fig. 14-9. The world in the Jurassic. Comparing this map with those for the Eocene (Fig. 14-7) and the Cretaceous (Fig. 14-8) we can see that the Atlantic is completely open in the former, and only open in the northern part in the latter; here, it is still virtually closed. But a large ocean separates Euroasia from Africa. This is the Tethys, an ocean which, in the Cretaceous period, still extends from the present-day Indian ocean to the present-day coast of Spain and even beyond towards the north Atlantic of those times. The present-day South Pole (indicated by the cross) is at a less advanced latitude. Key: 1. zones affected by orogenic movements in the Cainozoic; 2. extension of the continental shelf; 3. direction and course of residual magnetism, measured in the rocks of various regions; the number beside the arrow indicates the unit and sign of magnetic inclination; 4. regions affected by orogenic movements during the Mesozoic.

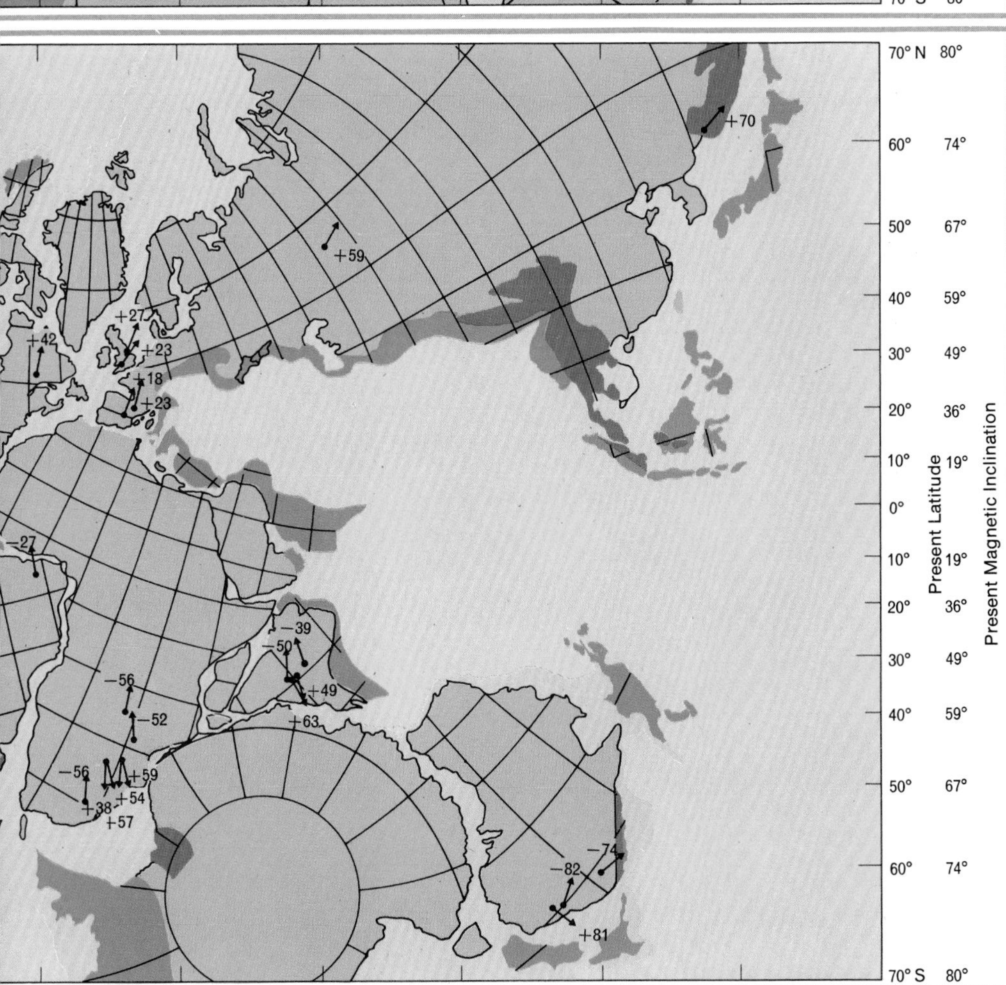

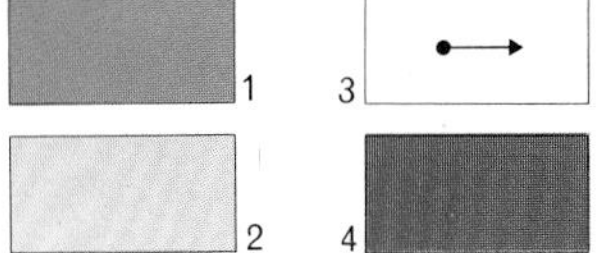

Fig. 14-10 The world in the Triassic, i.e. about 220 ± 20 million years ago. In practical terms we have just one continent, Wegener's Pangea, which lasts from the Permian and now starts to break up into those smaller parts which, much later, will become the continental masses which can be defined today. Note the latitudinal position of the Antarctic, the edges of which touch on Lat. 40° South. India is still positioned between the Antarctic coast and the coast of Madagascar and Africa. Only much later will it start its swift journey northwards, at the end of which it meets up with the Eurasian coast, to which it adjoins itself, forming the Himalayas as a result. For key see Fig. 14-9.

high latitudes and evaporites to be concentrated in former low latitudes straddling the old subtropical regions. If paleomagnetic latitudes are good estimates of paleogeographic latitudes, then these relationships should still hold. On the other hand, if the magnetic pole has remained in regions far from the geographic poles for long periods of time, then there should be little correlation between the positions of tillites and evaporites and those of former polar and subtropical regions inferred from the paleomagnetic data. This type of relationship between the magnetic and geographic poles could be regarded as a type of polar wandering in which the magnetic pole wanders relative to the Earth's spin axis. The apparent positions of the magnetic poles caused by plate motions—what we have referred to previously as polar wandering curves—would still occur, but superimposed on these curves would be the same additional movement caused by the wandering of the magnetic pole relative to the geographic pole.

Fig. 14-8 shows a plot of the inferred paleolatitudes of tillites and evaporites. There is a clear concentration of tillites in high southern latitudes and of evaporites in the subtropical regions, despite the fact that some of these tillites lie near the present-day equator, (as in the case of those in India) and some of the evaporites lie within the present-day Arctic circle (as in the case of some evaporites in Spitsbergen). The scarcity of tillites in the northern polar regions reflects the general absence of significant land areas in these regions throughout Phanerozoic time. The inferred sediment paleolatitude distributions confirm the view that to a good approximation the Earth's geographic poles have coincided with the magnetic poles. From the point of view of global map-making there is no evidence for significant long-term wandering of the magnetic pole relative to the Earth's spin axis. The average long-term geomagnetic field is a centered, coaxial dipole field.

Maps and Composites

Because the Earth's mean magnetic field approximates to a coaxial, centered, dipole field, the Earth's magnetic poles approximate to the geographic poles. The mean magnetic pole of a particular time period may be used to convert the reassemblies of Permian and later time into world maps. We simply equate the mean magnetic poles with the geographic poles of the period. A Euler rotation about a suitable point will bring the mean paleomagnetic poles of the reassembly into coincidence with the geographic poles. The rotated data points of the reassembly can then be projected as a paleogeographic map. This is how maps were made by Briden, Drewry and the writer (Figs. 14-9–14-14), of the Tertiary (Eocene), about 50 million years; Cretaceous, about 100 million years; Jurassic, about 170 million years; Triassic, about 220 million years and Permian, about 250 million years.

Different map projections are useful for different purposes. For example, if we are interested in the former equatorial regions, a Mercator projection has much to recommend it. The equatorial region occupies the central strip across the map, as on present-day Mercator maps. If we are interested in using the maps for the purposes of being able to make angular measurements, the stereographic projection is very useful. The whole globe may be portrayed on a Lambert equal-area projection (Fig. 14-9). The equal-area projection may be invaluable if we are interested in finding the relative areas of particular features at particular latitudes in past time. The map series shown here (Fig. 14-10–14) is a Mercator series. Its main disadvantages are that it is impossible to show the poles on a Mercator projection—they are at infinity—and it does grossly distort those areas near the poles. Nevertheless, it is a familiar projection, even if the shapes of the continents and their positions on these maps is unfamiliar.

Though paleomagnetism does not in general allow us to find the longitude separation of the continents, it does permit us to position any continental fragment into its former latitude range. Thus each of the pre-Permian continents can be repositioned into its former latitudinal extent. By projecting all such fragments onto a map frame that spans the whole globe, we may construct " composites" of the world at any time in the past.

To make composites we start with the Permian reassembly

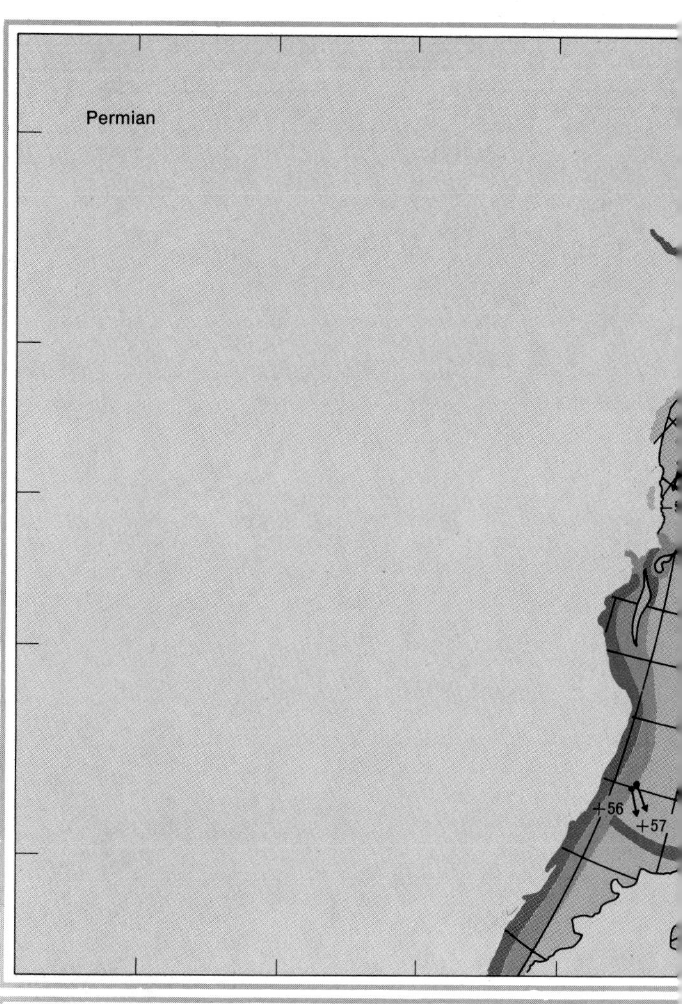

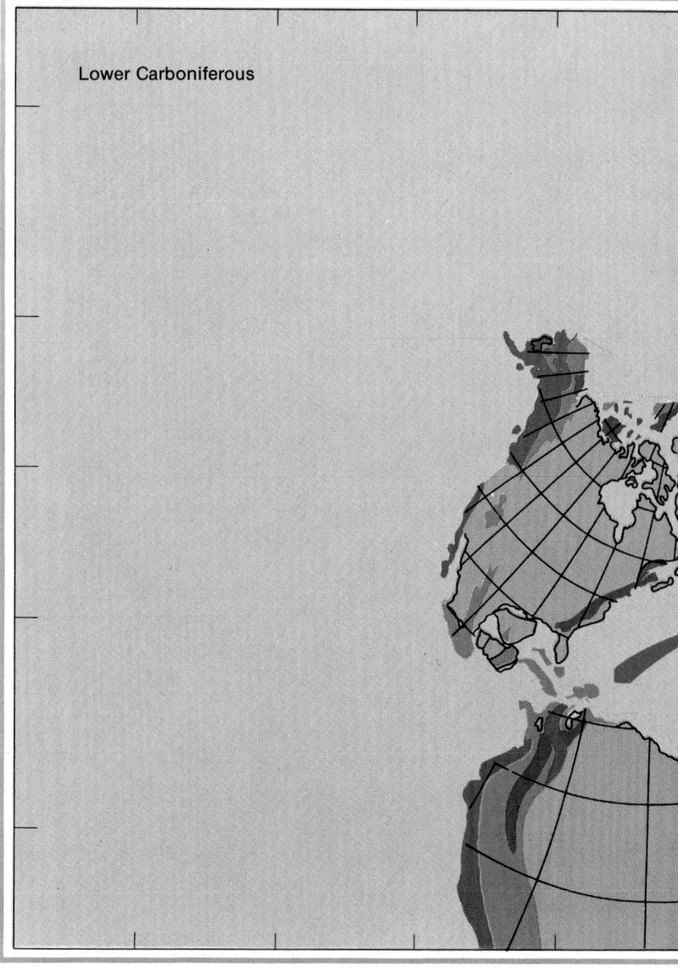

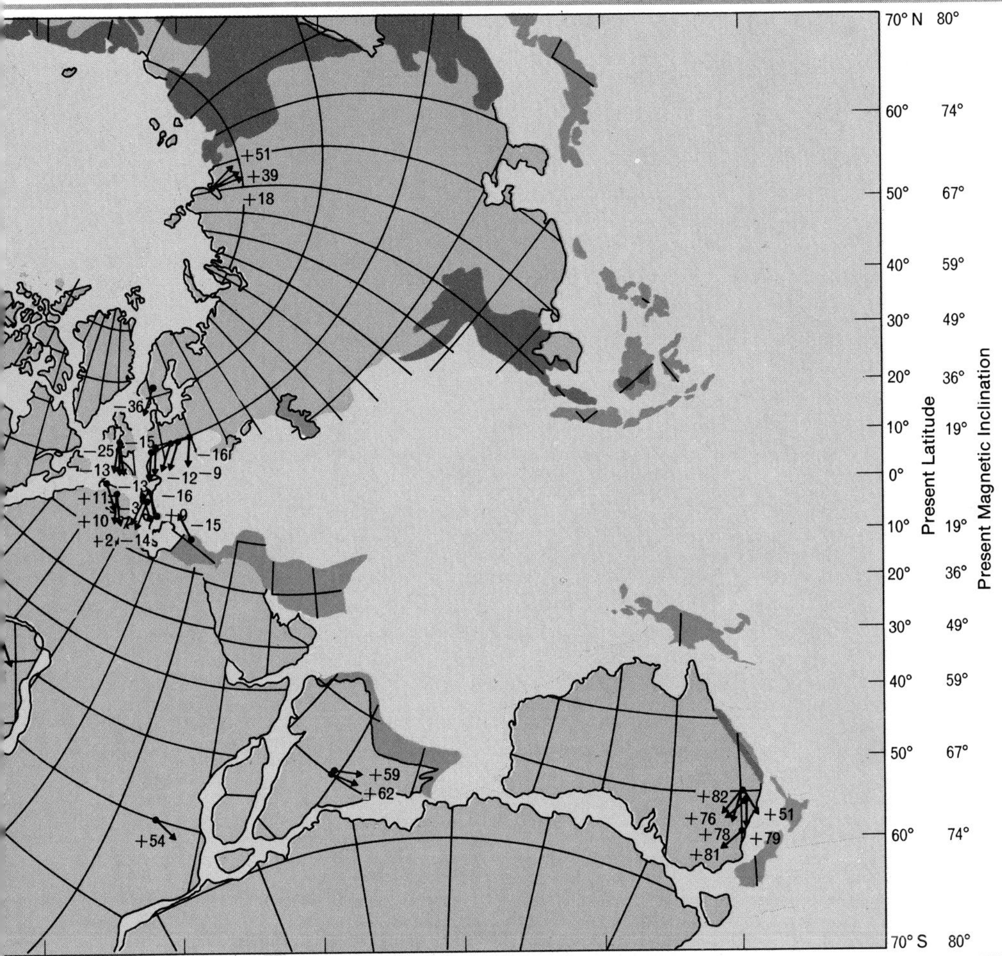

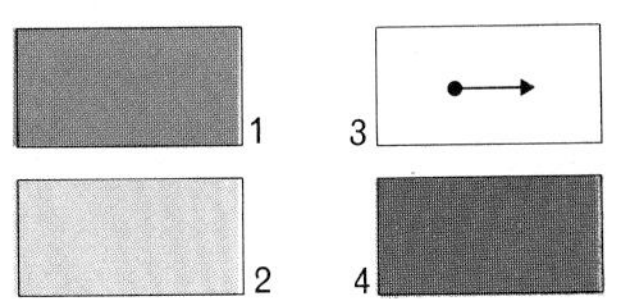

Fig. 14-11. The paleographic situation in the Permian. This is the most remote situation which can be reconstructed today. An impressive continental mass embraces the southern continents, the great continent of Gondwana, which was probably also linked with the northern continents, as indicated in the map, thus forming the large single landmass which Wegener called Pangea. In it the great oceanic gulf opens out which will later play a decisive part in the regions giving on to the Mediterranean basin as well. Key: 1. zones affected by orogenic movements in the Cainozoic; 2. extension of the continental shelf; 3. direction and course of residual magnetism, measured in the rocks of the various regions; 4. the numbers beside the arrows indicate the unit and sign of magnetic inclination; 5. zones affected by orogenic movements in the Mesozoic.

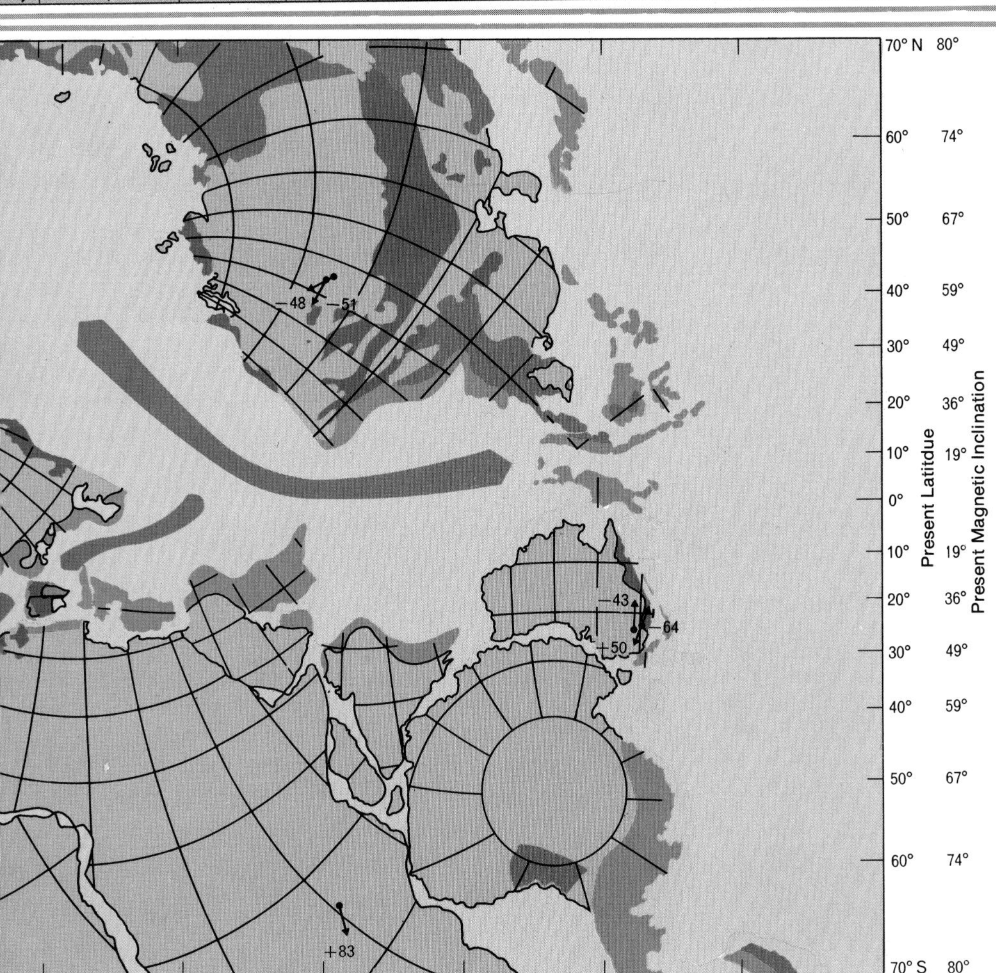

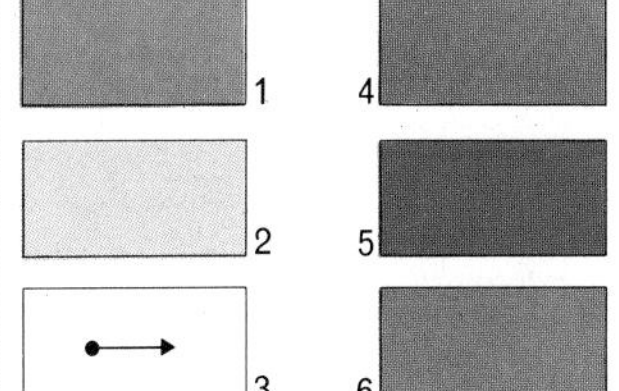

Fig. 14-12. Paleographic situation in the Lower Carboniferous, i.e. about 340 ± 30 million years ago. Even when it was established that present-day continents were part of a single continent (Pangea) it was realized that the mechanism only enabled us to go back as far as the Permian (Fig. 14-11), in other words, only to know the paleographic evolutionary process in the last 250 million years of the Earth's history. And before this? This reconstruction of the situation in the Carboniferous, and the reconstructions that follow, indicate how Pangea, too, was in turn the offspring of previous separate continents. Key, plus the four symbols used in Fig. 14-11: 5. regions affected by orogenic movements in the Upper Paleozoic; 6. axes of the principal oceanic areas.

of Pangea. Pangea is broken into three parts for the composite of the Lower Carboniferous, about 340 million years ago; and for the composite of the Lower Devonian, about 380 million years ago. The breaks are made along the orogenic belt running from the southern Appalachians to central Europe: the Appalachian-Hercynian orogenic belt. A second break is made along a belt of the same age: the Urals. The precise boundaries of the fragments is arbitrary, but we now have three major continental areas: Eurasia east of the Urals; Laurasia, excluding Eurasia east of the Urals; and the southern continents of Gondwanaland. The three fragments are treated independently and oriented according to the paleomagnetic data (Figs. 14-14–15). In the oldest composite, of Cambrian to Lower Ordovician age, about 510 million years ago, a further break is made. Laurasia west of the Urals is broken along the Caledonian-Appalachian belt into two pieces. The first consists of Europe, excluding northwest Scotland and western Spitsbergen; the second of northwest Scotland, western Spitsbergen, Greenland and most of North America. The four continental fragments are then oriented independently (Fig. 14-15).

The uncertainties in the older composites are considerable. Overlaps of one continent onto another are, of course, forbidden, but apart from this rule, the longitude separations of the fragments is arbitrary. Furthermore, the actual east-to-west sequence of the continents is uncertain. The positions of large areas involved in younger orogenics is not known. In the case of the Cambro-Ordovician composite the data used to orient the components span an interval of about 100 million years—as long as the interval from mid-Cretaceous time to the present-day.

The Phanerozoic maps and composites cover a time period of about 500 million years, 3 billion more years must elapse before we encounter the oldest known rocks. The Earth had probably been in existence for over 1 billion more years before that. In principle we would need perhaps another 60–70 maps, each as varied as the Phanerozoic maps before we could adequately portray the motions of the continents in geological time. About 15 of these maps would refer to a time period for which we have as yet discovered no record in the rocks! Bearing in mind the vast areas covered by Precambrian orogenic belts, we might conclude that the drawing of Precambrian maps, or even of Precambrian composites, is an intractable problem. But there is an intriguing possibility suggested by recent paleomagnetic work.

As noted above, one of the main supports for the plate tectonic interpretation of Phanerozoic orogenic belts is that the polar wandering curves of stable continental areas on either side of them, converge during the final phases of orogenesis. During the next episode of continental drift new continents form and the polar wandering paths diverge again. Africa is a continent made up of stable areas separated from one another by Precambrian orogenic belts of many different ages. We would expect the stable continental areas in Africa to show complex and independent polar wandering paths. On the contrary, it is possible to join in nearly all the known poles for the different stable areas into a single smooth polar wandering path for the whole of the Africa continent (Fig. 14-6). This result implies, but does not prove, that the older orogenic belts crossing Africa are not sites of continental collision, but have formed by some other process.

Whatever the explanation, the continuity of the polar wandering path does suggest that there were large, relatively stable continental fragments in much of Precambrian time. These fragments may have simply changed their orientation relative to the poles, rather than evolving as a continuously changing mosaic that is suggested by the Phanerozoic evidence. If these fragments really existed and can be identified, then the problem of drawing Precambrian composites may be soluble, at least for later Precambrian time.

Discussion

The Permian and later maps show how a large ocean that once lay between the northern and southern continents—the Tethys—has been progressively swallowed up. As it disappeared into the mantle, so other oceans came into existence; the Atlantic and Indian Oceans, and possibly the Mediterra-

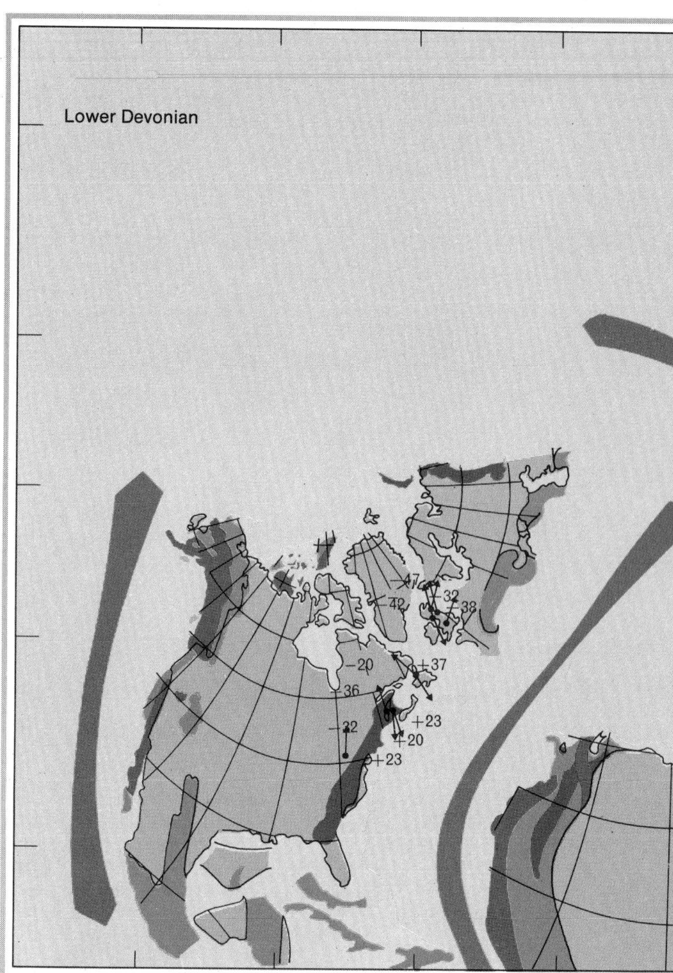

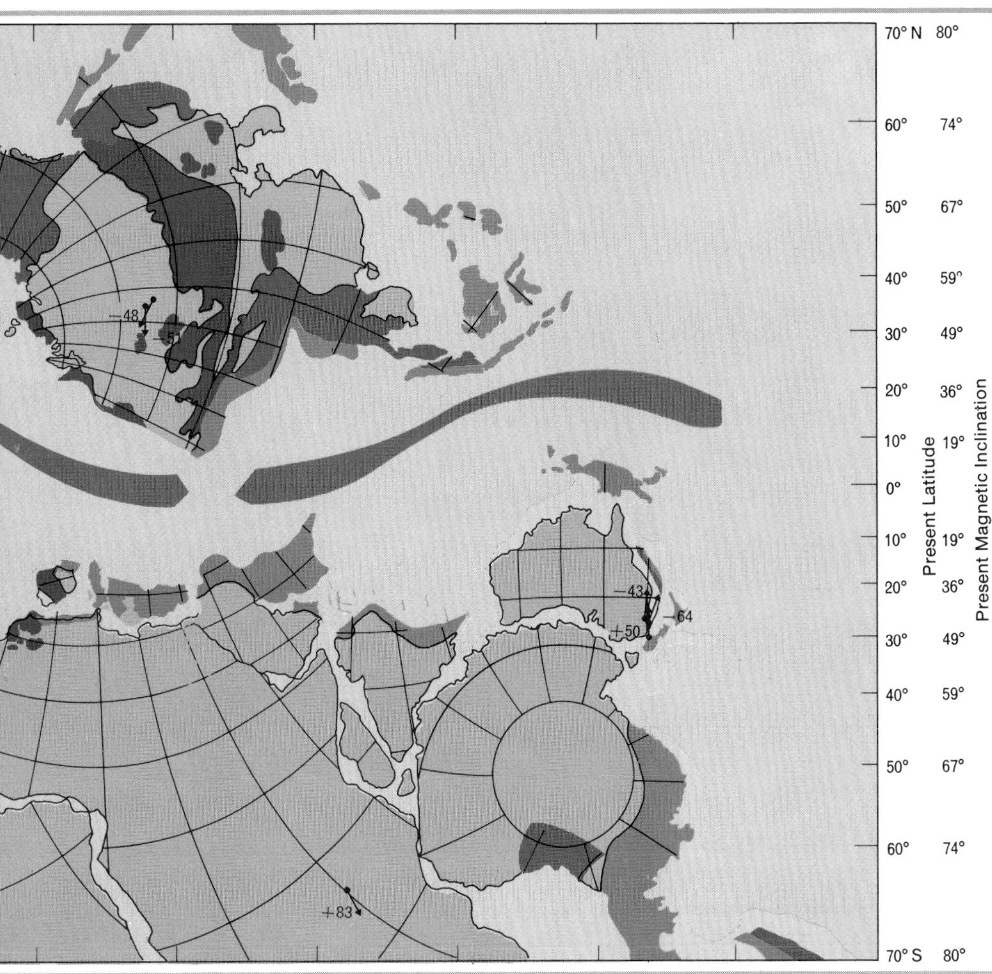

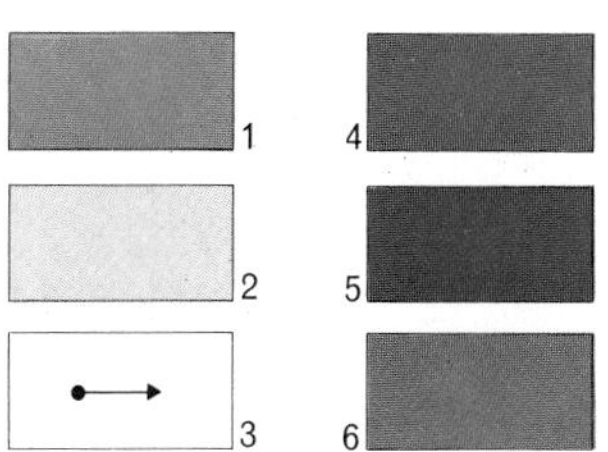

Fig. 14-13. Paleographic situation in the Lower Devonian. There are three continental cores: the southern one—Africa, South America, India, Australia and Antarctica; the western one—North America and Eurasia west of the Urals; and the northern one—Asia east of the Urals and north of the Himalayas. Key: 1. zones affected by orogenic movements in the Cainozoic; 2. extension of the continental shelf; 3. direction and course of residual magnetism, measured in rocks in the various regions (the number indicates the unit and sign of magnetic inclination); 4. zones affected by orogenic movements in the Mesozoic; 5. regions affected by orogenic movements in the Upper Paleozoic; 6. axial zone of the principal oceanic areas.

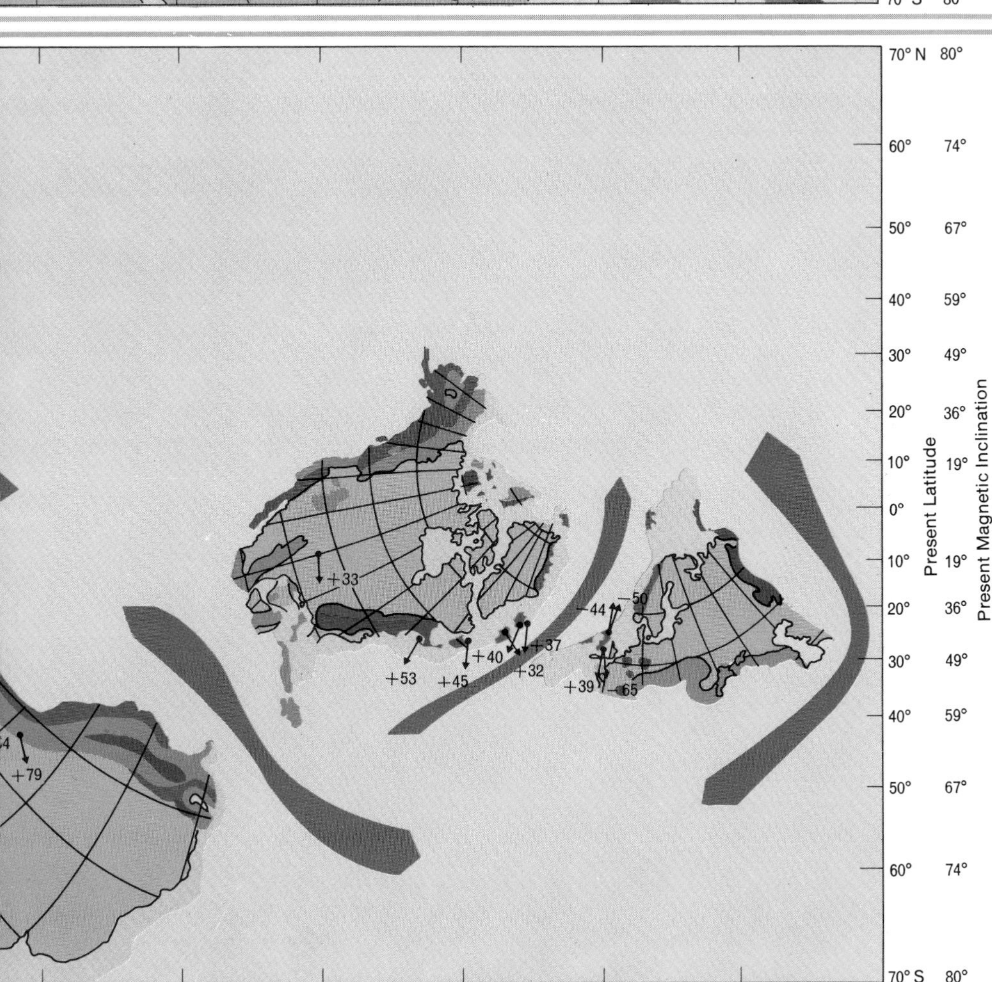

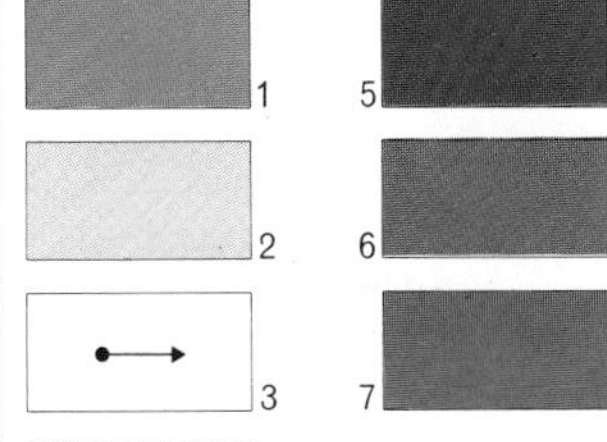

Fig. 14-14. Paleographic situation in the Cambrian and Lower Ordovician, i.e. about 510 ± 40 million years ago. This is the oldest projection and the most uncertain. But in addition to Gondwana, here oriented on the basis of paleomagnetic data so that Antarctica is quite near to the equator, while North Africa is near the South pole, alongside South America, there were, in this period, at least three other major continental masses; one embracing North America, Greenland, northern Scotland and western Spitzbergen; another embracing Europe, apart from the above areas; and the last one formed by the Asiatic continent east of the Urals. Key, plus symbols used in Fig.14-13: 7. regions affected by orogenic movements in the Lower Paleozoic.

nean. The former position of the Tethys is now marked roughly by the Alpine-Himalayan mountain belt. The evolution of this area during the past 250 million years shows quite clearly the inter-relationships among the vanishing of an ocean (the Tethys), continental collision (the northern and southern continents), mountain building (the Alpine-Himalaya chains), and the opening of new oceans (the Atlantic and Indian Oceans, and possibly the Mediterranean).

The main features of the earlier Phanerozoic movements may not yet have been discovered because we have only composites available: we have no maps. Nevertheless, we can see the extraordinary Paleozoic gyration of the southern continents as the south pole changed its position from somewhere in the Sahara in Ordovician time to somewhere in Antarctica in Permian time.

Similarly dramatic Precambrian gyrations are suggested by the polar wandering curves for Africa and North America.

What causes these motions? New global patterns are undoubtedly created by continental collisions. But why should particular continents collide with each other, and still others break up? Why should the motions occur at the rate of a few centimeters a year; neither faster nor slower? The deeper meaning of these motions is still a mystery. The movements of the continents are as incomprehensible as those of a silent dance. Yet the fundamental physical laws governing the patterns are probably all known. A predictive pattern of plate motions is one of the most fascinating but unsolved problems of classical physics. In the years to come, field finds and fluid flow will play their parts in revealing the music of the sphere.

ALAN G. SMITH

THE first 60 years of this century were characterized by a heated debate between the geologists who supported the "static" theory—that is, those who believed that the fundamental outlines of geography had remained practically unchanged throughout time—and those who supported the "mobile" theory, believing that even the continents were capable of motion, the continents being for many the symbol of all that is immobile and perpetual. In the last 10 years the latter have triumphed. Few scientists (but see also the article by Belousov) today dispute the fact that in the last 200 million years the continents have moved thousands of kilometers away from their original positions. In the face of the new data now available the majority of those who believed in the "static" hypothesis have changed their minds, accepting the existence of Pangea, the single continent of 200–230 million years ago, and its later splitting-up into various fragments.

The above debate, however, has not ended: it has just moved back to the period before 200–230 million years ago. Thus, as regards the whole of history before Pangea, two contrasting opinions are held: one, that of the "mobilists," tends to extend the plate-tectonics model and consequently the idea of continental drift, to the whole geological history of Earth, supposing that in the past the continents have repeatedly joined together and separated, making and unmaking so many Pangeas. The other, that of the "static" theory, considers continental drift to be a unique phenomenon in Earth's history, a series of events exclusively limited to the last 200 million years. At this point the question to be asked is whether or not continental drift took place over 200 million years ago. A. E. J. Engel, in the article that follows, develops an intermediate theory. While he does not deny the possibility of the formation and disappearance of entire ocean crusts, or the movement of continental blocks over the Earth's surface, he nonetheless underlines that at least one continental mass, Gondwana—which linked South America, Africa, Antarctica, India, Madagascar and Australia, and which 200 million years ago was joined to North America and Eurasia forming Pangea—moved, if at all, in a single block. It was formed at the beginning of the geological history of the planet, and while it may have moved, it has never since then broken up, re-formed again, as the diehard mobilist theorists propose. To support his views Engel shows that the reconstruction of Gondwana 200 million years ago is such that the alignment of the most ancient mountain ranges of the various continents coincides. According to Engel these ranges were formed 2.5 billion years ago as continous units on a single continent. Later the continents separated and each took its fragment with it; but when did this take place? If in a reconstruction of Gondwana as it was 200 million years ago these fragments join up again to form continuous mountain ranges, this must mean that such a splitting-up took place only upon that occasion, and not on other, earlier, dates. Thus the dispersion of continental masses which characterizes our more recent planetary history is a more or less unique phenomenon.

Not all students agree with this conclusion. For example, we have seen, in the preceding article, that Smith accepts Gondwana as united at least until 570 million years ago, while on the other hand he sets Asia and North America adrift. On this point nobody can as yet have the last word.

ALBERT E. J. ENGEL

Teaches earth science at the Scripps Institute of Oceanography of the University of California at San Diego. Born in 1916 in St. Louis, Missouri, he began his work in geology in the 1930s, first at the University of Missouri and later at Princeton University. A great part of Engel's scientific work is tied to that of his wife Celeste G. Engel, co-author of the following article, with whom, in 1964, he obtained and studied the first series of samples of deep oceanic crust. In 1967, in South Africa peridotites, he discovered evidence of the phenomena of ocean floor spreading during the primordial history of the Earth. He has worked on the reconstruction of the evolution of the continents during the period between the Pangea of Wegener and the beginning of the geological history of the Earth.

ALBERT E. J. ENGEL
CELESTE G. ENGEL

Continental Drift in the Early History of the Earth

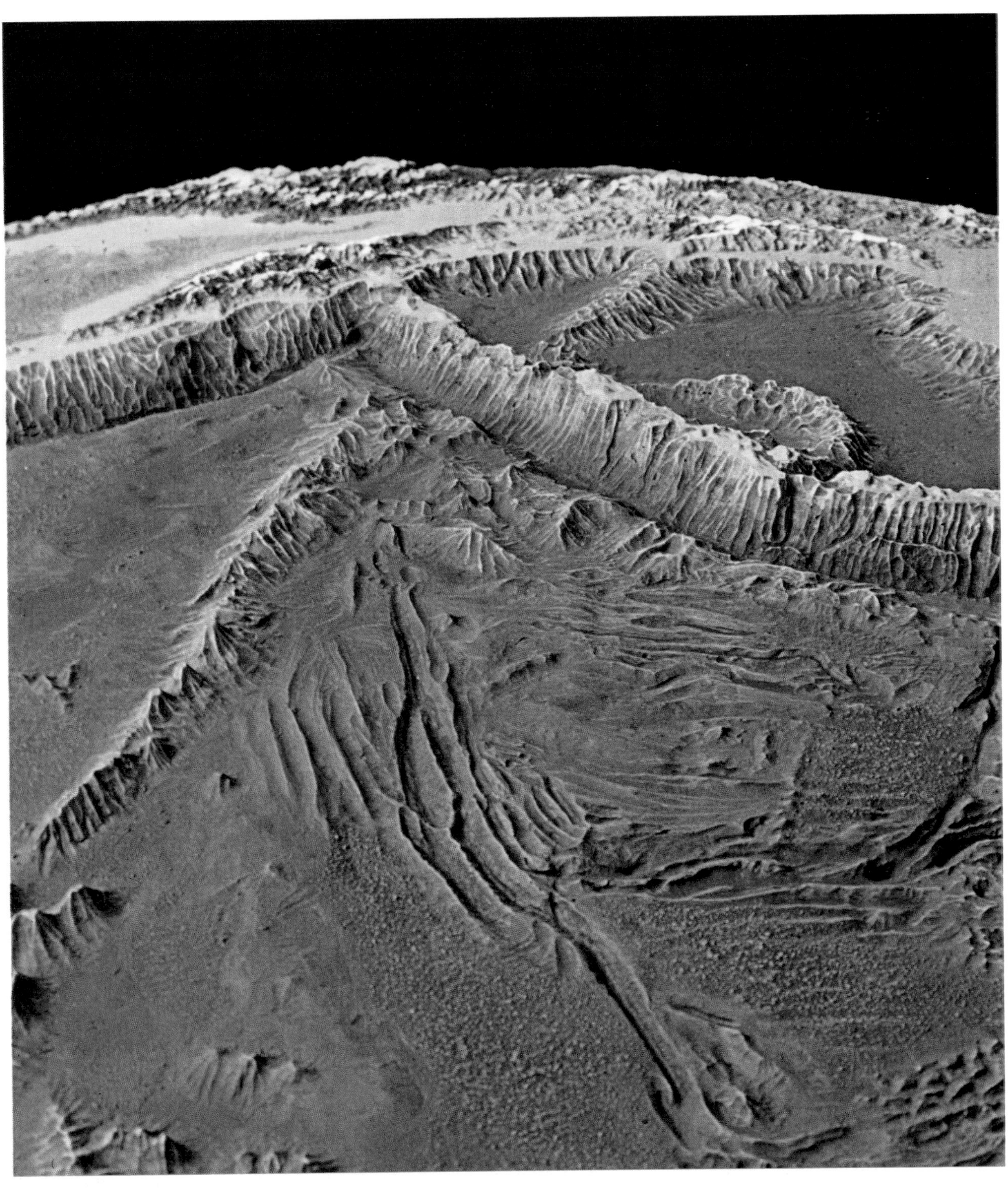

Overleaf: The Aleutian trench as it appears in a world map which also shows the morphological features of the ocean depths. Today, the oceanic trenches represent the structures along which the crust of the oceans sinks into the mantle and dissolves. In this phenomenon the plate-tectonics model has isolated a mechanism which may, as an outcome, cause the birth of a mountain chain. Conversely, the conclusion may be reached that most of the world's mountain chains may represent the result of subduction phenomena and evidence in the continents of the disappearance of an ancient continent and the consequent collision between continental masses. If this were proved, it would follow that continental drift is not a phenomenon restricted to the latter stages of the Earth's history, but that it is as old as the first continental masses. The debate surrounding this idea still continues (Rand McNally and Co., R.L. 75-GP 1-2).

Our earliest ancestors, almost 2 million years ago, were aware that the Earth seldom sleeps. Some were forcefully buried in a torrent of poisonous gas and volcanic ash. Many, more recent societies, as at Pompeii, Vesuvius, in Turkey, Iceland, Hawaii, Chile and Indonesia have been painfully reminded that the Earth quakes and erupts with frightening violence. Few of the ancients, however, ever realized that their homes on whatever continent or island, were more quietly adrift on great crustal rafts, or "plates."

Today, however, earth scientists have identified at least seven large crustal plates and other small ones. These jostle and migrate relative to each other, and move over the great mass of inner Earth at rates of 1–10 cm per year. Their motions are caused presumably by great currents within the Earth, much as great blocks of sea or river ice rift, drift and collide in response to the currents of water beneath them.

Plate Tectonics

The study of these features and movements of these plates called "Plate Tectonics" now preoccupies most earth scientists, and much has been written on the subject. The reasons are obvious. An understanding of the earthquakes, the volcanic eruptions, and continental drift associated with plate motions is fundamental to an understanding of the processes on the surface of the Earth and within it. The sites from which plates form, and diverge are now the great Earth-encircling oceanic ridge and rise complex, some 70,000 m long (Fig. 15-1). There, new oceanic crust is constantly forming as floods of basaltic lava issue from the inner Earth. But this is a relatively recent event in Earth history. Just 200 million years ago the Americas, Africa, India, Australia and Antarctica were one, a vast supercontinent. The famous, now sainted earth scientist, Alfred Wegener, called this continent Pangea (Fig. 15-2). For reasons not too clear, it is now known that Gondwana began to fragment along great rifts in the Earth's outer skin, the lithosphere. The Americas broke away for the first time some 125 to 180 million years ago and the Atlantic Ocean began to form at and along the great opening crack in the Earth's crust, the lithosphere. Subsequently, India and Africa broke away from Australia and Antarctica drifting northward toward Eurasia. Finally, only 50 million years ago Australia broke away from Antarctica and the Indian and Antarctic Oceans were born along the site of the fracture. This site is now marked by the great Mid-Indian Ocean ridge system (Fig. 15-1).

Continental Drift and Sea Floor Spreading

The rifting and drift of these great continental fragments as parts of even larger lithospheric plates, in the directions suggested in Fig. 15-1 is the subject of continental drift. The growth and motions of the new Atlantic and Indian Ocean floors from the emergent oceanic rises is the accompanying process called *sea floor spreading*. One possible ancient example very like some existing today in the western Pacific is shown in Fig. 15-3.

The major zones along which continental-oceanic plates collide, especially plates with continental edges, mark the sites of great mountain ranges of the world, such as the Alps and the Himalayas. These mountains formed as the drifting plates including Africa and India collided with plates containing Europe and Asia.

In the circum-Pacific, the plates contain the spreading Pacific Ocean crust. Its edges are thin, heavier oceanic crust. They tend to be thrust, or sink under the margins of the lighter, thicker granitic masses of Asia and the Americas. Here the sites of plate convergence are marked by the deep oceanic trenches, island arcs, and chains of active volcanic mountains as suggested for older examples in Fig. 15-3. The volcanoes are in fact what we often call the circum-Pacific "ring of fire," and of course "brimstone" from the devil's hearth beneath. This hearth we call the Earth's *mantle*.

The Earth's Crusts

The formation, and collision or underthrusting (consumption) of new crust has a fascinating impact on the nature of the Earth's crust. Oceanic crust is constantly being formed, but also largely being consumed, as it is thrust or sinks under the thicker, lighter, more elevated continents (Fig. 15-3). Hence, all of the crust under the contemporary oceans is very young relative to geologic time. It is largely born and consumed in periods of less than 200 million years. In contrast, the great continental fragments continue to age and thicken by collision and consolidation of island arcs and fragments of oceanic crust. All the continents contain rocks at least 2700 million years old. These ancient rocks form thick, stable "shield" regions in which granitic rocks between 3000 and 3700 million years have been found and dated by radiometric methods (Fig. 15-4). The general relations in age, thickness, and compositions of these several crustal types are shown in Fig. 15-5.

Looking Backward

It is necessary of course to read the history of the Earth largely from its rocks, and indirectly from meteorites, the moon, and other planets. But the great bulk of our knowledge of the Earth's history in times earlier than the date of the young ocean crust must come largely from studies of the older island arcs and continents. This means that all geologic history, including concepts of plate tectonics and continental drift prior to some 150 to 200 million years ago, back to the Earth's origin 4.6 billion years ago, must be read on land. It is to the land geologist, then, that we turn to learn of continental drift and seafloor spreading over the several eons (1 eon is 1 billion years) of geologic time before Gondwana and related supercontinents broke apart.

This is no simple task. The continents as any observant human knows are infinitely complex. But this complexity is a great benefit as well as a cause for controversy. Fortunately, the continents contain relic patches of various ages, of all the crusts—oceanic, arc, interarc basins, continental borderlands and the ancient, stable shields. And it is specially fortunate, if seemingly paradoxical, that some of the most informative and diverse crustal types and their constituent rocks occur in what are call *Archean* Shield terrains, formed 2.5 to almost 4 billion years ago (Fig. 15-4).

Reconstructing Supercontinents

The concept of fragmentation of earlier supercontinents such as Gondwana and the worldwide drift of the great continental fragments such as the Americas, Antarctica, Australia and India relative to Eurasia, was first proposed by A. Wegener some six decades ago. Wegener arrived at this conclusion from several lines of reasoning now well known. He noted, as had others before him, that the eastern coastal outlines of South America, fitted rather neatly into the great continental embayment of western Africa (Fig. 15-2). He also observed that he could refit Australia and Antarctica together with Africa. The resulting continental cluster forming this major supercontinent based largely upon their outlines was but a small part of Wegener's argument. He noted that when the continents were fitted back together again, that mountain belts formed in the periods just prior to 200 million years ago, between 300 to 600 million years ago, fitted together and formed great sinuous patterns across the supercontinent of Gondwana (Fig. 15-2). Wegener and others argued that both the matching continental outlines, and the matching mountain fabrics were hardly a coincidence. Wegener and his protagonists also perceived that many other geologic features made far more sense if the present continents were formerly parts of Gondwana. They pointed out that unique layers of sedimentary rocks, some containing related fossil organisms, all seemed to fit together into more coherent patterns in Gondwana. In fact, the fossil organisms in the refitted sedimentary rock strata all formed populations in Gondwana resulting from very special climates and environments. Again the "drifters" asked, could this occur by coincidence? The lines of argument that Wegener and others assembled, especially the morphological fit of continents, the matching of slightly older mountain belts, and the coherence of geologic strata bearing organisms from specific environments seemed to them overwhelming proof that the Americas, Africa, India, Australia and Antarctica were once part of a supercontinent.

Continental Drift and the Controversy

Wegener's ideas were immediately the subject of an enormous amount of discussion and controversy. Some geologists, especially those who had worked in the Southern Hemisphere, and including many European geologists, were inclined to accept Wegener's concept of widespread rifting and drifting of continents. Most other geologists, though, especially in North America, the U.S.S.R., and Europe were rather skeptical or contemptuous of the concept that all of the existing continents had broken apart and drifted on a worldwide scale in the last 200 million years. They ridiculed Wegener as well as other advocates of continental drift such as A. Du Toit in Africa, and Taylor in America. The skeptics assured each other that there were obvious and insurmountable reasons why continents could not drift across the face of the Earth.

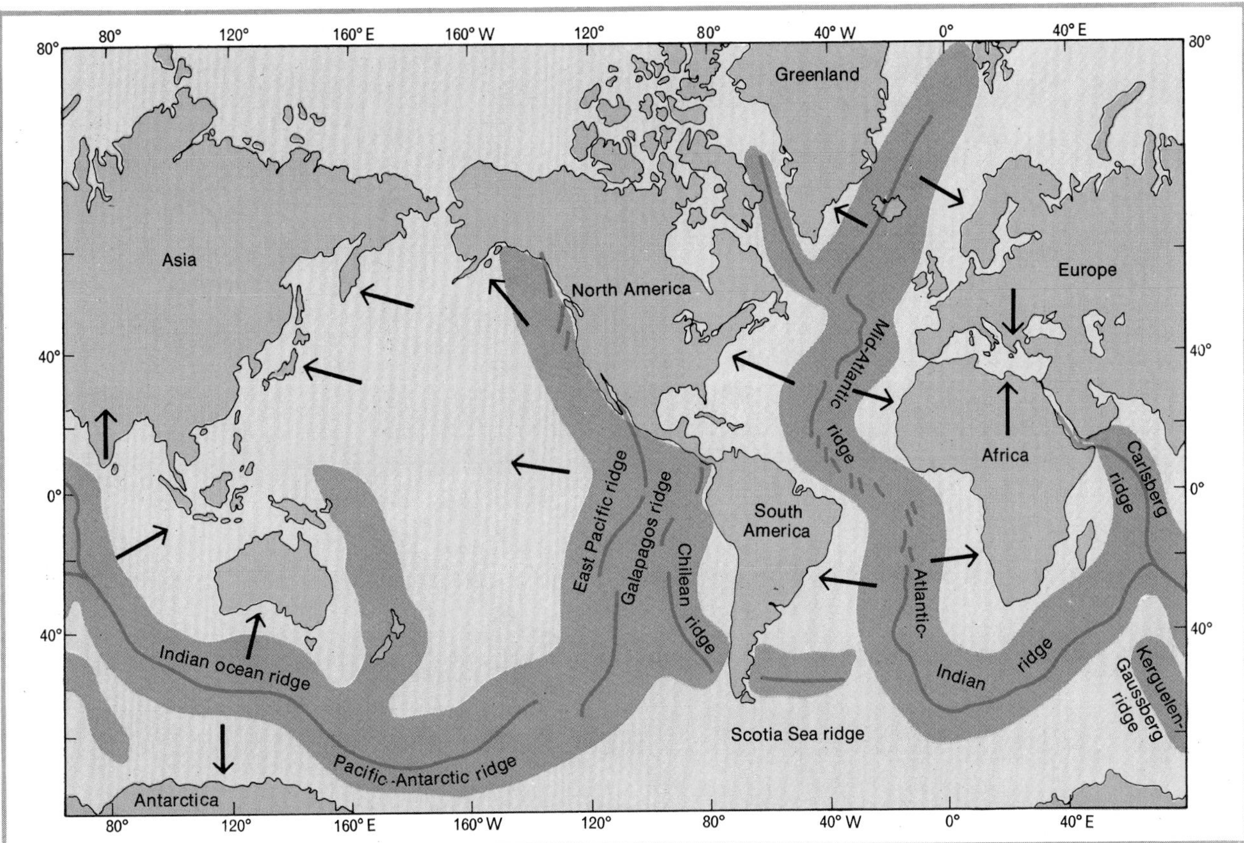

Fig. 15-1 Mid-oceanic ridges (blue). The thick blue lines indicate the zones in which currents of basaltic lava re-emerge from the Earth's interior, forming new oceanic crust. The arrows roughly indicate the direction of the movement or expansion of the oceans from the moment when the splitting-up process began and the drift movement of the large continental masses started, between 50 and 200 million years or so ago, with circum-Pacific and alpidic directions.The rate of drift varies from 1–10 cm per annum.

Perhaps the most powerful objections were advanced by scientists who studied the physics of the Earth. Wegener's suggestion that great continents had drifted implied to them the plowing of the continental rafts over the ocean floors themselves. Sea-floor spreading and the existence of lithospheric plates was not then understood. Hence, they argued, the ocean floors were composed of cold, very hard basalt. The continents were thick, generally cold and composed of brittle granitic rocks. The idea that the great continental rafts of granitic rocks could plow their way across or over the ocean floors was obviously impossible from a physical point of view. Their arguments were correct, but they did not think deeply enough. Had they thought in terms of plate tectonics, that is to say, of *thicker* plates, not just the thickness of the continents, or the thickness of ocean crust, but of the much thicker

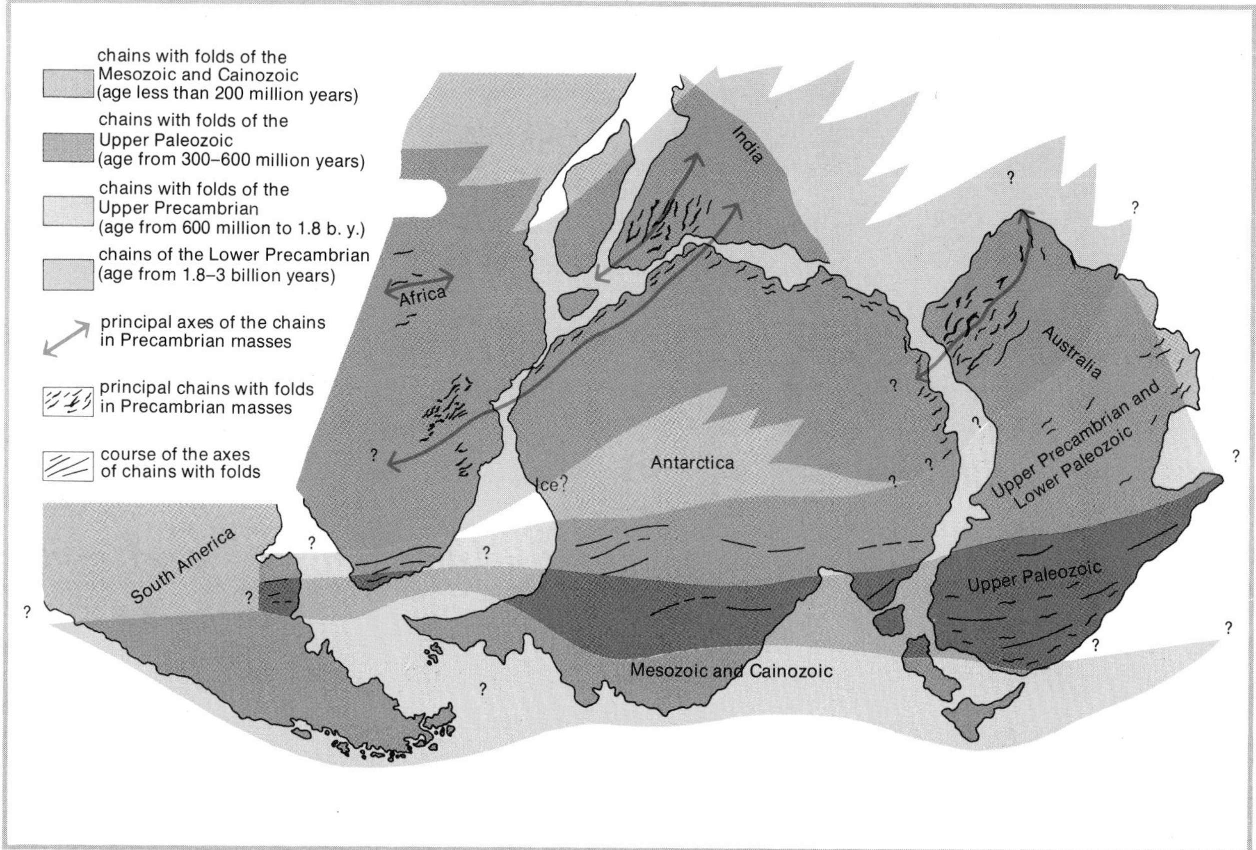

Fig. 15-2. The super-continent of Gondwana, about 200 million years ago. One of the tests of the permanence of this super-continent from the Lower Precambrian onwards derives from the confirmation that the oldest mountain chains present in the various continents are in alignment if related to present-day continents in the position that they occupied before the drift which began 200 million years ago.

lithospheric plates composing the Earth's hard skin as much as 200 km thick, the problem would have disappeared. These thick plates can move, or can be carried by currents in the hotter, softer solids and partly molten material in the Earth's mantle beneath (Fig. 15-3).

As geophysicists have looked more thoughtfully, with precise instruments, at the characteristics of the Earth's mantle they realize that at depths of 100 to 200 km, the velocities of elastic waves and the waves propagated by earthquakes, decrease. The slowing or attenuation of these waves is commonly indicated by their movement through layers of softened, incipiently molten rock of the upper mantle of the Earth immediately below the lithosphere. Here, then is an obvious mechanism for the movement of plates. Lithospheric plates probably are moved by currents within the Earth's mantle which convect hot, unstable rock from the Earth's interior towards its surface, and laterally in the mantle (Fig. 15-3).

But granting the worldwide rifting and drifting of continents, and the formation and spreading of new oceans in the last 200 million years, what about the preceding 4 billion years of geologic time? Have great plates of the outer Earth always moved over its mantle bearing with them fragments of evolving continents? Have sea floors always spread, been consumed, and renewed at intervals of 100 to 200 million years? Or is this a process that has evolved relatively recently? Answers to these questions can only come from the continents and the older volcanic arcs. And here is where we must search for evidence of ancient continental drift.

Reading the Record

But what types of evidence does one look for? Wegener and Du Toit, as well as others, have pointed to several. What about patterns of the mountain belts even older than those aligned by Wegener? Great mountain chains that were formed not merely 300 to 600 million years ago, but 3 billion years ago? Would they match if the great fragments of continents were pieced together as Wegener had done? What about distinctive older rock formations? Could great belts of some of the other distinctive rock formations such as anorthosite, a very curious rock composed largely of one mineral plagioclase feldspar, that now occurs in clusters on separate continents, be better matched if continents were placed together in various configurations such as Gondwana, and so on?

These questions in turn pose many other problems for any geologist who wishes to question when sea floor spreading and continental drift began. He must ask himself, when did the Earth differentiate into its present core, mantle, and the crusts? And at what rate did they differentiate? How hot was the inner and outer Earth 3 to 4 billion years ago. How unstable? And then he must ask, when did cooler, brittle plates of the type that are now moving over the surface of the inner Earth actually form? Let us look at just a few characteristics of the Earth that may provide clues to these intriguing questions. Perhaps a good place to start is at the beginning.

The Archean

The beginning is used in the sense that we can now read geologic history from the oldest rocks on each of the continents dating from 2.5 to those almost 4 billion years old. In this Archean era let us examine the characteristic features of the Archean rock terrains and see if they show evidence of rifting, drifting and collision of protocontinents. And are these Archean mountain belts something like the Alps and Himalayas? And have the oldest continental masses interacted with the sea floors, forming island arcs that were subsequently agglomerated into continents? Let us look to see if there are old ocean basins that form very different patterns from the existing ocean basins.

The patterns of preserved, old island arcs, ocean basins, as well as continental borderlands, and shields preserved on the existing continents necessarily are spotty (Fig. 15-4). But it is very amazing to find that the Archean record is consistent from continent to continent.

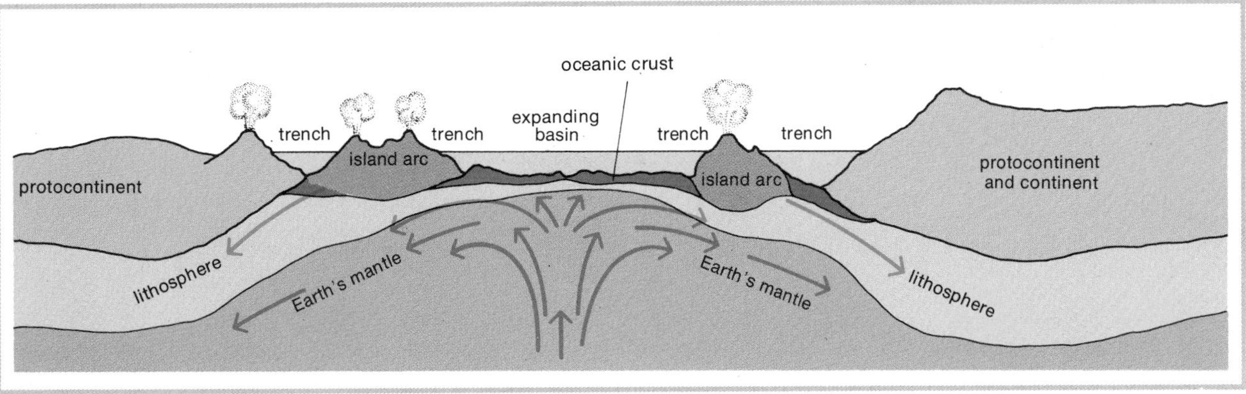

Fig. 15-3. A simplified diagram of the possible relations and interactions between island arcs and continents, either Precambrian or present-day; the theoretical continents under consideration here are separated by marine basins which are subject to expansion and bordered by volcanic island arcs associated with oceanic trenches.

Mountains Belts. For example, let us consider one aspect of the Archean era. Now, we know that mountains are probably always forming. But the great mountain belts formed during the history of the Earth appear to have done so in a series of six or seven major episodes. These orogenic episodes on Earth are spaced at several million years apart (Fig. 15-6). For example, prior to the presently recognized continental drift and sea floor spreading of 200 million years ago, the great Pan African-Appalachian-Caledonian-Hercyian mountain system of Africa, the Americas, Europe and Asia evolved. These mountains were formed in the period from 300 to 600 million years ago. They are the mountains that Wegener arranged into coherent patterns in his reconstruction of Gondwana. But the very careful and thoughtful collaboration of geochemists who

Fig. 15-4. The approximate form and distribution of the youngest mountain chains (i.e. those which formed during the last 200 million years) and the oldest (which formed earlier than 2.5 billion years ago). Both were caused by ocean-bed expansion and the drift of protocontinents or continents. Note that the youngest chains, around the Pacific and in the Alpine-Himalayan chain, have formed along the edges of drifting continents (Pacific), or in the collision areas between Africa and India on the one hand and Eurasia on the other, when the first moved northwards towards Eurasia. On the contrary, the old mountain chains of the Lower Precambrian now form the core of the main continental masses, cutting crosswise through the continents with axes directed NE and NNW, thus forming a marked angle between them. The arrows indicate the course of these chains.

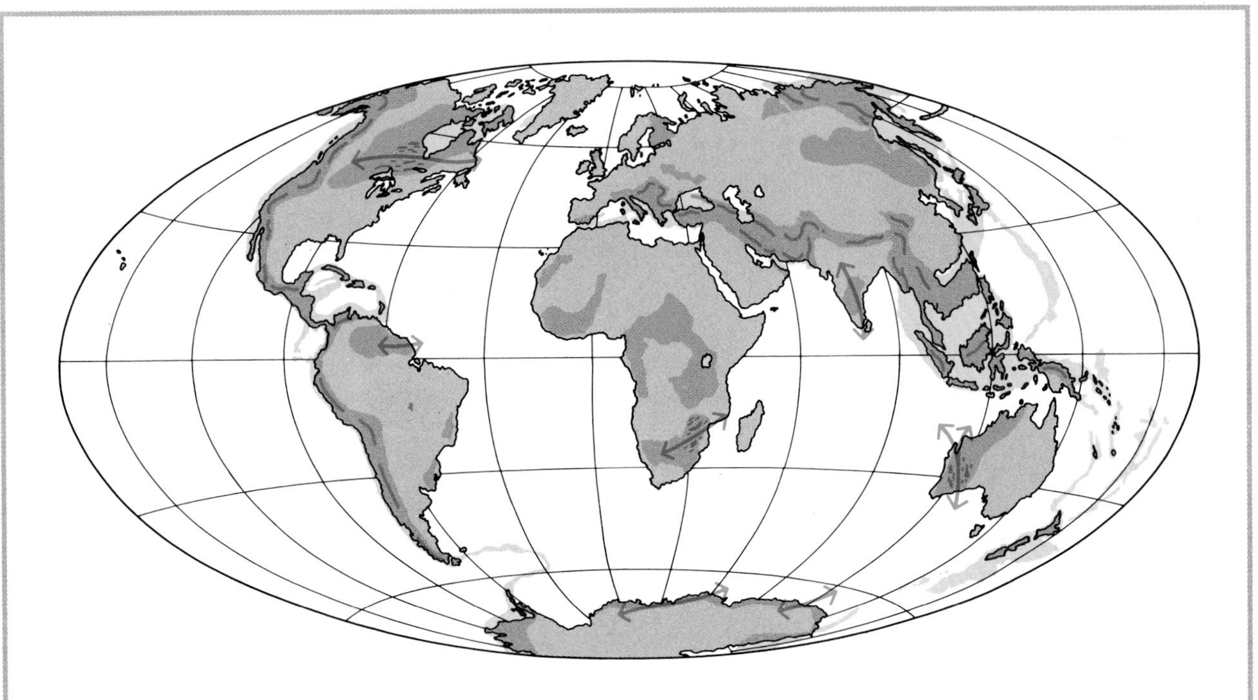

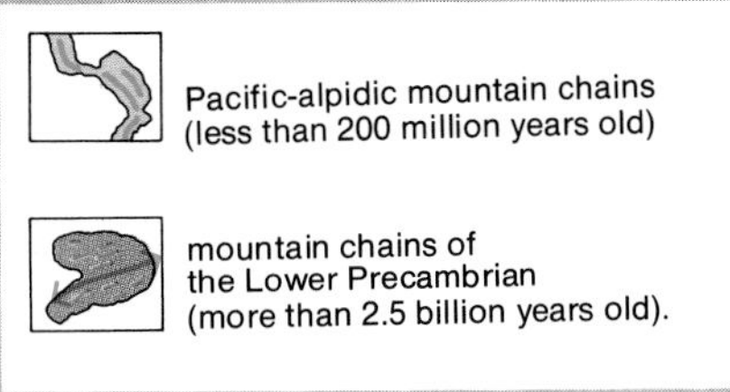

date rocks with geologists who study them in the field has shown that an older, and again, worldwide mountain-making episode occurred in the period between 1 to 1.2 billion years ago. We must emphasize that the great mountain-forming episode, sometimes called the *Grenville* episode, culminated then. This culmination is marked by the emplacement in the crust of the Earth of great masses of granitic rocks, as well as other rock types, including this curious rock anorthosite composed largely of plagioclase feldspar. In the Grenville orogeny, we have a unique mountain system and a sequence of rocks of almost worldwide distribution.

Looking backward from the Grenville event, we see another mountain-forming episode approximately 1.5 billion years ago, a fourth major event about 1.7 to 2 billion years ago, and then a gigantic series of episodes which began approximately 2.5 billion years ago and extended backward in time to at least 4 billion years ago. These Archean mountains contain the earliest remaining records of the Earth including the evidence of Archean life (Fig. 15-6). Prior to that time the record is far less clear. Fewer rocks remain with the record of this very early

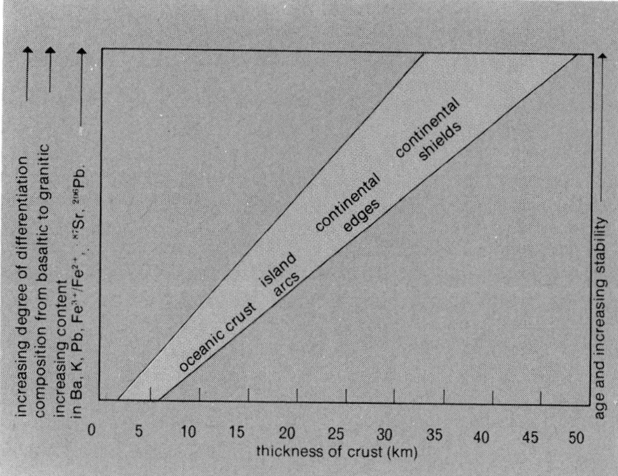

Fig. 15-5. Relations between the age, thickness and chemical features of certain types of the Earth's crust. Passing from the oceanic to the continental crust, the age, thickness and degree of differentiation increase progressively, while the composition varies from basaltic to essentially granitic. The island arcs and continental margins are included between the extremes represented by oceanic crust and continental shields.

history clearly written in them. Most of these older rocks have been so redeformed by later Earth motions that their origins and dates are somewhat obscure. But it is clear that in the early Archean, 3.5 to over 4 billion years ago, the outer Earth was a very warm and turbulent place. By this Archean time great masses of protocontinent had begun to form. The initial stages in continental formation as well as the evolution of oceanic basins were well advanced. For example, we find in South Africa remarkably well-preserved and little altered volcanic rocks as old as 3.5 billion years. Many of these rocks formed in ancient seas, for they are pillowed basalts which are known to form only in oceanic environments. There also are great masses of granitic rocks and iron and magnesium-rich *ultramafic* rocks essentially like those forming the Earth's upper mantle. And there is much evidence of the profound mountain-forming episodes involving extensive sea-floor spreading, and collisions of island arcs and evolving continents. The outer Earth was very hot and unstable. Looking backward even beyond that, a team of geologists and geochronologists have discovered in Greenland rocks as old as 3.7 billion years, which again show evidence of profound mountain-building episodes, evidence that both ocean floor and at least early vestiges of continental fragments formed and were largely destroyed at this time.

The Moon. The study of the moon gives further clues to the evolution of the early Earth. For example, it is clear that the moon's outer surface was largely molten in the period between its formation about 4.6 billion years ago and 4 billion years ago. Now, if the moon and the Earth were formed as companion planets, we may presume that the outer Earth was also a very hot and turbulent place in the period 4 to 4.6 billion years ago. This suggests that the differentiation of the Earth into its core, mantle, and crusts began essentially at the time of the formation of the Earth and that by 3.5 billion years ago,

Fig. 15-6. A diagram of the relations between marine transgressions, the evolution of life, atmosphere, and orogenic episodes. Note that the vertical time-scale is modified in such a way that the last billion years, which is the best known period, occupies a much larger space than the previous billion years. From the left: in the first column, the zone on the left indicates the percentage of dry land, and the zone on the right the percentage of land covered by sea; the second column shows the basic composition of the atmosphere and the appearance of the various types of organisms; in the third column we have the oxyen-enrichment curve as caused by the atmosphere until the oxygen has reached proportions similar to today's; on the right, the colored area indicates orogenic activity.

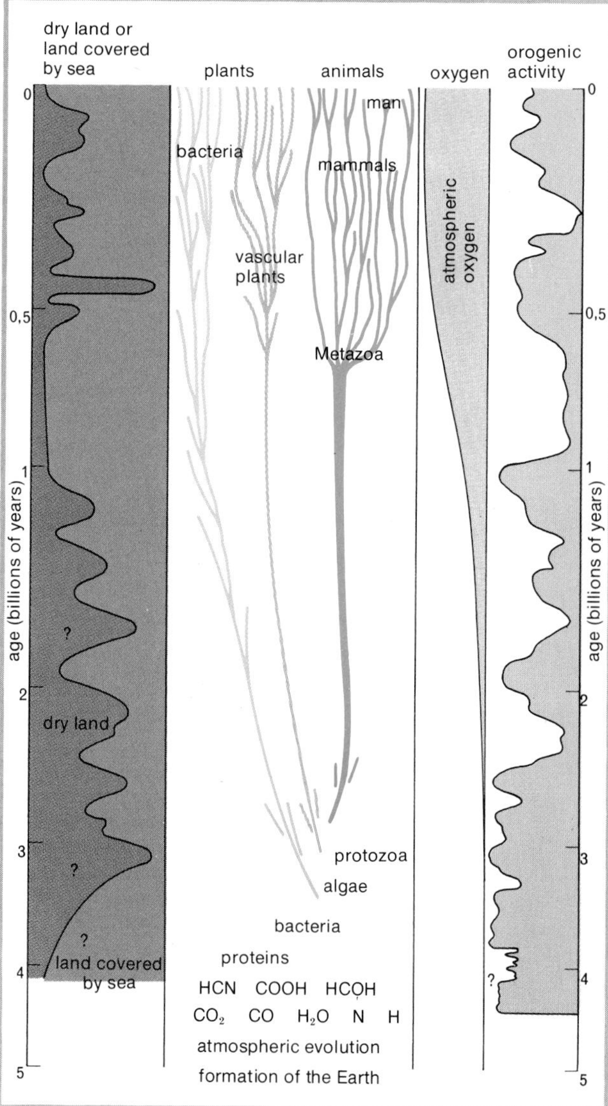

much of the differentiation of the Earth into core, mantle, and crust was complete.

Continental Drift through Time. By 3.5 billion years ago we recognize now folded segments of ocean floor, arcs, and protocontinents (Fig. 15-7). Hence, we may ask ourselves, have these fragments of the outer rind of the Earth, formed billions of years ago, been rifting and drifting throughout geologic time on the same scale as the present?

Surprisingly, the answer seems to be that the worldwide drift of large fragments of continents such as the Americas,

India, Australia and Africa relative to Europe and Asia in the last 200 million years is a relatively unique process in geologic history. In fact, it is only in the Archean between 2.5 billion years and the dim past at least 4 billion years ago that worldwide rifting and drifting of *protocontinents* or small continental fragments occurred again on a large scale. It should be noted, of course, that this does not preclude an almost continuous, or an episodic amount of sea-floor spreading or the *drift* of supercontinents throughout geologic time. But it does seem to preclude the possibility that great fragments of continents have either continuously or episodically *broken off* from other fragments of Gondwana and drifted widely over the Earth. We should, of course, note that smaller fragments such as tiny island arcs have clearly moved relative to continents that shoved against them, and became amalgamated. We know that continents have grown in thickness although probably not much in area, and that they have changed their average composition and that many other amazing features have occurred on the Earth. But the worldwide drift of great *fragments* of *continents* appears to have been confined entirely to

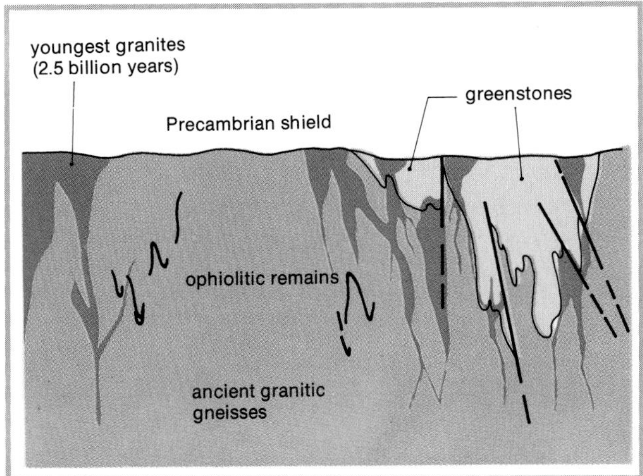

Fig. 15-7. A schematic view through a Precambrian shield region which shows the roots, exposed by erosion, of a Precambrian mountain chain, formed between 3 and 3.5 billion years ago. This ancient chain may have formed from an ocean bed and Precambrian island arcs crushed between the various fragments of a supercontinent forming at that time.

Fig. 15-8. Pillow structure in a peridotitic lava in the Barbeton Mountains in South Africa which contain rocks which were formed about 3.5 billion years ago. The lava emerged from the Earth's mantle into an underwater environment, as is decisively shown by the pillow structure: even then there were vast expanses of water which later became today's oceans.

the last 200 million years and on a different scale with somewhat different continental fragments and ocean basins. This is a rather startling conclusion and there are many competent earth scientists who would challenge it.

Let us look briefly at just one line of evidence that suggests that the conclusion seems correct. Following Wegener's lead, let us look at the patterns of mountain belts, but let us go backward in time far beyond 600 million years ago. Let us look at the mountain belts formed on the existing continents that are older than 2.5 billion years. The Archean mountains that are 2.6 to 3.5 billion years old. Fig. 15-4 is a map showing the present orientation of these old mountain belts together with those formed in the last 200 million years as the result of the recent worldwide continental drift and sea-floor spreading. Studies of the Archean Mountain belts show that in any particular continent such as Africa, or Australia or North America where they are clearly decipherable, these mountain belts are uniformly aligned in subparallel patterns as shown in Fig. 15-4. But looking at their orientation today from continent to continent we see striking divergences. For example, in North America the great Archean mountain belts formed over 2500 million years ago trend northeastward across the entire surface of northern North America in both the exposed and covered regions of the great *Canadian Shield*. This is also true in Africa (Fig. 15-4). But in much of western Australia, as shown in Fig. 15-4, the Archean mountains lie in crudely subparallel alignment trending north-northwest. This same pattern and trends are also true for India (Fig. 15-4).

It should be noted that because of the succeeding younger mountain-forming events on all of the continents, but specially in South America, Europe and Asia, the patterns of the Archean Mountains belts are not nearly as well defined or preserved. But studies of these mountain belts on all these now widely separated continents indicate that they are much alike in their characteristics and in the way they form. Most seem to be crushed ocean, and sequences of island arcs and interarc basins very much like those formed today in the circum-Pacific (Fig. 15-3). Thus, the Archean seems to have consisted of patches of protocontinent, evolving along and between great rifts composed of sea floor and arcs. Powerful convecting currents in the Earth's mantle clearly were at work spreading seas, colliding protocontinents, and creating great chains of island arcs. Then, about 2.5 to 2.7 billion years ago these currents carried most of the protocontinents together agglomerating them. In the process the Archean ocean basins, interarc basins and arcs were mashed into the existing Archean greenstone mountain belts shown in Fig. 15-7.

The reasons for these powerful convecting currents are obvious if one examines the rock types in the greenstone belts, especially at their roots. The roots of the greenstone belts are composed of great flows of ultramafic lavas. It was not until a few years ago that geologists and geochemists accepted the fact that such ultramafic lavas called *peridotites* could rise from the interior of the Earth and flow out onto the surface. The reasons that many doubted that this had ever occurred were that (1) the Archean terrain had never been studied carefully; and (2) the temperature of melting of these ultramafic or peridotitic lavas; at least 1400° to 1600° seemed far hotter than temperatures assumed for the early outer Earth.

Actually, because these lavas have a composition very much like that inferred for the entire Earth's upper mantle, the roots of the ancient greenstone belts we see are in reality oceanic rift zones and spreading centers crudely similar to those existing today (Fig. 15-3). Almost surely, however, the Archean rifts were sites of more widespread, high temperature igneous eruptions. Some of these ultramafic lavas also show pillowed forms, proof that they were erupted into seas. A photograph of ultramafic pillowed lavas, at the roots of one of these ancient greenstone belts, the Barbeton Mountain Belt in eastern South Africa is shown in Fig. 15-8. Pillowed lavas, as noted above, whether ultramafic, basaltic or felsic in composition, are unknown except where the lavas have flowed out into large bodies of water, probably evolving seas. This clearly tells us that there were large bodies of water on the surface of the Earth as long as 3.5 billion years. The ultramafic complex occurred beneath them, and they are overlaid by thick sequences of partly pillowed basalts, almost identical in composition to the basalts in the present ocean floors and in the roots of contemporary island arcs.

The Time Between

But what of the geologic record between 2.5 billion years ago and several hundred million years ago? In this vast span of time, some 2 billion years long, relics of ultramafic lavas, sea floors and island arcs are uncommon. There is no clear evidence of the rifting of most continental masses, or of the evolution of new ocean basins and chains of island arcs. Instead of great greenstone mountain belts, we find most of the great mountains formed between 2000 million years and 200 million years ago were formed by subordinate rifting of continents. Commonly they formed as great fold and fault zones upon, or between, *closely spaced* continents or protocontinents as indicated by their constituents, more granitic igneous rock types, and wedges of sediments eroded from and deposited upon, more granitic, not ultramafic to oceanic crust.

Now let us reexamine the interesting example illustrated in Fig. 15-2. Specifically the mashed and folded patterns of the Archean and successively younger mountain belts on a worldwide basis, if we recluster the continents into the supercontinent of Gondwana. This reclustering as illustrated in Fig. 15-2 clearly suggests that North America, Africa, India, Australia and Antarctica were clustered together 200 million years ago as the supercontinent Gondwana. It also tells us that the Archean mountain belts formed a vast subparallel mountain system, clearly the largest and most widespread mountain belts known on the Earth. One of the most interesting characteristics of this pattern is the enormous width, as well as length, of the Archean terrains in which these mountain belts occur. Note in Fig. 15-2, that in our reconstruction of the present continents into Gondwana, we must move India, Australia, the Americas and Antarctica across the spherical surface of the Earth. When we do this we cause these continents to rotate bringing the new, very divergent Archean mountain belts into one, or two subparallel belts. This rotation is necessary in any platelike mass as it moves over a sphere.

The alignment of the Archean fold belts into a Gondwana that existed about 250 million years ago has rather shocking implications. It suggests that Gondwana had existed more or less in the form it had 250 million years ago for the preceding 3 billion years. Let us suppose Gondwana had, in that 3 billion years, been broken apart and large fragments of it had drifted widely over the surface of the Earth, reclustering themselves into new continents. It seems too coincidental to conceive that in all these times of intermediate drift and collision, younger fragments could have rebroken and navigated back home again to proper positions in Gondwana; positions such that all of the Archean belts could be lined up. Such post-Archean, pre-Permian movements would imply that the continents had some skillful navigator who knew how to steer them about, and some very brilliant coordinator and dispenser who knew how to break apart each intermediately formed pattern of continental fragments and return them, fitting them again into the patterns of Gondwana that existed in the Archean.

The problem becomes even more complex as we look at the orientations of the mountain belts formed in intermediate periods between the Archean 2.5 billion years ago and the Paleozoic, 300 to 600 million years ago (Fig. 15-2). There are, for example the so-called Grenville belts, dated approximately one billion years in age, and the mountain belts formed at about 1800 to 2000 million years ago. Each of these belts also forms a coherent pattern in Gondwana when the Americas, India, Australia and Antarctica are refitted into Wegener's Gondwana.

The possibility that all of these beautiful alignments is simply accidental seems incredible. We are thus confronted with the possibility that the large-scale refitting and drifting of large fragments of Gondwana and any other megacontinent that occured 200 million years ago is relatively unique in geologic time. Prior to that Gondwana must have existed as a coherent entity, a megacontinent, subjected to innumerable mountain-forming episodes.

The Southward March

There are many other fascinating relations in conjunction with Gondwana, involving the mountain-forming episodes and the motions in the inner Earth that have caused the mountains. Note in Fig. 15-2 that, in general, the mountain belts decrease in age from the interior of Gondwana southward to its edge. This suggests that together with the parallelism of the belts, the motions in the Earth's mantle that made the mountains, maintained at least crudely the same orientation relative to Gondwana through 3 billion years of geologic time. Also that the central part of Gondwana was thickened and deformed first. Most subsequent deformation of Gondwana occurred along its southern parts, each mountain-forming episode thickening and refractionating Gondwana, and then consolidating it into a more stable shield region.

Such processes are readily visualized if we think that mountain-forming processes have been adding additional increments of more granitic material to the continent and thickening it in the process, and also refractionating it as suggested in Fig. 15-5. That the margins of the continents are more susceptible to deformation and mountain formation than the interior is, in effect, what we see today in the circum-Pacific plate tectonics. Most of the deformation throughout the circum-Pacific is at and within 1000 km of the continental borderland. It is here that the spreading Pacific sea floor is consumed. It is in these peripheral regions that the Pacific sea floor is thrust under Asia, Australia and the Americas creating in the process mountain-forming episodes very analogous in most respects to those we reconstruct for the Archean.

In Between

But *in between* the great analogies in Earth history tend to blur. Archean and post-Permian periods of geologic time show very similar crustal environments and very similar types of ocean floor, island arcs, interarc basins, continental borderlands and similar types of mountain belts. In contrast, however, between 2.5 billion years ago and 200 million years ago, the kinds of mountain belts, and the most common kinds of rock formed, are strikingly different. These differences offer a further powerful argument in defense of the interpretation that Gondwana and perhaps most of today's continents existed prior to 200 million years ago as one or at most two great megacontinents. For it was on the thinned interiors and peripheries of these continents that the mountains of intermediate ages were built, not in rifted ocean basins and interarc basins of the Archean and post-Permian type. It is impossible to summarize briefly these substantiating lines of evidence which further suggest fundamental differences through time in the nature of continental rifting and drifting; for we are talking of processes and rock products formed in over half of all geologic time. That is a subject in itself.

ALBERT E. J. ENGEL, CELESTE G. ENGEL

Bibliography: Engel A. E. J., et al., *Pre-permian global tectonics,* in Bull. Geol. Soc. of America, LXXXIII, 2325 (1972).

AFTER having seen how and why the continents move today, and after having seen their movements in the past described in the articles by Smith and the Engels, we note that two fundamental elements are still missing to complete our account of Earth's long history. These are the scenario and the actors on the stage. The next two articles deal with these aspects: the evolution of the climate and the organisms that lived upon the drifting continents. Nothing is more evanescent than a puff of wind or a shower of rain. Once they are over it seems that only human memory or instruments could bring them back to life. This, however, is not so: the Earth has highly accurate ways of recording all the events that take place on its surface. The puff of wind raises sand and dust and carries them from one place to another where they accumulate to form characteristic deposits; the shower of rain bites deep into a mountain, tearing away particle after particle of the material of which the mountain is formed, and creating once again deposits of an unmistakable kind. Thus the whole history of Earth's climate is written in rocks, and the only problem is learning how to read them.

The article by Rhodes Fairbridge which follows, first of all outlines those ways of reading rocks which form the basis of the science that reconstructs the character of the climate in the past (paleoclimatology). It then moves on to set out in chronological order, from the origins to the present day, the fundamental climatological events of the entire history of the Earth. It is a completely new vision because, for the first time this history can be reconstructed not in an abstract fashion but on the basis of the variable geographic features described in the preceding articles.

Within such a framework, even tales that have been told over and over again, such as that of the triumph and fall of the dinosaurs, or that of the ice ages, take on a new aspect and a new meaning.

The birth of terrestrial reptiles is closely linked to two other histories which, as a rule, have always been dealt with as if they were separate and independent. Between 350 and 250 million years ago the maps presented in the article by Smith show three great continents: to the south the gigantic continent of Gondwana including present-day Africa, South America, the Antarctic, India, Madagascar and Australia; and two others, much smaller, in the tropical-equatorial belt: on the one hand the block formed by North America and Europe, and on the other hand the Asiatic platform. During this period part of Gondwana was covered by gigantic glaciers while elsewhere the coal deposits were being formed that are largely responsible for the industrial revolution which had its roots in Europe; meanwhile the first reptiles began to multiply on the continents. Fairbridge relates these events to one another. The alternate advancement and withdrawal of the glaciers on the southern continents led to a rise and fall of the sea level which was in any case higher than today thanks to the different distribution of the continents, vast areas of which were covered by shallow water. This was ideal for the development of great equatorial forests which were subject to periodic flooding, and such events were responsible for the formation of coal. In these same forests land-based reptiles found a suitable ecological niche; here they increased and multiplied, spreading to all four corners of the Earth until, around 230 million years ago, the three continents joined together to give birth to the single great continent of Pangea.

RHODES W. FAIRBRIDGE

Teaches geology at Columbia University in New York. He was born in 1914 in Pinjarra, Australia, and was educated there and later in England and Canada. He has developed an extensive range of interests that extend over all areas of geology, oceanography and archeology. This has enabled him to examine closely an area in which the knowledge and the capacity to synthesize original data from various disciplines is indispensible to a study of paleoclimatology—that is, the reconstruction of the Earth's climate in the past, and the means of achieving it. He was among the first to integrate this reconstruction with that of continental drift, confirmed at the end of the 1960s.

RHODES W. FAIRBRIDGE

The Earth's Climate and Environment Through 4.5 Billion Years

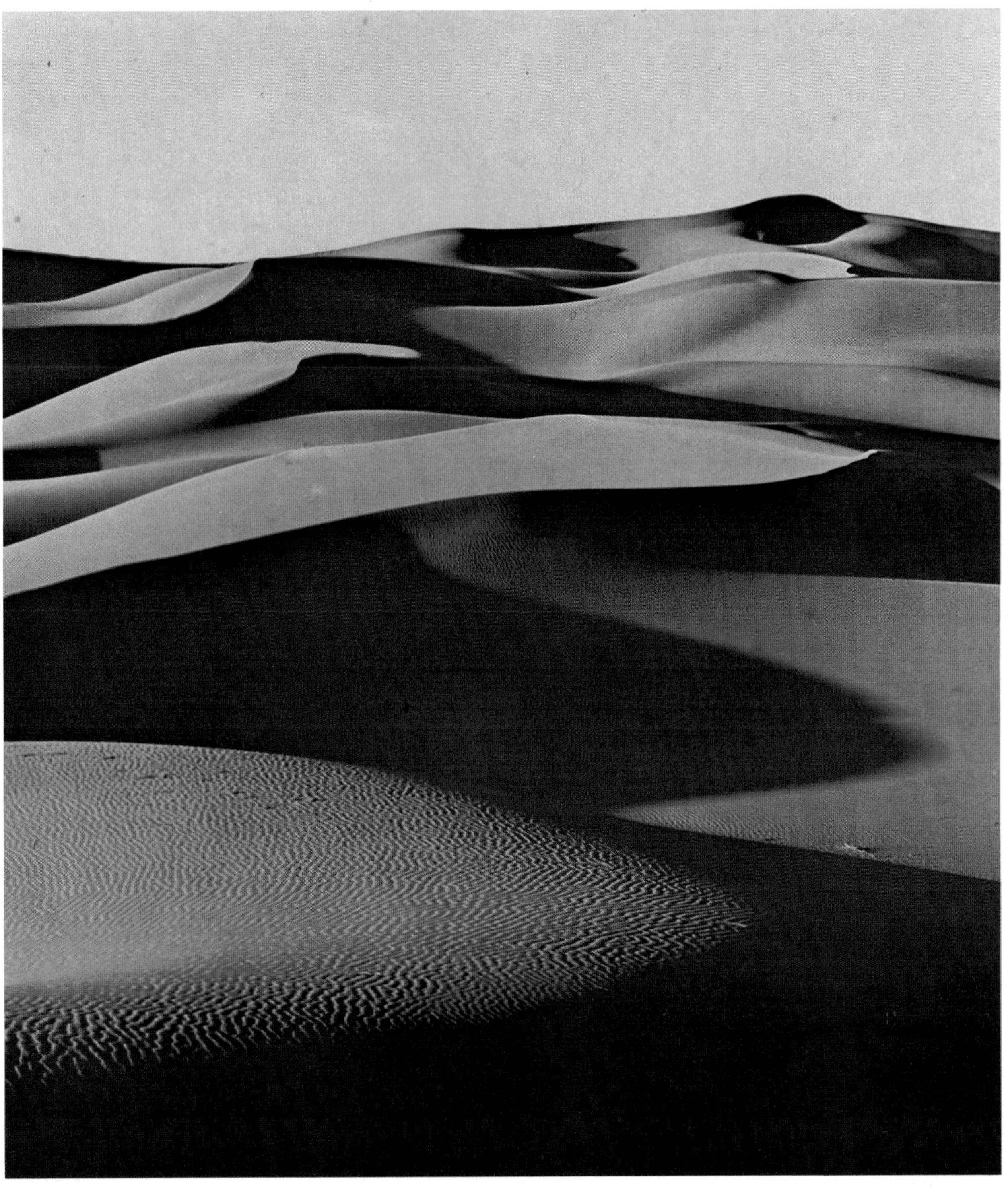

Overleaf: Sand-dune crests in the Sahara. Nowadays these structures are only formed in very precise environmental conditions and there is no reason to think that things were any different in the past. Present-day dunes contain very precise physical structures which enable us not only to recognize them independently by their outward shape but also to draw conclusions about the orientation and persistence or prevalence of the winds which have formed them. The same structures have also been identified in very ancient rocks, which thus represent a guide for reconstructing the environmental conditions of a given period. These and various other formations now enable us to reconstruct, at least along essential lines, the evolution of the complex of environmental conditions and, in particular, the climate in the remote past. (photo N. Cirani).

How does the historian reconstruct classical history? Largely from written documents. And how does the prehistorian do his work? He uses flint axeheads and arrow points, or the relics or ruins of old campsites and settlements. In both cases a *picture is reconstructed* verbally or by artistry, of a former time, its population, its ecology, the culture and the scenery.

But how does the geologist reconstruct his pictures of remote antiquity—not just over a few thousand years, but over many millions? He hammers rock; he digs; he drills deep wells; he employs geophysical instruments. Eventually he can collect sufficient data with which he can draw detailed geological maps that depict a three-dimensional picture of the rocks of the Earth's crust. Since geological maps indicate by color codes the ages of the rocks as well as their lithology, the map becomes a four-dimensional object, time being deduced from a reading of the code.

This complex documentation tells the casual observer nothing. What do we really know about the life and times and geography of the geological ages themselves that are represented in front of us by rocks, by minerals, by fossils, and synthesized for us on these beautifully colored but speechless maps?

In the Book of Job in the Old Testament one may read "Speak to the Earth and it shall teach thee." To the trained geologist, the stones and the fossils do indeed speak. At the present time a great deal of attention is being given by geologists to the reconstruction of logical coherent pictures of each stage of the Earth's history, descriptive and pictorial renderings of the paleogeography of any given moment in our immensely long evolution.

The science of paleoclimatology is one of the earth sciences, usually studied by geologists, but also to some extent by petrologists, botanists and zoologists, and sometimes also by physicists and meteorologists. Just as modern climatology relies on the statistical analysis of the direct observations and instrumental methods of meteorology, the study of ancient climates calls for the use of statistical analysis of indirect deductions from "climatic indicators." For example if one finds fossilized palm leaves in Greenland, one concludes that the climate was formerly warmer. If one finds reindeer bones in a cave in central France, one concludes that the climate was once more rigorous.

These qualitative deductions can be made precise by various refinements. For example, by comparing the ancient with the present ecological ranges of animals and plants, best grouped into assemblages, and handled by statistical methods of computer analysis, we can establish the mean annual summer and winter temperatures as well as the approximate seasonal precipitation.

Atmospheric pressure of the past cannot be established, but ancient wind systems may be demonstrated by structural analysis of former sand dunes, ripples and sediment textures. In this way the approximate centers of former anticyclonic cells together with the regular storm tracks may be identified. In desertic regions, a very characteristic distribution of longitudinal ("seif") dunes helps to identify the transition between westerly and trade winds (today, about 25°). The dunes in the northern Sahara, for example, advance from WNW towards the ESE, then gradually swing to S and then SW or WSW, describing a gigantic "hair-pin" swirl. The mean bisectrix of this swirl is sometimes called the "wheel round latitude," and if established for an ancient geological period, firmly demonstrates the tropical zone of that time.

The soil scientist is particularly interested in ancient climates because many of the processes of soil evolution, once established, can never be totally effaced. A soil is the complex product of physical and chemical weathering of the surface rocks or alluvial accumulations. Three specific trends are oriented according to latitude (or altitude): (a) *frost action* (cryoturbation gelifraction and solifluction), freeze-and-thaw mechanisms, which increase towards the poles in midlatitudes, but decrease once more in regions of permanent freezing; (b) *leaching* (chemical breakdown and solution of minerals), increases towards the equator because under humid conditions it is fostered by biological action that is positively orientated toward warm-wet latitudes; (c) *evaporation* (involving an upward, capillary motion of solutions and precipitation of chemical soil crusts) processes that relate to regions which have, at least during one season of the year, a positive evaporation/precipitation ratio, and thus afford an excellent indication of subtropical Mediterranean and Savanna type climates.

Soil is a dynamic medium, and with every change of climate a new biochemical and geochemical regime is initiated. But the reaction is often slow, so that the traces of the more stable minerals or structures introduced during a former regime is often "inherited" in the younger soil. The shapes and convolutions of a former termite nest or termitaria (favored by the humid subtropics or tropics) may be found in a soil which today is in a region of total aridity where termites cannot survive. These ancient soils are called paleosols and are tremendously helpful for identifying former climates.

In some parts of the world, e.g. West Africa, parts of South America, Australia and India, a special type of paleosol quite dominates the scenery. It is the so-called cuirasse de fer, ferricrete, or ferralite, an ironstone crust that developed during three distinct cycles: warm-wet "lateritic" (iron-rich) soil development; followed by a period of strong evaporation, concentrating the iron solution (as sesquioxides) near the surface as a massive crust, sometimes 1–3 m thick; and now today often seen either in places where there is no rainfall at all, or in very humid areas, lacking a season when vigorous evaporation takes place.

Another somewhat analogous evaporation (paleosol) process develops a calcareous crust, or calcrete in mediterranean latitudes; the Arabs call it "nari," the Spaniards and Americans "caliche." A third example, particularly notable in Australia, produces a paleosol crust of silica, silcrete (the famous opal gemstones of Australia are a biproduct of this paleosol process). A fourth, and economically very valuable product, is an aluminum-rich crust that is largely constituted of the ore mineral *bauxite;* small patches occur in southern France (formerly tropical), and elsewhere in Europe, but vast areas of it occur in the Guianas of South America, West Africa, the West Indies (Jamaica), and northern Australia. Yet a fifth, and also very valuable example of this crust-forming process under now past climatic conditions, is the nickel-rich crust that is responsible for the important nickel resources of New Caledonia, Cuba and the Philippines.

So, the sands of ancient deserts and the soils of the past tell us much about former climates and environments. But we cannot always discover these indicator rocks and paleosols. The role of chance in geology must be considered. The soils and sands of the land surfaces are the relics of an age most probable to be destroyed by subsequent erosion. Where then do we look? To the deposits of formerly marine sediments. These were laid down, grain by grain, on ocean floors, to become securely buried for long periods of time. They also contain evidence of old climates and environments: sands or dusts from deserts, "red beds" from tropical soils, ice-rafted debris from glacial centers; in constricted basins like the Mediterranean, there will be salt deposits to tell of periods of isolation and evaporation.

The new theory of plate tectonics tells us how ocean floors and their sediments slowly evolve, migrating apart, until at a critical revolutionary moment in geologic time, the crust between the ocean floor and continental margin cracks and equilibrium is shattered. The oceanic plate is usually overridden by the continental plate and the marine sediments disappear from sight, to be consumed, melted and "recycled" as granites and volcanic lavas. All traces of their former character and their buried fossils are now destroyed. Fortunately for us, not all of the marine sediments are consumed during this so-called subduction process. Some of the sediments are buckled up, folded and thrust up into mountain chains by the advancing continental mass. Here, then, in the deep valleys and mountainsides we may find the records that we have been seeking. The rocks are usually somewhat hardened and distorted, but the original material composition can be traced and the fossils of former flora and fauna can be patiently extracted. Now the paleoenvironmentalist and the paleoecologist can start their work.

The practical scientist works out many specialized techniques and makes numbers of assumptions that he constantly reviews and tests. Over the course of decades, some of his ideas change, but certain guiding principles do not change, and basic to them is the concept of actualism. We judge the past by using powers of human reasoning based upon evidence of actual experiences, experiments and processes that we observe today. We do not resort to the miraculous in geology—or we have not done so since the pre-Darwinian epoch, the days of Cuvier and Buckland of a century and a half ago. Actualism does not assume by any means that the world has always been exactly as it is today. Far from it, but changes have for the most part been evolutionary, only occasionally revolutionary. And always the change from one environment to the next follows strictly logical reasons of cause and effect.

Let us now review the history of global environments, as we read the story today. It must be constantly recalled, in all modesty, that man is not immortal and he cannot see through all the mists of time. Nevertheless, it is still one of the near miracles of man's triumphant march into science that we can indeed build up a picture of the ancient world that seems to meet most of the logical demands of our contemporaries and is susceptible to scientific testing and predictions.

The Dawn of History: 4.5 to 3.5 Billion Years Ago

According to modern theories of cosmology, the astronomic observations of other solar systems should provide a reasonable model for the origin of our planet. Generally accepted is the concept of a cold accretion of planetary dust, particles and asteriod-sided lumps of matter that collectively contained all the chemical ingredients commonly found today in meteorites and within the Earth itself. Our first problem is internal heat. Certainly a great deal of heating must be expected. Although the primeval dust cloud may have been cool, probably colder than −200° C, the multiple collisions and the trapping of short-lived, high energy radioactive particles must have heated the evolving planet. Then, either the Earth warmed rapidly, melting all components and then began to convect, slowly losing heat once more, or it warmed slowly, progressively developing heat from the interior.

The first model, the "quick-boil" concept, would be expected to leave a slag, such as develops in a steel blast furnace as a surface crust—the beginnings of the planetary crust as we know it today. The second model would not need to produce through melting, but rather melting by stages in response to the critical temperature/pressure relationships, while the crust would be of primary material intruded progressively from rising plumes of lighter molten ingredients from the furnaces below.

The oldest rocks we know are the Archean, and isotopic dating techniques have shown some of them to have been stabilized as far back as 3.7 billion years ago. Rocks of this age, or possibly even greater ages exist in Greenland, Russia and Antarctica. Very similar dates are reported from the continental nuclei of the whole world. It is evident, therefore, that the groundwork was laid for all of our continents during the first 1 billion years of geologic time. What do these oldest rocks tell us? First they contain certain conglomerates, stones rounded by rolling in streams or along the seashore. This tells us a great deal. There was already a hydrosphere. The water was in its liquid state, not ice or vapor. So the Earth was cool, or warm enough so that we can say thermal conditions must have been not too different from those of today—say within an average temperature range of 25–50° C.

What of the atmosphere? Some of the oldest conglomerates include pebbles of the minerals pyrite and pitchblende. These consist of compounds that are unstable in a moist, oxygen-rich air. Yet, they did not break down. It seems likely that there was a *reducing atmosphere* i.e. one free of oxygen. The geochemists have made experiments to demonstrate that the *first life* on this planet could only have evolved in a reducing atmosphere and that it consisted largely of hydrogen, helium, carbon dioxide, ammonia and methane. Now, H and He are light elements and would rapidly drift off, away from the Earth, because of the solar wind and their high velocities, and this would lead to eventual depletion of the unbound atoms of these elements. Then ammonia and methane are not stable without the hydrogen in the air, and these two would tend to dissociate, liberating further hydrogen.

First Life: About 3.5 ± 0.3 Billion Years Ago

In the early Archean (Precambrian) rocks we see clear evidence of erosion and sedimentation which together require a kind of landscape such as that we see today. But initially it was lifeless. Numbers of experiments have shown that under the energy of the Sun, possibly with the aid of thunderstorms with their electrical discharges, a synthesis of the primitive atmospheric molecules, in the presence of water that was evolving from below through hot springs and volcanoes, would lead to large, self-reproducing organic molecules. Thus was paved the way to the first life: *Revolution I* in Earth history.

Traces of bacteria are found in the Swaziland group in South Africa. More have been found in other places. The first megascopic life traces are the stromatolites that mimic some present-day formations found as far apart as Florida and Western Australia. These are lime crust that develop on tidal flats during the growth of primitive marine algae (see Glaessner, this volume). While the bacteria have a wide ecologic range, the environments favorable to modern stromatolitic algae are in the range 25 ± 15° C in shallow ocean water, not too saline, not too fresh. So we are gaining a picture of Archean times. The mineral types in the sediments of the day suggest a barren landscape without much life, subject to rainstorms, flood, river erosion and sedimentation. Volcanic rocks were more prevalent than today, but apart from that the picture is not unlike parts of the Sahara today.

The earliest land areas did not include high mountains and broad plateaus as today, but rather consisted of large numbers of small nuclear islands within a broad, shallow world-encircling ocean. These were no great ocean deeps. Climates were maritime and continentality effects were minimal, although day/night contrast was greater than today.

First Oxygen

About 2.9 ± 0.2 billion years ago, with the evolution of photosynthetic algae, a carbon dioxide rich atmosphere began to receive oxygen, the "waste product" of photosynthesis. This was *Revolution II* in Earth history. From the point of view of man and the animal world, this was almost as important as life itself, for without oxygen they could never have evolved.

Initially all the oxygen produced was immediately captured by chemical bonds because there were many reduced (oxygen-poor) chemicals in the early soil and water of the infant Earth. Iron, for example was mainly in the ferrous state and highly soluble, so that the early ocean was rich in dissolved iron. Liberation of oxygen began to produce ferric iron Fe_2O_3) which precipitated, in the middle Precambrian, a tremendous worldwide formation of iron ore deposits, largely in the form of Fe_2O_3. It is significant that all of the really enormous iron ore reserves of the world come from this specific time in Earth history. Certainly there are younger deposits, but they never again reached such a gigantic development.

Free oxygen evolved only slowly at first and not until about 1.6 billion years ago did the O_2 level reach 1% of its present level. In the meantime, water vapor and carbon dioxide continued to build up the atmosphere. As oxygen molecules began to appear they tended to be split up owing to the strong ultraviolet radiation from the Sun and to recombine as ozone (O_3). The land surface would have been fatal for any life. But gradually with the water vapor, CO_2 and oxygen a "greenhouse effect" was initiated that did two things: first it screened the land surface from potentially lethal radiation; second it helped to create milder climates. The second Earth Revolution thus saw the complete replacement of a primeval atmosphere by a new one that has gradually evolved through time.

One thing is of fundamental significance for anyone who wishes to understand and legislate sympathetically and knowledgably for the preservation of our existing ecology: the plants of this Earth that created the free oxygen. Prior to this gradual appearance of free oxygen there was no possibility of animal or human life.

First Mountains: About 2.5 Billion Years

As explained above, the first crust of the Earth supported low islands of continental nature within a broad shallow sea. The initial crust was rather thin and not very stable. It could not have supported high mountains or great plateaus, but there were innumerable small volcanoes. Scientists who specialize in the Precambrian Era (4.5 to 0.6 billion years ago) recognize an important change that occurred around 2.5 billion years. The older rocks, the Archean group, are associated only with the earliest crust. Subsequent to that time, the Proterozoic interval, troughs began to form along borders of the land masses and ringlike sedimentation sites evolved. These troughs have great linear extent, but are rarely more than a few 100 km across. They have become known as the geosynclines. Since long continued sedimentation in such troughs leads eventually to an unstable situation, the crust tends to fracture and buckle here. The overriding of one crustal segment by another generates complex structures, and possibly a doubling of the original thickness of crust, so that uplift results. Because of the hydrostatic nature of the Earth's outer skin (owing to its visco-elastic state), any doubling of thickness most eventually cause uplift. And following uplift we see erosion: valleys and peaks are sculptured and we have mountains.

One of the most remarkable features of mountain-building in geologic history is that it is definitely episodic. There have been many long, relatively quiet times during our protracted global history, interspersed by crescendoes of activity, marked by volcanicity, crumpling and mountain-building—orogeny. The discovery of "absolute" dating techniques (utilizing radioactive disintegration rates), produced statistical proof that there was indeed a long, rhythmic periodicity of Earth movements first recognized by Arthur Holmes of Edinburgh and carried further by Gordon Gastil in America. Once poetically described as the "mighty heart-beats of the Earth Mother," this periodicity is approximately 200 million years which is about the same as the period of the rotation of our galaxy, but no convincing proof of correlation has yet been offered. Nevertheless, it is tempting to speculate that our Earth moments are triggered gravitationally by our cosmic motion.

At the present time, there is a fundamental disagreement among geologists about the cause of Earth movements. One school, the endogeneticists, claim that heat flow from the Earth's interior must lead to massive internal convection currents that inevitably lead to coupling with the surface crustal plates causing their migration—as observed by continental drift. The other school, the exogeneticists believe that a steady-state convection would lead to continuous sea-floor spreading, continuous drift and continuous orogenies. But that is not so: spreading, drift and orogeny are all episodic and more or less cyclic. An external triggering, such as by solar system or galactic tides, would provide a more acceptable model. The two models are hard to separate, because undoubtedly, if the episodic stress peaks cause an occasional opening of fracture belts in the crust, one must expect an upsurge of heat and melted materials. How does one distinguish this upsurge from the rising "plume" of a hypothetical convection current?

We have something more tangible to discuss regarding the first major glaciers and icefields, and the first mountains. With small, low continental slabs of crust surrounded by the world ocean, the global climatic picture would be maritime. Under extreme oceanicity, there can be no ice ages. Without high mountains there cannot even be small mountain glaciers. In the geologic record of the Proterozoic there are numerous tillites. These are sediments that are composed of a random

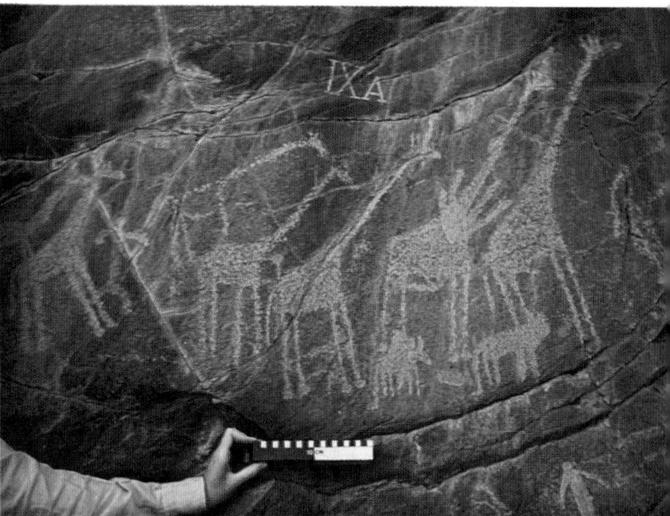

mixture of all sorts of rock types, unsorted by water, that characteristically are dumped by the melting ice of a glacier. The boulders are distinctively striated by the unique signature of the ice. Other slow, dragging processes in geology, e.g. landslides, also produce scratches, but their form and "signature" differ. The trained geologist can therefore recognize traces of ancient glaciations with a high degree of certainty.

Many of these glaciations occurred during the Proterozoic, evidence indeed of cold phases and polar land areas. The tillites occur only as relatively rare, periodic intercalations within which sequence of formations, e.g. the stromatolite "reefs," are contained subtropical indications. Examples of these glacial episodes are found today on *every single continent*. What does this mean? If the orogenic episodes of the Proterozoic are cyclic, is it not probable that the glacial phases are similarly episodic? This seems to be born out by the field evidence. But why are the tillites world wide? Was the glaciation global in effect? Surely that would be very hard to visualize in the equatorial belt. A more reasonable approach might be to imagine the beginnings of plate tectonics back in the early Proterozoic. With each break-up (taphrogenic) phase, seafloor spreading would be initiated and continents would have drifted to different latitudes. Proof of major changes in paleolatitude is strongly suggested by initial studies of paleomagnetism of Proterozoic rocks from different parts of the world, although the data at present are so sparse that we cannot work out any convincing picture of the general geographic relationships.

One of the clearest and best preserved representations of Precambrian glaciation is one that occurred close to its upper boundary, around 700 to 600 million years ago. This is often called the Eocambrian Glaciation, since it occurred shortly before the beginning of the Cambrian period. It occurred in every continent. First discovered in Norway and then elsewhere in northern Europe, it was later picked up in Australia, in China, in Spitzbergen, Greenland, in North America, and most recently, a very extensive development has been found in West Africa (almost coinciding with the distribution of the famous Ordovician glaciation that was to follow 200 million years later).

Clearly, if we accept climatic zones approximately as they are today (and every other geologic indicator suggests this was true), we must postulate fundamental and far-reaching notions of the various Eocambrian continents. This is of critical significance for the theory of plate tectonics, inasmuch as there are certain specialists who submit that only the plate tectonics of the last 200 million years have any sizable dimensions. Any earlier events, they say, must have been limited to quite miniature displacements. With that theory there is now clearly some solid opposition from the paleoclimatological viewpoints.

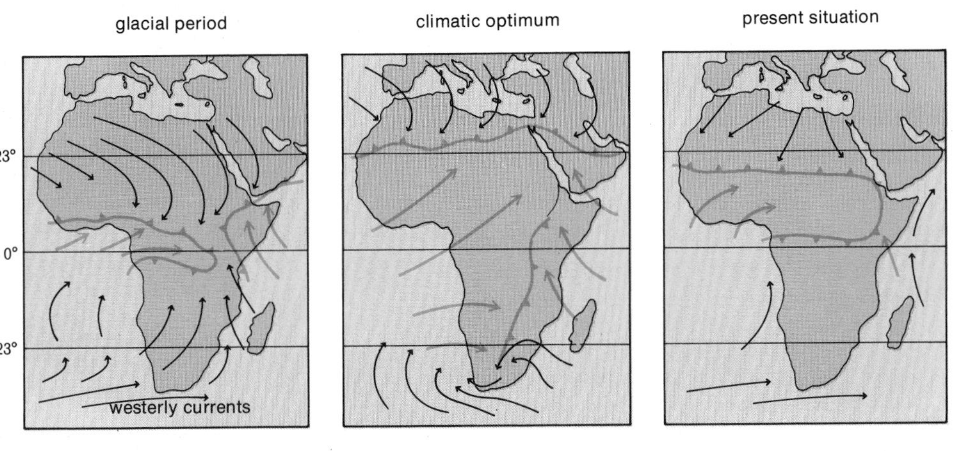

Fig. 16-1. On this page, Africa in three very different climatic conditions: on the left, the situation during the Quaternary glaciations, around 20,000 years ago; in the middle, during the climatically very favorable period which occurred around 6000 years ago; on the right, the present-day situation. The black arrows indicate the orientation of the prevailing winds: the westerly currents at the northern and southern tips and the tradewinds between the two tropics. The toothed line in color suggests the invasion limits of wet air carried by the monsoons (colored arrows). On the opposite page, a tale about the climate in the past: a drawing of giraffes made 5000–13,000 years ago at the Second Nile Cataract, where the climate today is too arid for these animals.

Paleoclimates

At some unknown time in the Proterozoic the first animal life began to evolve. Clearly a certain minimum oxygen level would be required, but the precise time and circumstances are elusive. All that we know is that by Eocambrian times a large and sophisticated marine biota of soft-bodied organisms had evolved. Their impressions and trails are widely distributed and are now well studied. They tell us a good deal about the environments: oxygen present—temperatures generally equable (say 15° ± 10° C)—marine conditions comparable with those today.

There is a popular theory that during the Proterozoic the moon was much closer to the Earth, which would mean a very short day, a high rate of rotation and immense tides. The sediments of the time should contain evidence of vast submarine sand dunes created by these violent tidal currents: there are none that we have been able to find. There should be belts of catastrophic scouring of the sea floor: we have not seen them. In contrast the stromatolites speak for relatively calm shallow-water conditions such as permit comparable structures to form today in Florida. The fossil impressions described by Glaessner (see this volume) are of soft-bodied organisms washed up onto sandy mud flats, but giant tides would have destroyed them. We suggest the astronomers should think again.

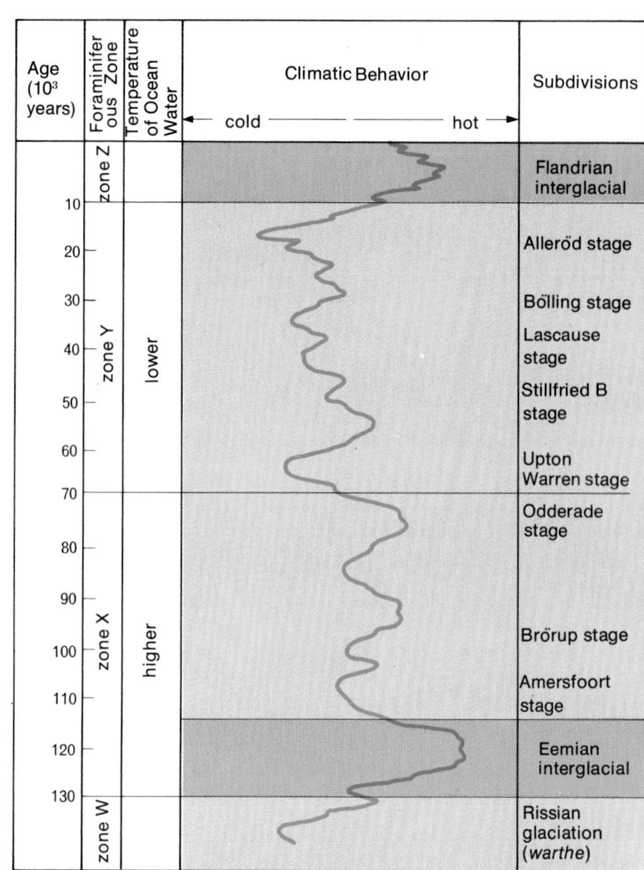

First Sea-Shells: c.570 Million Years Ago

Following shortly after the Eocambrian Ice Age came a long period of calm, mild conditions that saw one of the most exciting and obvious revolutions in Earth history—the first sea-shells: Revolution III. At this time, the *Cambrian Period,* there sprang into view the evidence of hard-shelled marine organisms, in great variety and complexity, ranging from the molluscs to the trilobites. These creatures were not simple unicellular blobs of primeval protoplasm but viable, sophisticated organisms complete with complex digestive systems, limbs, prehensile organisms, sexual equipment, nervous systems, eyes and other sensory cells.

The almost "instantaneous" appearance of this complex biota has astonished geologists and biologists for a century and more. Today the most likely explanation is partly climatic, rising temperatures; partly geochemical, rising alkalinity of the world ocean (making $CoCO_3$ excretion difficult); and partly biological: increasing population-pressure and predation which led to Darwinian selection of those organisms that developed hard carapaces.

From the end of the Eocambrian, throughout the Cambrian and until near the end of the Ordovician period there seem to have persisted generally mild to warm conditions on Earth. The near-universality of the major organisms, such as the trilobites is evidence of world-wide favorable conditions. Of the land plants, however, we know nothing. There were widespread carbonate-rich deposits (today only common between latitudes 30°N and 30°S). There were extensive stromatolites and spongelike organisms *Archaeocyathainae* that formed reeflike banks. In appropriate latitudes important evaporating basins permitted very thick deposits of rock salt and related minerals to accumulate. The general picture is for a long warm interval.

The early Paleozoic Era (Cambrian/Ordovician) closed with another great ice age. At around 450 million years ago, that is, 200 million after the Eocambrian refrigeration, continental ice conditions returned, and on a great scale, lasting probably at least 10 million years. At this time the African continent came, by pole motion and continental drifting, to lie in the south polar region. South America was then joined to western Africa, but most of Europe, Asia and North America were separated from Africa by a broad seaway sometimes called "Paleo Tethys." The glacier-covered regions were probably as large as modern Antarctica if not larger. The records are mostly clearly preserved in the hyperarid regions of the Sahara—in Algeria, Morocco, Mauritania, etc.

First Trees: c.400 Million Years Ago

Following upon the Late Ordovician "Sahara" Glaciation, the world saw a return to almost universally mild to warm con-

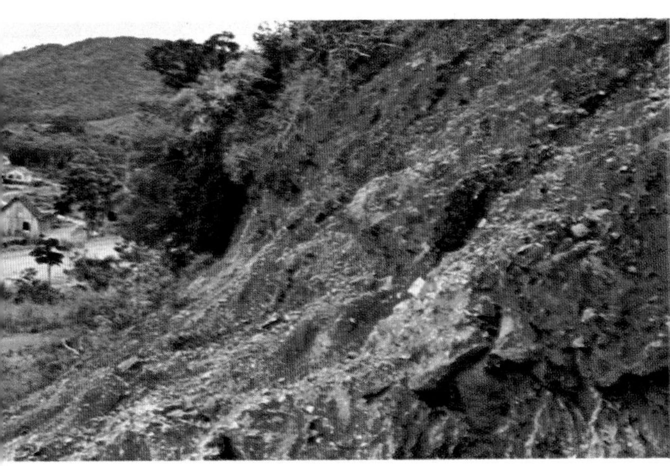

Fig. 16-2. The diagram shows the climatic variations relative to the last glaciation. The curve indicates temperature behavior related to average present-day temperatures which are used as a point of reference. Note how the last Interglacial (Riss-Würm or Eemian Interglacial) is followed by about 45,000 years of climatic oscillations characterized by marked advances of the ice over dry land but without any correspondence with an equivalent drop in the temperature of the oceans, as is indicated by the warm-water foraminifera in Zone X. Not until the second half of the glaciation did the ocean temperature also drop (Zone Y of the oceanic foraminifera). The photos on the page opposite and on this page present geological evidence about the climate of the recent geological past. On the opposite page, below, an erratic mass carried a great distance by a glacier 12,000 years ago (Quebec City, Canada). Above, evidence drawn from nature and the type of alteration of a silty-pebbly deposit (Florianopolis, Brazil): the deposit has become lateritized in today's wet, hot climate, but itself has features which require semiarid conditions in order to form. On the right, a loess deposit, an accumulation of windborne particles, evidence of the violent duststorms which must have raged over the nonglacialized part of the continents during the last glaciation. Below, left, the odd shapes of the Corsican *tafoni* near Filitosa, typical of the conditions which occurred in the Mediterranean belt during the glaciations. Below right, a series of high fluvial terraces on the Nile which are evidence of phases of intense fluvial sedimentation due to a reduced carrying capacity during the glaciations.

ditions. In this favorable setting the arthropods were replaced by the fish as the major denizens of the sea. The coral began to develop from a solitary polyp to form massive colonial associations, that with hard calcareous frameworks could rapidly build up fortresslike reefs, the bioherms, which by providing shelter and feeding, themselves created local environments favorable to a vast variety of symbiotes and associated creatures. The Silurian and Devonian reefs are almost worldwide in distribution so that any continental drift reconstruction does not materially alter their statistical scatter around the globe. In the middle latitudes wide spread rock salt and related evaporite deposits were laid down.

On land the great gesture of the age was the appearance of the first big nonaquatic plants. They were mainly of the primitive type—palms, cycads and so on—but they grew to become major trees. Fossils of this ancient vegetation are widespread and some trees grew up to 30 m in height. The forests in turn gave shelter (in the way that reefs did in the oceans) to a varied biota sharing this new habitat that now began to evolve.

Perhaps the most surprising feature of this scene is that as soon as the environment provided trees, evolution produced the flying organisms, the beetles, moths and dragonflies of the day, that could fly into those trees and presumably utilize their protection and exploit their flowers and fruits. Without predators, some of these flying organisms grew to giant sizes; fossil dragonflies with a 60 cm wing-span are known.

This mid-Paleozoic era, which included the Silurian and Devonian periods, was evidently a phase of generally mild climates, but continents were already on the move. Intercontinental collisions and plate-margin subduction led to tremendous mountain-building along certain belts—the most important being the so-called Caledonian Belt across N.W. Europe and eastern North America. This process of Ordovician to

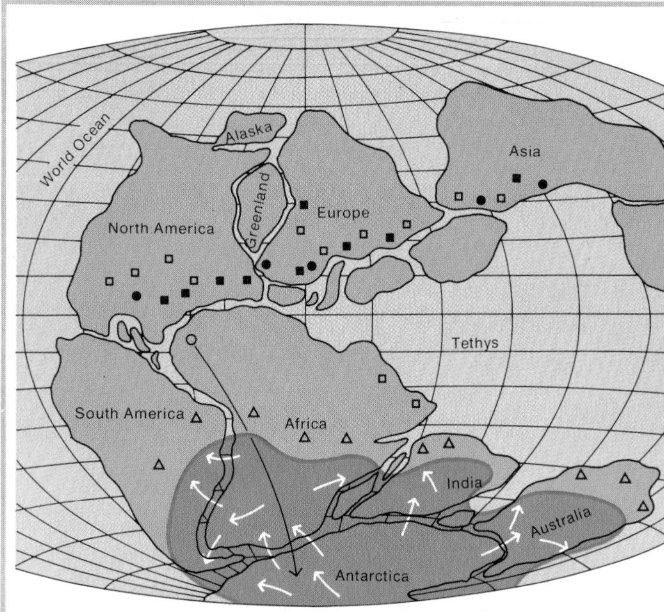

Fig. 16-3. The diagram above shows the now classic reconstruction of the geographical appearance of the continents in the Permian, some 230 million years ago. The group of southern continents, for a long time joined together to form the supercontinent of Gondwana, and the group of northern continents, called Laurasia, are, at this moment in the Earth's history, joined to form the huge supercontinent of Pangea. In this period, a great ice-mass (blue) spreads (white arrows) over the southern continents; this did not occur in the northern continents because in the northern hemisphere there are no land-masses at high latitudes. This is not the first glaciation to affect the land-masses in the South. The symbol ○ indicates the position of the South Pole about 450 million years ago and the arrow denotes its movement in time: the symbols □ indicate evaporitic deposits; the solid squares represent coral reefs; the symbols Δ indicate the deposits left by drifting icebergs. The photo on the right shows glacial structures in the Sahara which date back 450 million years.

Devonian age effectively closed the ancestral North Atlantic, creating a giant continent long recognized by geologists (with no concept of plate tectonics!) as the "Old Red Sandstone Continent." Its dry climate resulted from rain shadows and the climatic continentality factor, a consequence of the new mountains, and were also due to increased distances from the ocean. The "Old Red" basins accumulated desertic dune and lake sediments, characteristically red and isolated from the sea. Break-up of this continued towards the end of Devonian times permitting more maritime conditions to return to many of the land areas. This "continental" episode was accompanied by widespread development of evaporite lakes and lagoons, at which time atmospheric oxygen fell to a lower level than before. In general terms much of the world's land area enjoyed a warm dry climate (comparable, say, to Turkestan, in central Asia). In the oceans, separated into many provinces by the emergent lands, there was a widespread multiplication of organism varieties (though loss of *total* populations).

First Coal: c. 340 Million Years Ago

The first forests naturally grew in the dampest places—the low swampy delta areas close to the lakes or sea coasts. This has always been called the Carboniferous period, because of the widespread coal in the northern hemisphere. Continuing into the Permian period, another great ice age descended but this time in the southern hemisphere. While the northern hemisphere lands split up again after the Caledonian collisions, the southern lands (long known as Gondwanaland) had remained welded together as a single continent since the Cambrian time. In the intervening 200 million years the south pole had shifted progressively from West Africa, through Brazil to South Africa, thence through Antarctica to Australia. All these Gondwana continents (with India) were then one. Glaciation appears to have been only scattered and relatively minor at first, being limited to certain mountain ranges. But, in a cyclic way, with ever-increasing waves, the glacial phases grew greater and greater, until by the later Carboniferous and early Permian giant ice sheets were expanding over land areas ranging from South America to Australia, and including even India. The maximum accumulation of ice seems to have migrated in steps from west to east, accompanying a comparable shift of the south pole as evidenced by paleomagnetic measurements. This Permo-Carboniferous ice age evidence was first discovered by Blanford, a young geologist of the Official Geological Survey of India, over a century ago. What a dramatic discovery! At that time he could have had no inkling of continental drift or pole shift. But within a decade, comparable evidence had emerged in Australia, South Africa and South America.

At the same time as the South Pole lay over the southern continents, the North Pole lay somewhere in the north Pacific and meanwhile the North American, European and northern Asiatic continents lay scattered about as "islands" in the equatorial latitudes. The sediments and fossils tell the story. It was a junglelike environment. Dense swampy forests favored the terrestrial reptiles and later it was here that the earliest mammals were to evolve. From time to time gigantic oscillations of sea level took place, drowning the great forests and their populations, and burying them beneath blankets of muddy sediment. Thus were born the great coal deposits that we exploit today. Trapped organic material, when deprived of oxygen, tends to revert slowly to its primary nonvolatile component, carbon, with minor fractions of oxygen and hydrogen remaining in the lignin and cellulose material of the forest vegetation. This is what we now have, coal.

And what could have caused those giant inundations? It seems too much of a coincidence, that at precisely the same time as the coal swamps evolved there were occurring successive waves of glacial advances and retreats in the southern hemisphere. The sea furnishes the moisture to make ice caps and at the height of a glacial crescendo sea level will drop as much as 200 m. With the ensuing melt phase it will rise again, flooding the coastal plains that had been laid dry. The "glacio-eustatic" rise of sea-level resulting from a southern hemisphere glacier melt will drown the equatorial forests.

Thus it is that humanity, which still today so closely relies on its immense coal seams for its thermo-electric energy supplies, should thank the great Permo-Carboniferous ice age of the southern lands! In most of the northern hemisphere it was mainly Permian. (Additional coal sources are found in the Tertiary, but that is a later cycle.) The geochemists, studying the isotopes of sulfur, can now prove that during both Carboniferous and Permian periods there was a negative swing in the oxygen supply. Being a "continental" time, sea/levels were quite generally low, and outside of the glaciated regions and apart from equatorial or coastal coal-swamps the land climates tended to be extreme (hot summers, cold winters) and dry (like the Gobi Desert today), and there were appreciable evaporite-forming lakes—mainly ephemeral or seasonal.

The Mesozoic: A New Era c. 225–65 Million Years Ago

With the end of the Permian period came also the end of the Paleozoic Era, the first great era when shelly fossils and highly organized plants dominated the globe.

The first Mesozoic period is the Triassic, so called on account of the threefold division first observed in its characteristic red sediments in Central Europe—where it was first referred to as the "New Red Sandstone." The Triassic is indeed quite analogous to the "Old Red." The northern continents suffered another severe series of collisions. The Appalachian-Mauritaniede Revolutions document an Africa-American collision. Northern and southern Europe collided along a belt extending from southern Ireland to the Carpathians. Eastern Europe and Siberia collided along the Urals and have remained welded together ever since.

Again we see "continentality" as the dominant climatic feature of the time. Deserts and desert lakes began forming in the late Permian and persisted throughout the Triassic. Red colors dominate the rocks of these ages from the Rocky Mountains across Europe to northern Asia.

Against this backdrop the first of the giant reptiles, the dinosaurs, began to evolve. Their bones and footprints are often perfectly preserved in the muddy deposits of the former desert lake margins. Clearly many of these lakes were lush oases and isolated from the dangers and predators of the oceans. The dinosaurs could reign supreme. In the oceans the Triassic faunas suggest generally mild and warm conditions, with no traces of glaciation anywhere in the world. Almost the same marine organisms can be found in all parts of the world ocean. Clearly, only a universally equable ocean would afford this pattern. But the composition of this fauna is strongly different from that of the Permian. It is as if a fantastic plague had wiped out most of Paleozoic life, while completely new genetic strains suddenly emerged and proliferated. Some physicists have suggested a catastrophic dose of cosmic or solar radiation, causing irreparable gene damage and therefore extinction; a serious objection to this idea however, is that the major terrestrial families—reptiles and the earliest mammals—managed to survive, while the denizens of the oceans showed maximum evolutionary developments, although in this realm they are largely protected from cosmic radiation effects.

Other scientists have suggested that the climatic effects of the Permo-Carboniferous ice age played a key role in the end-of-Paleozoic extinctions. The evidence is clear: during the Permian one after another of the most vigorous Paleozoic families became mysteriously extinct. There was no apocalyptic mass destruction at the Permian/Triassic boundary. Worldwide studies by two experienced paleo-ontologists, Bernard Kummerl of Harvard and Curt Teichert of Kansas, have shown that boundary to be essentially transitional. Paleozoic forms gradually disappeared, to be replaced, in corresponding ecologic niches, by Mesozoic forms.

No gross climatic factor appears to be involved in the Paleozoic extinctions, although most of the later Permian lands seem to have been excessively arid (quite widespread desert deposits and evaporite accumulations have been found), but the seas, cold to begin with, ended with evidence of warmth, coral reefs, etc. Could the key factor of the biologic extension be a geochemical one? The progressive build-up of oxygen in the atmosphere during the Paleozoic was reversed by a tremendous withdrawal in Carboniferous and Permian times due to the development of vast forests, which, on being converted to coal, permanently withdrew carbon and oxygen (and sulfur) from the atmospheric carbon dioxide and marine sulfate systems? A return to hot dry climates in the early Mesozoic (Triassic) added appreciably to the partial pressure oxygen. Thus the coal swamps heralded the evolution of giant reptiles, birds and mammals.

Curiously enough, although the mammals evolved steadily throughout the Mesozoic, this time has become known as the age of reptiles, for it was the dinosaurs that dominated both the lands and seas of the entire era (225–65 million years ago). This reptilian paradise may have had a climatic explanation.

The Assessment of Climatic Variations in the Modern Period

Systematic meteorological records barely date back beyond the turn of this century, and long, uniform series are rare. The aim of this article is to draw up a concise list of the various means which enable us to reconstruct climatic conditions prevailing before the existence of calculated written observations.

Historians, have made frequent reference to climatic events which have precipitated economic catastrophes, such as the two appalling summers of 1315 and 1316, when it rained nonstop. They have also, though somewhat haphazardly, indicated texts they have consulted regarding the weather's caprices, such as the harsh winter of 1468 when the frozen wine in Paris had to be dispensed with an ax. But it is only recently, and thanks to the efforts of E. Le Roy Ladurie, that the idea of a continuous climatic history has been entertained. This author's work is most impressive, and anyone wishing to investigate this area owes a great deal to him, both because of his methodological thoroughness and because of the interesting conclusions he draws.[1] But we require more than a purely qualitative survey of long periods; we have to aim, from now on, for greater and greater accuracy about shorter and shorter periods of time. But the methods used call on very different disciplines which are hard for one scholar alone to embrace. There are three major areas of investigation: written accounts, glaciers and dendroclimatology.

Written Accounts

Two main categories of texts can be used: those which deal by design with the climate, and those which happen to touch upon this subject, by chance or indirectly.

1. Ancient meteorological observations do exist. Some of them, like those made by Vanderlinden about Belgium, are remarkable. Many of these documents have been lost, and some are certainly still waiting to be discovered. But they must be used with caution. This is because there are about 170 thermometers and barometers in relatively current use, and the readings they give do not always tally (because of inaccurate equipment, readings taken anywhere or anyhow, and readings recorded carelessly).

2. Chance observations tend to appear in five main types of document.

(a) Some of the most interesting occur in collections of letters. The letters of Mme. de Sévigné, for example, have already been used,[2] but this is a very special case because these letters are known the world over, and easy to get hold of. If it is important to explore what has already been published and classified, such as Diderot's correspondence (and his letters to Sophie Volland in particular), it is also important to go through national and provincial archives, those of religious institutions etc., and even to search in private collections.

(b) Diaries, especially those written in the 19th century when a keen love of nature prevailed, can be usefully perused, e.g. the diaries of Amiel, Maine de Biran etc.

(c) Ships' logs always contain very precise and valuable data about the weather and the state of the sea.

(d) Reports, records and registers to do with agriculture are important. These deal with indications about harvest dates, and in particular the dates of the grape harvest, which give a good idea of the quality of the vegetative or growing period. Marginal notes often refer to climatic calamities held responsible for poor yields and late harvests.

(e) Religious archives, lastly, sometimes give rise to unexpected results. We mention the Barcelona Rogations used by Giralt to list the years of drought in 16th century Catalonia, the summer drought being in proportion to the number of prayers. In Japan, for example, one must take account of the flowering dates of the cherry-trees and the appearance of the first snows in Tokyo, as well as the first ice on Lake Suwa, all these events being carefully recorded by monks for several centuries (the works of Arakawa).

Although, ordinarily, these written documents only yield qualitative data, it is possible to achieve a certain type of measurement by drawing a graph of the phenological series (the list of harvest or grape harvest dates) and other occasional or incidental series, like the Barcelona Rogations. Nanley and Garnier have shown an interesting case of parallelism between summer temperatures and the start of the grape harvest (the hotter the summer, the earlier the harvest starts). But it would appear to be difficult to extrapolate temperatures from phenological series.

Glaciers

The use of glaciers for paleoclimatological purposes has given rise to countless publications.

1. The Movement of Glacier Tongues. *It is well-known that the tips of glacier tongues do not stay in the same place over periods of time. Thus, as a general rule, present-day glaciers have been receding since 1870/1880. This phenomenon is easy to measure, and is in effect regularly measured. This kind of record-keeping did not exist in earlier times, so it is necessary to use a certain number of methods to work out movements occurring in the past. We shall mention five.*

(a) Written documents, from communities living near glacial regions have been widely and skillfully used by Le Roy Ladurie in the Alpine area.[3]

(b) Iconography.

(c) Lichenometry. A moraine is soon colonized by lichens. We now know the rate of growth of these plants: it is very uniform and very linear for example, 0.40 mm a year for Alectoria minuscula.[4] *It is also possible to date very accurately the positioning of a moraine.*

(d) Dendrochronology is a type of age-ring analysis which will be discussed below.[5] It enables us to date wood on the basis of thickness variations and age-ring density. Glacial invasions wreak havoc in lower forests, and trunks torn from the ground are embedded in the moraine, or even in the ice itself. By dating the death of such trees it is thus possible to draw up a chronological picture of the advance of a glacier. Conversely, a retreating glacier exposes ground which is quickly recolonized by vegetation; here again the age of trees enables us to trace the retreat of the ice very accurately.

(e) Periglacial peat-bogs. When glacier tongues happen to be in the vicinity of peat-bogs, we find sandy deposits which grow thicker nearer the tongue. Thus the glacial deposit which occurred from 1600 to 1850–1860 can be perfectly translated in the stratification of the Fernau peat-bog in the Tyrol; and in fact the deposit in question has been named the "Fernau oscillation." We should mention in passing that by means of pollen analysis peat-bogs can provide indications about the climate in past periods, in terms of many thousands of years. But this does not apply to our area of interest, which is limited to the modern historical period in which, at least since the year 3000 B.C., climatic variations have been relatively slight, and in any event not such as to alter the nature of the vegetation. Perhaps the only modification

There were almost no high mountains, but extensive plains. Sea level was high, conducive to a maximum effect of the ocean on climate. At no time, on no continent, was there any evidence of major glaciation throughout this 160-million-year interval. These "days of wine and roses" favored remarkable evolutionary developments; not only did the dinosaurs prosper and grow to gigantic dimensions, both on land and sea, but a branch of flying reptiles evolved that in turn became superceded by the birds, the first order of creatures elegantly specialized for the art of flying. As most of the lands became

has had to do with the distribution limits of certain species which might have shifted a little.[6]

2. The Stratification and Nature of the Ice. *In the heart of the ice-caps, nearest the poles, it only snows in summer, and rain is even rarer. In winter, all surface mechanisms are obliterated by the persistently high pressures caused by the intense cold which is in turn the result of permanent darkness.*

Each layer of summer snow is packed down to await next summer's falls. So when a vertical cut or section is made in the ice, one can see a series of very thin snowfalls which give a good idea of the precipitation (always low) occurring each year. But the lower layers become more and more compressed by the layers on top of them. So corrections must be made on the basis of the density of the ice. On the edge of the ice caps things are more complex because it also snows in winter. Here Lorius has delved much deeper in his glacial investigations. He has shown that the summer falls were richer in heavy isotopes. This led to the finding that the level of deuterium (heavy hydrogen), tritium and O^{18} *gives precise indications not only about seasonal precipitations, but also about the temperatures in which the snow fell, and lastly about the variations in the atmospheric structure.*[7]

3. Varves. *Varves are deposits from settling which can be observed in proglacial lakes.*[8] *A vertical section shows a succession of light and dark layers. The* summer varve *is light and thick because it contains detrital matter which becomes more abundant when the ablation process is at its strongest and fastest.* Winter varve, *on the contrary, is dark and fine; because there is almost no fusion, or only very slow and episodic fusion, there is a predominance of micro-organisms. The thickness and composition of the varves provide data about summers and winters. A long, hot summer is indicated by a very thick and sandy varve with almost no organic elements. Conversely, a cold winter is indicated by a very fine, clayey-muddy (thus dark) varve. Since the fine work done by Geer,*[9] *the Scandinavian varves are well-known and dated. Recently, Schove has attempted to compile a list of all the Baltic varves with a view to synchronizing them.*[10] *But there is more to be deduced from these lacustrine deposits. Taking a closer look, one sees that all varves, whether winter or summer, are themselves stratified.*

Dendroclimatology

We all know that tree-trunks grow by means of successive and concentric rings. Dendroclimatology *deals with the relationship between age-rings and climate. This science is inseparable from* dendrochronology, *to which I have already referred.*[11] *Dendrochronology and dendroclimatology constitute the analysis of age-rings.*

Age-ring analysis started to be practiced in geographical regions affected by a clearly defined climatic factor: drought in the western United States, cold in the northern regions. In the western United States, conifers which are affected by drought (including the famous sequoias, giant species of which are sometimes 2000 years old) show narrow rings when precipitation is insufficient: in a way these trees are natural rain-gauges. Fritts has used this characteristic to retrace variations in aridity in the Far West for the past 1000 years and more.[12] *At the present time, with his team from the University of Arizona, in Tucson, he is refining his methods and has just tried to give an overall view of the climatic fluctuations in North America since 1700.*[13]

At the northern forest limit, Slastad and Mikola have shown that the hotter the temperatures of the hottest month (July), the wider the age-rings.[14]

But in our temperate regions the age-rings embrace all the climatic elements, including precipitation, temperature and light. It is very hard to make a direct definition of the role played by each one of them in the growth process, which is based on a climatic ambience. To do so would mean comparing the growth of trees with different requirements differential disintegration. *This type of task has not yet been attempted. But circumstances have never been more favorable since Polge, at Nancy, developed a new method for exploring age-rings with X-rays.*[15]

The print-out from a radiodensitometer gives an immediate graphic record of the annual variations of the internal density of each ring (low density means spring wood, high density means summer wood). In addition one can follow the evolution of the metabolism throughout the vegetative period.

This procedure opens up huge possibilities. In effect, if the climate influences the rings, then conversely the rings express the climate (this is the reversibility *principle).*

So a good knowledge of present-day age-rings, particularly by the interpretation of the radio-pictures,[16] *should give us, by means of retropolation, the precise reconstruction of the climate in the vegetative periods of the past. One might even manage to obtain indications about the cold seasons, because it appears that certain essences, such as the Aleppo pine, work more slowly in winter. So there is considerable scope attached to dendroclimatology, and it is a good idea to stimulate research in this area.*

If the three major areas of investigation discussed are fundamental, there are additional areas which can also be used: variations in the level of lakes,[17] *movements of fishing sites (although this is not always reliable) associated with the transgression or regression of warm water, bird migrations and so on. But it is certain that no one on his own can reconstruct the climatic past. This will be achieved by matching results obtained by different methods, and by a process of complementarity.*

What is more, there must be considerable caution. The interpretation of given facts must be made with great care. For example, an advancing glacier does not necessarily indicate a general lowering of temperatures: winters with heavy snows, producing a superabundance, are milder than dry winters. In addition, mathematical methods must be used wisely and in most instances tempered by reflection and subtle analysis.

PIERRE DE MARTIN

Bibliography: 1. Le Roy Ladurie, E., *Histoire du climat depuis l'an mil,* Flammarion, Paris, 1967. The English edition: *Times of Feast, Times of Famine, a History of Climate since the Year 1000,* published by Allen and Unwin, London, 1971, is more complete. For those authors not mentioned in my text (the names are followed by an asterisk) see the bibliography in these volumes, which is very complete. 2. There are two publications on this subject: Dufour, L., *Chronique des événements météorologiques anciens d'après les lettres de Mme. de Sévigné,* Ciel et Terre, year LXXIV, no 11/12, Nov-Dec. 1958, Brussels. de Martin, P., *Le temps de 1670 à 1695 à travers les lettres de Mme. de Sévigné,* La Météorologie, V-21, pp. 67–68. Paris, 1972. 3. Le Roy Ladurie, E., *op. cit.,* 1967. 4. Andrews, J. T, Webber, P. J., *A Lichenometrical study of morphological technique,* Geographical Bulletin, 22, Ottawa, 1964. See also: Beschel, R. E., *Lichenometrical studies in West Greenland,* Arctic, vol. 11, no. 4, p. 254, 1958. 5. de Martin P., *Analyse des cernes, dendrochronologie et dendroclimatologie* Masson, Paris, 1974. 6. Modifications in the distribution areas of cultivated plants are just as well, if not better, explained by economic and human considerations than by possible climatic causes. As far as the forest is concerned, clearance has distorted the problem. In Brittany, however, the forest is not growing again; it is being replaced by moorland. This has still not been satisfactorily explained. 7. Lorius, C., *Le Deutérium. Possibilités d'application aux problèmes de recherche concernant la neige, le névé et la glace dans l'Antarctique.* Comité National Français des Recherches Antarctiques, no. 8, Imprimerie de l'Institut Géographique National, Paris, 1963. See also, in the present work, the observations of Lliboutry and Lorius, and Pontikis. 8. Brooks, C. E. P., *The problem of the varves,* Quat. Journ. Roy. Meteorology Soc., 54,34, 1928. 9. de Geer, G., Geochronologie Suecia. Principles. Kungl. Sv. Vet. Akad. Handl., series 3Bd, 18, no. 6, Stockholm 1940. 10. Schove, D. J., *Varve-teleconnection across the Baltic,* series A, Physical Geography, Vol. 53 A, no. 3/4, pp. 214–234, 1971. 11. de Martin P., *op. cit.,* 1974. 12. Fritts, H. C., *Tree-ring evidence for climatic causes in Western North America,* Monthly Weather Revue, 92, 7, pp. 421–443, 1965. 13. Fritts, H.C., Blasing T. J., Hayden B. P., Kutzbach J. E., *Multi-variate techniques specifying tree-growth and climate relationship and for reconstructing anomalies in paleoclimate,* Journal of Applied Meteorology (American Meteorological Society,) Vol. 10, no. 5, pp. 845–864, 1971. 14. Slastad, T., *Arringgunderskelser I Gudbrandsdalen. Tree-ring analysis in Gudbrandsdalen,* Meddellelser fra Det Norske Skogforsksvesen, 48, pp. 557–620, 1957. Mikola, P., *Temperature and tree-growth near the Northern Timber Line* in *Tree-growth,* ed. T. T. Koslowski, pp. 265–274, New York, 1962. 15. Polge, H., *Etablissement des courves de variations à la densité du bois par exploration densitométrique de radiographie d'echantillons prélévés à la tarière sur des arbres vivants. Application dans les domaines technologique et physiologique.* Annales des Sciences Forestières, vol. XXVIII, fasc. 1, Paris, 1969. 16. de Martin, P., *Croissance du chêne en Lorraine de 1946 à* 1970, Revue Géographique de l'Est, 1974, forthcoming. 17. Variations in the level of lakes reflect the pluviometry of their basin. Thus a drop in the level of Lake Chad, which is fed from the south, indicates a shortage of rain in the tropical wetlands to the south. Local evaporation must also be taken into account.

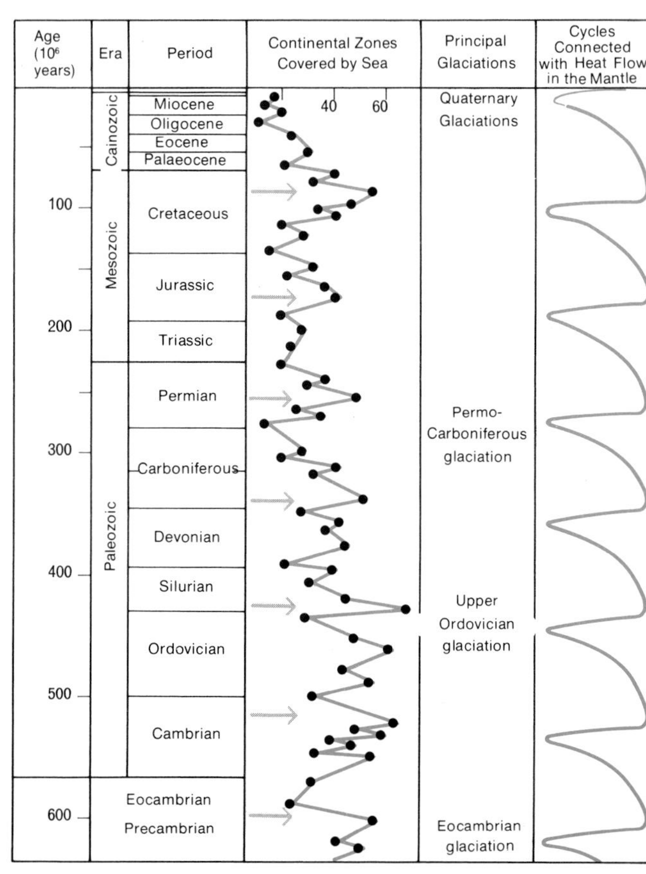

Fig. 16-4. In the Earth's long history the landscape of the various continents has changed many times. The diagram at right shows some of the elements of this history (the invasion of the continental surface by the sea, with the maximum points indicated by the gray arrows and the main glaciations), linking them to a galactic cycle which, over a period of 85 million years, becomes aligned to the cycle of the analogous period, which would seem to show the behavior of global tectonics and the expansion of the ocean-beds. The photos on this and the opposite page show some instants in this history. Above left, a present-day landscape: the Tassili in the Sahara, which, according to some geologists, is a good representation of the Earth's appearance as it must have looked 3 billion years ago, with no vegetation and so liable to active erosion. Top right, dried mud dating back 450 million years formed in lagoons liable to periodic drying-up periods which were plentiful in the Devonian. Above right, center, sedimentary cycles of the Carboniferous which indicate periodic variations of the sea level which occurred about 300 million years ago. In this changeable environment life emerged from the water, where it had remained from the start for more than 2.5 billion years. On the opposite page, an indisputable trace of life: the footprint of a dinosaur, 200 million years old, on the shores of a lake in Connecticut.

forested a net loss of oxygen from the atmosphere developed.

Environmental and climatic conditions requisite to the livelihood and prosperity of Mesozoic life call for luxuriant vegetation, trees and swampy areas. On land there is also proof of this warm climatic interpretation in subtropical paleosols—laterites, bauxites and so on. For the marine life of the times, the same is true: the corals formed reefs (but a different class from the Paleozoic ones), and even certain mollusca and sponges contributed reeflike communities. The sea temperatures so indicated would be in the range 15–35° C (expressed as monthly averages—greater extremes are acceptable, but generally only for a few hours).

In the last quarter of the Mesozoic, the Upper Cretaceous Period, a remarkable "flowering" of small floating marine organisms occurred. These are the nannoplankton, unicellular plants (coccolithophoridae) and animals (foraminifera), the shelly parts largely constructed of calcium carbonate. Previously most of marine life was benthonic (bottom-livers) or nektonic (free swimmers). But now, for the first time, a vast proliferation of pelatic (floating) forms started to populate the surface waters of the world ocean. This was only possible under conditions of worldwide warmth and equable seasonal changes. The result of this "flowering" was a rain of dead shells into the seafloor—to form what we know as chalk, and now found all over the world, from Dover to Texas, from the Ukraine to Australia. Of special significance to the climatologist is the semipermanent withdrawal of carbon dioxide (as $CaCO_3$) from the atmosphere. So our atmosphere lost CO_2 to gain still more oxygen. This would lead to a slight reduction of the "greenhouse effect," favoring a cooling trend that was perceptible throughout the next era.

The Cenozoic: Our Own Era c. 65 Million Years Ago to the Present

As remarked above, the end of the Paleozoic was marked by great extinctions and the beginning of the Mesozoic was marked by the emergence of a vast and flourishing new biota. A comparable "re-tooling" of the great bulk of the biota of the late Mesozoic is visible in the early Cenozoic. Again, the mechanism of this phenomenon is clouded in deep mystery. From the evidence of sediments and soils we are satisfied that no great ice age or violent climatic revolution marked either the transition from the Paleozoic or that from the Mesozoic.

Numerous ad hoc theories have been proposed to explain these events: some of the suggestions are environmental, some climatic, or atmospheric, some are biological. For example, the geological record discloses that there was a crescendo of mountain-building events in the late Mesozoic. Continental splitting and drifting apart of the major continental blocks—Gondwanaland in the southern hemisphere and Eurasia in the north—had begun in the early Mesozoic, and the critical and final break-apart events did not come till the late Mesozoic. Could the newly developing mountain-belts inhibit the migrations of organisms? It is possible, but more important, could the final separation of continents, e.g. West Africa and South America parted about 80 million years ago, cause an isolation of the gene pools for the major reptiles and thus bring about their extinction? It is an interesting idea.

The crescendo of sea-floor spreading activity must have led to an accelerated heat flow along the global spreading axes, and this would cause crustal expansion, leading to a universal rise of sea level. Around 80 million years ago sea level reached over 500 m above its present level, greatly reducing the world's available land areas, and in the late Cretaceous Period, only 15% of the Earth's surface was dry land. The population crowding must have been Malthusian. (There is perhaps an object lesson here for humanity, take care!) Furthermore, the lack of natural catastrophes, such as ice ages, during the Mesozoic did nothing to trim the population growth. There is abundant evidence from the fossils, that predators, for example, *Tyrannosaurus rex,* developed on an immense scale—a phenomenon suggestive of food shortage and desperate competition.

An ingenious theory couples the rise of the angiosperm plants with the demise of the dinosaurs. The high eustatic sea-levels of the late Mesozoic created immensely wide shallow seas and lagoons—most favorable sites for the growth of giant mangrove swamps. But it is said the dinosaurs could not negotiate the impenetrable thickets of mangrove and thus became progressively cut off and died out.

Yet another idea suggests that the mountain-building of the late Mesozoic was associated with prodigious volcanic activity, which led to blankets of dust being distributed over the whole Earth (illustrated in Walt Disney's classic film "Fantasia"—to the music of Stravinsky; the dinosaurs fell down giant cracks in the Earth, were engulfed by lava streams and choked to death on the volcanic fumes and dust).

Or yet again, it has been proposed that the gradual rise of the mammals led to the appearance of the rodents, that as a group are known to be fond of eating eggs. If the dinosaur eggs were subject to constant predation by hordes of late Mesozoic rats (or their equivalent), it would not take long before all the giant reptiles were wiped out.

A feature of the reptile metabolism is that they thrive in a carbon dioxide rich atmosphere, whereas mammals do not flourish under these conditions. The biogenic oxygen build-up of the global atmosphere was presumably increasing steadily throughout the Mesozoic and toward the close the environments were becoming more attractive to the reptile's competitors. Removal of carbon dioxide was in progress at the same time due to formation of deep-sea chalks ($CaCO_3$). Sulfur-isotope studies show that a serious oxygen deficit was building up during the Mesozoic and this was reversed at the start of the Cenozoic.

But there is one serious flaw in all these arguments: they only extinguish one class of organisms—the dinosaurs. What about all the precisely contemporary changes that occurred amongst marine animals and plants? These too exhibited the same sort of violent evolutionary upheavals.

What about a geophysical explanation that would treat all the organisms known? For example, it is known the Earth's magnetic field from time to time becomes weaker and may reverse itself. Could the faunas of the time become so disoriented that they could not survive? It is an imaginative suggestion but alas, it forgets that the plant kingdom also evolved at this time, and plants are very little concerned with locomotion.

The Earth's magnetic field also controls the global charged particle screen, popularly known as the Van Allen Belt. This screen effectively cuts down on a great deal of potentially dangerous or even lethal solar and cosmic radiation. Organisms, both plant and animal, would be equally affected, although those that live underground or in the deep sea would be protected. Those living in open spaces and in the shallow surface waters of the ocean would be most vulnerable. This is the most viable explanation at present, but clearly the evidence is meager. Perhaps it is more satisfactory simply because the alternative hypotheses are clearly less than satisfactory.

The climatic history of the globe during the Cenozoic Era is very illuminating, for it shows us how an ice age is initiated. At the beginning of this era we have evidence of worldwide

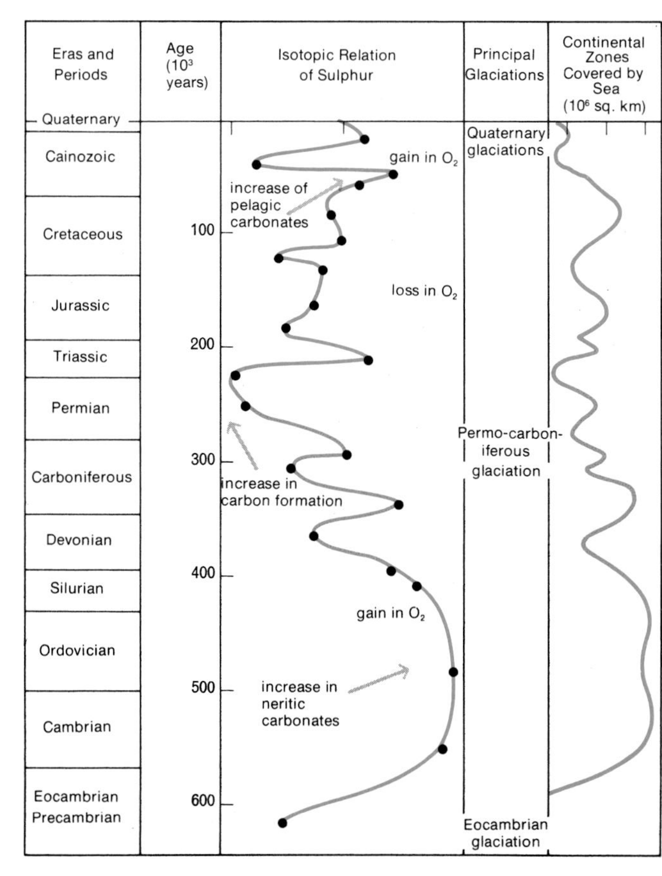

Fig. 16-5. The diagram shows the oscillations in the relation between the sulphur isotopes in the last 600 million years, comparing them with other events in the Earth's history: the major oscillations of the sea level, the major glaciations, the development of carbon on the continents and of carbonatic deposits in the oceans. In effect the comparison is significant, because the state of the atmospheric oxygen, which is so important for our planet, is linked by a reverse relation with the isotopic relation of the sulphur and with the atmospheric carbon dioxide. The formation of impressive carbonatic deposits, like those shown in the photo on the page opposite, in fact represents the result of the impoverishment of the atmosphere in carbon dioxide, caused by organisms which lived in the oceans.

mild semitropical conditions, that is warm and moist. The sizes of the continents were initially very small, and the oceanicity factor was dominant. It was a "thalassocratic" era of low climatic extremes. Continental drifting and sea-floor spreading during the Cenozoic progressively led to a series of intercontinental collisions from Spain to the Himalayas. What had been, in Mesozoic times, a wide equatorial seaway, the "Tethys," or ancestral Mediterranean, that joined the North Atlantic to the South Pacific, now became blocked. By about 30 million years ago no further east-west migration was possible. By the late Miocene (12 million years ago) the Mediterranean became limited to a salt lake. Other barriers were also rising: in the East Indies the uplift of island chains almost connected Australia to south-east Asia; in Panama, volcanic chains eventually joined North America to its southern partner.

What happens to an automobile engine when its temperature regulation system breaks down? What if its radiator becomes blocked with rust? The engine boils. The Earth's heat balance is mainly related to *heat loss;* so what happens when there is a blockage in the global temperature regulating system? There is an ice age.

Guided at first by mountain-building and the closing of oceanographic temperature regulation systems, and with continental drift bringing continents into polar latitudes, the snows of winter began to stay all through the summers. Minor climatic cycles were superimposed on the major trends so that episodically the glaciated regions grew larger. As long ago as 30 million years there were scattered signs. These grew more and more extensive—from Antarctica to Alaska—until around 2 million years ago the present ice age (the "Quaternary Period") is deemed to have begun in earnest. Icelandic faunas began to appear in the Mediterranean. In ever greater cycles of about 100,000 years each, the great glacial stages closed down over the globe. We are now in one of the mild interglacial phases, but how long will it last?

Throughout the Cenozoic Era, as we reach closer and closer to our own time, the geologic record becomes more and more complete and precise. This permits the testing of numbers of key concepts. For example, do the minor extinctions and faunal appearances of the Cenozoic coincide with paleomagnetic reversals? There are strong indications that they do. And, assuming they do, what biologic mechanisms are involved?

Furthermore there is a growing weight of evidence to suggest there is a definite relationship between the Earth's magnetism and contemporary climate changes. Intuitively it might appear that both are controlled ultimately by radiation from the Sun. Now there is a substantial body of data. The Russian climatologist M. I. Budyko (*Tellus,* 21, p. 612, 1969) has shown that solar radiation received in Northern Hemisphere ground stations has been dropping since 1938, and Murray Mitchell (of ESSA, Washington, D.C.) has demonstrated that since 1940 the mean world temperature has been dropping. Conversely, the mean magnetic intensity has been rising. However, there are also regional reversed trends: in some parts of North America and the Southern Hemisphere there is a warming climatic trend during the same period, paralleled by a falling magnetic intensity. One reason for regional anomalies in geomagnetism is the eccentric distribution of the magnetic dipole field of the Earth and its secular (westward) drift. Since the global climate average is likely to be steered by effects that coincide with the world's largest Hemisphere, the Pacific and Southern, it is the strengthening magnetic field and dropping temperature over Eurasia today that are most significant.

It has been possible to measure past solar radiation by an indirect technique. The unstable isotope C^{14} is generated by

solar (or cosmic) radiation in the upper atmosphere. It then sinks as CO_2 and is incorporated in plants and other living things. If one counts tree rings back from the present, one can calibrate them to an exact sidereal calendar. But if one determines the C^{14} ratio in the wood of each year, one finds that there are anomalies. The rate of C^{14} generation appears to be variable; the writer of this article showed that there is a correlation between this and another world indicator of the temperature—the oscillation of sea level. This was extended back for 1000 years (in 1961), but since then systematic dendrochronologic work with radiocarbon dating now takes the record back over 8000 years. During this epoch there has been a notable and systematic variation that rose to a peak around 5000 years ago, the time of the warmest mean world temperature in nearly 100,000 years.

World climatic patterns are much affected by "feed-back" phenomena. For example, if Eurasia becomes 1–2° C cooler over a few decades, the snows of winter will last a few days to a few weeks longer each spring. The incoming solar radiation at that time, instead of warming the soil, is largely lost by reflection off the white surface (the albedo effect) and consequently some of the *net heat* intake of the globe is lost. Thus the general climate becomes colder and colder.

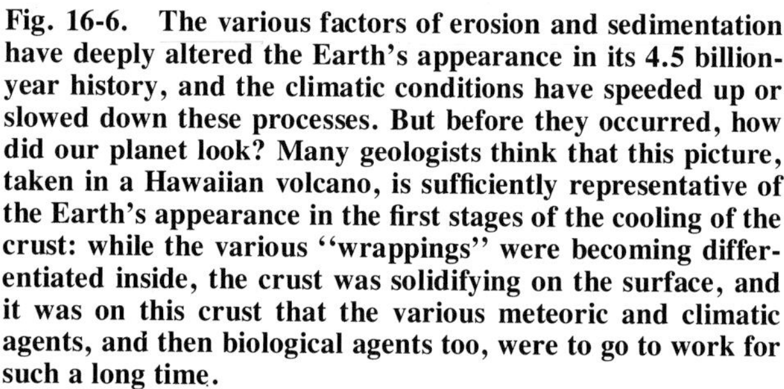

Fig. 16-6. The various factors of erosion and sedimentation have deeply altered the Earth's appearance in its 4.5 billion-year history, and the climatic conditions have speeded up or slowed down these processes. But before they occurred, how did our planet look? Many geologists think that this picture, taken in a Hawaiian volcano, is sufficiently representative of the Earth's appearance in the first stages of the cooling of the crust: while the various "wrappings" were becoming differentiated inside, the crust was solidifying on the surface, and it was on this crust that the various meteoric and climatic agents, and then biological agents too, were to go to work for such a long time.

During a glacial phase there is an additional feedback effect of stupendous consequence. The transfer of moisture from the oceans to continental glaciers lowers the ocean level. At the last glacial maximum (about 18,000 years ago) approximately 80 million km^3 of ice covered the high latitudes of the Earth, thereby lowering sea level by around 200 m from its pre-ice age stand. What does this do to global geography? It exposes about 12% more of the continents to the atmosphere and reduces the size of the oceans. The global "continentality" increases, a phenomenon that favors extremes of climate, specifically cold extremes inasmuch as the largest continental areas are located in high latitudes.

Yet a third feedback mechanism now begins to operate. Shallowing of the northern seas, e.g. around Arctic Canada and the Siberian continental shelf, permits them to cool and freeze more easily, so that the sea-level drop favors increased sea ice. But the sea ice is white and this contributes still further to the albedo heat loss.

And there is a fourth feedback mechanism. As glaciers grow in the higher latitudes, every spring and summer they tend to calve icebergs into the ocean; these giant floating lumps of ice are carried by ocean currents into lower latitudes where they melt, cooling the intermediate latitudes as well.

Lastly, we mention a fifth consequence of global cooling and glacier-building. The writer of this article presented this argument to a meeting on Climate Change held in Rome in 1961, organized jointly by F.A.P. and the World Meteorological Organization. The reasoning was as follows: a general cooling of the ocean surface coupled with a reduction of solar radiation should lower the mean evaporation rate, and this in turn should cause a drop in rainfall. Glaciations should coincide with global aridity. The eustatic drop of sea level and the increased areas of sea ice would amplify this effect by reducing the surface area of oceanic water available for evaporation. A contrasting effect would be generated by the increased equator-pole temperature gradiate. This should raise wind velocities and *raise* the evaporation rate but only, of course, in limited regions. According to Hubert Lamb, the British meteorologist, the glacial climatic regime would be marked by "blocking," that is to say, there would be suppression of the normal zonal winds, the prevailing westerlies, which bring the bulk of the moisture into the midlatitude continental regions. In contrast there would be much north-south turbulence, but since the northern land masses are largely oriented north-south, this turbulence would not favor increased precipitation on a global scale. The matter is controversial because the traditional theory claims that glaciations were marked by increased precipitation.

The ice-age aridity problem can be tested by geological evidence. During the last glacial maximum, the land areas of the globe were intensely arid. Violent duststorms raged across the whole of Eurasia from the coast of France to the Yellow Sea, and across the Midwest of North America. The Sahara and Kalahari Deserts of Africa both expanded catastrophically, invading the once fertile tropical savannas and even penetrating the region of the Congo. The same thing happened in South America, sand dunes reaching from Argentina into Brazil and much of the Amazon basin became semi-arid. Such specialists in pollen analysis as Van der Hammen and his associates from the Netherlands have been systematically testing deposits of the last glacial period. They have put down test borings in suit-

able lake basins—in Italy, Spain, Greece and elsewhere—where continuous sedimentation would preserve an uninterrupted record of events. In regions where today there are forests and moist climates, during the glacial period there were semi-arid grasslands, or, in the mountains, even tundra conditons. Numbers of great rivers, that were not fed by melting glaciers, almost dried up during the glacial maximum—the Nile and the Niger suffered great loss of discharge. Many equatorial lakes suffered a change of regime, e.g. Lake Victoria became an evaporating, carbonate basin and did not overflow.

The picture of ice age desiccation contrasts with the classical view that ice ages were somehow connected to the Biblical Flood. In the last century, when evidences of glaciation first began to be recognized, the Flood dogma was rather universally believed. It became then only a short intellectual leap to think of the "forty days and forty nights" precipitation falling as snow in the high latitudes. Nowadays it is not even felt necessary to provide increased precipitation for glaciers to expand. Cooling may decrease the melting rate. Indeed most of the present glaciated regions are "ice deserts." They remain as glaciers simply because the annual budget of snowfall is not exceeded by the annual melting or ablation; the controlling factor is the continued low temperature.

The correlation of glaciation with excessive precipitation has readily led to this remarkably universal error of correlation. There are widespread evidences of giant lakes and former rivers in the subtropical regions, particularly in Africa. Lake Chad for example once covered an area half the size of Western Europe. It was *presumed* that this correlated with the last glaciation. Evidence of repeated "pluvial" stages in many parts of the world has led to slavish correlation with glacials. But radiocarbon dating of the former lake terraces has brought about a revolutionary change in view. The highest lake level of the Paleo-Chad was about 5000 years ago. Since the time of the Pharoahs, North Africa has been drying up. At that time all sorts of wild animals roamed the extensive savanna, and Neolithic man drew pictures on the rocky outcrops to dramatize the scene.

During the last 5000 years there has been a progressive drying up of the Sahara landscapes. Even in the classical times of Herodotus and Ptolemy there were vivid descriptions of the "great lake" (the Paleo-Chad); the fossils of crocodiles and hippopotamus, found today, attest to its former dimensions. Desiccation has accelerated since the Roman era and, although the advance of the deserts is oscillatory, the long-term prediction for the future of low-latitude savanna countries (e.g. Mauritania to Ethiopia, parts of India, northern Australia, northern Mexico, etc.) is very serious indeed. A conference held at Providence, Rhode Island in 1972 concluded that the next glacial age is to be expected within a few thousand years, at most, but possibly within a few hundred years. It is important to realize that these climatic changes do not proceed steadily. Rather they advance with strong oscillations, sometimes warmer, sometimes colder, but ultimately reach a threshold limit, when the climatic deterioration is rapid and absolute.

Epilogue

In this climatic and environmental review of the history of our planet, events have succeeded one another as in the turning of a carousel (see Glaessner's diagram in this volume). As the carousel sluggishly turns more and more passengers climb on, but no one gets off. Some seem to get buried under the crowd, but the effect in time is that everything on the globe becomes progressively more and more complex. This is an interesting concept for physicists and mathematical philosophers. A basic law of physics is that of *Entropy,* which states that in a physicochemical process with the passage of time, there is an increase in entropy, the measure of disorder. With progressive loss of energy there is a trend toward breakdown or disorder. But in long-term processes connected with geology and biology, entropy seems generally to be negative (i.e. exactly the opposite), because the crust of the Earth and its populations become progressively more complex. That may not be true in cosmic time, but it is true for the 4.5 billion-year-record we have before us.

While the tendency toward equilibrium or maximum entropy is evidenced in isolated systems, the global complex is not isolated and is even subject to many external radiational and gravitational influences. *We have a nonsteady-state history.* Only short-term processes may be viewed in terms of steady-state.

To recapitulate briefly, the following evolutionary trends in Earth history are now considered as basic:

1. *The Atmosphere* has been progressively generated by emanations from the Earth (H_2O, CO_2, SO_2, etc.) and fundamentally altered by plant photosynthetic generation of oxygen, and in part modified by sediment burial (e.g. of carbon in coal and hydrocarbons, gypsum and anhydrite, limestone and chalk).
2. The *Hydrosphere* has grown progressively like the atmosphere (at an unknown rate), while its chemistry has changed through time, partially in response to that of the atmosphere. The trend has been from a more acid beginning, rich in carbon dioxide, towards a more alkaline condition.
3. The *Sediment Chemistry* has evolved in response to changes of the crust, containing progressively less iron, and magnesium but more calcium and sodium.
4. The *Sediment Lithology* has been more and more sorted, i.e. organized through time. In the primeval crust, all components were mixed. Today vast formations of quartz sand, clays or limestones speak for long periods of sorting and segregation.
5. The *Continents* have become progressively higher and thicker, each continent today containing representative rocks of every single geological period.
6. The *Ocean Basins,* have become steadily deeper and

more segregated from one another, while continental shelves have become smaller.

7. The *Biosphere*—organized life, both plant and animal—has become successively more complex through time. As speciation increases, total populations of each tend to decrease. In spite of numerous extinctions, no phylum has ever been completely terminated and the total number of classes, groups and species increases with time.

Evolution of this planet and its organic populations has only been secular if viewed from afar, in a highly generalized way, just as human history and culture tends to evolve in a series of jumps, alternating threshhold levels or plateaus of achievement, or even regressions. This humanistic language might, at first sight, appear to be man's wishful thinking, to interpret everything in terms of the familiar. Far from it, the geological history of this planet is being hammered out with painful care against a background of constant questioning, rechecking and re-evaluation. With each new instrumental discovery a new area becomes available for study. But the basic laws of our science were well and truly laid more than a century ago. New discoveries from day to day expand our horizons but for the most part they do not reverse the conclusions of the pioneers.

RHODES W. FAIRBRIDGE

Bibliography: Cloud P., *Evolution of ecosystems,* in Amer. Sci., LXI, **1**, 54 (1974); Van den Henvel E. P. J., Buurman P., *Possible causes of glaciations,* in: Herman Y. (ed.), *Marine geology and oceanography in the Arctic Seas,* New York (1974); Condie K. C., *Archean magmatism and crustal thickening,* in Geol. Soc. Amer. Bull., LXXXIV, 2981 (1973); Bigarella J. J., *Paleocurrents and the problem of continental drift,* in Geol. Rundisch, LXI, **2**, 447 (1973); Fairbridge R. W., *Glaciation and plate migration,* in: Tarling F., Runcorn S. K. (ed.), *Paleoclimatic implications of glaciation and continental drift,* New York (1973); Holland H. D., *Systematics of the isotopic composition of sulfur in the oceans during the Phanerozoic and its implications for atmospheric oxygen,* in Geochim. Cosmochim. Acta, XXXVII, 2605 (1973); Pitrat C. W., *Vertebrates and the Permo-Triassic extinction,* in Paleogeogr., Paleoclimat., Paleoecol., XIV, 249 (1973); Wollin G. et al., *Magnetic intensity and climatic changes 1925-70,* in Nature, CCXLII, **5392**, 34 (1973); Woodrow D. L., et al., *Paleogeography and paleoclimate at the deposition sites of the Devonian catskill and old red facies,* in Geol. Soc. Amer. Bull, LXXXIV, 3051 (1973); Fairbridge R. W., *Climatology of a glacial cycle,* in Quat. Res., II, 283 (1972); Goldreich P., *Tides and the Earth-moon system,* in Sci. Amer., CCXXVI, **4**, 42 (1972); Holland H. D., *The geologic history of seawater—an attempt to solve the problem,* in Geochem. Cosmochim., XXXVI, 637 (1972); Lamb H. H., *Climates and circulation regimes developed over the Northern hemisphere during and since the last ice age,* in Paleogeogr., Palaeoclimat., Palaeoecol., X, 125 (1972); Nairn A. E. M., *Paleoclimatology: present status,* in Naturwissenschaften, LIX, 388 (1972); Newell N. D., *L'evoluzione delle scogliere,* in Le Scienze, 52 (1972); Trendall A. F., *Revolution in Earth history,* in J. Geol. Soc. Australia, XIX, 287 (1972); Valentine J. W., Moores E. M., *Global tectonics and the fossil record,* in J. Geol., LXXX, 167 (1972); Beuf S. et al., *Les gres du Paleozoique inferieur au Sahara,* Parigi (1971); Fairbridge R. W., *Upper Ordovician in Northwest Africa? Reply,* in Geol. Soc. Amer. Bull., LXXXII, 269 (1971); Van der Hammen T., Wijmstra T. A., Zagwijn W. H., *The floral record of the Late Cenozoic of Europe,* in: Turekian K. K. (ed.), *Late Cenozoic glacial ages,* New Haven (1971); Kukla J., *Correlations between loesses and deep-sea sediments,* in Geologiska Föreningen i Stockholm Förhandlingar, XCII, 148 (1970); McAlester A. L., *Animal extinctions, oxygen consumption and atmospheric history,* in J. Paleontol., XLIV, 405 (1970); Merifield P. M., Lamar D. L., *Paleotides and the geologic record,* in: Runcorn S. M. (ed.), *Paleogeophysics,* London (1970); Budyko M. I.; *The effect of solar radiation variations on the climate of the Earth,* in Tellus, XXI, **5**, 611 (1969); Mesolella K. J., Matthews R. K., Broecker W. S., Thurber D. L., *The astronomical theory of climatic change, Barbados data,* in J. of Geol., LXVI, 250 (1969); Bray J. R., *Glaciation and solar activity since the fifth century B. C. and the solar cycle,* in Nature, CCXX, 672 (1968); Cloud P., *Atmospheric and hydrospheric evolution on the primitive Earth,* in Science, CLX, 729 (1968); Fairbridge R. W., *Carbonate rocks and paleoclimatology in the biogeochemical history of the planet,* in: Chilingar G. V., Bissell H. G., Fairbridge R. W. (ed.), *Carbonate rocks,* Amsterdam (1967); Newell N. D., *Revolutions in the history of life,* in Geol. Soc. Spec. Pap., LXXXIX, 63 (1967); Harland W. B., Rudwick M. J. S., *The great infra-Cambrian ice age,* in Sci. Am., CCXI, **2**, 28 (1964); Nairn A. E. M., *Problems of paleoclimatology,* in *Proceedings of the NATO paleoclimates conference,* London (1963); Schwarzbach M., *Climates of the Past - An introduction to paleoclimatology,* London (1963); Fairbridge R. W. (ed.), *Solar variations, climatic change, and related geophysical problems,* in New York Acad. Sci. Annals, XCV (1961); Barghoorn E. S., *Evidence of climatic change in the geological record of plant life,* in: Shapley H. (ed.), *Climatic change,* Cambridge (1953).

IN December 1969, in the middle of the austral summer, a small group of geologists working under a National Science Foundation grant reached a point about 700 km from the south pole on the slopes of the Transantarctic Mountains, the great mountain range that divides the Antarctic in two. This expedition was organized by Edwin H. Colbert, author of the next article, and the objective was a search for reptiles. Colbert, an internationally renowned paleontologist, was, of course looking not for living reptiles, but for fossils. In the Antarctic, in fact, until that moment nobody had reported finds of fossil reptiles. This appeared to be a contradiction in paleobiogeographical terms since, if it were true, as the supporters of continental drift claimed, that for a long period in the past the Antarctic had been in close contact with Africa, South America, India and the other southern continents, the terrestrial reptiles that had been abundant in these other regions could hardly have kept out of the Antarctic. If no reptile remains were found in the few rocks rising out of the ice masses which today cover the white continent, much of the paleogeographical reconstructions based on the continental drift hypothesis would have had to be revised.

Colbert had not set out across the ice by mere chance. A couple of years before an Australian student P. J. Parrett, had brought him from the Antarctic a more or less unrecognizable pebble that might just be interpreted as a fossil fragment. This small but important clue was destined to enter into the history of geology. A few days after the Colbert expedition's arrival in Antarctica, they found abundant fossil remains of amphibians and reptiles. These reptiles included *Lystrosaurus,* which was known to have been widespread in Africa and South America about 200 million year ago. It was not a good swimmer, and if it were present in these three continents it must have been able to travel overland, thus confirming the continental drift hypothesis, that these three continents had indeed been linked together.

As for *Lystrosaurus*, all land and marine life has been widely affected by the arrangement of the continents in the past, as we have seen in the articles by Smith and Engel, and by the climate as described in the article by Fairbridge. The distribution of plants and animals in the past cannot be understood unless we suppose that the continents were joined together and separated several times. It thus becomes clear that, like many other geological data, all the paleontological evidence must be reviewed in the light of this new perspective, which bears witness to the constant evolution of Earth's environment. This is the theme as it relates to the main groups of animals, which is dealt with in this article by Colbert.

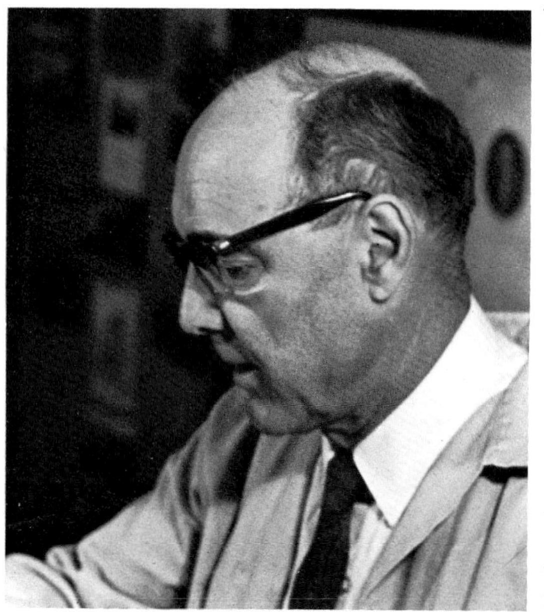

EDWIN H. COLBERT

Curator of Vertebrate Paleontology at the Museum of Northern Arizona's Society of Science and Art in Flagstaff, Arizona. Born in Clarinda, Iowa in 1905, he was one of the most active organizers of the American Museum of Natural History in New York, where he has spent most of his scientific career. Initially an opponent of the concept of continental drift, he became one of its most decisive supporters when, in 1969, he discovered in Antarctica the remains of terrestrial vertebrate fossils identical to the fauna of today in South America and South Africa. He was thus forced to reanalyze his long experience in the field of paleontology in light of the new dynamic model of the Earth, tackling first the problem of the evolution of terrestrial life on continents in reciprocal motion.

EDWIN H. COLBERT

Life on Wandering Continents

Overleaf: The footprints of a dinosaur found in the impressive deposit of fossil vertebrate remains at Gadufaoua in the Sahara. Dinosaurs are commonly found in continents which are now further apart than they used to be; the same applies to many other typically land-dwelling animals of the past, which were poorly adapted to swimming long distances in water. For a long time this phenomenon remained a mystery until it was realized that the configuration of the world's geography has altered in time and that the continents have not always been where they are now. From that moment on—i.e. since the definite confirmation of continental drift in the 1960s—it has become essential to reconsider the whole of biological evolution in the light of the changing configuration and position of the continents, because this was the basic condition which enabled or prevented exchanges between organisms living on parts of the land which were differently arranged than they are today.(photo L. Pizzi).

Every continent is distinguished by the life it supports. Perhaps it would be more correct to say that until a few hundred years ago life on each continent was distinctive to a greater or a lesser degree. But within the past few centuries man has obscured the natural biological differences between the continents by introducing plants and animals (consciously or inadvertently), and by exterminating many of the native organisms once peculiar to the several continents. However, let us keep in mind the continents as they were, antecedent to the interfering hand of man.

On these continents the plants and animals of North America, Europe and Asia north of the Himalayas are closely related. They are rather different in many respects from the biota of India and southern China. Africa south of the Sahara is a great botanical-zoological paradise with a rich flora and fauna. South America has its own characteristic animals and plants, in part distinctive and in part related to the organisms of North America. Australia is a land of marsupials and in the plant world, of a host of eucalyptus species. Antarctica is a continent of ice and lifeless mountains.

The biological resemblances and differences between the continents are the result of organic evolution, combined with various interrelationships between the continental masses through the long duration of geologic time. The resemblances of the modern biota of Europe to North America are owing to the rather close physical relationships of the two continents during the past 60 million years of Earth history. The very distinctive Australian fauna reflects the isolation of that island continent during the same 60 million years.

But are the continental relationships as we know them necessarily similar to the arrangements of the continents in past geologic ages? Until about 20 years ago a majority of geologists accepted this idea. And relationships were based upon this premise, in spite of certain anomalies.

For example similar Paleozoic amphibians and reptiles, about 300 million years old, are found in central Europe and in North America. On the premise of fixed continents it was postulated that their similarities were the result of long intercontinental migrations between the two land masses, by way of eastern Asia and a Bering connection. Likewise, closely related early Mesozoic reptiles, some 200 million years in age, are found in South Africa and South America, necessitating, on the theory of fixed continents, a tremendously long journey involving the length of Africa and Asia, a Bering crossing, the latitudinal length of North America and a Panamanian crossing into South America, or vice versa. In any event a possible but a tortuous explanation. Are not such fossil occurrences more logically explained upon the basis of past continental arrangements different from those of the present?

The New Geology

This question leads to the subject of plate tectonics and continental drift, or in simpler language, the New Geology. During

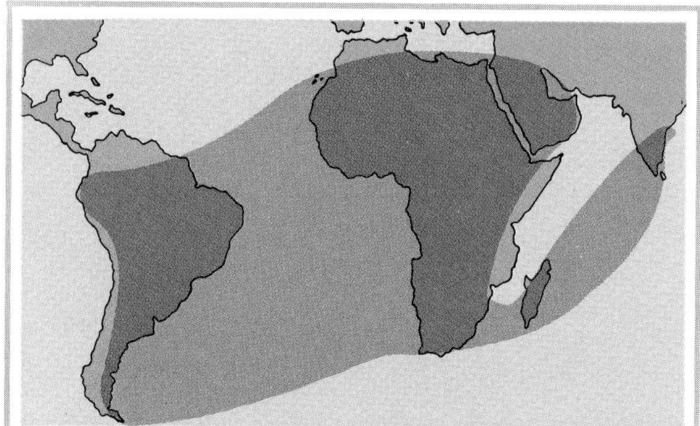

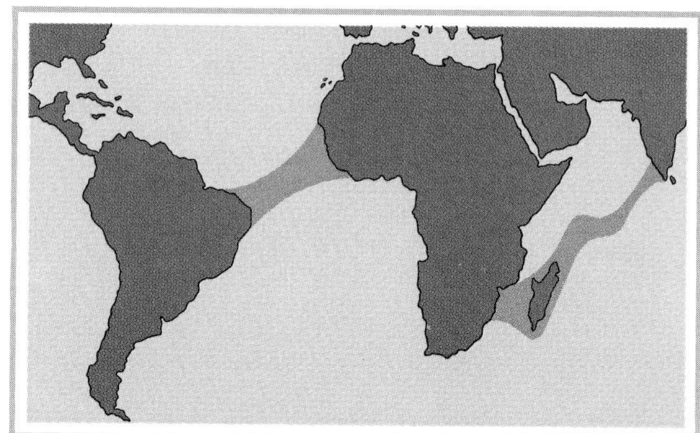

Fig. 17-1. Paleontological data gathered from the last century onwards showed undisputably that in given past periods various dry land-masses now separated by vast oceans must have been in some way connected because they yield the fossil remains of animals and plants which could not conceivably have crossed the oceans. For those who did not believe that the continents had moved, there was no other alternative but to think in terms of an overland "bridge," either across vast continental areas which have now disappeared and left behind ocean basins (left) or narrower continental bridges (right).

these past two decades the science of geology has undergone a profound revolution, as significant as the Darwinian revolution in biology that took place more than a century ago. Darwin showed that species are mutable—that life has been evolving and changing through geologic time. The new revolution in geology is showing, as a result of cumulative evidence, that the continents have been mobile and mutable through geologic time. In short, studies of the Earth during these past 20 years show it to be a dynamic planet.

The theory of plate tectonics postulates that the crust of the Earth is composed of several large plates, constantly moving in relation to each other. Some of the plates contain the continents as we know them; some contain ocean basins. The boundaries between the plates may lie within the ocean basins or in some cases may transect parts of continents.

coast of Africa, eventually to collide with Asia. Antarctica and Australia, remaining conjoined during Mesozoic time, separated from their original position at the southern tip of Africa. Eventually there was a split—Antarctica drifting southward to its present position, Australia drifting to the northeast. In the meantime Laurasia rotated in relation to Gondwanaland, opening an ancient seaway, Tethys, which was ancestral in part to the Mediterranean, while eventually North America pulled away from the Mauritanian border of Africa and finally split from Europe, thus opening the North Atlantic.

These tremendous events occurred over a long span of time; indeed they are still progressing. And as they progressed the continental blocks, carried on their respective plates, moved to their present positions. Old connections between continents were broken; new connections were established.

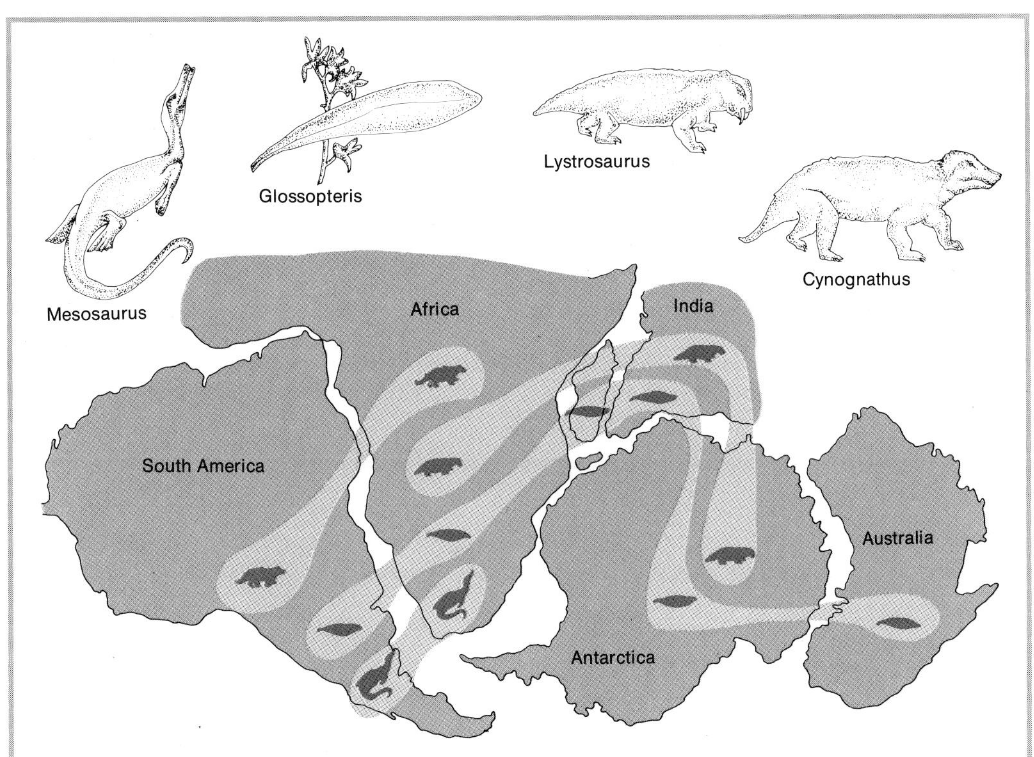

Fig.17-2. Some of the animals and plants whose remains are today found on continents some distance from each other: in South America, Africa, Madagascar and India, Antarctica and Australia, then joined to form the great continent of Gondwana; in gray, the distribution area. These are: *Mesosaurus,* a Permian reptile present in Brazil and Southern Africa; *Glossopteris,* a Permian plant found in all the zones shown: *Lystrosaurus,* a Lower Triassic reptile which is found in Southern Africa, India and Antarctica; *Cynognathus,* a Lower Triassic reptile which lived slightly longer than *Lystrosaurus* and is found in Argentina and Southern Africa.

According to this theory a single supercontinent, known as Pangea, existed in Paleozoic time, more than 400 million years ago. It seems that Pangea consisted of two conjoined moieties, Laurasia in the northern hemisphere and Gondwanaland in the southern hemisphere. With the close of the Paleozoic Era and the advent of the Mesozoic Era, about 200 million years ago, this ancient supercontinent began to break apart, and as it fragmented the several component portions, each carried on a tectonic plate, drifted away from each other.

The break probably began in Gondwanaland, as South America split away from Africa and drifted to the west, thus opening the South Atlantic Ocean. At about the same time another portion of Gondwanaland, which was to become peninsular India, broke away to slide northward along the east

The geophysical evidence for this physical evolution of the Earth becomes increasingly convincing with each passing year. Does the paleontological evidence—the evidence of the fossils—accord with the geophysical evidence? And if so, how do we now interpret the evolution of life on the drifting continents?

Some Early Inhabitants of Pangea

We return to the examples of late Paleozoic amphibians and reptiles in central Europe and North America, and of early Mesozoic reptiles in South Africa and South America. As already mentioned, we must invoke long intercontinental movements covering half the circumference of the globe or more,

Fig. 17-3. Relations between Laurasia and Gondwana at the end of the Permian, i.e. about 230 million years ago. In that period there must have been definite communication routes between the group of northern continents (Laurasia), here represented by the Eurasia-Greenland block, and the group of southern continents (Gondwana) here represented by Africa. This is the conclusion that can be drawn from the presence in southern Africa and in central-northern Russia of dinosaur remains of that period. They were essentially large herbivorous dinosaurs (pareiasaurs) and their enemies were carnivorous. The morphological relationship between the Russian and African groups are evident from a comparison of the skulls shown here: compare the African carnivore (1) with the Russian carnivore (2), and then the African herbivore (3) with the Euroasiatic counterpart (4). The arrow indicates the probable communication routes: the light colored area indicates dry land as it is today, the darker area the present-day continental shelves.

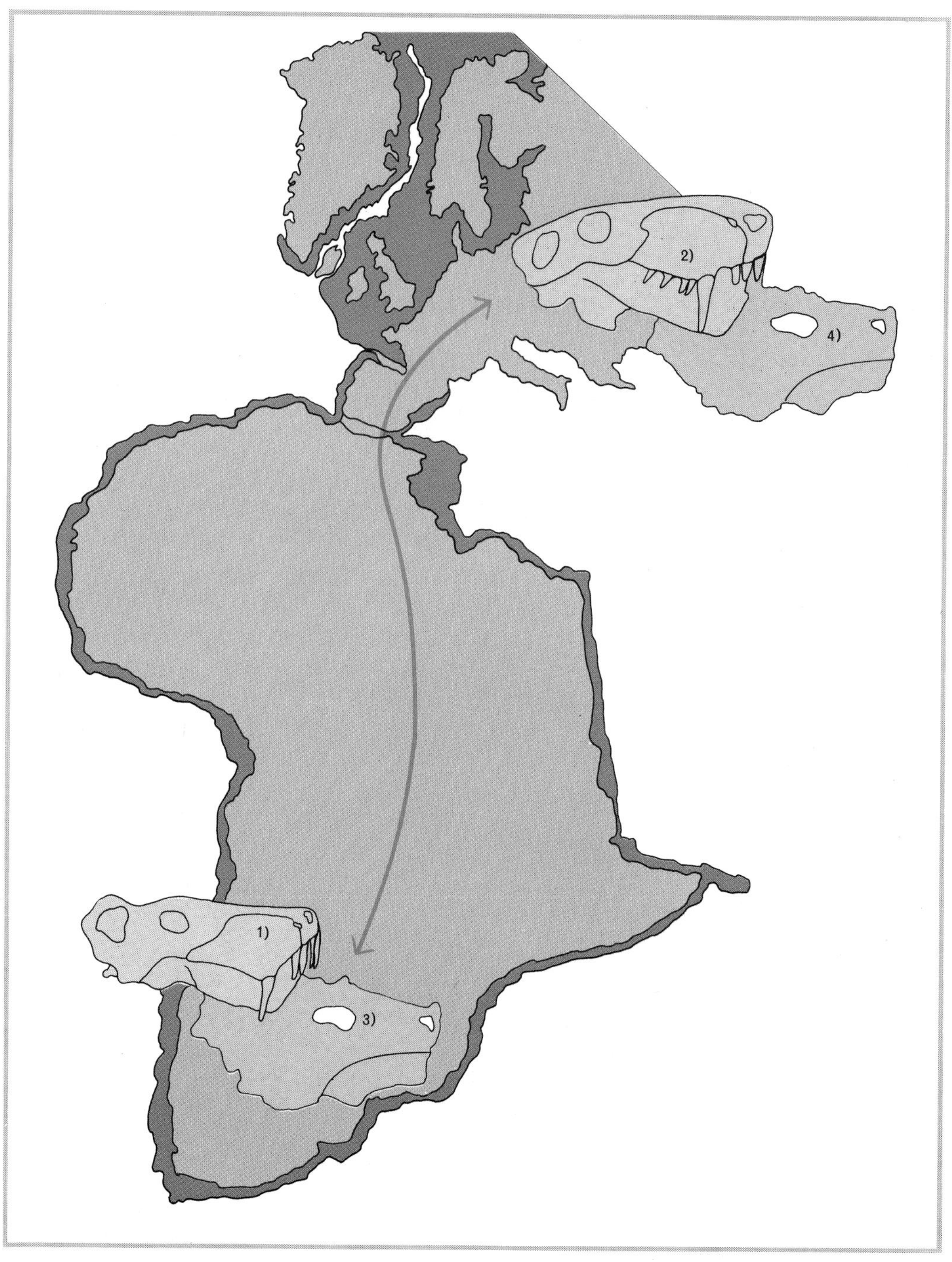

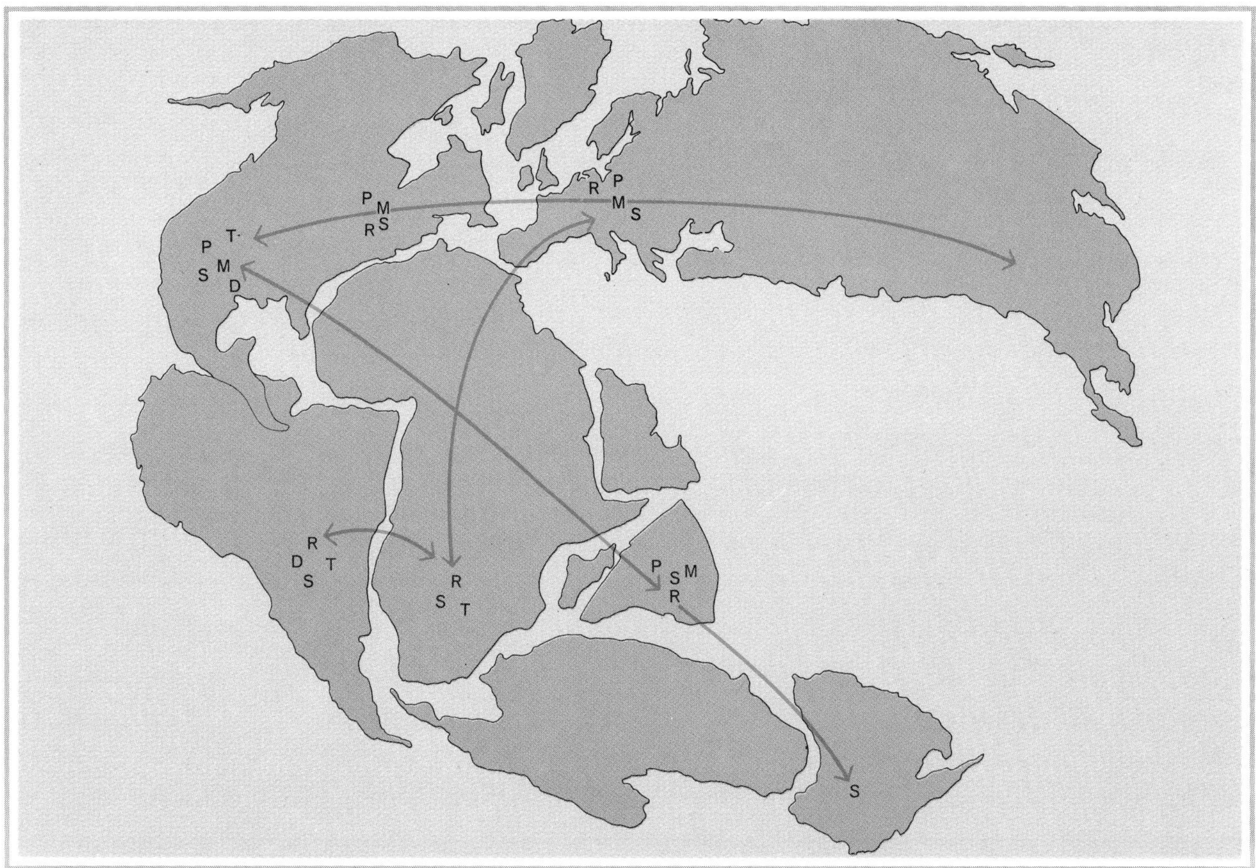

Fig. 17-4. Tetrapod migrations in the Upper Triassic, some 200 million years ago; these were made possible by the merging of all the continents into a single large land-mass, Pangea. The remains of these dinosaurs are found over a very wide latitude range today, from northern Europe and North America and from eastern Asia to Australia. In the key: D, dicynodonts; M, metoposaurs; P, phytosaurs; R, rhynchosaurs; S, saurischia; T, tritylodonts.

if the obviously close relationships of these ancient land-living vertebrates are to be explained upon the basis of fixed continents with connections more or less as they exist today. But if there had been a former Pangea, extant in the Paleozoic Era and just beginning to rift apart during the early part of the Mesozoic Era, the occurrences of these early vertebrates are brought within reasonable distances of each other. Thus the late Paleozoic amphibians and reptiles of central Europe and middle-western North America, instead of being separated by a migratory distance of 15,000 km or more, would have been perhaps 5000 km or less apart. Such a distance might very well be within the limits of a faunal range, as we understand the ranges of modern land-living vertebrates. Likewise the early Mesozoic reptiles of South Africa and South America, separated by a migratory route of at least 30,000 km on a fixed-continent globe, would have been within about 2000 km of each other on the Gondwanaland portion of Pangea. This latter distance certainly is encompassed within the range of a vertebrate fauna, as based upon modern distributions. Moreover the mere 2000 km that might have intervened between the early Mesozoic reptiles of South America and South Africa offers a possible clue as to their very close relationships. In many instances these reptiles were in essence the members of a single fauna, inhabiting a continuous range in Gondwanaland. These are cogent considerations, indeed.

Yet it still might be argued that these early land-living vertebrates inhabited a fixed-continent globe, reaching the extremities of their occurrences along extended routes of intercontinental migration.

Within recent years this rather tenuous argument has been effectively refuted by discoveries in Antarctica.

The Lystrosaurus Fauna. The *Lystrosaurus* fauna is typically found in the lower Triassic rocks of the South Africa Karroo, the great semidesert basin occupying thousands of square kilometers between the folded mountains of the Cape, and Johannesburg. The fossil reptile, *Lystrosaurus* (which gives its name to the fauna) is so very characteristic of the lowest sediments in South Africa, that they have been designed as the *Lystrosaurus* Zone. Fossils are plentiful and among the fossils by far the most numerous are the remains of *Lystrosaurus*.

Lystrosaurus was a rather bizarre mammal-like reptile, as small as a cat or as large as a sheep, with a robust body, rather short, thick legs, a very short tail, and a peculiar skull in which there were just two teeth—a large tusk on each side. The toothless jaws were obviously sheathed with a hornylike covering, like the beak of a turtle.

Along with *Lystrosaurus* in South Africa are found carnivorous mammal-like reptiles, particularly the genus *Thrinaxodon,* which may be compared ecologically with a very small fox, or perhaps a civet. Also, there is a small, primitive reptile, *Procolophon,* lizardlike in form and probable habits, but not in relationships. In addition there are small truly lizardlike reptiles known as prolacertilians, obvious forerunners of the lizards that were to evolve during late Triassic times. And there are small labyrinthodont amphibians—members of a large and diverse group of amphibians that lived during late Paleozoic and Triassic time. There are also other elements in the *Lystrosaurus* fauna, but these are the common and characteristic forms.

During the austral summers of 1969–1970 and 1970–1971 the *Lystrosaurus* fauna was discovered in the Transantarctic Mountains, about 700 km from the South Pole. It is a characteristic assemblage, showing frequent specific identities with the *Lystrosaurus* fauna of South Africa. *Lystrosaurus* is abundantly represented, while *Thrinaxodon* and *Procolophon* are commonly present. And the same species typical of the African forms represent these genera in Antarctica.

Antarctica is now an isolated island continent, distant by

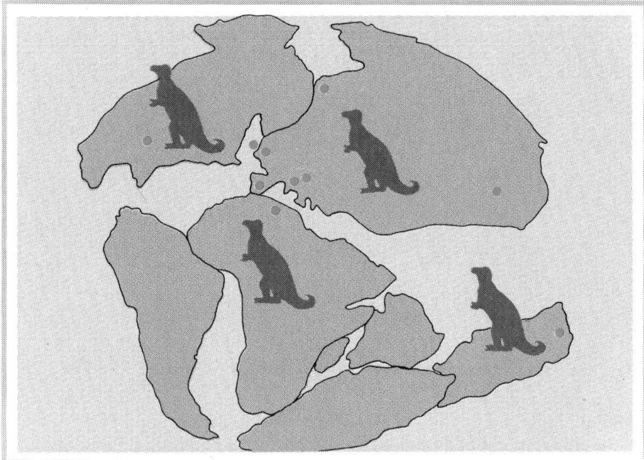

Fig. 17-5. Distribution of Iguanodonts at the beginning of the Cretaceous, some 136 million years ago. Their remains are found from the Spitzbergen islands over a very wide latitude range. This indicates not only a different distribution of the dry land but also a more uniform climate.

thousands of kilometers from the other continents of the world. If continents were fixed, then the amphibians and reptiles composing the *Lystrosaurus* fauna, all of them terrestrial animals, could only have reached the antarctic continent by an enormously elongated land bridge which has since foundered in the depths of the Antarctic Ocean. There is no geological evidence for such an impossibly long bridge. Moreover, the *Lystrosaurus* fauna bespeaks a tropical or subtropical environment, which can hardly be supposed to have prevailed at the southern axis of the globe.

Therefore, the *Lystrosaurus* fauna in Antarctica would seem to offer the strongest kind of paleontological evidence for the close litigation of the south polar continent to southern Africa, in latitudes where tropical plants, amphibians and reptiles could flourish. And since the evidence for the close connection of Antarctica to Africa, as attested by the occurrences of the *Lystrosaurus* fauna, is so overwhelmingly convincing, there is no longer any reason to doubt the close connection of Africa to South America, thus accounting for the very close relationships of early Triassic reptiles in these two continents; or for the earlier close connection between Europe and North America, thus nicely explaining the close relationships of Paleozoic amphibians and reptiles of central Europe and the middle United States. It therefore follows that the paleontological evidence appears to be in full accord with the geological and geophysical evidence, to indicate the probable reality of a former Pangea.

It should be added that the *Lystrosaurus* fauna is also known from peninsular India and from southeast Asia; which taken in conjunction with recently adduced geological and geophysical criteria, points to the possibility that these regions also were portions of ancient Gondwanaland, united with southern Africa and Antarctica. In fact, the presence of the *Lystrosaurus* fauna in southern Africa, Antarctica, peninsular India and southeast Asia would seem to represent the scattered remnants of what was once a continuous faunal range—a range subsequently fragmented and dispersed by the process of continental drift, acting through the long years of geological time.

The Vertebrates of Pangea. With the obvious coincidence among the several lines of evidence—geological, geophysical and paleontological in particular—there is now good reason to think that the land-living backboned animals, conveniently designated as the tetrapods, (as well as other forms of terrestrial life) evolved on mobile continents. How is the evolutionary progress of these vertebrates on shifting continental blocks to be interpreted?

At the beginning of tetrapod evolution it would appear that Pangea was a pristine supercontinent, a great continuous landmass as yet unbroken by the rifting that later was to take place. Consequently there were broad overland avenues for the dispersal of tetrapod faunas. Consequently the amphibians and reptiles of central Europe were closely related to those of

Fig. 17-6. Migration of dinosaurs in the Upper Cretaceous in the northern continents (dark). The presence of identical dinosaurs in Mongolia and in the northwestern part of America forms a strong indication in favor of the existence of a bridge which straddled the Bering Strait and must have linked the two areas, enabling these giant reptiles which could not swim, to cross on dry land. The distribution of these dinosaurs was possibly restricted by the shallow seas then present in the center of North America and Eurasia.

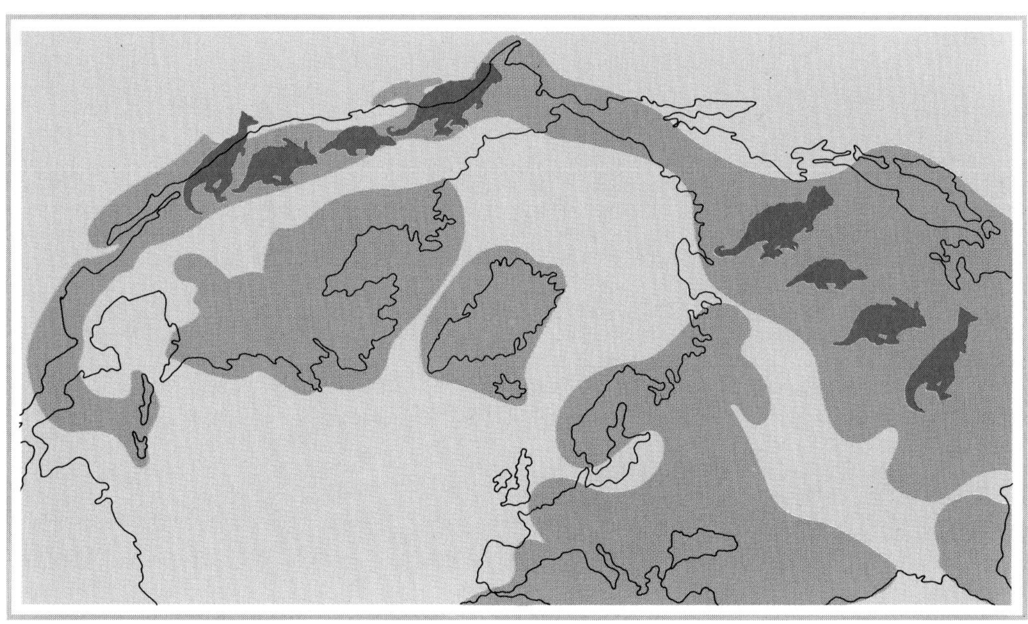

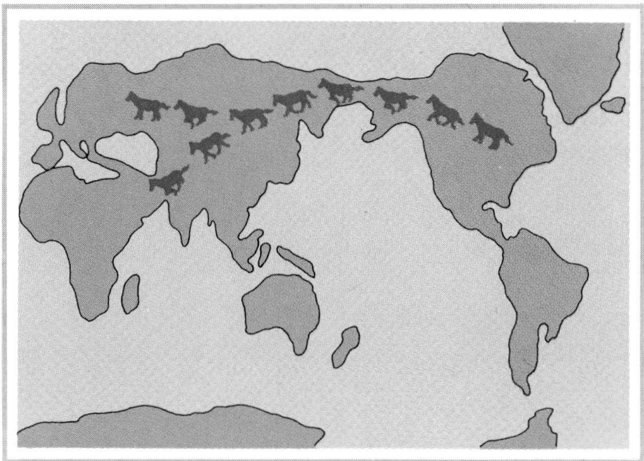

Fig. 17-7. Migrations of *Hipparion*, ancestor of the horse, in the Pliocene; from North America, where Upper Miocene forebears had appeared, it moved to Eurasia and Africa, crossing a land bridge corresponding to the present-day Bering Strait.

North America, as has been pointed out. Furthermore certain fossils show that there was an early interchange of Paleozoic tetrapods between South Africa and South America, as might be expected. Indeed the presence of the little Permian reptile, *Mesosaurus,* in South Africa and in southern Brazil, and nowhere else in the world, was a fact avidly seized upon by the early proponents of continental drift in support of the theory. Now were the faunal relationships confined to longitudinal lines during Paleozoic time? Many of the Permian reptiles of Russia are very closely related to their counterparts in southern Africa, indicative of broad faunal movements along latitudanal lines. Additionally, relationships are to be seen between the Permian reptiles of Russia and some fragmentary but nonetheless definitive Permian fossils found in the central part of the United States. In short, Pangea would seem to have been one world during late Paleozoic time—at least so far as the distributional migrations of tetrapods was concerned.

The Permian and Triassic periods in certain respects constitute a rather distinctive phase of Earth history, marked by the continuity of life on the land. There were many extinctions at the close of Permian time, yet in broad aspects many of the tetrapods so characteristic of the Permian continued their evolutionary histories into the Triassic period.

However, the Triassic was a period of change, so that many new tetrapods appeared within Triassic faunas, and these newcomers, combined with the Permian holdovers, gave the tetrapod assemblages of the Triassic a very distinctive cast.

The early Triassic *Lystrosaurus fauna,* so characteristic of Gondwanaland, has already been described. A somewhat later fauna was widely dispersed in Gondwanaland, its various constituents having been found in South America, southern Africa, peninsular India and southern China. This assemblage is typified especially by advanced mammal-like reptiles, such as *Cynognathus* and by large, tusked mammal-like reptiles, such as *Kannemeyeria* and its relatives.

At the same time, rather different tetrapod faunas inhabited the northern hemisphere or Laurasian portion of Pangea. Here there were various labyrinthodont amphibians and early thecodont reptiles—the thecondonts being specialized reptiles, some of which gave rise to dinosaurs, others to crocodiles, others to flying reptiles. So it would seem that there may have been a sort of faunal dichotomy between the northern and southern segments of Pangea at this stage of earth history. But the dichotomy, if it was real, disappeared with the advent of late Triassic time. By the late Triassic the early dinosaurs, descended from thecodont ancestors, had become established and ranged widely across the lands. Similar upper Triassic dinosaurs are found in all portions of Pangea (except Australia and Antarctica, where the lack is probably a result of insufficient fossil evidence at present), indicating the continued availability of overland migration routes.

At this time there was significant wave of extinction among the land-living tetrapods. The old holdovers from the Permian largely disappeared. The new reptiles that arose in the Triassic became dominant. This was the beginning of the reign of the dinosaurs.

The Age of Dinosaurs. Although the dinosaurs arose and spread throughout Pangea during late Triassic history, it was with the advent of the Jurassic period and from then until the close of the Cretaceous period that the dinosaurs were truly dominant. They ruled as the undisputed lords of the lands during a time span of 100 million years, when these lands were being disrupted and shifted from the ancestral Pangean pattern to something approaching the arrangement of continents in the modern world.

The rifting of Pangea had commenced in late Triassic time, as we have seen. During Jurassic history this rifting continued. South America, continued to pull away from Africa, possibly in part by a counterclockwise rotating motion, which opened the lower part of a narrow South Atlantic ocean, leaving the northern portions of the two continents connected. Likewise Laurasia was rotating in a counterclockwise manner, thus separating the eastern coast of North America from the Mauretanian coast of Africa and opening the lower portion of a narrow North Atlantic ocean. But the North Atlantic remained closed at its upper end by the firm junction between Europe and North America across the region of Greenland. And Laurasia maintained a contact with Gondwanaland across the Spanish-Moroccan "hinge." Antarctica and Australia, still firmly connected to each other, had not as yet moved away from Africa.

It is possible that peninsular India had commenced the long journey to the north, that eventually was to bring about its collision against what was then the southern border of Asia. Some authorities picture the Indian peninsula as an isolated island during Jurassic and Cretaceous time, floating away from its original position next to Africa. But there are large dinosaurs of Jurassic and Cretaceous age in the sediments of the Indian penisula, indicating that this land mass somehow maintained a firm contact with the rest of the world during these two Mesozoic periods. Perhaps peninsular India ground its way up the east coast of Africa, in the manner that today California is grinding its way up the west coast of North America.

However that may be, there were connections whereby dinosaurs roamed far and wide across the surface of the rifting continents. This is particularly well documented by the closely related dinosaurs of late Jurassic time—most of them giants—which are found in western North America, Europe and Africa.

Such wide dinosaurian distribution continued into and through the Cretaceous period, when these great reptiles had reached the zenith of their evolutionary development. The advent of Cretaceous time was marked by an evolutionary "explosion" of the angiosperms—the flowering plants. This sudden multiplication of plant life opened vast new pastures, so to speak, for plant-eating animals. So the herbivorous dinosaurs responded to the new ecological opportunities by evolving through rich and intricate patterns of adaptive radiation. Never had the dinosaurs been so numerous or so various.

As the dinosaurs reached this apogee in their evolutionary history, the separating blocks of a rifting Pangea continued to draw apart. South America was virtually surrounded by oceans, but it possibly retained a narrow connection with northern Africa—as attested by the late Cretaceous dinosaurs of these two continents. Northeastern North America and northwestern Europe seemingly were still connected. At the same time further rotation had carried the other extremities of the two continents close together in the region of Beringea, so that there was an interchange of dinosaurs between western North America and eastern Asia. At this time a north to south seaway through the middle states divided the eastern from the western portions of North America, while another seaway in the Caucasus region similarly divided Eurasia into eastern and western segments. Africa probably retained a connection with Eurasia—again in the Spanish-Moroccan area, while Antarctica—Australia was still connected with Africa.

This was the world in which the dinosaurs were widely distributed—in part by way of ancient Pangean connections, but in part, as in the case of the Bering crossing, by connections that were to prevail in later ages.

Then, with the close of Cretaceous time, all of the dinosaurs became extinct. Other great reptiles, also characteristic of the Mesozoic Era, likewise disappeared. Why there was such widespread extinction at the end of Cretaceous time has been interminably argued. There is no point here in exploring the possible reasons for the great Cretaceous extinctions. Suffice

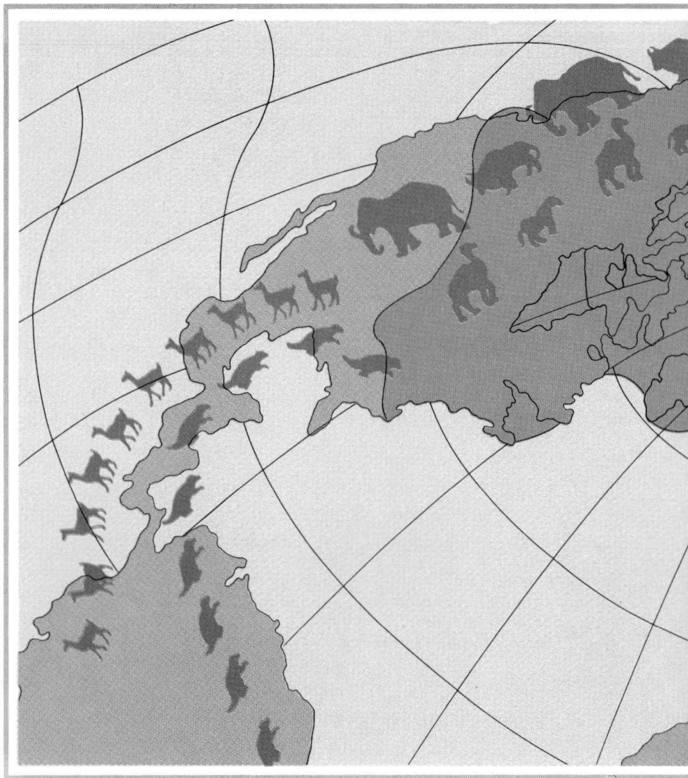

Fig. 17-8. Large intercontinental migrations of certain types of mammals at the beginning of the Pleistocene. Horses, of the genus *Equus*, and camelids crossed the bridge which joined North America, their place of origin, to Eurasia. At the same time the first mammoths and bison, using the same route, reached North America, but from the opposite direction. Even before the beginning of the Pleistocene, there were impressive movements of mammals on both sides of the Panama bridge between North and South America: small camels (llamas) pushed south from North America, their place of origin; and certain huge bradypodids made the crossing south to north. But there are various different cases: the woolly rhinoceros, which was very common in Eurasia, never reached North America, and the American antelope never reached Eurasia. In blue, the Quaternary ice-caps which conditioned these and other migrations.

it to say that the marvelous reptiles which had ruled the world for a 100 million years suddenly (in geological terms) disappeared. The Age of Dinosaurs had come to an end. The Age of Mammals was beginning.

The Age of Mammals. The first primitive mammals had arisen from reptilian ancestors during Triassic time, and haired mammals had evolved during the 100-million-year span of the Age of Dinosaurs. But as long as the dinosaurs were dominant, the mammals were small and insignificant. It seems that the very presence of dinosaurs on the land inhibited mammalian evolution.

With the disappearance of the dinosaurs at the close of Cretaceous times there was literally an explosion of warm-blooded mammals to occupy the lands vacated by the dinosaurs. The mammals evolved in a multiplicity of directions to fill the available ecological niches. Many of the mammals were small, in decided contrast to the giant dinosaurs, but some of the mammals quickly evolved into giants of sorts. These large mammals did not attain the overwhelming proportions of some of the dinosaurs; nevertheless they were giants in their own right.

There were in effect two successive waves of mammalian evolution during Cenozoic time. The first evolutionary wave, lasting through the Paleocene and Eocene epochs, was marked by the development of various groups of archaic mammals—especially heavy and clumsy herbivores and rather unspecialized carnivores. Then during Oligocene time there was a widespread replacement of the archaic mammals by modern groups of mammals—the ancestors of which had arisen during the Eocene epoch. The archaic herbivores were replaced by early representatives of the herbivores familiar to us—tapirs, horses, rhinoceroses, pigs and peccaries, camels, and the early ruminating herbivores. Likewise the archaic carnivores, often known as creodonts, were replaced by civets, cats, mustelids and dogs (in the broad sense). During Cenozoic history these groups of mammals evolved into their modern representatives, and were augmented by many other mammalian orders which inhabit the world of today. Such were the rodents, rabbits, proboscideans (elephants and their predecessors), bats, insectivores and other orders that need not be listed in detail. Not the least among the orders of evolving mammals were the primates, the group to which we belong.

One of the most significant facts of Cenozoic history was the long-enduring trans-Bering land bridge, foreshadowed in Cretaceous time. This connection between northeastern Asia and northwestern North America did not persist continuously through Cenozoic time; it was subject to periodic interruptions. But its existence was of longer term than its absence, thereby affording an avenue for intercontinental migrations that determined the general homogeneity of north circumpolar life—the life of Holarctica, the name give to modern North America and Eurasia north of the Himalayas.

Here we see the importance of a new connection between continents, a connection unrelated to the old Pangean land mass, a connection brought about by the drift of the continents into new positions, relative to each other. Several such new intercontinental relationships were to characterize the Cenozoic world, setting off from the Pangean world of Paleozoic and early Mesozoic time.

But some of the old connections persisted into the beginning of Cenozoic time. For instance, northern Europe and northeastern North America still formed a wide bridge closing the

northern end of the Atlantic ocean and affording a pathway for mammalian migrations. So it is that the Eocene mammalian fauna of western Europe and North America show very close relationships, typified for instance by the presence of the ancestral horse, *Eohippus* (more properly *Hyracotherium*) in the two regions. The connection was broken after the Eocene, never to be reestablished.

South America had drifted far from Africa by the beginning of Cenozoic time, and evidently had become joined to North America by an isthmian link. Here again we see a new intercontinental bridge, unrelated to the connections that prevailed in earlier geologic time. For a comparatively brief geologic interval there would seem to have been a flow of archaic mammals between North and South America. Then the Panamanian link was broken and South America became an island continent, to remain so until near the end of Cenozoic time. On this island continent a succession of unique faunas evolved, dominated by indigenous groups largely descended from the archaic mammals that had populated South America prior to its isolation. Marsupials—pouched mammals—were prominent in the Cenozoic faunas of South America. By Eocene time the Antarctic-Australian land mass had separated from Africa, and in due course these two blocks separated from each other, drifting through time to their present positions. It seems very possible, however, that before the separation and isolation of Australia it may have received a primitive marsupial fauna—perhaps by way of Antarctica, which in Cretaceous or very early Cenozoic time may have been connected with South America by what is now the Scotia Arc. However that may be, the history of life in Australia during Cenozoic time is a story of evolution on an isolated island-continent. This was a theater for the unique development of the marsupials, and of many distinctive forms of plant life, among which the hundreds of species of eucalyptus are especially impressive.

Antarctica, drifting to the south, became a frigid and essentially lifeless continent.

In the meantime peninsular India had been drifting toward Laurasia. Eventually it collided with the northern continent, pushing up the Himalayas to form a mighty barrier between this subcontinent and the rest of Asia. And on the Indian subcontinent life evolved along distinctive patterns. Much of the Cenozoic mammalian life of India evolved in concert with the developing mammals of Africa. There obviously were movements back and forth. And Africa, the core of ancient Gondwanaland, was the vast locale for mammalian evolution on a grand scale during the extent of Cenozoic history. Today we can see in Africa a legacy of life, unparalleled in the rest of the world.

The Age of Man. As the Age of Mammals progressed inexorably toward our own time there was a steady and perceptible cooling of climates in the middle and high latitudes. The Earth, which generally speaking had enjoyed benign climates during much of the Cenozoic Era, was gradually subjected to more varied and extreme climates, with an increasingly sharp definition of climatic belts, from the equator to the poles. Great chains of mountains arose, in part as the result of pressures engendered by the drift of continents. The collision of peninsular India against Asia caused the upward wrinkling of the Himalayas. The Rockies, the Sierras and the Andes were pushed up as North and South America drifted to the west, exerting pressures on the Pacific tectonic plate. The Alps arose in Europe. And the polar ice caps grew.

In the north there were great southward advances of continental glaciers across northern Europe and North America during the last of the Cenozoic epochs—the Pleistocene, often called the Great Ice Age. Four times the glaciers advanced; four times they retreated. And we are living in the age of the fourth retreat.

However, the Pleistocene Epoch had its beginnings well before the southward advance of the first northern glaciation. This beginning was marked by widespread migrations of modern mammals in the northern hemisphere—particularly back and forth across the trans-Bering bridge. Horses and camels flowed from North America, the center of their origins, into the Old World, while elephants and bison crossed the northern bridge into the New World.

South America, an island continent during much of Cenozoic time, became reunited with North America by a rejuvenated Panamanian Isthmus, so that during late Pliocene and Pleistocene times there were processions of mammals back and forth. Horses and little camels (which we call llamas), and various carnivores including bears and large cats, invaded South America to overwhelm much of the indigenous fauna that had evolved on that continent through so many millions of years. Giant ground sloths and glyptodonts, and armadillos (the little cousins of glyptodonts), and porcupines, entered North America from the south.

India and Africa were biotic gardens of Eden, where mammals flourished on a scale and in numbers perhaps never before realized on the Earth. (We see the remnants of this remarkable flowering of the mammals in the modern African fauna.)

It seemingly was in India and Africa where man had his beginnings. Manlike "apes" are found in the late Cenozoic of India. "Apelike" men—the australopithecines—are found in Africa. And from the australopithecines early men evolved in Africa and Asia. Pithecanthropoids, who used primitive tools and fire, were widely dispersed through the lands of Asia and Africa. From them later men evolved, to inhabit all of the Old World. The Neanderthaloids were skilled hunters and makers of chipped stone tools, who inhabited much of the Old World, pushing into northern Europe when that land was a subarctic realm, refrigerated by the last of the great continental ice sheets. At this time Eurasia was also inhabited by men of modern aspect, the so-called Cro-Magnons, who embellished caves in France and Spain with marvelous murals depicting the animals of that time and place.

Until perhaps 20,000 years ago the Americas were to man a terra incognita. Then stone age hunters came out of Asia and in a succession of migratory waves crossed the trans-Bering bridge into North America, and wandered down through the Isthmus into South America.

When man entered the Americas there were still many large mammals living in this land—mammoths, horses, camels, ground sloths, and various others. But within a relatively short time after man's entrance into the Americas there was a widespread extinction of these mammals. Was man the prime agent in this extinction? This is a question widely debated at the present time.

Interestingly there were many extinctions in Europe and Asia at the same time, involving animals that had long lived in association with man. For example, the woolly mammoth and the woolly rhinoceros, abundantly depicted on the walls of European caves, disappeared, as did various other mammals.

Australia, the floating island continent was also invaded by man only a few thousand years ago. The Australian aborigines, coming out of southern Asia, entered a strange world of marsupials, to make this their home. They brought with them one domestic animal, the yellow dog of Asia, which ran wild in Australia to become the Dingo.

Thus the modern phase of earth history began, the time when man had spread to all of the continents, the time when various large mammals so characteristic of the Pleistocene had become extinct. This was the last act in the long drama of life on wandering continents.

Conclusion

The history of evolving life on wandering continents is remarkably complex, a story of such dimensions that its salient features have been barely outlined in this foregoing discussion. Yet perhaps this brief account will give some impression of the closely entwined courses of organic evolution and continental evolution—the former dependent upon and affected by the latter. It is a story of the adaptations among plants and animals to the profound changes taking place through geologic time on the evolving, drifting continents.

Some 400 million years ago it would seem that there was literally a single world continent, Pangea, composed of a Laurasian portion in the northern hemisphere and a Gondwanaland portion in the southern hemisphere. The tetrapods of Paleozoic time evolved on this great supercontinent.

During the progess of Mesozoic history there was a rifting of Pangea, with the several resultant continental blocks drifting apart. And as the continents moved apart, there was an increasing diversity among the tetrapods that inhabited the moving segments.

With the advent of Cenozoic time the continental blocks approached their present positions and new intercontinental connections were established, notably the trans-Bering bridge and the Panamanian isthmus. These new conditions inevitably affected the course of tetrapod evolution on the continents, so that the faunas of our modern world became established.

Today we live on a globe in which Africa is in a sense a great faunal heartland. To the north is the oriental region, showing faunal relationships to Africa. And in the northern hemisphere the dominant land is Holarctica the great circumpolar Eurasian-North American land mass in which tetrapods, especially mammals, show the close relationships resulting from millions of years of longitudinal movement. In the southern hemisphere are the two island continents, Australia and Antarctica, the one with unique fauna, the other essentially lifeless, as well as the former island continent South America, joined to the rest of the world only at a geologically recent time, and showing in its fauna the result of long isolation and subsequent invasions.

As we view the history of life on the wandering continents we see in effect a geological-biological palimpsest. It is the story of the evolution of mammal on dispersed continents, written large over the earlier story of reptilian evolution on rifting continents, this in turn written over the ancient story of early tetrapods on a great ancestral supercontinent. It is the long extended background for the world in which we live.

EDWIN H. COLBERT

Bibliography: Hallam A., *Una rivoluzione nelle scienze della Terra* (1974); Ippolito F. (ed.), *Tettonica a zolle e continenti alla deriva,* Milano (1974); Sullivan W., *Continents in motion. The new Earth debate,* New York (1974); Calder N., *La terra inquieta,* Bologna (1973); Colbert E. H., *Wandering lands and animals,* New York (1973); Brouwer A., *Paleontologia generale,* Milano (1972); Tarling D. H., Tarling M. P., *Continental drift. A study of the Earth's moving surface,* London (1971); George W., *Animal geography,* London (1966); Darlington Ph. J. jr., *Biogeography of the southern end of the world,* Cambridge, Mass. (1965); Simpson G. G., *Evolution and geography,* Portland (1953).

To conclude this rediscovery of our planet we must not omit a dissident opinion—indeed the most authoritative of them all. Vladimir Belousov, amid the general enthusiasm of geologists over the last 15 years, has remained apart from those who suddenly saw in the plate-tectonics model a solution, and an accurate scale in which to place the global problems of the life and history of the planet. He has always asserted that the main features of the planet are not the result of imposing forces acting parallel to Earth's crust, but are to be explained by radial (that is to say, vertical) movements. In his view the only movements of which the Earth is capable are up-and-down ones. This view, substantially close to the fixed-continents model, had met, even in the times of traditional geology, with a number of apparent contradictions. Since the end of the last century geologists had recognized, in mountain ranges, the traces of great movements in the so-called coverage strata, great masses of rock which had without any doubt moved several hundreds of kilometers from their most probable place of origin. How could such movements, parallel to the Earth's surface, be reconciled with the conviction that the Earth's only movements were instead vertical? The answer came from a number of laboratory experiments carried out not only by Belousov but also by the Norwegian scientist, Ramberg. They arranged a series of strata formed of materials that could be accepted as simulating those forming the Earth's crust, and the upper part of the mantel, over a source of heat. Immediately above this heat source the material expanded and rose: here was the vertical motion. On the sides of this raised area, by the effect of gravity, portions of material slipped away, becoming superimposed one upon the other and causing lateral folding of the strata: hence the horizontal movement and hence too the origin, according to Belousov, of mountain ranges with folded strata.

Belousov's view is that the plate-tectonics model explains many problems but leaves as many again unexplained. His article, which now follows, puts forward an alternative to the plate-tectonics view. On the one hand it is a document of interest in the history of geology, and on the other it is a concrete and general admonition not to consider the plate-tectonics formula as a solution to all problems, because the Earth certainly has more surprises in store for us than we can imagine.

VLADIMIR VLADIMIROV BELOUSOV

Director of the Institute of Earth Physics of the USSR Academy of Sciences. He was born in Moscow in 1907, and became interested in geology at the beginning of the 1930s. Belousov's main contribution, however, was made to global tectonics in which he has always been a supporter of the pre-eminence of vertical movement, as opposed to the model of plate tectonics that foresaw the development of great horizontal movement. The following article is presented by Roberto Malaroda, Director of the Institute of Geology at Turin, who in the course of his scientific activities has been able to measure the interest of several of the opinions expressed by Belousov, even those on the reconstruction of the history of the Alpine mountain chain.

VLADIMIR VLADIMIROV BELOUSOV

An Alternate Viewpoint

Overleaf: A general view of a large part of the Alpine range moving eastward from the ample arc which it forms starting from the Gulf of Genoa. The Alpine range can be called the birthplace of geology; it was here, at the end of the past century, that the first tectonic hypotheses were born and the presence of large horizontal forces and movements was predicted.

The Alps are one of those typical folded mountain ranges about whose origin geologists are divided, some arguing in favor of ample horizontal movements, and some against. The hypothesis of plate tectonics implies the intervention of large horizontal forces. V. V. Belousov and R. Malaroda express herein a sharply differing opinion (NASA).

The deformations of the Earth's crust attest to the fact that both subvertical and subhorizontal internal movements have taken place. Certain structures, such as faults, horsts, grabens, diapirs, and ascent of magma indicate the prevalence of the former; other structures, such as slip planes, transcurrent faults, and continental drifts indicate the latter.

Until now the primary cause of the forces which act on the crust and on the interior of the mantle has been unknown; one hears of contraction and dilatation of the entire Earth or of some parts of it, of effects of astronomical nature (effects of tides, of spin rate, of direct or near collision with other celestial bodies), of forces due to magma or to the weight of sediments, of isostatic equilibria due to density differences, or of convective currents produced by thermal gradients. Very little is known of the physical, chemical, and physicochemical equilibria of the more internal parts of the Earth or of their stability; in particular, little is known of the distribution of radioactivity which can be a cause of local and perhaps rapid thermal gradients. Deep ignorance also exists about reactions of hydratation, dishydratation and ionic migration which can create variable density gradients in materials located at different depths in the Earth's interior.

Therefore, it can be understood how either one of the two above-mentioned types of movement (horizontal and vertical) has been, on occasion, considered prevalent according to the assumed motive force. Either movement can be considered as the only real one since vertical forces can be viewed as simple, local consequences of horizontal compressions or extensions, and horizontal forces can be reduced to particular gravity-induced accidents connected with uplift or subsidence of more or less vast portions of the Earth's crust. To the geologists who first became interested in structural problems, the existence of radial (or vertical) forces should have appeared more logically obvious. These forces could be attributed to faults (invariably represented as vertical as in the first real geological profile published in 1719 by W. Strachey) or to subsidence (as in the first ideal geological profile drawn in 1669 by N. Stenone in order to illustrate his concept of "erosional mountains") or to the thrust of magma (as in the plutonistic ideas and in the first transverse geological profile of the Alps published in 1800 by P. S. Pallas). The hypothesis became sanctioned in L. von Buch's now classic theory of the "Uplifting Cratons." E. Suess had the distinction of being the first to attract attention to horizontal, or tangential, movements which he attributed to compressions between mutually approaching rigid crustal platforms (cratons) on more plastic and unstable intervening areas (geosynclines). The sediments of the geosynclines are folded and squeezed out of the basin on which they deposited and are theorized to be possibly subjected to important vertical movements which are chronologically and quantitatively secondary to the horizontal ones. Other vertical displacements, such as horsts (columns) and grabens (trenches), are connected with the ascent of magma and are related to distensional activites also of horizontal nature.

The theory of continental drift, formulated in 1912 by A. Wegener, maintained the idea that orogenesis stems from horizontal displacements of cratonic areas. The Wegenerian hypothesis has had alternating fortunes and has been totally abandoned in some of its details (the origin of the Mid-Atlantic Ridge, the origin of the South American Cordilleras), and in revised form, it is essentially contained in the modern theories

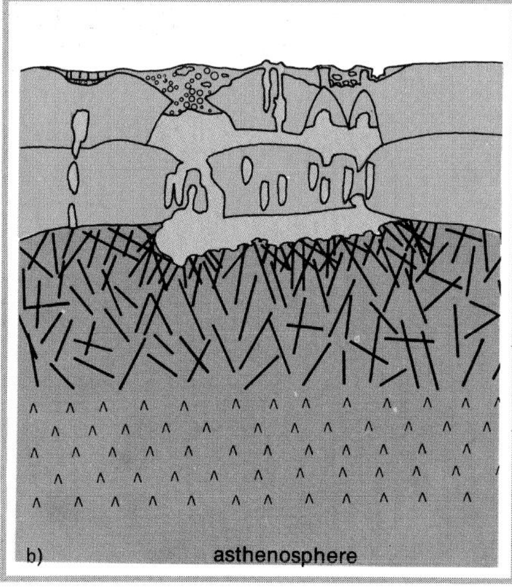

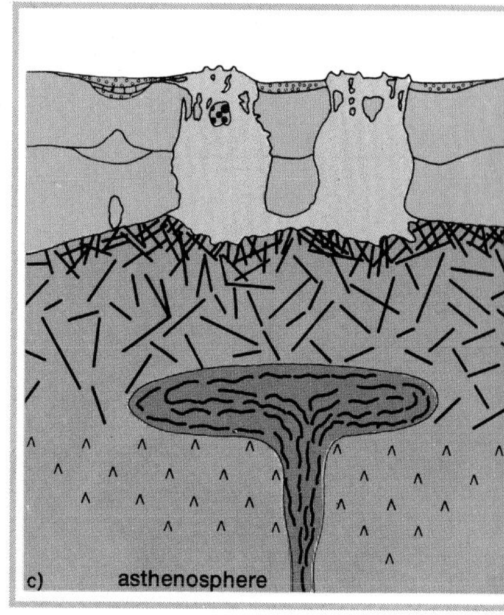

of ocean crust spreading and of plate tectonics. Numerous pieces of evidence and in particular the knowledge of the paleomagnetism of rocks of the oceanic crust and of the age of its overlying sediments, furnish a satisfactory explanation of the evolution of the Atlantic Ocean, of the genesis of new oceanic crust and of the origin of the oceanic ridge. In cases such as the Andes and Indonesia the centrifugal motion of the ocean crust gives a convincing explanation of such morphological and structural features as oceanic trenches bordering orogenic areas, and their connected magmatic phenomena.

These local successes led to the application of this theory to the entire surface of the Earth and in particular, to all orogenic areas and to all locations in which the oceanic crust or fragments of it come to the surface. The tangential motions generated by plate displacements are now considered by many geologists to explain by themselves all tectonic processes. The plate tectonic theory has thus been given a universal value and the name of global tectonics. This process has resulted in the complete disregard of, or the reduction to a role of mere accident, of all geological, geophysical and experimental data which support essentially vertical movements. Nevertheless, particularly on the vast continental cratons, there is ample evidence of the existence of purely vertical movements which are often repetitive and which present inversions of an oscillatory character; they extend over areas which can be large and isolated (basal folds), or small and associated (intercratonic geosynclines) with subcircular contours or in elongated bands and which seem to be unexplainable by the spreading of the ocean crust even if aborted.

Stratigraphic and paleoecological studies point to great differences in the thicknesses of contemporary sediments over contiguous and often small cratonic regions. Geodesic and geophysical arguments for recent cases and geomorphological arguments for ancient ones, demonstrate the considerable speed of vertical movements, their sharp boundaries and their inversion and frequent oscillatory character. The causes and mechanism of isostatic vertical displacements can be considered perfectly known in particular cases such as in areas which have undergone recent deglaciation. Even in the interior of continents, geothermometry identifies areas with conspicuous thermal gradients and with large heat flux (geothermic spots). This leads to the consideration of metamorphism, anatexis, local volume and density variations, and local uplifting of matter in plastic or more or less completely molten state. Finally, seismology proposes for the crust and the mantle models in which layers of greater and lesser density are interspersed, with the inevitable consequence that vertical forces of diapiric type are produced according to the laws of isostasy.

In spite of the clamorous attention presently being accorded to plate tectonics it should not be forgotten that, although with less fanfare, earth science is also progressing in the analysis of prevalently or exclusively vertical displacements. In many cases these displacements can explain not only local structures of modest dimensions but also large-scale phenomena such as the formation of geosynclines and orogeny, the genesis of sedimentary deposits, the phenomenon of continuously subsiding regions being often found adjacent to continuously uplifting ones, the location of granite and granitoid domes, and the upward convection of basal portions of the lithosphere. Moreover, geophysics and geochemistry indicate that the crust and the mantle below it must be closely interdependent. These considerations prevent some authors from conceding to plate tectonics the primary mechanism which propels the geodynamic evolution of the globe.

Among the studies of orogenic phenomena with prevalently vertical movement one should recognize those which led E. Haarmann to formulate in 1930 his "Oszillationstheorie" and R. van Bemmelen in 1933 his "undation" theory. In 1963 H. Ramberg and his coworkers at the University of Uppsala initiated experimental geology studies on models in which the density ratios of various Earth strata are simulated by the use of various substances (clays, waxes, putties, silicone putties, mixtures of colophony or paraffin with different oils) and in which the action of gravity is simulated in centrifuges. Similar studies were intitiated in U.S. universities. The results have repeatedly confirmed in increasing detail that gravity is by itself sufficient to cause in the mantle and crust uplifting displacements of diapiric type in the shape of columns or mushrooms with expanded top. These diapirs may or may not reach the surface, but they always produce dome-shaped insertions of granitic and gneissic base material in the overlying soils; folded overlying structures or thin strata covering the upper reaches of the diapir; downward pull and subsidence of the areas which surround the uplifted material. Experiments of the same type have been conducted by research workers at the University of Moscow with analogous results.

Vladimir V. Belousov deserves to be recognized at the forefront of the proponents of research which emphasizes the role of vertical movements. Aside from the personal experience he acquired particularly in the Caucasus, as director of the Department of Geodynamics of the Geophysical Institute of the U.S.S.R. Academy of Sciences, he was able to synthesize the precious data gathered by geological research over the entire Russian territory and especially in cratonic areas and in regions located at the border between cratons and geosynclines. Throughout his career, Belousov developed ideas which are particularly interesting owing to his careful evaluation of available geophysical data in the above-mentioned regions, to his vast knowledge of other sections of the globe

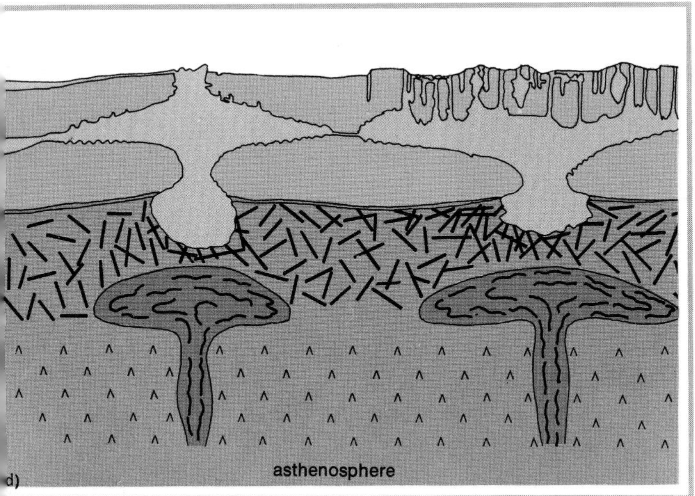

Fig. 18-1. Four conditions of the Earth's crust thought to correspond to four different geodynamic regimes. (a) the situation in the geosynclinal regime, that is in long and deep trenches subjected to rapid and continuous subsidence and to a corresponding influx of detritic material; (b) the situation during the folding and metamorphic phase which occurs when the subsidence characteristic to geosynclines is inverted; (c) the situation during an orogenic phase, that is during the uplift of a mountain range; (d) the situation in a regime of tectonic activation, that is of thrust from below which causes at the surface either simple tectonic fractures of the appearance of the pillars and tectonic trenches characteristic of vertical displacement of portions of the crust. In (a) the asthenosphere is warm and greatly mobilized, the lithosphere is penetrated in a diffused manner and consequently the above-mentioned generalized subsidence occurs. In (b), instead, the asthenosphere undergoes cooling while the thermic wave reaches the surface producing metamorphic actions and granitization, strong lithospheric oscillations and, consequently, vast folding phenomena. In (c) new molten basaltic rocks rise from the asthenosphere and the crust is uplifted. In (d) the rigidity of the crust is such that the thrust of the asthenospheric masses can only produce fractures or limited uplifts.

and to the position he has occupied since 1960 within the International Union of Geology Geophysics and in the Upper Mantle Project. Although Belousov's ideas have already been made partially known in the West through the treatise, "Basic Problems in Geotectonics" and other works, they are more precisely expressed in the pages that follow.

ROBERTO MALARODA

Evolution of Continental Areas

Geological history shows that, in continental areas, all endogenous geological processes (tectonic, metamorphic, and magmatic) take place in orderly successions thus revealing their intrinsically common character. Therefore, they can be grouped into regimes with specific characteristics (endogenous regimes).

Different regimes can be distinguished on the basis of the following characteristics:

1. Character and degree of penetrability of the Earth's crusts by magma. When high, it can be of two types: diffused and concentrated;
2. Character and intesity of magmatism;
3. Regional metamorphism and granitization: these phenomena determine the origin of crystalline crust and create impenetrable carapaces; thereafter, the crust becomes again penetrable only if it is fractured by new tectonic displacements;
4. Degree of contrast existing among various vertical crustal displacements;
5. Relation between crustal uplifts and subsidences;
6. Folding characteristics.

Typical examples of endogenous regimes are the following: geosynclinal, orogenic, cratonic, rift, horst, and graben (Fig. 18-1).

The geosynclinal regime has the following characteristics: great diffused penetrability of the crust; very accentuated magmatism; great contrast between vertical displacements; subsidence sharply predominant over uplifting.

The orogenic regime is also characterized by high crustal penetrability but in this case it is sharply concentrated; vertical displacements are sharply contrasting and uplifts are prevalent over subsidences.

The cratonic regime is associated with very reduced crustal penetrability and presents vertical displacements which are only feebly contrasting.

The rift regime constitutes a very particular manifestation of tectonic and magmatic activation of a craton in conditions of high impenetrability of the crust. Before fracturing, the crust offers a strong resistance to the pressure of the mantle material in its ascending motion.

Since metamorphic, granitization, and tectonic processes took place everywhere during the Archean (Precambrian) period, the geosynclinal regime must have dominated during that time over the entire continental area. The prevailing structures were then dome-shaped rather than linear. This "pageosynclinal stage," or "hypermobile stage" terminated approximately 3 billion years ago. At ttat time, the "transitional stage" began with the subdivision of the continental crust into

Fig. 18-2. Undulations of the Earth's crust during the Alpine orogenic cycle. The upper diagram shows the average value of the undulation velocity for the different periods. The lower diagram shows a statistical treatment of the same data and illustrates the average dispersion of the velocities. In both cases the values corresponding to two active zones are compared to those corresponding to plate zones. In dark color, the Caucasus geosyncline; in light color, the Crimean parageosyncline; in black, the Scythian plate; in white, the Sarmatian plate.

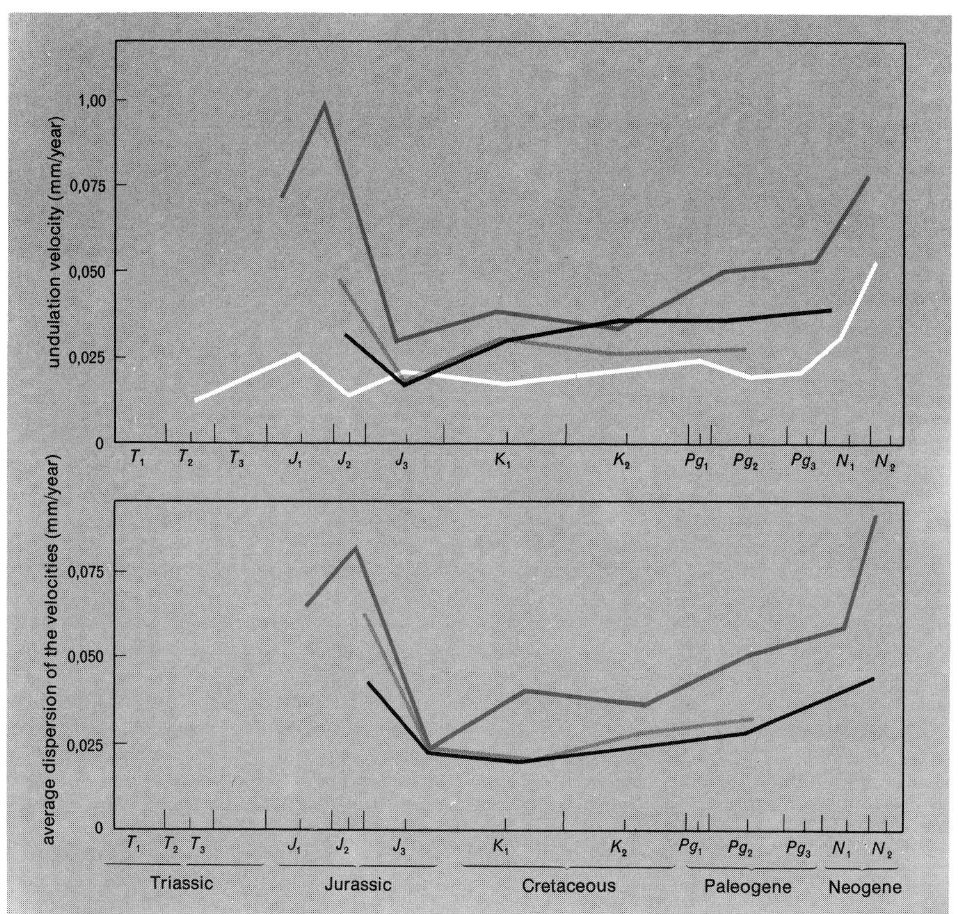

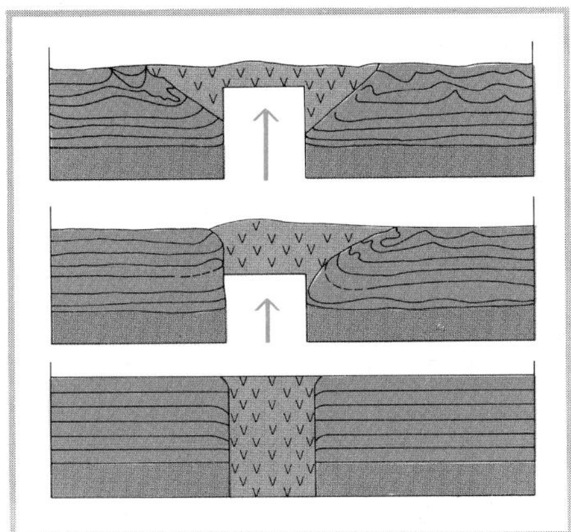

Fig. 18-3. Laboratory tectonics: a solid block of resin (center) expands and uplifts. As a consequence, a whole series of various types of folds forms on the stratified surface resins.

protogeosynclines and protocratons, whose spatial distribution remained unstable. The transitional stage normally extends from 1.6–1.8 billion years ago, that is until the end of the Middle Archeozoic Era. This marked the advent of the "stable stage" of the geosynclinal and cratonic regimes which continues at present. In certain regions the transitional stage had a

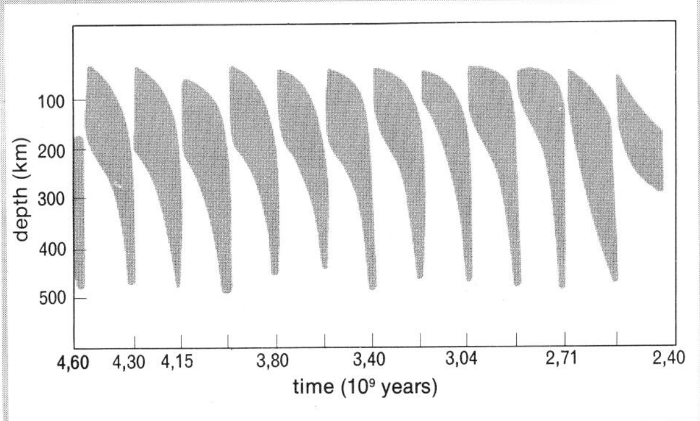

Fig. 18-4. Cyclic melting which has supposedly taken place in the Earth's upper mantle from the Middle Archeozoic Era (right) until today (left). The abscissa indicates the age of the Earth in billions of years from its formation, 4.6 billion years ago. Melting always starts in a deep location and expands rapidly upwards. The average duration of each period is indicated by the width of the color bands and it is approximately of 170 million years.

longer duration, extending in certain cases until the conclusion of the Archeozoic Era. At the end of the transitional stage, the protocratons amalgamated into the presently known cratons; this obviously occurred as a consequence of the fact that some zones comprised between groups of protocratons passed from a geosynclinal regime to a more stable one.

During the Phanerozoic Era the number of geosynclines was further reduced. The process was not random but followed a clear trend: the cratons assumed the role of "foci of stabilization" and increasingly more recent plates formed along their edges. As a consequence, during the Mesozoic and Cenozoic Eras, the reduction of geosynclines led to the formation of two elongated zones, the Circumpacific one and the Mediterranean one (or Tethys).

Over this background of a general trend toward decreasing endogenous activity in continental areas, some opposite trends were taking place with a temporary or local character. This latter class of phenomena is identified with the name of "tectonic and magmatic activation." During the Archeozoic Era regional metamorphisms and granitization phenomena were generally localized in well-defined periods which were separated by time intervals between 300 and 600 million years.

In Phanerozoic times, these endogenous cycles are normally called tectonic cycles. They are taken as the basis for the cyclic repetition of great general oscillations or undulations, i.e., subsidences or uplifts, which spread over entire continents or considerable portions of them. These undulations produced transgression and regression phenomena on a continental scale over periods of 150–200 million years.

In the geosynclines the tectonic cycles are characterized by the coordinated association of tectonic, magmatic, and metamorphic phenomena. The two Paleozoic cycles, the Caledonian and the Hercynian, took place everywhere with relative simultaneity. In contrast to this, the history of the Mesozoic and of the Cenozoic Eras was more complicated: a group of geosynclines developed following the "Atlantic Model" with a continuous Alpine cycle. Although these geosynclines are mostly related to the Atlantic Ocean, an analogous development could be observed over a portion of the Circumpacific geosyncline (from Japan to New Guinea). Other geosynclines follow the "Pacific Model" and manifest either a Mesozoic or a Cenozoic cycle.

The internal subdivision of geosynclines and cratons is connected to vertical movements of the crust, that is to undulations. The contrast between positive undulations (uplifts) and negative ones (subsidence) of both the geosynclines and the cratons undergoes modifications even during the time of a single cycle. It is at a maximum at the beginning and the end, while it is at a minimum during the middle portion of the cycle (Fig. 18-2). In the Phanerozoic Era, it is usually the rule that the geosyncline of each cycle is located in the interior of the region occupied by the geosyncline of the preceding cycle (although on a reduced area). The general direction of the geosyncline and that of its interior zones may however not correspond to the direction of the structures of the preceding geosyncline. There is reason to believe that at the end of each tectonic cycle regional metamorphism and granitization produced scarring of pre-existing deep faults and attenuation of structural inhomogeneities. Thus, at the beginning of each new cycle, the formation of a new system of faults not corresponding to the preceding ones determined the orientation of the new geosyncline. For a long time folding was considered to be the principal tectonic phenomenon. In reality, in comparison with vertical crustal movements, which are permanent and universal and in comparison with regional metamorphism and magmatism, folding possesses a considerably smaller scale and a much shorter period of formation.

Folds were and still are often considered a proof of horizontal compressions of the Earth's crust. In reality several types of folds exist: basal folds and injection folds do not depend upon horizontal compressions; only holomorphic folds do. On the other hand, there is reason to suppose that this compression is not applied externally to the entire geosyncline but that it develops internally in some zones of it. The most probable explanation is that these phenomena are of a gravitational nature, i.e. they are due to pressures conditioned by the dilatation of the upper portions of uplifted crustal slabs or by "deep diapirism." The latter consists of the tendency of granitic, granitoid, and metamorphic masses to emerge through the overlying rock strata (Fig. 18-3).

When a temperature increase causes partial melting of the rock or only the transformation of lattice water into porosity water, the rock tends to become less dense and consequently to move toward higher levels of the lithosphere. During the ascending motion of deep diapirs, both the uplifted materials and the surrounding rocks undergo complex deformations.

Furthermore, a correlation exists between endogenous regimes and heat flux. The latter reaches minimum values in the shield areas, [approximately 0.98 μcal/(sec.cm^2)], it is larger in areas which have undergone recent strong uplifting [1.80 μcal/(sec. cm^2) in the Tien Shan] and it is maximum in rifts (approximately 2.0) and in volcanic regions (up to 3.6). The above-mentioned differences should be related to the depth and the character of the asthenosphere which is essentially expressed by the greater velocity of seismic waves and by the greater height of the roof of the asthenosphere in correspondence to the regions with high heat flux. Even in geosynclines, the heat flux is high (from 3 to 5 times normal) whenever regional metamorphism develops. It follows that different endogenous regimes must be related to differences in heat flux and in the degree of partial melting of the mantle. It also appears obvious that endogenous regimes are connected with very deep heterogeneities (at a depth of 200 km and more). At this point we must recognize that the rhythmic character of the changes in the degree of contrast of vertical crustal movements and of the magmatic and metamorphic phenomena implies that the responsible internal processes should also have a rhythmic character.

Evidently, in the exterior part of the mantle, periods of excitation alternate with periods of quiescence. The former ones can be made to correspond to the times in which the highest temperatures are achieved in the asthenosphere. In such conditions both density and viscosity decrease while the density inversion existing between the asthenosphere and the lithosphere causes greater mechanical instability and mobility. Fracturing of the lithosphere into slabs implies asthenospheric diapirism and strong contrasts between uplift and subsidence of lithospheric slabs. On the contrary, whenever a regime of asthenospheric excitation exists under a lithosphere with high but concentrated penetrability, an orogenic regime will ensue. Finally, if the same conditions of asthenospheric excitation are combined with high but diffused lithospheric penetrability, a geosynclinal regime is created due to the fact that the diffused penetration of dense basaltic magma into the lithosphere forced the latter to subside. Therefore, it is possible to assert that the endogenous regimes of given areas depend on the state of the asthenosphere and on the behavior of the overlying lithosphere.

A. N. Tikhonov and coworkers demonstrated in 1970 the possibility of periodic, partial remelting within the mantle at a depth of 400 to 500 km. This zone of partial melting would move upwards thus melting overlying rocks. After a period of time, cooling forces the situation to stabilize until the formation of a new deep zone of partial melting begins a new cycle of upward propagation. Since each of these zones convects to the asthenosphere a certain volume of overheated material of deep origin, this may constitute the cause of periodic asthenospheric excitations (Fig. 18-4).

Asthenospheric diapirism can explain not only the contrast in vertical crustal movements, but also their magmatic and metamorphic evolution. Indeed, with each remelting the affected zones lose a part of their light elements and, gradually but irreversibly, active endogenous regimes are transformed into more stable ones. Therefore, tectonic and magmatic surface activity would indicate that from time to time the melting of deep rocks is intensified. As an example, this could be due to the fact that this process affects deeper strata of the mantle than had been previously involved.

On the other hand, let us consider the major folding regions, whose origin is usually attributed to horizontal compression forces. Valid arguments oppose the concept of external horizontal compressive forces; that is, forces applied by rigid cratons bordering the folding zone. These arguments are:

1. The shape of the corrugated areas; such areas are frequently subdivided into arcs which penetrate the cratons; this would require that the craton in question exercise on the corrugated zone pressures with various orientations. This is impossible due to the rigidity of the craton itself.

2. The chronological development of corrugated zones: initially, uplifting takes place in the axial portions and subsidence at the periphery; later, the intumescence migrates toward the sides while the uplifting of the interior portions continues; the migration of the intumescence, propagating as a wave toward the boundaries of the geosyncline, is not in agreement with genesis by lateral compression.

3. The alternation of different fold types within the folding zone; corrugation folds are separated from one another and surrounded by folds of other types such as basal and diapiric ones; since these are not caused by horizontal compression it

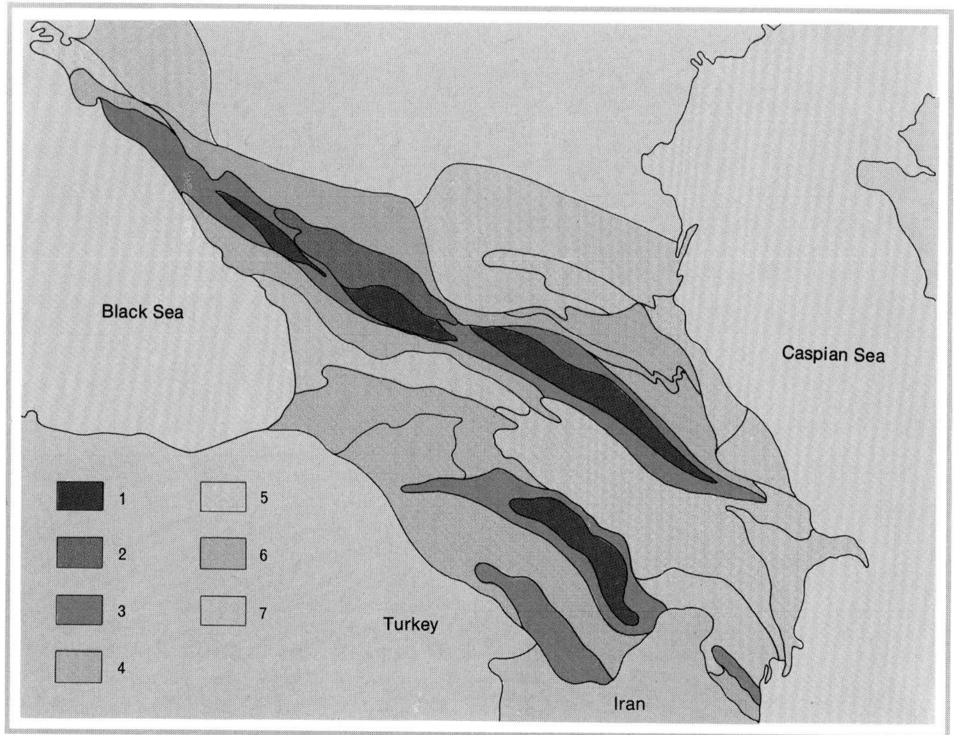

Fig. 18-5. Distribution of the different types of folding in the Greater Caucasus. Key: 1, stable unfolded area; 2, prealpidic base structured pillars or with raised listric wedges; 3, zones characterized by compression folding; 4, zones with brachianticlinal folds and domes; 5, areas with diapiric structures and folds; 6, Greater Caucasus, northern monocline; 7, post-tectonic deposits.

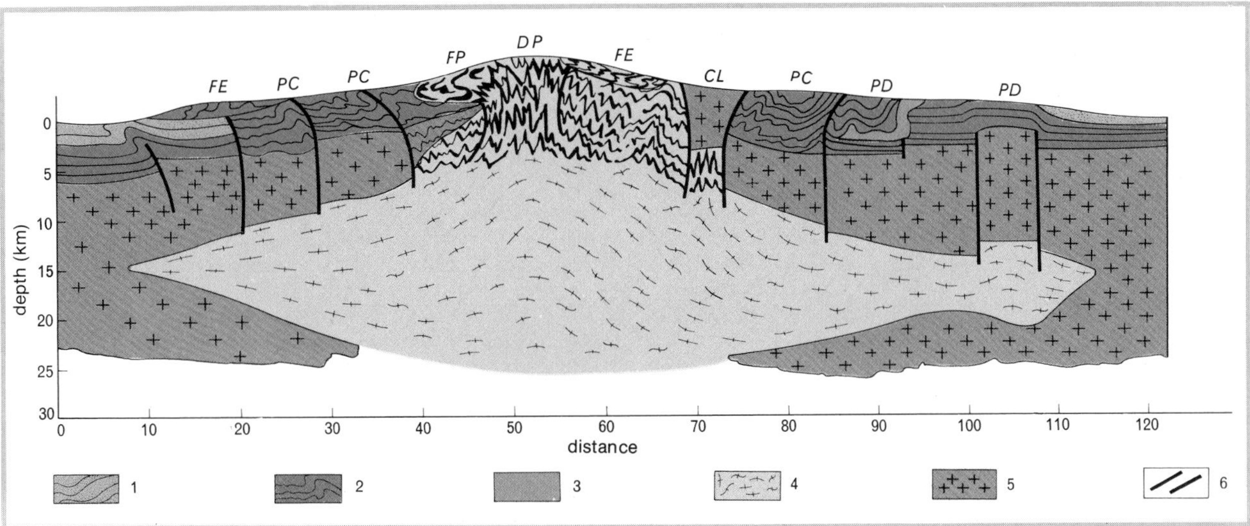

Fig. 18-6. Diagram of an orogenic phenomenon in which several types of folds are generated by a thrust from below due to a deep diapir. Key: 1, mostly detritic sedimentary deposits (flysch and molasse) which formed during the inversion stage of the geosynclinal displacement; 2, layers of evaporites which generate surface diapirs; 3, sedimentary deposits formed during a prevalently subsident phase; 4, activated base; 5, nonactivated base; 6, faults. The initials indicate: DP, deep diapir; FE, strata of Helvetidic type; FP, strata of Pennidic type; PC, compression folds; PD, diapiric folds; CL, wedge dilated in its upper portions.

is impossible to justify the propagation of external compression through zones containing basal and diapiric folds void of any trace of tangential compressive phenomena (Fig. 18-5).

From the above-mentioned considerations one can deduce that the horizontal compression causing the general corrugation folds must have internal and local character associated to processes which take place within the geosyncline itself. These processes are probably of a gravitational nature and are unleashed by the differential vertical movements of the crust which are peculiar to geosynclines. Usually, gravitational processes are interpreted solely in the form of rock displacements in the proximity of the surface and along inclined planes, a mechanism which is typified by the one which generated the Helvetidic type of fault. But gravitational displacements are not limited to surface motions; they are also evident in the dilatation of the upper parts of uplifted crustal slabs in which the lateral stress is not due directly to gravity but to the pressure exercised by upward flowing deep diapiric masses and the associated lateral sinking of relatively dense rock (Fig. 18-6).

The above-described conditions arise in numerous examples spread all over the world, from the Rocky Mountains to the Southern Alps. A profile of the Polar Urals by N. P. Kheraskov and A. S. Perfiliev (1963) shows that the external dilatation of the highest part of the Harbeisk anticline is associated with corrugation folds in the zone located along its western border (Fig. 18-7).

W. Bucher (1957) has explained the Appalachian folds as due to the ascent of the central metamorphic massif which is presently, for the most part, buried under the Piedmont depression, to its outward dilatation and over adjacent sediments and to the resulting thrust exerted on them. According to Ramberg (1966), an analogous role was played in the Norwegian Caledonides by gneissic massifs uplifting in the form of domes and surrounded by arcs of folds of metamorphic and sedimentary rocks from the Inferior Paleozoic Era. Phenomena of the same class are the actions exerted on surrounding rocks by the gneiss granite and "mushroom" migmatite domes or "complexes" described by Haller (1964) in the Greenland Caledonides. It is appropriate to mention at this point the metamorphic masses of the axial Alpine zone and the bands of corrugated and fractured Jura shales (argillitis) which are characteristic of the axial zone of the eastern half of the Greater Caucasus.

The action of slabs of denser rocks being subducted through underlying lighter and more plastic materials is evident in the interaction between certain Helvetidic strata in the Alpine autochthon. The movements of the Helvetidic strata have locally

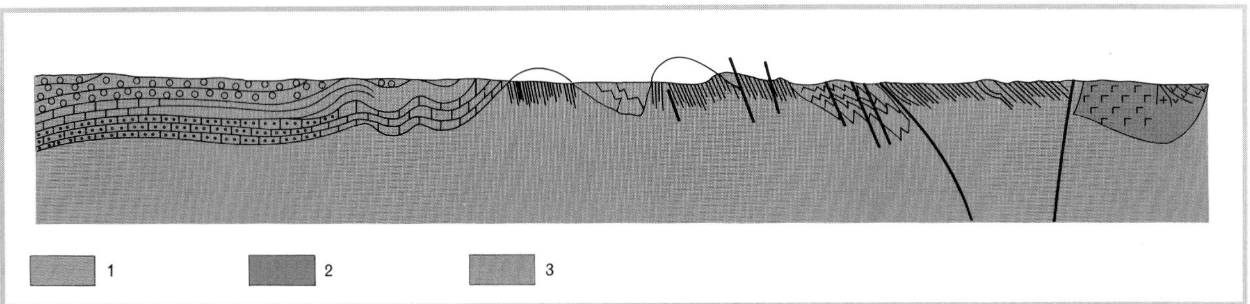

Fig. 18-7. Geological section through the polar Urals. Key: 1, nonmetamorphic deposits; 2, series of synclinal folds consisting of metamorphic formations also comprising green rocks or ophiolites; 3, basal complex which precedes the formation of the range. The ophiolites which are abundantly present in range (2) represent, according to supporters of plate tectonics, the rest of the basaltic oceanic crust which 270 million years ago divided the European and the Asiatic portions of the U.S.S.R. In the present interpretation, the ophiolitic formations are simply considered to be the fruit of metamorphism which accompanied the uplift of the mountain range.

caused density inversions. As a consequence, certain heavy sectors of the overlying strata "drown" in the rocks of the autochthon. Within the space intervening between different strata, the autochthonous plastic and relatively light material is squeezed upwards in the form of injection folds. Even in the reciprocal relations between certain Pennidic strata of the central Alps and the Helvetides one can observe heavy tectonic elements pressing on lighter ones; consequently, the latter are squeezed upwards and laterally.

The above explanations imply that in the geosynclines the masses which are uplifted and which exert deforming action on surrounding rocks can be of two types. Some are formed by rocks which are older than the underlying ones and of which they determine the corrugation; in such a case during the uplifting motion, their materials are neither folded nor metamorphized, nor granitized. They are, so to speak, passive, basal slabs, with a tectonic, magmatic and metamorphic history which was already concluded in preceding tectonic cycles. They present only deformations associated with their gravitational expansion.

A second type of massifs which produce deformations in surrounding rocks is constituted: (a) by rocks of the same age as the surrounding ones; (b) by rocks older than the surrounding ones, but which have undergone metamorphosis or granitization during their uplifting; (c) by granites, gneiss-granites, or gneiss, whose origin (or at least their current position) is contemporary to the uplifting. In contrast to the massifs of the first type, all these are active; their tectonic, magmatic and metamorphic history continues during the course of their uplift. This phenomenon is identified as "deep diapirism."

It is obvious that from a kinematic point of view, many common characteristics exist between deep and surface diapirism. The analogy begins with the external shape of intrusions. The shape of the massif of gneiss-granite or migmatite described by Haller (1956) in the Greenland Caledonides is similar to that of diapiric domes, i.e., it manifests compression in its lower reaches and mushroom expansion in its upper parts, evident indication of an identical genetic process. Furthermore, their internal structures are similar. In ordinary diapirs, "salt tectonics" consists of extremely complex folds which are strongly compressed, mostly vertical and fan-shaped. The internal structure of alpine metamorphic massifs is very similar. Even the corrugating pressure exerted on surrounding rocks is identical in deep diapirs and in surface ones.

The concept of "active" massif explains the complicated fold structure of the axial portions of corrugated zones. The ordinary concept of gravitational folding does not contain such an explanation since it hypothesizes a zone of tension at the axis of the range where the strata are either absent due to tectonic action or deformed by distensive forces. The fact that, on the contrary, the axial zone of mountain ranges often develops isoclinal folds has been the principle obstacle against the hypothesis of gravitational folding.

What is the mechanism of deep diapirism? Nobody doubts that granites, gneiss-granites, and migmatites can be uplifted during their genesis; indeed, the granitization process is associated with a reduction of density and an increase of the volume of the rocks. Less understood is the uplifting mechanism of the metamorphic rocks which constitute many active diapiric masses.

In these rocks the degree of metamorphism varies from the initial one (slates, phyllites) to that of schists and gneiss of amphibolic facies. Actually, the density of these rocks is normally higher than that of sedimentary nonmetamorphic rocks. Nevertheless, active metamorphic blocks must have been lifted through sedimentary, nonmetamorphic rocks.

In spite of the fact that the problem as a whole has not been sufficiently studied, Rutland pointed out the possibility that temporary excesses of fluid pressure could develop "whenever the rate or production of water or carbon dioxide during a metamorphic reaction is larger than the speed with which these compounds are expelled from the system." Metamorphism is always associated with deployment of water which is either bonded to or absorbed by minerals and which initially fills intergranular pores and then begins to migrate toward the exterior of the metamorphosing rock mass. This process requires time and, therefore, it is very probable that during certain time periods the process of water expulsion from massifs proceeded more slowly than that of water elimination from the individual minerals. In the event of such a delay the rock can be compared to a sponge saturated with water and carbon dioxide.

Although the density of the minerals increases with the deployment of water, the average density of the rock decreases. Indeed, the volume occupied by porosity water will be greater than that of the water absorbed at the surface of the minerals or of the water which is chemically bonded to them. Approximate calculations show that this density decrease can be of the order of 10–15%. It is exactly during this phase of water accumulation in the pores of metamorphizing rocks and during the accompanying phenomena of their volume increase and density decrease that the uplift of "active" metamorphic masses should take place. Afterwards, when they have lost the water content, the rocks become denser and lose the capacity to emerge.

Naturally, many of the above ideas are only hypotheses. However, they correspond to realistic, working hypotheses for those interested in investigating the relation between the readily observable endogenous regimes and the much more mysterious deep processes. However, it must be observed that if, indeed, the Earth's crust is subdivided in areas with different endogenous regimes and if that subdivision is permanent and to be related to inhomogeneities hundreds of kilometers deep, it is then absolutely impossible to accept any possibility of ample horizontal displacements of the lithosphere and asthenosphere. Such displacements would have destroyed the indicated vertical connections and their most apparent manifestation which is the relation which exists between the endogenous regime, the heat flux, and the state of the asthenosphere.

VLADIMIR V. BELOUSOV

Bibliography: Belousov V. V., Reisner G. I., Rudich E. M., Sholpo V. M., *Vertical movements of the Earth's crust on the continents*, in Geoph. Surv., I, 245 (1974); Ramberg H., Sjöström H., *Experimental geodynamical models relating to continental drift and orogenesis*, in Tectonophysics, XIX, 105 (1973); Scheinmann Yu. M., *Continent-ocean differences and a differentiation of the Earth*, in Tectonophysics, XIX, 21 (1973); van Bemmelen R. W., *Geodynamic models for the Alpine type of orogeny (Test-case II: The Alps in Central Europe)*, in Tectonophysics, XVIII, 33 (1973); Belousov V. V., *Modern concepts of the structure and development of the Earth's crust and the upper mantle of continents*, in Quart. Journ., CXXII, 293 (1966); Beloussov V. V., *Basic problems in geotectonics*, New York (1962); van Bemmelen R. W., *New view on the Alpine orogenesis*, in Proc. of XXI Int. Geol. Congr., XVIII, 99 (1960); van Bemmelen R. W., *The undation theory and the development of the Earth's crust* in Proc. of Int. Geol. Congr., XVI, **2**, 965 (1933); Haarmann E., *Die Oszillationstheorie*, Stoccarda (1930).

CONCLUSION

The revolution which jolted the earth sciences in the 1960s and 1970s has still not come full-circle. We have tried to give a survey of it up to the beginning of the last quarter of this century, during which we shall probably see the completion of this new model of the Earth and its past, with the no less impressive results achieved by physics, astronomy, astrophysics, oceanography and biology. This model will find more tangible applications in the quest for new sources of raw materials and energy, and in the prediction of natural disasters.

Where resources are concerned, the model has already made its basic contribution by eliminating the distinction between "renewable" resources and "nonrenewable" resources, and replacing it with the concept of renewal-rate.

In fact the fundamental feature of the plate-tectonics model is the continuous two-way link-up between the Earth's surface and its interior: the mid-oceanic ridges are the places along which new matter from within the Earth continually emerges; the oceanic trenches are the places where, just as continually, the old part of the oceanic crust sinks into the Earth, and is once more available for a later cycle. So the concept of inevitability surrounding the exhaustion of "nonrenewable" resources—a concept which has played a fairly important part in the construction of the first global models of resource dynamics—appears to have been superseded by the concept of the renewal-rate of the resources themselves: there is no substantial difference between water—a typical "renewable" resource—and copper, which is typically considered to be "nonrenewable," except in the different rates at which each of these commodities occurs in possible consumption areas. Both are governed by the concept that the absolute consumption rate must not exceed the inflow rate.

As we have seen in this volume, the model is far from being definitively consolidated and all the problems are far from solved. The last years of this century will see further in-depth discussion about the reality of the model, but also about some of its aspects which today seem less well supported. In particular: the deep mechanisms which support the dynamics of the Earth's crust; its application to the most remote past; the role played by the various forms of activity in the Earth's crust. What we have learned in recent years about the universe and the stars, the ocean, life, and the structure of matter is no less intriguing than what we know about the Earth. The next step is the completion of an integrated and coherent model of the entire natural world.

FUNDAMENTAL STAGES IN THE EARTH'S HISTORY

Era	Period	Epoch	Major Events	Millions of Years Ago
QUATERNARY		Holocene	*End of glaciations. Rise in sea level. Man farming	0.011
		Pleistocene	*Major Quaternary glaciations. Appearance of man	1.8
CAINOZOIC	Neogene	Pliocene	*Appearance of Australopithecids in Africa	7
		Miocene	*Emergence of the Andes	26
	Paleogene	Oligocene	*Collision between Africa and Eurasia; Alps and Appennines start to develop	37–38
		Eocene	*Collision between India and Eurasia; growth of Himalayas	53–54
		Paleocene	*Emergence of Rocky Mountains; mammal kingdom born	65
MESOZOIC	Cretaceous		*Culmination of reign of dinosaurs, which then die out; at the same time the Ammonites and microscopic *Globotruncata* disappear in the oceans	136
	Jurassic		*Pangea starts to split up; the Atlantic starts to open up. First birds	190
	Triassic		*Reptile kingdom established	225
PALEOZOIC	Permian		*Collision between Asia and Europe; Urals born; all the continents merged into one: Pangea *Collision between Africa and North America; Appalachians born	280
	Carboniferous		*Great forests and diffusion of winged insects	345
	Devonian		*Collision between Europe and North America; development of Caledonian chains	395
	Silurian		*Life emerges from the water, invading the continents	440
	Ordovician		*First vertebrates, the fishes, appear	500
	Cambrian		*First animals with shells and skeletons	570
ARCHEOZOIC or PRECAMBRIAN			*Great glaciation *First certain association of pluricellular soft-bodied animals *First soft-bodied animals *First certain fossils: algae *First stromatolites *First structures of possible biogenic origin (unicellular algae or bacteria) *First sedimentary rocks *First rains; formation of expanses of water; erosion and sediment accumulation *Solidification of first Earth's crust	650 700 1500 2200 2900 3200 3750 3800 4000
PREGEOLOGICAL ERA			*Cooling of outer part of planet *Resettlement by gravity of the planet's interior and subdivision into "wrappings" with increasing density *Progressive reheating until almost complete melting *Formation of the Earth from the cold solar nebula	4600

Index

Q

R

S

T

U

V

W

Y